Remote Sensing of Natural Hazards

This book presents a comprehensive coverage of remote sensing technology used to gather information on 12 types of natural hazards in the terrestrial sphere, biosphere, hydrosphere, and atmosphere. It clarifies in detail how to yield spatial and quantitative information on a natural hazard, including its spatial distribution, severity, causes, and the likelihood of occurrence. Presented in the book are detailed explanations of multiple methods of data acquisition, elaborated description of the pros and cons of each method, so that readers are informed to choose the best method applicable to their case. The book is written with a practical leaning towards data analysis using the most appropriate data, methods, and software.

- Covers all major natural hazards, including hurricanes, tornadoes, wildfires, and avalanches.
- Studies each natural hazard holistically, ranging from spatial extent, severity, impact assessment, causes, and prediction of occurrence.
- Explains different remotely sensed data and the most appropriate method to use.
- Compares different ways of sensing and the pros and cons of the selected data and their analysis.
- Provides ample examples of each aspect of a natural hazard studied, augmented with graphic illustrations and quality assurance information.

All professionals working in the field of natural hazards, as well as senior undergraduate and graduate students will find in-depth approaches and sufficient information to become knowledgeable in the methods of analyzing remotely sensed data and yielding the desired information, ultimately providing a deeper understanding of how to sense natural hazards.

Remote Sensing of Natural Hazards

Jay Gao

CRC Press is an imprint of the
Taylor & Francis Group, an informa business

Designed cover image: Reprinted by permission from Springer Nature Customer Service Centre GmbH: Springer Nature | Natural Hazards, Morgenroth, J., Hughes, M.W. & Cubrinovski, M. Object-based image analysis for mapping earthquake-induced liquefaction ejecta in Christchurch, New Zealand. *Nat Hazards* **82**, 763–775 (2016). https://doi.org/10.1007/s11069-016-2217-0.

First edition published 2023
by CRC Press
6000 Broken Sound Parkway NW, Suite 300, Boca Raton, FL 33487-2742

and by CRC Press
4 Park Square, Milton Park, Abingdon, Oxon, OX14 4RN

CRC Press is an imprint of Taylor & Francis Group, LLC

Library of Congress Cataloging-in-Publication Data

Names: Gao, Jay, author.
Title: Remote sensing of natural hazards / Jay Gao.
Description: Boca Raton, FL : CRC Press, 2023. | Includes bibliographical references and index.
Identifiers: LCCN 2022047279 (print) | LCCN 2022047280 (ebook) | ISBN 9781032404028 (hardback) | ISBN 9781032406909 (paperback) | ISBN 9781003354321 (ebook)
Subjects: LCSH: Natural disasters--Remote sensing. | Hazard mitigation--Remote sensing.
Classification: LCC GB5014 .G36 2023 (print) | LCC GB5014 (ebook) | DDC 363.34/630284--dc23/eng20230324
LC record available at https://lccn.loc.gov/2022047279
LC ebook record available at https://lccn.loc.gov/2022047280

ISBN: 978-1-032-40402-8 (hbk)
ISBN: 978-1-032-40690-9 (pbk)
ISBN: 978-1-003-35432-1 (ebk)

DOI: 10.1201/9781003354321

Typeset in Times
by KnowledgeWorks Global Ltd.

I would like to dedicate this book to my parents, who gave me unreserved care and love, without which I would never be able to achieve what I have done.

Contents

Preface

The world we are living in is in a state of constant change, not necessarily for the better. In fact, with the continued warming of the global climate, all types of natural hazards are striking the earth at an alarming and unprecedented pace around the world each year. This century has witnessed the occurrence of such natural disasters as floods, wildfires, and droughts in record frequency and magnitude. They have posed a grave threat to the natural environment and its elements (e.g., biodiversity and its eco-service value) and caused immeasurable damage, a huge economic loss, and the extinction of species. Suddenly, there is an awareness and realization that natural hazards are closely related to each of us. Nearly everyone on earth is impacted by natural hazards in one way or another. How to cope with Mother Nature has received increasing attention in the media and from the general public. There is an ever urgent need to understand natural hazards, predict their occurrence, and minimize their devastating effects by implementing mitigation measures before they strike again.

In recent decades, remote sensing has found increasing applications to the study of natural hazards owing to the growing interest in environmental issues and the values offered by geospatial technologies. In particular, the ability of remote sensing to supply near-real-time imagery and auxiliary information about natural hazards has proved irreplaceable in monitoring and mapping them as they evolve via diverse means, including multi-media and the Internet. This book presents a timely and comprehensive coverage of the topic. In total, it expounds how to sense 12 types of natural hazards in the terrestrial sphere, atmosphere, biosphere, and hydrosphere. Here the term "hazard" is used instead of the common "disaster," as some hazards covered in this book are exerting an impact only gradually, such as drought and land subsidence. They do not occur overnight but can cause enormous destruction in the long term. Some of the covered hazards take the form of mass movement, such as landslides, avalanches, and lahar flows, all of which will be discussed in different chapters because of the materials involved are quite different. More importantly, the target of remote sensing, the suitable remote sensing data, and the method of data analysis all differ from each other. Specifically, avalanches apply to (mostly) snow, while flows are reserved for lahar in this book. Not covered in this book is acid rain which is caused mostly by human activities. Although acid rain can also be formed by sulfur in volcanic ash, the contribution of volcanism to acid rain is highly confined to certain parts of the world, and the acidity level is easily diluted by wind and rain, so the rain is not so acid and cannot do much damage to the environment to be considered as a hazard. Also not covered in this book are oil spills as they are purely human-made disasters.

This book consists of 13 chapters. After the introductory chapter, a unique type of the 12 natural hazards is covered in each chapter, starting with earthquakes that can trigger secondary hazards, such as landslides and avalanches. The next group of chapters is all related to land-based hazards, such as landslides (Chapter 3), land degradation (Chapter 4), desertification (Chapter 5), and land subsidence (Chapter 6). The third group of hazards pertains to climate, including droughts (Chapter 7), dust storms (Chapter 8), hurricanes and tornadoes (Chapter 9), floods (Chapter 10), and, to some degree, fires (Chapter 11). The last haphazard group of hazards includes volcanoes (Chapter 12) and (snow) avalanches (Chapter 13). In spite of this sequence of organization, the reader does not have to follow it and is at liberty to choose to read whichever chapters interest them without losing comprehension, except Chapter 1. It provides crucial background materials for the comprehension of the remaining chapters. All chapters are structured in the rough pattern of articulating the feasibility of sensing the concerned hazard first, followed by an examination of the best data/method to use, and an explanation of how to study its different aspects or components. They encompass spatial extent mapping, severity and impact/damage assessment, and prediction/modeling of risk and likely occurrence, all lavishly supplemented with examples and graphic analytical outputs. References are provided at the end of each chapter for those who would like to find further information on the sensing of a particular type of hazard.

Increasingly, a huge number of abbreviations have been used in remote sensing publications to describe remotely sensed data and data processing methods. Some of them are commonly accepted in the English language, such as radar, LiDAR, and GPS. Others are novel to those not so familiar with the subject. It is almost impossible to avoid acronyms when reading the remote sensing literature. There is no attempt to circumvent these abbreviations in this book. After an acronym is first introduced, it will be used subsequently throughout the remainder of the book without full spelling. An alphabetic list of all major abbreviations contained in this book is provided at the beginning of the book for reference.

Written for upper-level undergraduate and postgraduate students, this book can be used as a supplementary textbook on remote sensing of natural hazards. It can also function as a handy and valuable reference for those professionals working in the field of natural hazards, and anyone who would like to study a particular natural hazard using remote sensing in their work. Through reading it, the reader can expect to become knowledgeable in how to sense widely ranging natural hazards qualitatively and quantitatively and gain a clear understanding of how to undertake certain applications of interest to them. This book paints a holistic picture about the capability of remote sensing in mapping, detecting, analyzing, and modeling natural hazards and assessing their risk and severity using diverse analytical methods. All covered topics are discussed in sufficient depth and breadth, backed by the latest research findings. The content of the book showcases the type of information that can be retrieved from both imagery and non-imagery remotely sensed data using the most effective approaches. Throughout the book, all discussions emphasize how a particular hazard element is sensed quantitatively, the achievable accuracy, and its influencing factors. Apart from a broad theoretical exposure, this book also contains a pragmatic dimension to elucidate how to select and process the most relevant remote sensing data to derive the desired information about the concerned hazard. Whenever relevant, it also elaborates on the practical implementation of the analysis using the most appropriate software/computational tools. When multiple methods and tools are available, they are compared and contrasted for their pros and cons, with their capability comprehensively and critically assessed so that the reader can decide the most appropriat one to choose.

Acknowledgments

It has been more than three decades since I first started to research natural hazards using remote sensing and spatial analysis for my PhD thesis. Later when I was preparing for lectures on natural hazard research using GIS, this book was conceived more than a decade ago. Back then I did not have the time to actually put my thoughts down in words. It was always in my mind that I would complete the project someday toward the late stage of my academic career. The advent of the Covid-19 pandemic and the immeasurable disruption it caused to higher education accelerated the arrival of this stage. In the process of transitioning from a full-time academic to a freelance researcher, I could finally work on this book full-time and see its completion within several months. During the writing of this book, I received generous assistance from numerous parties whose contribution toward this book is hereby heartily acknowledged. I am deeply indebted to the University of Auckland for granting me an honorary position that enables me to access all of the university's library and computing resources, without which this book could not have been completed.

In addition to my own research results, this book also benefits from the projects my postgraduate students completed over the years. I am especially appreciative of the contributions of Simon Chen and Sesa Wiguna to this book. A few figures contained in the book are based on their theses. In comparison, many more tables and figures are sourced from recently published literature to showcase the breadth of remotely sensing natural hazards. The generosity of a large number of fellow researchers who have made their works openly accessible is hereby acknowledged. There are too numerous of them to list individually here. All figures and tables with a cited source but without "used with permission" fall into this category of publication.

Next, I would like to express my gratitude toward a few publishers and professional organizations for permitting me to reuse their illustrations and tables in this book. I am deeply impressed by the ease of acquiring permission using the online systems of Elsevier, followed by Springer and John Wiley & Sons. The granting of permission was nearly instantaneous. In addition, the American Meteorological Association, the Geological Society of America, and the Coastal Education and Research Foundation, Inc. all graciously granted me the permission to reuse their published figures and tables in this book, for which I am deeply thankful.

Finally, I owe a special thank-you to my siblings who cared for my mother throughout my academic career when I was thousands of miles away from home to perform my filial duty, and for doing everything they could to ensure that she finally rested in peace with maximal comfort at a time when I was barred from international travel. Their selflessness enabled me to concentrate on my work and pursue what I am always passionate about in life.

Jay Gao
Auckland, New Zealand
August 2022

About the Author

Jay Gao used to be an associate professor affiliated with the School of Environment, University of Auckland. He received his Bachelor of Engineering degree in the field of photogrammetry and remote sensing from the Wuhan Technical University of Surveying and Mapping in 1984. Four years later he obtained his MSc in geography from the University of Toronto, Canada, and his PhD in geography from the University of Georgia in the US in 1992. Upon graduation, he became a faculty member of the current university. His research interest spreads widely among different disciplines of geoinformatics, including remote sensing, digital image analysis, and spatial analysis and modeling. Over his academic career, he has completed numerous projects on the applications of remote sensing to the management of natural resources and the studies of natural hazards using remote sensing and GIS. He has published nearly 200 papers in international journals, authored two books, and edited several more. His solely authored book on *Digital Analysis of Remotely Sensed Data* was published by McGraw Hill in 2009, and his second book, *Fundamentals of Spatial Analysis and Modelling* was published by CRC Press in 2021.

List of Acronyms

ABI	Advanced Baseline Imager
AHI	Advanced Himawari Imager
AHP	analytical hierarchical process
AIRS	atmospheric infrared sounder
ALOS	advanced land observing satellite
ALS	airborne laser scanning
AMSR	advanced microwave scanning radiometer
ANFIS	adaptive neuro-fuzzy inference system
ANN	artificial neural network
AOD	aerosol optical depth
AOT	aerosol optical thickness
ARVI	atmospherically resistant vegetation index
ASI	avalanche susceptibility index
ASTER	Advanced Space-borne Thermal Emission and Reflection Radiometer
ATI	apparent thermal inertia
ATMS	advanced technology microwave sounder
AUC	area under curve
AVHRR	Advanced Very High Resolution Radiometer
AWC	available water-holding capacity
BADI	brightness temperature adjusted dust index
BP	backpropagation
BRDF	bidirectional reflectance distribution function
BSI	bare soil index
BT	brightness temperature
BTD	brightness temperature difference
CART	classification and regression trees
CCD	coupled charge device
CCM	convergent cross-mapping
CDIR	comprehensive drought index of remote sensing
CFRISK	cumulative fire risk index
CMORPH	CPC (Climate Prediction Center) morphing technique
CRSI	canopy response salinity index
CSI	critical success index
CSI	coherent scatterer InSAR
CV	cross-validation
CVI	coastal vulnerability index
CWSI	crop water stress index
DAR	damage area ratio
DEM	digital elevation model
DGPS	differential global positioning system
DI	degradation index
DRBTD	dynamic reference BTD
DRI	degradation risk index
DS	distributed scatterers
DSI	drought severity index
DSM	digital surface model
DSSI	dust spectral similarity index

DTRI	desertification trend risk index
EC	electrical conductivity (for soil salt content)
ECMWF	European Centre for Medium-Range Weather Forecasts
EDI	enhanced dust index
EDII	enhanced dust intensity index
ENDVI	enhanced normalized difference vegetation index
EO	Earth observation
ERS	Earth resources satellite
ERS	European Remote Sensing
ESA	European Space Agency
ESP	exchangeable sodium percentage (for soil alkalinization)
ETCI	evapotranspiration condition index
ETM+	enhanced thematic mapper plus
EVI	enhanced vegetation index
FAR	false alarm ratio
FHI	flood hazard index
FOM	frequency of misses
FPAR	fraction of photosynthetically active radiation
FR	frequency ratio
FRP	fire radiative power
FTA	fire thermal anomaly
GAM	generalized additive model
GBRT	gradient-boosted regression tree
GDDI	global dust detection index
GDVI	generalized difference vegetation index
GFDS	global flood detection system
GLCM	gray-level co-occurrence matrix
GMF	geophysical model function
GNSS	global navigation satellite system
GOES	geostationary operational environmental satellites
GPM	global precipitation measurement
GPS	global positioning system
GRA	gray relational analysis
GSD	ground sampling distance
GSI	grain size index
HAI	human activities impact
HII	human impact index
HJ	Huan Jing (environment)
HYSPLIT	Hybrid Single-Particle Lagrangian Integrated Trajectory
IDII	improved dust identification index
IMU	Inertial Measurement Unit
InSAR	interferometric SAR
INSAT	Indian National Satellite System
IRS	Indian remote sensing
JERS	Japanese Earth Resources Satellite
LAI	leaf area index
LCC	land capability classes
LDAS	land data assimilation system
LiDAR	light detection and ranging
LISS	linear independent self-scanner
LNS	local net primary production scaling

LOS	line of sight
LR	linear regression
LR	logistic regression
LS	length-slope
LSeI	liquefaction sensitivity index
LSI	liquefaction susceptibility index
LSM	linear spectral mixing
LSM	land surface model
LST	land surface temperature
MAPE	mean absolute percentage error
MEDI	Middle East dust index
MIRBI	mid-infrared burn index
MLR	multivariate linear regression
MODIS	Moderate Resolution Imaging Spectroradiometer
MPI	microwave polarized index
MPI	Max Planck Institute
MSAVI	modified soil-adjusted vegetation index
MSDI	moving standard deviation index
MSDI	multivariate standardized drought index
MSI	multispectral instrument
MSS	multispectral scanner
NASA	National Aeronautics and Space Administration
NBR	normalized burn ratio
NBT	Naïve Bayes Trees
NDAI	normalized difference angle index
NDDI	normalized difference desertification index
NDDI	normalized difference drought index
NDDI	normalized dust difference index
NDI	normalized degradation index
NDI	normalized difference index
NDII	normalized difference infrared index
NDLI	normalized dust layer index
NDSI	normalized difference salinity index
NDVI	normalized difference vegetation index
NDWI	normalized difference water index
NIR	near infrared
NMDI	normalized multiband drought index
NOAA	National Oceanic and Atmospheric Administration
NPP	National Polar-Orbiting Partnership
NPP	net primary production
NPV	non-photosynthetic vegetation
NRCS	normalized radar cross-section
NSBAS	new small baseline subset
NSS	normalized skill score
NTI	normalized thermal index
OBIA	object-oriented image analysis
OLI	Operational Land Imager
OMI	ozone monitoring instrument
PALSAR	Phased Array type L-band SAR
PCA	principal component analysis
PCC	Pearson correlation coefficient

PCI	precipitation condition index
PDSI	Palmer drought severity index
PET	potential evapotranspiration
PERSIANN	Precipitation Estimation from Remotely Sensed Information using Artificial Neural Networks
PFSM	parameterized first-guess spectrum method
PLSR	partial least-squares regression
POD	probability of detection
POFD	probability of false detection
P-PET	precipitation and potential evapotranspiration
PPI	population pressure index
PRISM	panchromatic remote-sensing instrument for stereo mapping
PS	permanent/persistent scatterers
RBF	radial basis function
RDI	reconnaissance drought index
RF	random forest
RFR	random forest regression
RGB	red, green, blue
RI	ratio index
RMSE	root mean square error
RPD	residual predictive deviation
RSP	relative slope position
RTK	real-time kinematic
RUE	rain use efficiency
SAR	synthetic aperture radar
SARVI	soil-adjusted and atmospherically resistant vegetation index
SAVI	soil-adjusted vegetation index
SBAS	small baseline subset
SDCI	scaled drought condition index
SDDA	simplified dust detection algorithm
SDFP	slowly decorrelating filtered phase
SDI	synthesized drought index
SEVIRI	Spinning Enhanced Visible and Infrared Imager
SFMR	stepped-frequency microwave radiometer
SI	salinity index
SLA	sea level anomaly
SLC	single-look complex
SMAP	soil moisture active passive
SMSI	soil moisture saturation index
SNAPHU	statistical-cost, network-flow algorithm for phase unwrapping
SNR	single-to-noise ratio
SPEI	standardized precipitation evapotranspiration index
SPI	standardized precipitation index
SPI	stream power index
SPOT	Satellite Pour l'Observation de la Terre
SRTM	shuttle radar topographic mission
SSI	standardized soil (storage) index
SST	sea surface temperature
StaMPS	Stanford Method for Persistent Scatterers
SVM	support vector machine
SVR	support vector regression

SWDI soil water deficit index
SWIR shortwave near infrared
SWSI surface water supply index
TBI three-band indices
TDI thermal infrared dust index
TDLI temporal difference liquefaction index
TEI thermal eruption index
TIIDI thermal infrared integrated dust index
VIIRS Visible Infrared Imaging Radiometer Suite
TIR thermal infrared
TM thematic mapper
TMII tidal marsh inundation index
TOA top of atmosphere
TPI topographic position index
TRI topographic ruggedness index
TRMM Tropical Rainfall Measuring Mission
TS time series
TSS total suspended sediments
TSS true skill statistics
TVDI temperature vegetation drought index
TWI topographic wetness index
UAS/V unpiloted aerial system/vehicle
USGS U.S. Geological Survey
USLE universal soil loss equation
VH vertical transmission and horizontal reception
VHR very high resolution
VIF variance inflation factor
VIIRS Visible Infrared Imaging Radiometer Suite
VNIR visible light – near infrared
VRM vector ruggedness measure
VSSI vegetation soil salinity index
VSWI vegetation supply water index
VV vertical transmission and vertical reception
WBDI water budget drought index
WDI water deficit index
WDVI weighted difference VI
WEI wind exposition index

1 Introduction

1.1 INTRODUCTION

A natural hazard is defined as "any atmospheric, earth, or water related occurrences, the action of which adversely affects or may adversely affect human life, property or other aspect of the environment" (RMA [Resource Management Act], 1991). As the world we are living in is struck by more frequent occurrences of all types of natural hazards, including earthquakes, tsunamis, volcanic and geothermal activity, landslides, subsidence, wildfires, droughts, hurricanes, and floods, they have received increasing attention from the general public and been studied extensively using various means, especially remote sensing. Remote sensing in its various forms refers to the technology of acquiring information about natural hazards without being in physical contact with them. In its broad definition, remote sensing also encompasses the methods by which the acquired data are analyzed, either in isolation or in conjunction with data from other sources, to derive the desired information about a natural hazard of interest. Remote sensing of natural disasters refers primarily to the use of remotely sensed data to study natural disasters, including their detection, analysis, quantification, modeling, prediction, and impact assessment and mitigation (Chien and Tanpipat, 2013). These data used to be acquired exclusively from airborne and space-borne platforms in the graphic format. As remote sensing has been evolving, especially over the last two decades, a far wider range of data has been acquired at an ever finer temporal resolution from an ever expanding fleet of sensors deployed on the ground, just above it, in the air, and from space. They have opened the floodgates for data acquisition. These data enable natural hazards to be studied at an unprecedented detail level that is unimaginable only years ago. Besides, the results generated from these remotely sensed data, facilitated by powerful machine learning algorithms, convey an amazingly diverse range of information about natural hazards. Both the fine details and short temporal resolution of the generated results make it feasible to study the process of some natural hazards and predict their likely spot of occurrence.

This chapter lays the foundation of the book to facilitate the comprehension of the subsequent chapters. It first overviews all types of hazards, paying particular attention to those natural hazards that are best studied via remote sensing. Then it differentiates a few terms commonly used to depict natural hazards, including vulnerability/susceptibility, hazard, and risk, articulates their relationship, and spells out the role of remote sensing in hazard research. Next, this chapter presents a brief but comprehensive overview of the remote sensing data that have found applications in studying natural hazards. The emphasis of the discussion is placed on the main features of each type of data and their best uses. Whenever relevant, the limitations of and the accuracy level achievable with each type of data are acknowledged and evaluated so that the reader can judge the appropriate type of data to use for a given application. Afterward, this chapter introduces methods by which remotely sensed data are commonly analyzed and converted to the desired information about natural hazards. Some of the analyses are concerned with the mapping of natural hazards, while others target the prediction of hazard occurrences that may not seem to bear a close relationship with remote sensing, but they are still included in the discussion as remotely sensed data play an indispensable role in them. Finally, this chapter summarizes the accuracy measures that have been used to indicate the reliability of remote sensing-generated hazard maps, the predicted hazard severity level, and the likelihood of future hazard spatial occurrence.

DOI: 10.1201/9781003354321-1

1.2 NATURAL HAZARDS AND RISK

1.2.1 Classification of Hazards

There is no universal agreement on what constitutes a natural hazard among scientists. Some disasters such as earthquakes are purely natural as humans play no role in their initiation. Pure natural disasters include earthquakes, tsunamis, snow avalanches, hurricane-induced floods, volcanic eruptions, and dust storms. Other hazards may not be so clear-cut. Take hurricanes as an example. Strictly speaking, tropical storms and cyclones are not disasters but natural meteorological phenomena. Only when they impose a threat to properties and human lives, they become disasters in the form of landslides, floods, and avalanches. Hurricanes are regarded as purely natural disasters by some (Table 1.1), but there is still an element of human influence in their initiation. In this day and age of climate warming, humans alter the atmospheric composition through the emissions of greenhouse gases into the atmosphere, indirectly impacting hurricanes. Climate change-related disasters include droughts, cyclones, and wildfires. Some hazards, such as wildfires and coalmine fires, may be caused by lightning naturally or accidentally set off by humans. These hazards are still covered in this book in order to predict their occurrences and provide early warnings, such as the monitoring of the hurricane eye and its movement to evacuate those at risk prior to the eventuation of the natural disaster. However, air pollution, acid rain, oil spills, and land mines are beyond the scope of this book as they are purely anthropogenic in origin (Table 1.1).

Natural hazards can be studied differently using remote sensing to fulfill various objectives. The most common application is mapping that aims to illustrate the spatial distribution of a disaster and the affected area, and hazard severity. Analysis of a hazardous event concentrates on the process of change by generating time-series results. All hazards can be studied in terms of damage and the effectiveness of disaster mitigation measures, such as building damage in an earthquake. It is also possible to issue early warnings for some disasters before they strike and do damage (e.g., forecast of forest fires and identification of fire hotspots). Monitoring is one step further than analysis by assessing the current state of a disaster as it is happening, such as tracking hurricanes. Some of the disasters listed in Table 1.1 take place within a matter of minutes, such as flash floods and earthquakes, so it is impossible to track them. Only the aftermath damage is mapped and assessed.

TABLE 1.1

Classification of Hazards in a Graduate Scale between Purely Natural and Purely Human-Made

Purely Natural	Partial Natural Influence	Mixed Natural/ Human Impacts	Partial Human Impact	Purely Human-Induced
Earthquake	Crop disease	Landslides	Flood	Land mines
Tsunami	Insect outbreak	Subsidence	Dust storm	Oil spill
Volcano/lava flow	Wildfire	Soil erosion	Drought	Water/soil/air pollution
Snow avalanche	Mangrove/kauri dieback	Desertification		
Glacial lake outburst	Coral reef decline	Coal fires		
Heatwaves	Acid rain	Coastal erosion		
Windstorm	Ozone depletion	Greenhouse effect		
Hailstorm/tornado		Sea-level rise		
Cyclone/hurricane				

Source: Modified from van Westen (2000).

1.2.2 Vulnerability and Risk

Hazard vulnerability or susceptibility refers to the inherent likelihood that an area is going to be impacted by a hazardous event. Vulnerability is synonymous with the propensity or condition of an area or individual subject to hazard-triggered damage and suffering from the adverse effects of physical, social, economic, and environmental factors or processes. It is related to the geographic location and its settings in the immediate spatial context. For instance, those coastal areas lying slightly above the current sea level are vulnerable to flooding during a severe storm coupled with a king tide as a consequence of the projected sea-level rise in the long run. Low-lying areas are vulnerable to flooding after lasting downpours. Those areas with a low elevation, a gentle slope, and a small curvature have the highest chances of being flooded, much more vulnerable to flooding than elsewhere (Mojaddadi et al., 2017). Vulnerability (V_L) is frequently evaluated from a set of site-specific parameters and expressed mathematically as follows:

$$V_L = P\left[D_L \geq 0/L\right]\left(0 \leq D_L \leq 1\right) \tag{1.1}$$

where D_L = the assessed or the expected (forecast) damage to an element in light of a hazardous event L. Vulnerability is measured by the probability of total loss to a specific element or the portion of damage to an element caused by a hazardous event (Mojaddadi et al., 2017). Vulnerability is usually expressed on a scale of 0–1, with 0 meaning no damage and 1 indicating total destruction or loss. Vulnerability to a given type of natural hazard is expressed quantitatively in monetary terms or qualitatively in heuristic terms.

Hazard is sometimes indiscriminately regarded as vulnerability, but the two have subtly differing connotations. Hazard refers to the risk of a hazardous event that may include the area being affected (e.g., inundation area, depth, duration, and frequency in case of flood). This hazard is jointly controlled by the local physical environment (e.g., topography and land use) and the event itself (e.g., severity, duration, and frequency), in contrast to vulnerability that is subject only to the environmental settings. Hazard is also related to external factors such as infrastructure and population distribution. The same hazard certainly poses a higher risk if more people live in the hazard-prone zone. An area may be vulnerable to the occurrence of a hazardous event but its hazard or risk can be low, such as the slow rise of water level just above its normal height during a flood. On the other hand, an area normally not subject to flood can face a rather high hazard in light of flash floods that can do considerable damage within minutes. Hazards can be single, consequential, or a combination of both. For instance, an earthquake itself is a single hazardous event. If it triggers landslides, more secondary hazards are evoked in addition to the damage done by the quake itself, such as landslides damaging buildings.

Risk is defined as the probability of suffering from loss, damage, and other negative consequences of a hazardous event or process in the future. It differs from hazards that are potentially damaging events or activities and that may lead to negative consequences to human life and safety, and the built and natural environments. Risk is expressed as the quantity of the potential or expected damage (loss) in a particular area within a given period (in general one year). It may be estimated for damage assessment in terms of hazard (i.e., the affected area, severity, frequency, and even duration and depth for floods), exposure, and vulnerability (i.e., sensitivity to economic damage). Exposure is related to the contact of the vulnerable cohort with a natural disaster, such as people and property. Disaster risk can be perceived as a function of hazard, exposure, and vulnerability, or

$$Disaster\ Risk = function\left(Hazard,\ Exposure,\ Vulnerability\right) \tag{1.2}$$

The risk of a hazardous event is commonly calculated from potential adverse consequences multiplied by the probability of a hazardous event, or

$$Risk = C \times p_h = Element\ at\ risk \times Hazard \times Vulnerability \tag{1.3}$$

where p_h = the probability of a hazardous event, C = the potential adverse consequence incorporating factors such as exposure and vulnerability. It is calculated as

$$C = V \times S(mh) \times E \tag{1.4}$$

where V = the value of the element at risk, in monetary terms or human life; S = susceptibility (vulnerability): damaging effect on the element at risk. It is a function of the magnitude of influence (e.g., severity/intensity and duration, as well as volume in case of rainfall). S ranges from 0 to 1. E = exposure: the probability of the element at risk to be present during the hazardous event. Identical to vulnerability, exposure is also expressed over a range of 0–1. V, S, and E are the three vulnerability parameters.

1.2.3 Remote Sensing and Natural Hazards

Remote sensing plays a significant role in disaster mitigation, preparedness, response, and recovery, such as risk modeling, vulnerability analysis, early warning, and post-event damage assessment (Table 1.2). Mitigation relies on remotely sensed results prior to the eventuation of a hazard, such as

TABLE 1.2
Role of Remote Sensing in Mitigation, Preparedness, Response, and Recovery of Seven Types of Natural Hazards

Hazard	Mitigation	Preparedness	Response	Recovery
Cyclone	Risk modeling; Vulnerability analysis	Early warning Long-term climate modeling	Identifying escape routes; crisis mapping; impact assessment; cyclone monitoring; storm surge predictions	Damage assessment; Spatial planning
Drought	Risk modeling; Vulnerability analysis; Land and water management planning	Weather forecasting; Vegetation monitoring; crop water requirement mapping; early warning	Monitoring vegetation; damage assessment	Informing drought mitigation
Earthquake	Building stock assessment; Hazard mapping	Measuring strain accumulation	Planning routes for search and rescue; damage assessment; evacuation planning; deformation mapping	Damage assessment; Identification of sites for rehabilitation
Fire	Mapping fire-prone areas; monitoring fuel load; risk modeling	Fire detection; predicting spread/direction of fire; early warning	Coordinating firefighting efforts	Damage assessment
Flood	Mapping flood-prone areas; delineating floodplains; land use mapping	Flood detection; Early warning; Rainfall mapping	Flood mapping; Evacuation planning	Damage assessment; Spatial planning
Landslide	Risk modeling; hazard mapping; DEM creation	Monitoring rainfall and slope stability	Mapping-affected areas; Debris volume and displacement distance estimation	Damage assessment; Spatial planning; Suggesting remedy measures
Volcano	Risk modeling; hazard mapping; DEM generation	Emissions monitoring; plume tracking; Thermal alerts	Mapping lava/lahar flows; evacuation planning; Diversion of air traffic	Damage assessment; Spatial planning

Source: Modified from Lewis (2009).

the identification of likely spots where flooding is going to take place so that they are barred from human settlement. Preparedness takes place much closer to the eventuation of an imminent hazardous event, such as a hurricane. Those residents dwelling in its path must be alerted and prepared to evacuate. Response pertains to the reaction immediately before or during the hazard strike, such as the most effective deployment of resources to combat the hazard as in wildfire fighting. Recovery takes place in the aftermath of the hazardous event in the form of damage assessment, such as whether the liquefaction-affected area is still fit for human habitation.

In studying natural hazards, remote sensing brings numerous advantages, such as synoptic view and continuous coverage of an extensive ground area per image that allows the impact of a hazard to be assessed (e.g., dust storm and hurricane tracking), a feature that is impossible to retain otherwise (Table 1.3). Multi-sensor data of complementary strengths allow several aspects of a hazard to be studied. For instance, detailed information about its damage may be gleaned from fine-resolution imagery (e.g., volcanic lava flows), while the process of the hazardous event is studied from fine-temporal images recorded every 15 minutes (e.g., volcano plume spreading). The use of multi-sensor data also prolongs the span over which the same natural hazard may be studied or monitored over multi-decades. Such studies have been eased by the availability of a variety of remote sensing data products of environmental variables. The desired hazard information is derivable from remotely sensed data on a routine and operational basis using existing software packages. Although remote sensing is critical to natural hazard mitigation and response, it is not without limitations. The main limitation is related to data unavailability and delivery as there is no centralized inventory of data except Google Earth Engine that serves as a clearinghouse for only public domain data. Some data are available for certain periods of time, and for only certain parts of the world.

TABLE 1.3
Strengths and Limitations of Satellite Remote Sensing in Studying Natural Hazards

Strengths	Limitations
Synoptic, periodic coverage	• Insufficient spatial and temporal resolution in many cases
Availability of pre-event, reference imagery	• Delayed initial image acquisition, especially using non-pointable sensors
Variety of complementary sensors (radar and optical, high and low spatial resolution)	• High cost of imagery
Availability of increasingly sophisticated and user-friendly software and algorithms	• Low number of operational radar satellites
Easy and free accessibility of certain images	▪ Typically slow data dissemination
Availability of a wide range of data products	▪ Lack of central inventory of available satellites and their current location
Minimized fieldwork required, increased safety	▪ Frequent data incompatibility
Almost unlimited number of sample points (area-specific sampling rather than location-specific sampling)	▪ Large image-file size, which makes electronic use/dissemination in the field difficult
Easy integration of diverse data to generate image-derived maps	▪ Lack of global coverage by ground receiving stations for some satellites, resulting in incomplete coverage
Acquisition of stereo images or elevation data via pointable sensors	Unfamiliarity of disaster managers with satellite images and their use
Visual, yet quantitative data	Variable equatorial crossing times (and, consequently, variable illumination) of some polar orbiters
Repeatability of analytical methods, and/or testing of a new theory using archived images	
Rapidly increasing sophistication of satellites (e.g., multi-sensors)	

Source: Modified from Kerle and Oppenheimer (2002). Used with permission.
Note: Dots (•): will diminish with future satellites; squares (▪): likely to lose relevance with future satellites

1.3 REMOTE SENSING DATA

Based on their nature, remote sensing data fall into two categories of imagery and non-imagery. Imagery data graphically illustrate the ground cover at the time of imaging, while non-imagery data such as global positioning system (GPS) and light (laser) detection and ranging (LiDAR) capture the 3D position of surface points.

1.3.1 Imagery Data

In terms of the wavelength of the energy used for sensing, imagery data are broadly classified as optical and microwave. Optical images are acquired passively using the solar radiation reflected or emitted from the target to the sensor over the ultraviolent (UV), visible light and near infrared (VNIR), shortwave infrared (SWIR), and thermal infrared (TIR) portions of the spectrum (Table 1.4). Mostly multispectral, they faithfully preserve the ground features under cloud-free conditions. Multispectral images are rather good at revealing the surface cover. Optical images can be further classified as medium resolution or very high resolution (VHR) in terms of the details they show. Medium resolution images have a spatial resolution of 5 m or coarser. There is no precise definition of what spatial resolution is considered very high. In general, it is understood to range from sub-meter to less than 5 m. Each type of imagery can be acquired from different satellite missions, either commercial or public-funded. Radar images are recorded using microwave radiation supplied by the sensor itself, in a process known as active sensing. Since microwave radiation is able to penetrate clouds, radar sensing is operational under all weather conditions unless the cloud is rather thick. Each type of remote sensing image has its best uses in studying natural hazards (Table 1.4). For instance, UV is the only mode of sensing that can detect the composition of volcanic plumes, while radar imagery is the best at studying the 3D structure of hurricanes.

TABLE 1.4
Overview of Remote Sensing Data and Their Utility in Studying Natural Hazards

Type of Sensing	Wavelength	Exemplary Images	Best Uses
UV	0.3–0.4 μm	OMI (UV–vis) sensor	Volcanic plume composition
Visible light	0.4–0.7 μm	ASTER, SPOT, OLI, Sentinel-2	Impact assessment;
		MODIS, IKONOS	Flood extent mapping;
			DEM generation
NIR	0.7–1.0 μm	SPOT, ETM+, OLI, Sentinel-2	Desertification
		MODIS	Dust storms
		WorldView-3	Landslides
			Lava temperature
SWIR	0.7–3.0 μm	Landsat OLI	Drought
TIR	3.0–14.0 μm	MODIS	Active fire detection
		ETM+	Burn scar mapping
		Hyperion	Hotspot identification
			Soil salinization
Microwave (radar)	0.1–100 cm	RadarSat-2, Sky-Med	Ground subsidence/deformation monitoring
		AMSR-E	Flood forecasting
		TRMM	Hurricane tracking
		QuickSCAT radar	3D storm structure
			Surface wind

1.3.1.1 VHR Imagery

VHR images may be acquired from diverse sensors, all of which make use of radiation, the wavelength of which does not extend beyond NIR. In most cases, both panchromatic (P) and red, green, and blue (RGB, i.e., true color) bands are acquired, with the spatial resolution of the latter being four times coarser than that of the former (Table 1.5). The earliest VHR imagery is exemplified by IKONOS emerging in 1999, and it acquires the finest resolution of 1 m (panchromatic) and 4 m (multispectral) bands. Since this century, many more VHR images have been acquired by commercial satellites that have a life expectancy of three to five years, such as WorldView-3 and QuickBird. WorldView-3 images have a very fine resolution up to 0.31 m. The ground sampling distance (GSD) of QuickBird images ranges from 0.60 m for the panchromatic band to 2.40 m for the multispectral bands. Associated with the fine resolution is a narrow swath width smaller than 20 km for all VHR images except SPOT6/7 (Table 1.5). Nevertheless, the temporal resolution or revisit period of some VHR images can be shortened to a few days by tilting the scanning mirror, even though the exact period varies with the latitude of the area being sensed.

Owing to their superb fine spatial resolution, VHR images allow detailed hazard features, such as landslide paths and lahar debris, to be mapped at high accuracy. Limited by their spectral paucity, such images are not good at studying vegetation in hazard-affected areas or other ground features, the spectral response of which is rather distinctive in the SWIR region of the spectrum. Besides, they are rather expensive and are not economic for large areas of study. In contrast, medium-resolution earth resources satellite images such as Satellite Pour l'Observation de la Terre (SPOT), Landsat, and Sentinel-2 are cheaper or mostly free.

1.3.1.2 Medium Resolution Optical Imagery

Medium resolution optical images have the longest history of existence apart from meteorological satellite images, dating back to the early 1970s or mid-1980s, which makes them the ideal choice for long-term monitoring of natural hazards. Over the years, the spatial resolution of these

TABLE 1.5
Main Properties of VHR Images that Can Support the Monitoring of Natural Hazards

Satellite	Life Span	Spatial Resolution (m) Pan	Multispectral	Swath Width (km)	Revisit Period[a] (days)	Bands
Air-borne optical[b]	N/A	0.25				NIR, R,G, B
IKONOS	1999	1	4	11	1.5–3	NIR,R,G,B
QuickBird	2001	0.60	2.40	16.5	1.5–3	NIR,R,G,B, P
KOMPSAT-2	2006	1	4	15	2–3	NIR,R,G,B,P
GeoEye-1	2008	0.46	1.84	15.2	2–8	NIR,R,G,B
WorldView-2	2009	0.46	1.85	16.4	1–4	P, 8 XS[c]
Pléiades-1/2	2011	0.5	2.0	20	1	NIR,R,G,B,P
KOMPSAT-3	2012	0.7	2.8	15	2–3	NIR,R,G,B,P
KOMPSAT-3A	2012	0.55	2.2	15	2–3	NIR,R,G,B,P
SPOT-6/7	2012/2014	1.5	6	60	1	NIR,R,G,B,P
WorldView-3/4	2014/2019	0.31	1.24	13.1	1	P, 8 NVIR, 8 SWIR, 12 CAVIS[d]
Gaofen-2	2014	0.8	3.2	45 (2 cameras)	60	NIR,R,G,B,P

[a] It varies with the latitude of sensing. The given resolution is the best possible
[b] Images taken with Leica ADS (an extra panchromatic band), Vexcel Ultracam, and DMC digital cameras
[c] P = panchromatic; XS = multispectral
[d] CAVIS = clouds, aerosols, vapors, ice, and snow

TABLE 1.6

Medium-Resolution Images that Can Support the Monitoring of Natural Hazards and Their Main Properties

Satellite	Life Span	Spatial Resolution (m) PAN	Spatial Resolution (m) Multispectral	Swath Width (km)	Revisit Period (day)	Spectral Bands
ASTER	1999–2008		15, 30, 90	60	16	VNIR, TIR
SPOT-5	2002–2015	2.5, 5	10	60	2–3	NIR, SWIR,R,G,B,P
Formosat-2	2002–2016	2	8	24	1	NIR, R,G,B,P
Landsat-8[a]	2013–present	15	30	185	16	NIR,TOR,R,G.,B,P
Sentinel-2A/2B	2015/2016–present	10	10	290	5	R,G,B, NIR,P
HJ-2A/2B	2020–present		16 (48[b])	200	Up to 2	R,G,B, NIR, P, plus more

[a] Landsat-8 carries the OLI (Operational Land Imager) sensor, while its predecessor carries the TM (thematic mapper) and ETM+ (enhanced TM plus) sensors

[b] Hyperspectral bands

images has been gradually refined with the exception of Landsat series images, the spatial resolution of which remains unchanged at 30 m (Table 1.6). Nearly all medium-resolution images have a spatial resolution coarser than 10 m with a few exceptions. Corresponding to such a medium resolution is a much large image swath width on the order of tens of kilometers, which makes medium-resolution images the ideal choice for studying regional hazards. The temporal resolution of medium-resolution imagery is long, usually more than ten days, but can be shortened by deploying two identical satellites (e.g., Sentinel-2A and 2B satellites). Most medium-resolution images are freely accessible. A good source of such satellite images is Google Earth Engine (https://earthengine.google.com). It continuously provides geo-referenced, high-quality, and mostly multi-temporal images, such as 8-m FormoSat-2 and SPOT-5 images. Two of the medium satellite images listed in Table 1.6 merit special mention, SPOT and Advanced Space-borne Thermal Emission and Reflection Radiometer (ASTER), because they can acquire 3D images. SPOT images provide across-track stereoscopic coverage of the ground via sideway viewing by tilting the scanning mirror off the nadir. SPOT-5 carried one High-Resolution Stereoscopic sensor able to generate 10 m stereo images, and two High-Resolution Geometric sensors able to capture stereo images at a resolution of 2 m in the panchromatic mode and 10 m in the multispectral mode. ASTER bands 3N and 3B of the same wavelength view the ground from two slightly different angles simultaneously, producing a stereo view of the ground from the same track. The spatial resolution of ASTER imagery ranges from 15 m to 30 m to 90 m (TIR band) (Table 1.6). Such images can potentially be used to study natural hazards involving a change in surface elevation such as landslides and volcanic lava flows.

Compared to their VHR counterparts, medium-resolution images have a much broader spectral and spatial range of sensing, which makes them ideal to study all types of ground features on the regional scale. Such images are particularly suited to study certain natural hazards such as land degradation and desertification and assess the likely spot of earthquake occurrence via active faults and hazard recovery (e.g., impact assessment). However, they are ill-suited to study natural hazards of small features such as landslides or spatially extensive features such as hurricanes. Their temporal resolution is too coarse to study fast-changing natural hazards such as dust storms.

1.3.1.3 Radar Imagery

Radar is an acronym standing for RAdio Detection And Ranging. It is an active means of sensing by first transmitting polarized electromagnetic waves toward the target and subsequently recording the backscattered radar signals off a target (area) and measuring the two-way travel time between the target and the sensor. The distance between them is determined in terms of the radiation's velocity of propagation which is constant. A radar sensor must encompass two components: an antenna and a transmitter. The antenna array is a device that operates in two alternative modes of transmission and reception. The transmitter transmits microwave radiation pulses over a very short duration on the order of micro-seconds and receives the radiation scattered back from the target of sensing. The echoes (signal) from the target are time-dependent, and their amplitude is proportional to the strength of radar return.

Radar images are commonly acquired by satellites launched by public institutions. The earliest radar satellites were launched experimentally in the 1990s, including the European Remote Sensing (ERS-1 and -2) satellites. Now they have been superseded by new generations of satellites with better quality. Among the more recent radar satellites is EnviSat, an environmental satellite launched by the European Space Agency (ESA) on 1 March 2002. It has an altitude of 790 km, a revolution period of 101 minutes, and a revisit period of 35 days (Table 1.7). EnviSat exemplifies multi-sensor sensing, with the payload comprising nine sensors, one of which is a C-band (5.331 GHz) Advanced Synthetic Aperture Radar (ASAR). ASAR images have a spatial resolution of 30 m, identical polarization of VV (vertical transmission and vertical reception) or HH (horizontal transmission and horizontal reception), and a swath width varying between 56 and 105 km, depending on the incidence angle. Other space-borne SAR data are acquired from ALOS (Advanced Land Observing Satellite) PALSAR (Phased Array type L-band SAR), ALOS-2, and the Japanese Earth Resources Satellite (JERS).

The most recent additions to the existing fleet of radar satellites are Sentinel-1, RadarSat-2, and TerraSAR-X. Launched in April 2014, Sentinel-1A significantly increases the volume of publically available radar data. Its constellation comprises two identical satellites, A and B. They halved the normal revisit period of 12 days to just 6 days. Such a short temporal resolution enables the acquisition of interferograms with a six-day interval and is ideal for monitoring height-related hazards (e.g., ground subsidence) and characterizing its evolution, and rapidly estimating the volume of landslide debris. The Sentinel-1A satellite carries a C-band sensor that operates at two incidence angles of 23° and 36°, alternating along the satellite orbital direction at 100 km intervals. Sentinel-1 image swath width is a function of the scanning mode. It is 250 km in the interferometric wide-swath (IW) mode. Each IW-SLC (single-look complex) image has three sub-swaths, stemming from image acquisition in successive bursts by steering the radar antenna between adjacent sub-swaths. In this way, images of a large swath width are acquired with reduced scalloping and minimum noise. At present, Sentinel-1A/1B is the sole operational mission able to provide global coverage in near real-time. Sentinel-1 data with pre-defined observational scenarios are archived and freely accessible through the Copernicus Open Access Hub (https://scihub.copernicus.eu/). These data produce potentially better coherence than other radar images owing to the enhanced spatial (2.7 m × 22 m) (range × azimuth) and radiometric resolutions.

Launched on 14 December 2007, RadarSat-2 operates in three imaging modes of single-, dual-, and quad-polarization. It started acquiring SAR images in 2011, sensing anywhere around the globe between 81.4°N and 81.4°S. RadarSat-2 can operate in 18 imaging modes, 2 of which worth mentioning are polarimetric and ultra-fine. The former allows retrieval of full-vector polarization information, while the latter can record images of 3 m resolution. In the cross-polarization ScanSAR mode, the acquired images cover a nominal ground area of approximately 500 km × 500 km at a pixel size of 50 m.

Two of the radar images listed in Table 1.7 are VHR images, COSMO-SkyMed (COnstellation of small Satellites for the Mediterranean basin Observation), and TerraSAR-X. The former's constellation (X-band) consists of four identical satellites launched between 2007 and 2010 by the Italian Space Agency, recording images at a nominal spatial resolution of 2.5 m. As the name implies, this satellite's images are not available outside the Mediterranean region. TerraSAR-X is a German radar satellite successfully launched on 15 June 2007. It acquires radar images of the finest, sub-meter spatial resolution. The exact resolution varies with the mode of scanning. SpotLight images have a resolution up to 1 m, covering a ground area of 10 km (width) × 5 km (length). The resolution

TABLE 1.7
Major SAR Satellites and Their Image Properties

Satellite	RadarSat 1/2		EnviSat	Sentinel-1	JERS-1	ALOS-1	ALOS-2	TerraSAR	COSMO-SkyMed
Sensor	SAR	SAR	ASAR	CSAR	SAR	PALSAR	PALSAR-2	TSX-1	SAR2000
Band	C	C	C	C	L	L	L	X	X
Wavelength (cm)	5.6	5.6	5.6	5.6	23.5	23.5	22.9	3.1	3.1
Polarization	HH	All	HH/VV		HH	All		All	HH/VV
Incident angle	20–50	10–60	15–45	30–46	35	8–60	8–70	20–55	20–60
Mode	Fine	Fine Ultra fine ML fine	IM	IWS/SM		FBS	HDS/HSD[a] FBS/FBD		Spot Strip scan
Range resolution (m)	10–100	3, 5, 10	30	5–20	18	7–100	3, 10, 100	1–16	2.5, 1
Azimuth resolution (m)	9–100	3, 5, 10	30	5–20	18	5–100	1, 3, 100	1–16	2.5, 1
Swath width (km)	45–500	50–500	56–400	250	75	40–70	25 spot 50 strip 350 scan	10 30 200	10 40 200
Repeat cycle (days)	24	24	35	6[b], 12	44	46	14	2–11	2–16
Data duration	1995–	2007–	2003–2010	2014–present	1992–1998	2006–2011	2013–2017	2007–2018	2007–2014
Orbit height (km)	798	798	800	693	568	692	639	514	619
Max. velocity (cm·year^{-1})	20.4	20.4	14.6	85	48.7	46.8	149.2	25.7	17.7 spot 35.4 strip 70.7 scan
Max. detectable subsidence D_{LOS} (mm·m^{-1})[c]	1.8	4.6 ultra-fine 2.8 ML fine 1.4 fine	0.5	0.7 IWS 2.8 SM		5.9	19.1 UBS/UBD[a] 9.5 HBS/HSD 5.7 FBS/FBD		

Source: Modified from Metternicht et al. (2005). Used with permission.

[a] UBS/D = Ultrafine (3 m) Beam Single/Dual (polarization), HBS = High-sensitive (6 m) Beam Single, FB = Fine Beam (10 m); HSD = High-sensitive, StripMap Dual

[b] Two identical satellites of 1A and 1B can half the return period

[c] Assuming an incidence angle of 37°; RadarSat-2: spot mode – 3 m resolution, 10 km swath width; strip mode – 8 m resolution, 40 km swath width; scan mode – 25 m resolution, 200 km swath width; RadarSat-1: fine mode – 45 km swath width; strip mode – 100 km swath width; scan mode – 200 km swath width. ML – multi-look

drops to 3 m in the StripMap mode that enables a ground area of 30 km (width) × 50 km (length) to be covered in one image. The resolution decreases further to 16 m in the ScanSAR mode, even though each image covers a much larger ground area of 100 km (width) × 150 km (length).

Compared with their optical counterparts, radar images are advantageous in that they are not subject to cloud contamination. However, these images suffer from excessive geometric distortion, such as slant range distortion, radar layover, and radar shadow, due to the scanning range far exceeding the sensor altitude. In spite of these limitations, they are still useful in studying certain hazards such as ground subsidence. In comparison with optical images, they are more sensitive to surface relief. Since airborne optical images tend to be one-off, they do not allow subsidence to be detected. Thus, it is space-borne SAR images that are exclusively used for this purpose.

1.3.2 LiDAR

Non-imagery LiDAR sensing, also known as airborne laser scanning (ALS), is an active means of sensing the surface height in dense 3D points. Its principle of sensing is identical to that of active radar sensing except the energy used being laser light of 0.8–1.6 μm wavelengths. The sensing is grounded on transmitting laser signals to the target at short durations (e.g., up to 25,000 times per second) and receiving the signals bounced back from the target (Figure 1.1). Through timing, the duration of laser signal transmission, the distance between the sensor and the target is calculated from the velocity of light, or

$$Distance = 0.5c \times \Delta t \quad (1.5)$$

where c = velocity of light (299,792,458 m·s^{-1}); Δt = duration of the pulse transmission from the laser scanner to the target and back. The target area is sampled at a density up to 1 point·m^{-2}.

During the transmission and reception of LiDAR signals, both the orientational and positional information of the scanner at any given moment is precisely known via two instruments, the IMU and GPS. The former records the orientational information of the scanner (φ, ω, κ), while its

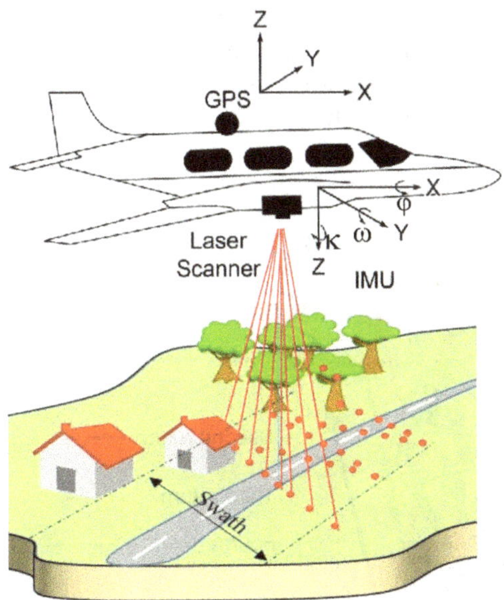

FIGURE 1.1 Principle of airborne laser scanning in determining the position of ground features. The aircraft must be equipped with a GPS and an Inertial Measurement Unit (IMU) to measure its position (X_s, Y_s, Z_s) and orientation (ϕ, ω, κ) at any given moment, respectively.

positional information (X_s, Y_s, Z_s) at the time of transmission is obtainable from the onboard GPS unit (Figure 1.1). During scanning, both the range of a given target from the sensor and its associated scanning angle at the time of laser pulse transmission are captured and recorded. Virtually, raw LiDAR data comprise just a huge collection of range measurements and sensor orientation parameters. The laser range (slant range) calculated using Eq. 1.5 is subsequently converted to three-dimensional increments between the scanner and the target in a global coordinate system through trigonometry. The modification of the sensor position by these increments enables the determination of the 3D position (X_t, Y_t, Z_t) of the target. Although LiDAR was invented in the 1960s, it did not find wide applications until the advent of GPS coupled with IMU in the 1990s when commercial LiDAR systems became available. At present, it is used mostly for generating elevational information in hazard studies.

The positioning accuracy of the ground surface is affected by a number of factors, such as the GPS and IMU accuracy, the atmospheric conditions, and the LiDAR signals bounced back from the target. In order to improve the accuracy of LiDAR data, it is desirable to derive the absolute aircraft position (trajectory) from a differential, geodetic GPS network so as to eradicate any consistent bias in the GPS signals. Once striking the target, the LiDAR signals may be bounced back a number of times. The first return is the signal bounced back from the closest target (Figure 1.2). It measures the range of the first object encountered (e.g., building roof and tree top). The last return is the signal bounced back after penetrating the canopy to hit the ground as in a forest. The subtraction of the last return from the first return leads to the bare earth height. This elevational differencing effectively separates ground points from non-ground points after spatial filtering. It is the ground points, the height of which at different times is indicative of surface change caused by natural hazards. In comparison, the intensity of LiDAR return that may enable the separation of the bare earth surface, building footprints, vegetation structure, roads, signs, and lamp posts is not so useful in studying most natural hazards.

LiDAR is a fast and accurate means of sensing. ALS potentially offers a few distinct advantages over slow and expensive ground-based leveling in mapping surface deformation induced by natural hazards, especially in remote and inaccessible areas. It is able to yield a spatial perspective into the vertical movement over an area. The detection is implemented by converting the randomly distributed, dense points of 3D coordinates into a regular grid digital elevation model (DEM), the grid cell size of which can range from 1 to 10 m. Hazard-triggered surface deformation is detected by subtracting a recent DEM from an earlier one of the same area, spatial extent, and grid size. The net difference represents the vertical change (e.g., ground subsidence) in elevation.

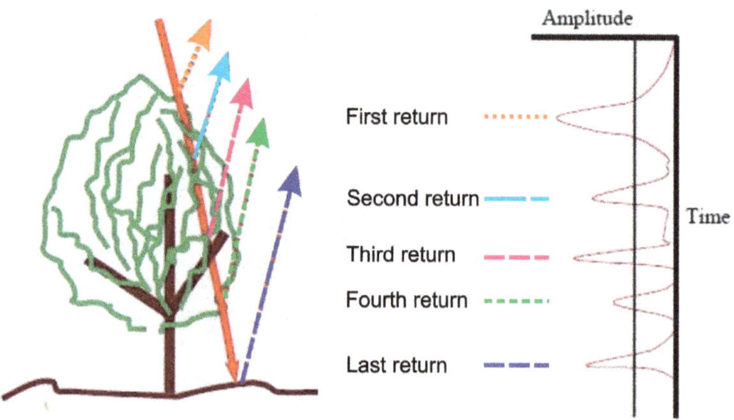

FIGURE 1.2 Multi-returns of laser scanning from the ground cover and their amplitude.

1.3.3 GPS

1.3.3.1 GPS Fundamentals

GPS is a worldwide navigation system launched by the US Department of Defense. This system comprises a constellation of 24 satellites (currently numbered 31) and ground stations, the space segment. Of the 24 satellites in the operational mode, 21 (28) are in use, 3 are spares, and 3 are used for testing (Figure 1.3). All satellites have an altitude of 20,200 km with an orbital period of 12 hours. Their geometric configuration guarantees that a minimum of four satellites are always visible from any point on the Earth's surface. These satellites all transmit radio signals of a unique wavelength that are received by GPS data loggers, the user segment. The GPS control segment that the user seldom interacts consists of 12 ground stations. They continually monitor all satellites, track their signals, and determine their precise position. The Master Control Station calculates orbit and clock information every 15 minutes. Ground Antennas broadcast updated navigation information to satellites once or twice a day. Satellites use updates in their broadcasts.

FIGURE 1.3 Constellation of GPS satellites in space. (Source: https://seos-project.eu/GPS/GPS-c01-p03.html.)

Virtually, GPS is a ranging system just like LiDAR. A GPS receiver continually tracks the radio signals transmitted from several satellites simultaneously. Radio signals are waves of electromagnetic radiation traveling at a mostly constant speed of 299,792,458 m·s^{-1}. The internal microprocessor of the GPS receiver contains "an almanac" of satellite positions, calculates the distance from each satellite, and trilaterates its position on the Earth's surface. The distance between the satellites and the GPS receiver is calculated based on the duration of signal propagation from the satellite to the receiver using Eq. 1.5 without the 0.5 coefficient. Thus, the distance between a GPS receiver and a satellite is known if the duration of signal propagation is timed very accurately. In this way, the determination of the GPS receiver's position is reduced to the determination of its distances (ranges) to multiple satellites at the same time. Each range puts the GPS receiver on a sphere about the satellite, and the intersection of several ranges yields a unique position. For instance, if the distance from the receiver to one satellite of a known position is known, the receiver's possible location is confined to a theoretical sphere, the radius of which is equal to the distance. If the receiver's distance to two satellites is known, it must lie somewhere on a theoretical circle formed via the intersection of the two spheres. If the receiver's distance to a third satellite is known, its position is reduced to only two possible points, formed via the intersection of the theoretical circle with the third sphere. One of them is easily eliminated because it is neither on the surface of the Earth nor close to it. Intersection with the fourth sphere uniquely locates the position of the receiver at higher accuracy. The more satellites are tracked, the higher the positioning accuracy.

The timing of radio signal transmission from the tracked satellite to the GPS receiver is based on pseudo-random code (Figure 1.4). Each transmission is time-tagged and contains the satellite's position. The code sent from a satellite at t_0 matches the code recorded at the receiver at t_1. The time of arrival at a receiver is compared to the time of transmission based on the correlation of the signal's unique phase delay (Figure 1.4). The propogation duration Δt (t_1–t_0) is determined by locking into the same phase.

1.3.3.2 GPS Accuracy

Fundamentally, GPS can supply three types of end results: time, velocity, and position, of which position is relevant to most natural hazards, while velocity applies to certain hazards such as debris creeping rate. The GPS-logged position is subject to the reliability of timing and other factors, the chief ones being satellite constellation, the atmosphere, the logging environment, and the GPS receiver's functionality (Table 1.8). Timing (clock) inaccuracy and atmospheric delay are minimal with the GPS satellites that are equipped with hydrogen maser atomic clocks. They lose 1 second every 2,739,000 million years; hence, they are extremely accurate to 1 × 10^{-9} seconds. At this

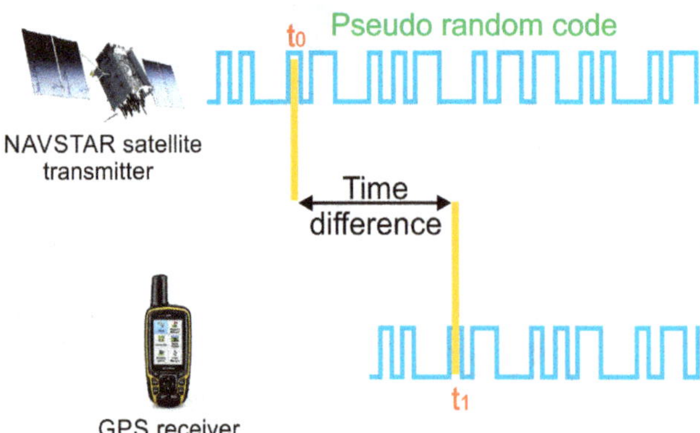

FIGURE 1.4 The duration of signal transmission from the tracked satellite to the GPS receiver determined from the same phase of the pseudo-random code sent from the satellite at t_0 and the identical phase at the receiver at time t_1.

accuracy, timing-induced inaccuracy is negligible. However, the clock embedded in a GPS receiver is far less accurate. Its timing inaccuracy degrades the ranging accuracy. Vertically, the atmosphere is heterogeneous (e.g., amount of ions) in different layers, making the velocity (v) of electromagnetic radiation waves in Eq. 1.5 inconsistent and deviating from 299,792,458 $m \cdot s^{-1}$. The ionosphere and troposphere delay the propagation of GPS satellite signals and cause them to bounce around when traveling through them. This prolongs the time it takes for the signals to reach the GPS receiver on the Earth's surface, which alters the trilaterated position.

The relative position of the tracked satellites exerts an impact on the accuracy of GPS-logged coordinates due to the uncertainty in the calculated distance between the satellites and the receiver caused by the inconsistent signal speed and timing inaccuracy. In trilaterating the GPS position from the signals from two satellites, the uncertainty zone of trilateration is minimized if the two satellites are distant from each other, leading to accurate positioning, and vice versa. This issue is known as positional dilution of precision, a measure of the geometric configuration of satellites used to calculate a fix. The uncertainty in intersecting the position of the receiver from various ranges is magnified if the tracked satellites have an ill-formed constellation (e.g., they are too close to each other). This uncertainty can be reduced by adopting a positional dilution threshold, below which the signal from one of the tracked satellites is excluded from trilaterating the GPS position, provided there are still four satellites available for trilateration.

Logging conditions refer to the surrounding environs that may affect the propogation path of satellite signals in areas of high relief. When the signals eventually reach the Earth's surface from the satellites, instead of reaching the GPS receiver directly, they may strike tall buildings in the vicinity of the receiver before being bounced back to it along multiple pathways. The net effect of multi-path is a prolonged time taken for the signals to reach the receiver, leading to an exaggerated range between the satellite and the receiver. Predictably, the positioning accuracy is noticeably lowered in areas with too many obstacles such as forests, in close proximity of high-rise buildings in urban areas, and in tall mountainous areas. In the worst case, the obstruction of GPS signals by tall features in the landscape may completely disable the functioning of a GPS receiver because insufficient signals are received, as in a dense forest. As shown in Table 1.8, most of these sources result in an inaccuracy smaller than 2 m except for positional dilution of precision. Its exact value depends on the geometry formed by the tracked satellites. So far GPS vertical inaccuracy has not been budgeted in detail, only relative positioning accuracy. Under optimal circumstances, over short lines

TABLE 1.8

Sources of GPS Positioning Inaccuracies and Their Typical Ranges (Unit: m) for Three Types of GPS According to Trimble (2007)

Sources of GPS Inaccuracies	Autonomous	Differential	RTK
Range inaccuracy – satellites			
Clocks	1–1.2	Removed	Removed
Ephemeris errors	0.4–0.5	Removed	Removed
Range inaccuracy – atmosphere			
Ionosphere (charged particles)	0.5–5	Mostly removed	Almost all removed
Troposphere (the dense part)	0.2–0.7	Removed	Removed
Subtotal	**1.7–7.0**[a]	**0.2–2.0**	**0.005–0.01**
Receiver inaccuracy	0.1–3	0.1–3	Almost all removed
Data logging environment			
Multi-path	0–10	0–10	Markedly reduced
Positional dilution of precision	3–5	Mostly removed	Almost all removed

[a] Ephemeris and clock inaccuracies are correlated with each other, and they total typically less than their sum

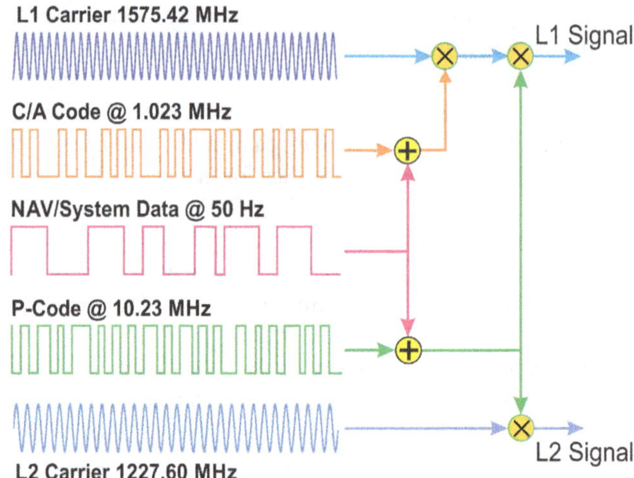

FIGURE 1.5 Types of GPS signal codes and their frequencies (+: modulo 2 sum; ×: mixer).

(e.g., <10 km), the error budget is predominated by antenna centering and height errors, antenna phase center motion, and multipath inaccuracies (Table 1.8). Differential troposphere influences primarily the relative height component. If the baseline is shorter than 100 km, GPS inaccuracies are typically within 2–3 cm (Table 1.8). Of the three types of GPS, RTK (real-time kinematic) is the most accurate, followed by differential GPS (DGPS).

Finally, GPS accuracy is also related to the receiver's functionality, such as the frequencies of GPS signals it can track or whether it uses a single channel or dual-channel. GPS signals fall into several types, such as L1, L2, P, Y, and coarse acquisition (C/A) (Figure 1.5). L1 and L2 are carrier frequencies at 1,575.42 and 1,227.60 MHz transmitted by each satellite. Both are modulated with pseudo-random codes to be picked up by the receiver. The C/A code of 1.023 Mb·s^{-1} is freely available. The P (precision) code at 10.23 Mb·s^{-1} is ten times faster than the C/A code but has restricted access. Y code is encrypted for military use. GPS receivers with dual-frequency carrier phase observables can double differencing wide lane signals between the satellite and the receiver. Multi-channel, multi-frequency receivers are able to achieve a vertical accuracy of ±5 mm +1 ppm root-mean-square error (RMSE) in the static mode (Marin et al., 1998).

GPS positioning accuracy can be improved via three means. The first method is to average multiple readings logged at the same stationary spot (e.g., fixed station). This method is applicable when the GPS receiver is immobile and keeps logging data continuously. The more readings are used for averaging, the higher the accuracy. Averaging increases positioning accuracy to around 2–4 m. This method has limited applicability to hazard studies except in monitoring deformation or subsidence. The second method is via Kalman filtering, a process of converting satellite-to-receiver pseudo-ranges to receiver position estimates. This complex processing requires specification of the receiver status, such as stationary, low dynamics, and high dynamics. Kalman filtering is very demanding and impractical for hazard studies. In comparison, the third method of differential correction is much simpler and easy to implement.

DGPS involves the deployment of two receivers, one placed at a base station or reference point, the precise coordinates of which are known through long-term repetitive observations, and another placed in the field known as the rover GPS (Figure 1.6). The roving receiver is moved around to survey the hazard-affected target in the field. Both receivers track the signals transmitted from the same set of satellites simultaneously. The station's true coordinates are then compared with its GPS-logged coordinates, and their discrepancy is used to offset the rover GPS readings via logging time to derive the truer coordinates (Figure 1.7). This correction requires that the two receivers always lie

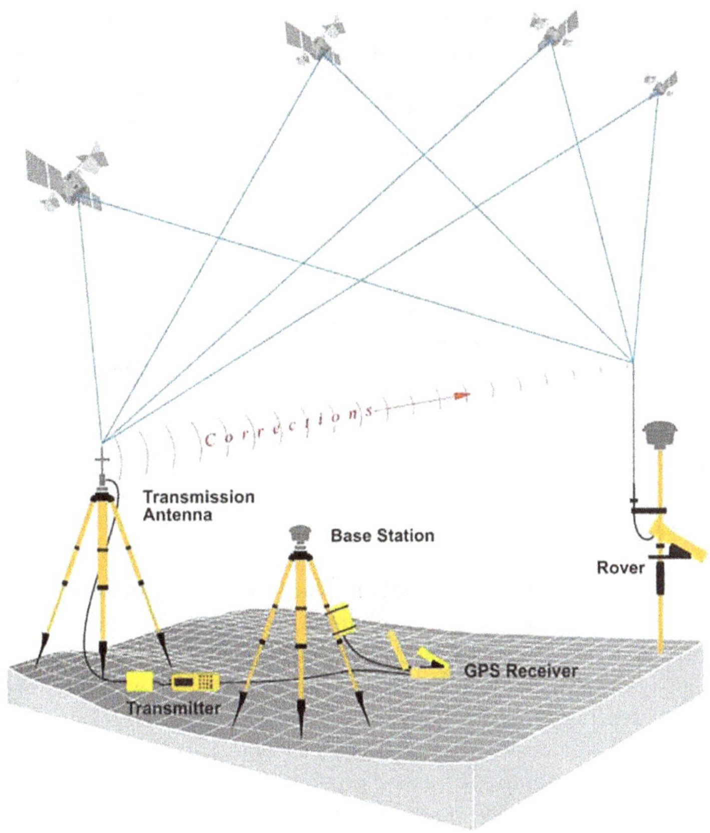

FIGURE 1.6 Principle of differential GPS using two receivers tracking the signals from the same set of satellites simultaneously, one placed at the fixed base station, and one rover in the field. (Source: Casciati et al., 2016.)

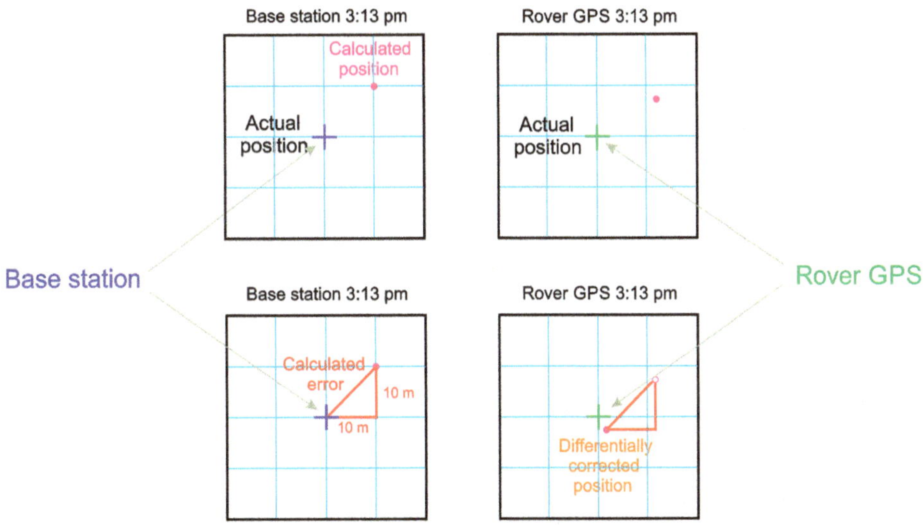

FIGURE 1.7 An example illustrating how differential GPS is implemented to correct positional coordinates logged by the rover GPS to improve its positioning accuracy.

within a certain distance from each other (e.g., <350 km). Under this condition, signal inaccuracies of the two receivers are assumed to be identical at the same time and are completely canceled out via the correction. The positions calculated by the mobile GPS receiver are adjusted according to the broadcast differential information either in real-time or during post-logging processing.

This method of correction works if the errors (e.g., pseudo-range noise and multipath) are virtually identical within a baseline of 350 km around the base station. Apparently, the accuracy of DGPS decreases as the baseline lengthens because of larger tropospheric, ionospheric, and orbit uncertainties. Differential correction can reduce vertical inaccuracy from ±10 to ±3 m (Okuda et al., 1990), still not accurate enough to study most natural hazards such as estimating debris volume. More precise coordinates can be computed using the precise ephemeris and a Saastamoinen tropospheric model, and via correcting the tropospheric delay by recording water vapor radiometry using Raman LiDAR. It is able to retrieve temperature profiles from total air density. After correction for the path delay, GPS-derived heights can be improved to a vertical accuracy of 5–10 mm (Chen et al., 2003), adequate for studying most hazards.

1.3.4 DEM

DEM data are vital to studying natural hazards in which topography plays a role, such as (flash) floods, landslides, snow avalanches, and even wildfires. From a DEM, a number of topographic and hydrological parameters and indices can be derived, including slope, aspect, topographic position index (TPI), and topographic wetness index (TWI). DEM data come with different spatial resolutions and accuracies, depending on the data sources used for their creation. Primarily, they are created from three types of data, stereoscopic images using the photogrammetric method, LiDAR data, and InSAR. The photogrammetric method produces a 3D view of the area of interest from a pair of stereoscopic aerial photographs or satellite images of a roughly equal scale. Through proper orientation and manipulation, a 3D view of the ground is established, from which contours are drawn first and later converted to a DEM via spatial interpolation digitally. While this method is able to produce detailed and highly accurate DEMs from aerial photographs, however, it requires precise ground control and complex computation, so is rather slow and has been gradually replaced by other methods.

Apart from aerial photographs, DEMs can also be produced from space-borne images such as IKONOS, ASTER, and SPOT based on the same photogrammetric principle. These images are able to sense the surface stereoscopically by viewing the earth from two slightly different perspectives or from adjacent paths. The resolution of the DEMs produced from such satellite images and their accuracy are inferior to those of photo-derived DEMs, but these DEMs cover most of the Earth's surface and are freely available as a data product. For instance, the ASTER global DEM (GDEM) product has a spatial resolution of 30 m or 1 arc, freely downloadable as 1° × 1° tiles at https://search.earthdata.nasa.gov/search/. Personal experience with GDEM suggests that the quality of such DEM data is low in areas prone to cloud coverage. Although the latest version (version 3) is significantly improved over the previous release, users are advised that the data still suffer from anomalies and artifacts that impede their effectiveness in certain hazard studies.

Another fine resolution DEM data product is TanDEM-X. This mission called TerraSAR-X add-on for Digital Elevation Measurement (https://tandemx-science.dlr.de/) was launched in June 2010 by the German Aerospace Center (DLR) to produce accurate DEMs. TanDEM-X acquired in the StripMap mode has a spatial resolution of 2–3 m, much finer than that of previous SAR sensors. In comparison with GPS-logged elevations, TanDEM-X has an absolute vertical accuracy of 1.6 m (Albino et al., 2015).

GDEMs have also been produced from Shuttle Radar Topography Mission (SRTM) at 90 m spatial resolution. SRTM DEMs of 30 m are available at the national level. Horizontally, the

SRTM 1 arc sec DEM has a global mean accuracy of about 20 m. This uncertainty is nearly halved with TanDEM-X (Bagnardi et al., 2016). More DEMs can be produced from similar radar data such as IFSAR, GeoSAR, and AIRSAR using the interferometric technology or InSAR. In Arctic areas, the best DEMs available are ArcticDEM produced from high-resolution (2 m) images. If completed, ArcticDEM will encompass all land areas north of 60°N, including all territory of Greenland, the State of Alaska in entirety, and the Kamchatka Peninsula of Russia. This data product is produced from 0.32 to 0.5 m WorldView-1 (since 2007), WorldView-2 (2009), WorldView-3 (2014), and GeoEye-1 (2008) panchromatic band. This data product is available freely at the Polar Geospatial Center, University of Minnesota (https://www.pgc.umn.edu/data/arcticdem/). Included in the data product are time-dependent 2 m digital surface model (DSM) strips along a satellite single pass, covering a typical scene of 17 km × 110 km. The DSMs have an internal (pixel to pixel) accuracy of 0.2 m, and a vertical and horizontal geolocation offset of 3–5 m (Dai and Howat, 2017). Similar accuracies have been achieved from VHR images such as IKONOS and QuickBird (Table 1.9). Since the ArcticDEM DSM data date back to 2007 only,

TABLE 1.9

Vertical Accuracies Achieved with Various Stereo Images Using the Photogrammetric Method

Imagery	Ground Resolution	Swath Width (km)	Accuracy	Affecting Factors[a]	Pros	Cons
Aerial photographs	Varying with camera focal length	Varying with flying height	3 cm–14.3 m	Flying height, image resolution Accuracy of ground control	Highly accurate; DSM possible	Slow; expensive; available for small areas.
IKONOS	1 m (PAN) 4 m (XL)	11.3	0.2–0.7 m (DEM accuracy: 1–4 m at 68% LE)	Sensor orientation; ground cover; Extra ground control	Images available at short return period; large B/H ratio; along-track and across-track 3D possible	Expensive, limited ground cover
QuickBird Geo Stereo	2.4 m	16.5	0.2–1.38 m (DEM accuracy: 10 m)	3D model used, ground control quality; external DEMs	DSM possible; stereo images obtained at the same time	Expensive, sophisticated software needed
SPOT 1–4 SPOT-5	0–20 m Up to 4 m	60	5–10 m 5 (5.2–6.8) m 2.38 m (2.89–2.60)	Geometric model adopted; GCP quality;	More accurate than ASTER images A large ground area is covered	Expensive; Across-track 3D only; time lapse between stereopair
ASTER	15 m	60	10–39 m	Surface relief	Inexpensive; data available globally	Not so accurate, mostly for DEM; long repeat cycle (16 days)
TK-350 space photograph	10 m	200–300	17 m (27–39 m for DEM of 40 m cell)	Surface cover; DEM cell size; ground control	Wide area coverage	Low accuracy; images not widely available

Source: Gao (2007).

[a] All photogrammetric methods are subject to the B/H (base/height) ratio

other data with a longer history of existence (e.g., SRTM) may have to be used to detect pre-2007 vertical changes caused by natural hazards.

The spatial resolution and vertical accuracy on the order of meters for some DEMs (see Table 1.9) may be insufficient for studying small-sized natural hazards, such as lava flows and landslides, or warranting the derivation of change in surface elevation (e.g., detection of ground subsidence) to a creditable accuracy level. They require more current, accurate, finer resolution sub-meter DEMs that are commonly produced from airborne LiDAR. LiDAR data allow reasonably accurate DEMs to be produced via spatial interpolation, during which dense, irregularly distributed LiDAR points are converted to regular grids. LiDAR (and InSAR) data offer repeated measurements from periodic images at a fine temporal resolution of around tens of days, whatever height inaccuracy of the raw LiDAR data ultimately propagates to the DEM vertical inaccuracy. LiDAR data can have a planimetric and vertical RMSE of 0.83 and 0.11 m, respectively, that are reduced to 0.60 and 0.09 m after correction for systematic errors (Behncke et al., 2016). LiDAR DEMs are useful for local-scale hazard studies, such as the estimation of lava flow volume. They are especially important if secondary parameters (e.g., surface curvature) are to be derived from them.

Different methods of generating DEMs have different pros, cons, and achievable accuracy levels. They are compared in Table 1.10 among GPS, interferometry, and LiDAR. Of the three types of GPS, RTK is the most reliable and can generate DEMs in real time, but it is operational within a short range. Radar is much less accurate, achieving a vertical accuracy of up to 1 m. Radar-generated DEMs are good for regional or global-scale studies. InSAR data offer densely spaced measurements from repeated images at a fine temporal resolution for tens of days. ALS can produce DEMs of dm-range vertical accuracy quickly but very expensively, and a huge volume of data must be processed, so it is suitable for local-scale applications.

As illustrated in Table 1.11, the vertical accuracy of those DEMs produced from large-scale photos and topographic maps is the highest, followed by LiDAR. Pléiades-1 DEMs have a mean offset of −7.6 m (standard deviation = 0.4 m) and −1.3 m ($\sigma = 0.3$ m) in the east and north directions, respectively (Bagnardi et al., 2016). Of all the DEM sources, only LiDAR, topographic maps, and large-scale photos produced DEMs with a cm-level vertical accuracy. They are the best at studying small-sized hazard features. Even so, the vertical accuracy may not be sufficiently high to monitor subtle changes induced by natural hazards. One way of improving the reliability of monitoring is to make use of DEMs produced from the same sensor in change detection. In this way, any offset can be canceled out through DEM differencing.

1.3.5 Data Selection Considerations

So far a large variety of remotely sensed data is available for studying natural hazards. Which type of data is the best for studying a given type of hazard requires careful consideration of three important data properties: spatial resolution, temporal resolution, and spectral resolution in this order. Spatial resolution is the most important to consider because it governs the detectable detail of the target. Fine resolution images are critical to recognizing very small hazard features on the ground, such as landslide paths, avalanches, and lava flows. It must be borne in mind that images of a very fine spatial resolution cover only a small ground area (Figure 1.8), so many frames of images are essential to cover an extensive study area, and they take a lengthy time to process. Before deciding on the best imagery to use, readers need to ask themselves about the smallest features to be recognized from the remote sensing images. Then the image resolution must be 5–10 times finer than this dimension to be successful. No matter which type of image is selected, there is always a marked trade-off between data spatial resolution, processing time, and areal coverage (Figure 1.8). Since most of the data are complementary to each other in their strengths, naturally, a multi-sensor approach of combining lower resolution data at a larger scale with those VHR images is advantageous in studying natural hazards as it takes advantage of each's strengths to the maximum. As an example, hazard damage may be mapped over a large area from a coarse resolution image rapidly

TABLE 1.10
Comparison of the Main Features of Three Types of Height-Finding Methods

Method	GPS			Radar		LiDAR	
	Standard	**DGPS**	**RTK**	**InSAR**	**SRTM**	**ALS**	**SLA**
Accuracy level	10 m	Sub-meter	2–4 cm	Meter-level	16 m	Dm-range	Up to 1 m
Affecting factors	Logging environment; Satellite constellation	Receiver functionality; Surface vegetation	Baseline length; receiver functionality	Slope gradient; slant range	Baseline; surface relief and cover	Topographic relief, surface cover, soil moisture	Same as ALS
Strengths	Accurate	Improved accuracy over standard GPS	Real-time results; highly accurate	Cost-effective; Sub-surface height and surface structure possible	Readily available; inexpensive	Fast, areal-based, sub-surface height possible	Wide extent; cheap
Limitations	Not possible to measure canopy height; site access essential	Two receivers required; results not immediately available	Two receivers essential; short baselines in hummocky terrain	Narrow swath width	Not so accurate; accuracy difficult to improve	Expensive Small area covered	Resolution and accuracy not suitable for certain uses
Best applications	Spot height Navigation, LiDAR	Accurate DEM over accessible areas	National leveling; real-time	Tectonic and Volcanic monitoring DEM	Regional to global DEMs	3D surface, DEM, and contouring	Regional DEM

Source: Gao (2007).

TABLE 1.11
Comparison of the Vertical and Horizontal Accuracy of DEMs Produced from Various Sources Using Different Methods

Data Source	Ground Pixel Size (m)	Vertical Accuracy (m)	Authors
Photos (1:5,000–1:800)	0.13	0.05	Baldi et al. (2008)
Topographic maps	0.20	0.09	
LiDAR	N/A	0.09	Behncke et al. (2016)
Pléiades-1	1	0.51	
TanDEM-X	5	1.12	Bagnardi et al. (2016)
SRTM	30	3.64	
ASTER-GDEM	30	5.74	

in a preliminary study. Later the initial damage estimates in specific areas of severe damage are refined and improved using finer resolution data.

Temporal resolution is very important to studying natural hazards that change quickly with time, such as wildfires, hurricanes, volcano eruptions, dust storms, and floods. These events take place so fast that multiple images must be acquired at a short temporal interval frequently to track the change and predict their likely spread and movement. Since images of a fine temporal resolution tend to have a coarse spatial resolution, it is impossible to study the hazard in detail, such as the type of vegetation burned in a wildfire. In most cases, only the location and/or extent of the hazard can be tracked. However, it is still possible to study the composition of volcanic plumes from coarse resolution images. Multi-temporal data are essential to assessing the effects of a hazard, such as an area burned by a fire or affected by a landslide. In comparison, spectral resolution is not a quality critical to the study of most natural hazards except land salinization and soil moisture as in mapping floods.

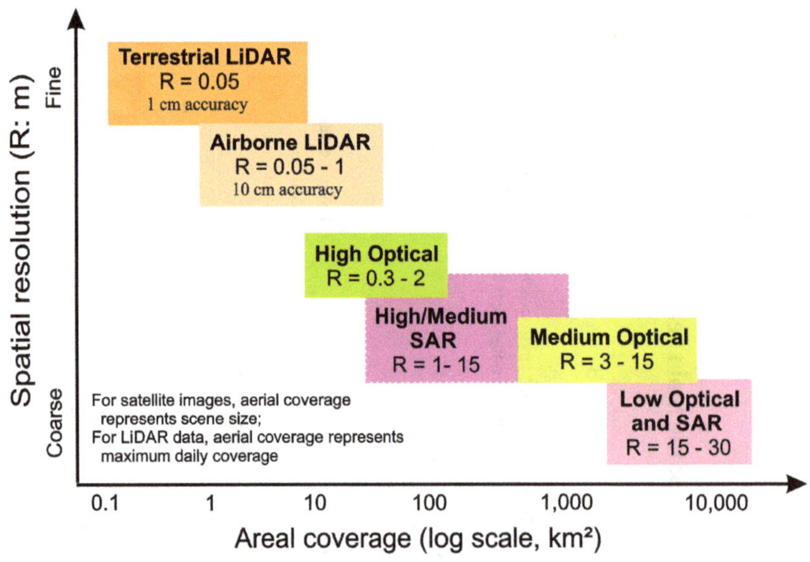

FIGURE 1.8 Relationship between spatial resolutions of typical types of remotely sensed data and their areal coverage.

1.4 DATA ANALYTICAL METHODS

While most hazards are directly observable on remote sensing images, their severity is not so easy to judge or assess visually. Besides, the assessment of hazard severity and prediction of its likely spot of occurrence are not possible via the analysis of remotely sensed data alone. Such tasks require sophisticated methods of data analysis, in which remotely sensed data are analyzed jointly with data from other sources. So far a few analytical methods have found wide applications in assessing hazard risk and predicting hazard occurrence, the most successful ones being machine learning algorithms. Machine learning is a type of artificial intelligence able to generate the optimal output from a given set of inputs based on the experience gained from the training samples in a process known as supervised learning. This suite of automatic methods predicts the output from a set of the most important factors based on a prediction model constructed from the training dataset. This section introduces six of them: support vector machine (SVM), random forest (RF), artificial neural network (ANN), classification and regression trees (CART), Naïve Bayes (NB) Trees (NBT), and generalized additive model (GAM), together with two image classification paradigms.

1.4.1 Image Analysis Methods

There are two paradigms of digital image classification, per-pixel and object-oriented, both undertaken in the multispectral domain. Per-pixel classification is usually performed on medium- or high-resolution images, during which pixels of a similar value are lumped together to form categorical classes of ground cover. Instead of classifying pixels individually, object-oriented image analysis classifies groups of pixels on VHR images or even aerial photographs. The images may be PAN-sharpened and contrast stretched, before the panchromatic and multispectral bands are co-registered with each other. Then the image is converted to clusters of pixels or "objects" that are treated as the unit of decision-making in the classification. Object-based image analysis (OBIA) allows more non-spectral features to be incorporated into the analysis to compensate for VHR images' lack of spectral richness.

1.4.1.1 Per-Pixel Image Classification

This paradigm of image classification is underpinned by an implicit assumption that pixels of the same or a similar value in the same spectral band represent the same ground cover, which is mostly true with medium-resolution optical images. Conversely, different ground covers have different pixel values in the same spectral band. Thus, all the pixels of the same or a similar value are grouped together in an effort to produce a map of qualitative hazard severity, based on subjective judgment. Per-pixel classification may be implemented as unsupervised or supervised. Unsupervised classification does not assume any prior knowledge about the image to be classified. It is also known as spectral clustering analysis in which all pixels in the input image are grouped into clusters based on their spectral similarity. The identity of these spectral clusters is ascertained in a post-classification session. This implementation is fast, but the identity of a cluster may not be informative or unique. Besides, it is unlikely that there is always a one-to-one correspondence between a spectral cluster and an information class, which is not the case with supervised classification. It can be undertaken using one of three methods: parallelepiped, minimum distance to mean, and maximum likelihood. They all require feeding training samples selected by the image analyst to the computer. They represent all the land cover types or information classes to be classified. Then their statistical parameters in each of the multispectral bands are calculated, such as mean, standard deviation, and covariance between multispectral bands. The computer then evaluates each individual pixel against these statistical parameters to decide which of the pre-defined information classes it should be assigned to optimally. Both unsupervised and supervised classifiers treat individual pixels in isolation, while the rich spatial information of the pixels such as context and texture is completely ignored in the decision-making process. Understandably, the classification accuracy is rather limited in some cases, especially with VHR images, which can be overcome using OBIA.

1.4.1.2 Object-Oriented Image Classification

With the wide availability of VHR images such as QuickBird, IKONOS, WorldView-3, GeoEye-1, and drone-acquired digital aerial photographs of a sub-meter pixel size, the assumption of the same ground cover having the same pixel value in the same spectral band is frequently violated. For instance, the moisture content of a denudated surface created by a landslide is variable spatially and will not be classified as a single cover in a per-pixel classification. Besides, those landslide paths falling in topographic shadow have a pixel value different from sun-lit landslides. Furthermore, VHR images are not rich in their spectral bands that are usually numbered four or fewer (three visible light and one NIR bands, see Table 1.5). They are ill-suited for spectral-based per-pixel classification. These images are better classified using OBIA, in which decisions regarding the identity of a pixel are made for a group of them called patches collectively. They are formed from neighboring pixels in a process known as image segmentation, during which any variations among the member pixels are homogenized locally according to the pre-defined criteria. Similarity criteria are devised from maximum spectral homogeneity within a segment, smoothness, and area-to-perimeter ratio (i.e., shape). Critical to the formation of patches is the specification of a proper scale at which local spectral heterogeneity is measured and compared. To find out the most suitable scale of segmentation, the same image may be repeatedly segmented at multiple scales, and all the segmented results are compared with one another to reveal the most suitable scale to use. After image "objects" are formed, their classification is implemented as the usual nearest neighbor analysis. In deciding the identity of these patches, consideration is given to their geometry and value, so more image parameters are involved in the decision-making than in per-pixel image classification.

The main features of per-pixel and object-oriented image analysis in automatically mapping natural hazards are compared in Table 1.12. Per-pixel classification is simple, especially with unsupervised classification, but it involves assumptions that are frequently violated, resulting in low accuracy in high topographic relief areas. Besides, it does not allow the incorporation of non-pixel data such as texture and contexture in the decision-making. This method is best suited to medium-resolution images. In contrast, OBIA is complex as objects must be formed first via image segmentation. It does allow non-spectral information to be used in the decision-making, such as shape and texture. The use of more evidence in decision-making tends to produce more reliable results. However, neither per-pixel nor object-oriented image classification allows the incorporation of non-imagery, auxillary data in the decision-making, a drawback that can be avoided by machine learning methods.

TABLE 1.12

Comparison of Per-Pixel and Object-Oriented Methods in Automatically Mapping Natural Hazards from Multispectral Remote Sensing Data

Classification Paradigm	Per-Pixel	Object-Oriented
Unit of decision-making	Individual pixels	A group of pixels
Evidence of decision-making	Pixel value	Object's geometry (e.g., size and shape) and texture
Complexity	Simple, statistical-based	Complex, logical-based
Nature	Supervised or unsupervised	Always supervised
Main problems	Assumptions frequently violated	Scale of segmentation uniform for all features
Ability to incorporate external data	No	Yes
Best use	Medium-resolution imagery	Very high-resolution imagery

1.4.2 Machine Learning Methods

In recent years, machine learning methods have gained popularity in mapping certain hazards. Machine learning algorithms have various forms, the most common being SVM, ANN, decision trees, RF, and CART, all of which are introduced in this section.

1.4.2.1 Support Vector Machine

SVM maps the input data onto a high-dimensional feature space and is more suitable to handle a huge set of variables than other regression methods. It is particularly good at addressing the non-linear relationship between the dependent and independent variables. The strength of this algorithm lies in its global optimality, generalization capability, simplicity, ease of calculation and implementation, and high learning efficiency. It is particularly strong at handling a small training set common in the traditional classification method. The SVM algorithm involves two critical parameters: C (the penalty parameter) and ε (residual), both of which are estimated from a kernel. It has various forms. The four most popular types of kernel functions are linear, polynomial of various degrees, (Gaussian) Radial Basis Function (RBF), and sigmoid (Table 1.13), all of which affect the selection of the candidate variables from a large pool. Whether a variable is selected depends on whether it minimizes the RMSE and maximizes the coefficient of determination (R^2) of the model to be constructed. The selection of a given kernel function affects the generalizability of SVM. The RBF kernel should be selected when the problem at hand is non-linear, such as natural hazards.

The constructed non-linear SVM regression function takes the following form:

$$f(x)=\sum_{i=1}^{N}\left(\alpha_i-\alpha_i^*\right)\cdot K\left(x,y\right)+b \tag{1.6}$$

where N = the number of (training) samples, $(\alpha_i-\alpha_i^*)$ = the Lagrange multipliers (i = 1, 2, …, n), b = bias (a constant), $K(x, y)$ = the kernel function (Table 1.13), via which the non-linear variable of natural hazard is solved. In Table 1.13, d, σ, and b are all kernel parameters. They affect the efficiency of SVM, together with C for the error. In particular, σ controls the shape of the clustering hyperplane, and hence the overall classification accuracy. The C value dictates SVM success and effectiveness. Whatever value is selected for C, it always represents a trade-off between the complexity of the adjustment function and the tolerance of the empirical material. A larger C value requires more training and leads to an over-fitted model, while a smaller C value may not produce a model of sufficient accuracy. Both C and σ can be set via optimization.

SVM algorithms fall into two categories – classification and regression. As a kernel-based maximum-margin method, SVM classification is grounded on the statistical learning theory and aims to partition the domain of the input data into clusters using an optimal regression-established hyperplane (Figure 1.9). It transforms the input data space into a higher dimensional feature space

TABLE 1.13
Four Common Types of SVM Kernel Function $K(x, y)$

Function	Formula
Linear	$x^T\cdot y$
Polynomial	$[x^T\cdot y+1]^d$
Radial basis	$e^{\frac{\lvert x-y\rvert^2}{2\sigma^2}}$
Sigmoid	$\tanh(x^T\cdot y+b)$

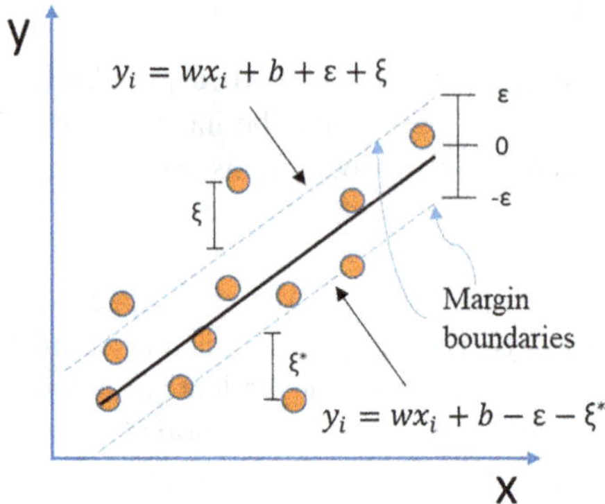

FIGURE 1.9 The partitioning of the input data space by a hyperplane in SVM-based image classification. (Source: Shi, 2022.)

non-linearly. With non-linear models, the data must be projected into a feature space via one of the four kernel functions, in which the data are classified linearly. The best separation (e.g., minimized estimation errors and model complexity) is achieved via supervised learning. The hyperplane is considered optimal if the Euclidean distance of the separating margins between the defined classes of the problem is maximized.

The shortest distance (ε) between the predicted and observed values is determined by minimizing the prediction error (Eq. 1.7) and imposing three constraints (Eq. 1.8).

$$\text{Minimization: } \frac{1}{2}\|w^2\| + C\sum_{i=1}^{N}(\xi + \xi^*) \tag{1.7}$$

$$\text{Constraints: } \begin{cases} y_i - wx_i - b \le \varepsilon + \xi \\ wx_i + b - y_i \le \varepsilon + \xi^* \\ \xi, \xi^* \ge 0 \end{cases} \tag{1.8}$$

where ξ = distance of samples to the margin boundary (Figure 1.9), i = the ith training sample. Thus, the final output decision function is depicted mathematically as follows:

$$\begin{cases} z = \sum_{i=1}^{N}(\alpha_i - \alpha_i^*) \times K(x, y) + b \\ K(x, y) = (xy + r)^d \\ 0 \le \alpha_i, \alpha_i^* \le C, \; i = 1, \ldots, n \end{cases} \tag{1.9}$$

where α_i = dual coefficients that are upper-bounded by C, $K(x, y)$ = the polynomial kernel function, d = degree parameter (d = 1, 2, 3), r is a constant. In running SVM classifications, three parameters must be specified: d, ε, and regularization, in addition to the type of kernel function $K(x,y)$.

1.4.2.2 Random Forest

As an assemble machine learning algorithm, RF classifies the input data using a combination of multiple decision trees. This set of trees iteratively resamples the input data based on the training dataset, from which n bootstrap samples are created. Each of them is used to establish an unpruned classification or regression tree. The tree is then applied to the input data, and the classification results from all the *n* trees are averaged and output as the final outcome. Each time a random set of evidence features is selected to predict the output. The output is weighted by the value of the votes received. The majority voting is based on the outputs of the estimated trees converging to a single tree. In implementing RF classification, a set of samples in the input data are reserved for validation. These out-of-bag samples are not used for tree growth, but for estimating prediction accuracy and determining feature importance. Apart from the number of out-of-bag samples, three more training parameters must be specified: the number of trees in the forest (*ntree*), the number of predictors at each split node (*mtry*), and the minimal size of the terminal nodes of the trees (*nodesize*). It is important to select appropriate *ntree* and *mtry* to achieve a satisfactory classification. A small *ntree* value causes the results to deteriorate considerably. Conversely, if its value is above the optimal threshold, it increases the general computation cost without improving the results notably. Similarly, the use of a small *mtry* debilitates the predictive power of each tree due to insufficient predictor variables available. The determination of the optimal values for these two parameters may require experimentation with a range of values.

The RF regression predictor is expressed mathematically as follows:

$$\hat{f}_{rf}^{K}(x) = \frac{1}{K}\sum_{k=1}^{K} T(x) \tag{1.10}$$

where $\mathbf{x}$ = the input data vector comprising various evidential features, K = the number of decision trees. RF-based classification is implemented in four main steps of "tree planting" and "tree pruning":

i. The training dataset is divided into N samples via repeated multiple sampling with reset.
ii. Each node in the tree selects m variables from a total of M input variables ($m < M$). The selected variables are used to determine the best split point and produce m trees.
iii. Each decision tree is allowed to grow as deep as possible without pruning.
iv. The entire dataset is used for prediction by summing the results from all trees using majority voting in classification and averaging in regression.

RF is a bagging algorithm that makes use of the integrated learning method by training multiple weak models to form a strong model. It has a superior performance to that of a single weak model and can produce highly accurate results without suffering overfitting, a problem of trying to produce an accurate prediction from the input based on samples only, with a poor ability to generalize the unseen data. The success of RF classification lies in the derivation of the high variance from different decision trees.

1.4.2.3 ANN

ANN is a machine learning paradigm that mimics the human brain in its decision-making. It aims to automatically predict the dependent variable (e.g., hazard occurrence) accurately by constructing a quantitative model from the training samples fed to it. The simplest network comprises three layers: the input layer, the hidden layer, and the output layer. Each layer contains a variable number of nodes, and all nodes in the same layer are linked to all other nodes in the immediate layers. The strength of a link is indicated by its weight. The input layer contains all predictor variables, and the output layer shows the predicted outcome. During training, the weight is constantly adjusted so as to reach the best match between the observed and the predicted values. The huge number of processing nodes or artificial neurons is equivalent to biological neurons in the human brain. They fulfill different functions. The nodes in the input layer represent the features considered in the decision-making, and the nodes in the output layer denote the outcome of analysis (e.g., presence or absence of a natural hazard) based on the output from the hidden layer. Each node fulfills two functions of summation (sum of weighted input from the previous layer) and activation. These nodes must have non-linear activation functions to be of any use. This simple model of input-output layers can be made more complex by inserting a hidden layer, thus turning the linear network into a non-linear one. Non-linear neural network models are distinctly advantageous over the traditional statistic technique as they are able to perform multivariate logistic analysis. All inputs to the output nodes are summed and passed to their counterparts in the next layer via the non-linear sigmoid function.

ANN can be implemented as either feed-forward or backpropagation (BP). In feed-forward networks, data always flow from the input layer to the hidden layer and ultimately to the output layer unidirectionally (Figure 1.10a). The output (y) and the input (z) conform to the following relationship:

$$y = \alpha \tan\mathrm{h}\left(\sum_{k=1}^{j} w_k z_k + c\right) \tag{1.11}$$

This kind of network lacks the feedback mechanism from the output to the input nodes and is able to handle only linear relationships through the linear learning discriminants. A set of weights (w_k) and biases (c) associated with each node must be known prior to classification, usually determined during network training. In reality, the relationship between natural hazards and their influential factors is unknown and most likely non-linear. It is better handled by BP networks, in which the difference (error) between the predicted and observed values at the output nodes is sent back to the input nodes (Figure 1.10b). A BP network has a feed-forward topology and a supervised learning algorithm. Data are sent forward from the input layer to the output layer via the hidden layer(s) just as with the feed-forward network. After the input is presented to the input nodes, an output is produced from the input via randomly assigned connection strengths initially. The calculated output is then compared with the desired output. Their discrepancy or the error signal is subsequently propagated backward from the output nodes to the input nodes (dashed arrows in Figure 1.10b) iteratively. In each iteration, the synaptic strengths of connection or weights between two nodes in different layers are continuously adjusted until the predicted and observed values match each other within the defined tolerance threshold. However, there is a risk in over-training the network in that it may be able to produce a good prediction based on the training samples, but its ability to generalize the rules for the unseen data is compromised.

Owing to its self-adapting and self-organizing abilities, ANN is the best at processing a huge volume of data of various natures and outputs the results that best match the given input (e.g., training samples) through the self-learning process, independent of any particular functions, or involving

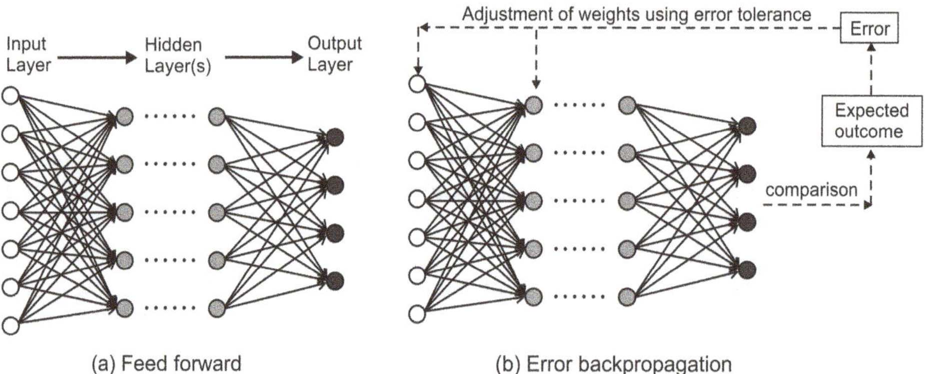

FIGURE 1.10 Comparison of the difference between feed-forward (a) and backpropagation – BP (b) artificial neural networks. Empty circles: input nodes; grey circles: hidden nodes; dark circles: output nodes. Solid arrows: direction of data flow; dashed arrows: direction of error propagation.

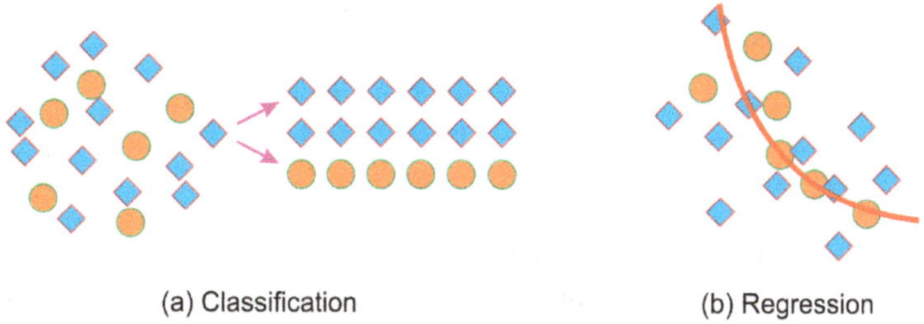

FIGURE 1.11 The distinction between classification (a) and regression (b) trees in a CART.

any assumptions regarding the data distribution. As such, it is perfectly suited to model the severity of natural hazards from a diverse range of physical, environmental, and climate variables.

1.4.2.4 CART

As a supervised, non-parametric, and non-linear learning algorithm, CART is a classification method resistant to missing data and does not require the data to be normally distributed. This binary classifier recursively partitions the input data of continuous or nominal attributes to form clusters in an effort to predict values of the dependent variable from predictor variables. Of the two types of trees, classification trees are used to categorize independent, categorical, and mutually exclusive variables such as land covers that can be binary (Figure 1.11a). As the input dataset is split into binary subsets, the classified subsets become increasingly homogenized. A regression tree is used to predict the value of the continuous and independent target variable (Figure 1.11b). CART is virtually a sequence of trees with terminal nodes. These hierarchical nodes are joined together by branches. How a classification is executed is governed by three elements: (i) rules about how the data should be split at a node based on the value of one of the variables; (ii) rules governing when a tree branch should be terminated as it can no longer be split further; and (iii) a prediction for the target variable in each terminal node. While it is straightforward to build a tree, it is not an easy task to prune it to produce a lean and functional tree.

CART analysis comprises three phases. The first is classification to produce a tree with several branches, the quantity of which is related to the level of data dispersion. Besides, tree size is also subject to such parameters as the minimum number of successive nodes, the minimum size of children, the maximum number of levels, and the maximum number of nodes, even though the size of a tree bears no relationship with classification accuracy. Correct classifications can be produced even with reduced overfitting of the training set. The second phase is pruning. Since the preceding phase aims to generate the largest trees possible, they must be pruned to reduce the total number of leaves, and in so doing to improve the classification accuracy. Whether a branch should be pruned depends on whether doing so lowers misclassifications and hence leads to higher accuracy, determined via cross-validation. In cross-validation, the linear dependency between the complexity of the tree and the cost of misclassifications R(T) is expressed as

$$R\alpha(T) = R(T) + \alpha T \Leftrightarrow \alpha = R\alpha(T) - R(T)/|T| \tag{1.12}$$

where $R\alpha(T)$ = the cost-complexity; $|T|$ = tree complexity with a value between 0 (maximal tree) and 1 (minimal tree), indicated by the number of terminal nodes in the tree. The generated regression rule set is then applied to all the data to map the targeted natural hazard (susceptibility).

1.4.2.5 NBT

NB is a machine learning classifier that creates a probability model based on Bayes Theorem (Eq. 1.13). It involves an implicit assumption that all the attributes are fully independent of each other, called class conditional independence. This trait enables the network to be trained faster.

NBT has a decision-tree structure and organizes an NB model on every leaf node of the constructed decision tree:

$$P(A/B) = \frac{P(B/A) \cdot P(A)}{P(B)} \tag{1.13}$$

where $P(A/B)$ = conditional probability of encountering A among B; $P(B/A)$ = conditional probability of B existing among A; $P(A)$ = probability of event A occurrence; $P(B)$ = probability of event B occurrence.

NBT-based classification starts with the estimation of the probability of each class to be mapped, calculation of the covariance and variance matrices, and construction of the discriminant function for each class. This model is easy to construct and simple to run. Nevertheless, the model is robust and not so much affected by noise and irrelevant features.

1.4.2.6 Generalized Additive Model

GAM is a machine learning method by extending the generalized linear model. As a semi-parameter model, it combines both linear and non-linear relationships among the predictor variables, their response, and the dependent variable (target) of analysis. Non-linear relationships are based on smooth functions of covariance to transform the predictor variables. The response of the predictors is the summation of all smoothers. GAM flexibly classifies the dependence of the response on the covariates. Able to analyze the non-linear response to changing conditions, it has found applications in predicting the location of natural hazard occurrence. In the GAM model, the additive predictor is an alternative to the common linear predictor in the generalized linear model. In its simplest form, a GAM is expressed as

$$g\left(E\left(Y_i\right)\right) = \beta_0 + \sum_{i=1}^{m} f_i(X_i) \tag{1.14}$$

where $E(Y)$ = the expectation of variable Y, $g()$ = the link function; β_0 = the intercept; $f_i(X_i)$ = the smoothing function of the input variable X_i; m = the number of features or predictor variables considered. The running of the GAM method requires three steps: (i) determination of the distribution of the response variable and the corresponding link function $g()$; (ii) determination of the smoothing function $f_i(X_i)$; and (iii) model optimization using contour plots (Elmahdy et al., 2020).

1.4.3 Vegetation Indices

In remote sensing, it is a common practice to derive spectral indices from multispectral bands of contrasting wavelengths to detect certain features related to natural hazards, such as the effect of drought on vegetation conditions. Some of the very commonly used indices are the normalized difference vegetation index (NDVI) from pixel values on multispectral bands, or

$$NDVI = \frac{\rho_{NIR} - \rho_R}{\rho_{NIR} + \rho_R} \tag{1.15}$$

where ρ_{NIR} and ρ_R = spectral reflectance of the NIR and red bands, respectively. The use of these two bands is justified by the fact that vegetation spectral reflectance peaks in the infrared spectrum but is much subdued in visible light wavelengths. It is particularly low in the red wavelength of 0.6–0.7 μm. In contrast, both soil and water have much less variations in their reflectance at these two wavebands. The differencing of the NIR and red bands considerably maximizes the spectral disparity between vegetation and these two covers, thus accentuating the presence of vegetation on the derived index layer. NDVI is able to indicate quantitatively biomass on the ground and has been used to monitor vegetation biomass and its decline over time, which is indicative of desertification.

Owing to its usefulness, NDVI products have been produced from coarse resolution images, such as Global Inventory Modeling and Mapping Studies (GIMMS) and Advanced Very High Resolution Radiometer (AVHRR), and are available to the public freely. For instance, bi-monthly GIMMS data are available from http://ecocast.arc.nasa.gov/data/pub/gimms/3g.v1/. Usually, the data are temporally aggregated weekly or bi-monthly to iron out the influence of clouds. They are good at detecting degradation at the continental and global scales over a long period as the data date back to the 1980s. The change in the vegetative cover is ideally monitored from time-series vegetation indices derived from multi-temporal images.

NDVI works well with dense vegetation cover as is common in the tropics but is less reliable in areas where vegetative cover is patchy and bare ground is prevalent, or in areas where surface biomass is low, as in grassland. In such environments, NDVI and other biological indicators-based methods tend to overestimate the degree of desertification due to seasonal fluctuation in vegetation cover, the severe effect of rainfall, and other factors. Consequently, several alternative VIs have been developed, including enhanced VI (EVI), soil-adjusted vegetation index (SAVI) and its variants of SAVI1 and SAVI2, and modified SAVI (MSAVI). EVI measures the greenness and health of vegetation and vegetation productivity using three bands of blue, red, and NIR. They are calculated as follows:

$$EVI = \frac{2.5(\rho_{NIR} - \rho_R)}{\rho_{NIR} + C_{1\rho R} - C_{2\rho B} + L} \tag{1.16}$$

$$SAVI = \frac{(\rho_{NIR} - \rho_R)\cdot(1+L)}{\rho_{NIR} + \rho_R + L} \tag{1.17}$$

$$MSAVI = \rho_{NIR} + 0.5 - 0.5\sqrt{(2\rho_{NIR}+1)^2 - 8(\rho_{NIR} - \rho_R)} \tag{1.18}$$

where C_1 = atmosphere resistance red correction coefficient (default = 6); C_2 = atmosphere resistance blue correction coefficient (default = 7.5); L = canopy background brightness correction factor (default = 1), or soil-adjusted factor. EVI requires a blue band to derive that is missing from some optical satellite images such as SPOT. Another formula for calculating MSAVI is

$$MSAVI = \frac{\rho_{NIR} - \rho_{red}}{\rho_{NIR} + \rho_{red} + L}(1+L) \tag{1.19}$$

where $L = 1 - 2\gamma$DNVI × WDVI, in which γ = slope of the soil line in the scatterplot of red vs NIR reflectance and is commonly taken as 1.06 (Qi et al., 1994), WDVI = weighted difference VI or $\rho_{NIR} - \gamma\rho_{red}$.

1.5 ACCURACY MEASURES OF SENSING

1.5.1 (Cross) Validation

The remote sensing-estimated hazard risk or severity is validated for its accuracy using a number of methods. If the sample size is large enough, a portion of it can be reserved as the validation sample, while the remaining ones are used to construct the model to predict hazard or its likely spot of occurrence. A comparison of the modeled hazard with the observed hazard reveals the reliability of prediction. However, when the sample size is very small, which is the norm with certain hazards such as volcano lava flows, all available samples must be used to construct the model. Otherwise, the prediction from the model suffers low confidence. In this case, it is still possible to validate the prediction model by leaving out one of the samples from constructing the model, and then comparing the predicted value with the observed value. The use of a single observation for verifying the accuracy is risky as there may be a large divergence between the observed and the predicted values. This can be prevented by repeating the same process tens of times, each time one of the observations is randomly selected and excluded from model construction. The discrepancy between all the observed and predicted values is analyzed statistically (e.g., averaged) to derive the accuracy. This method is known as leave-one-out cross-validation.

1.5.2 Accuracy Measures

The quality of remotely sensed hazard results is assured by different accuracy indicators, depending on their nature. For image classification using either per-pixel or object-oriented methods, the results are assessed against the ground truth, and the assessment outcomes are expressed as the user's accuracy, the producer's accuracy, and the overall accuracy, plus errors of omission and errors of commission. The overall accuracy indicates the proportion of correctly classified pixels out of the total number of evaluation pixels used. Omission errors are true misses, such as those genuine landslide pixels not included in the classification results. Commission errors are just the opposite, representing false alarms (e.g., the inclusion of non-genuine pixels as landslides). The effectiveness of image classification is measured by Kappa statistics. It indicates the effectiveness of image classification in comparison with random assignment of pixel identity. The higher the Kappa value, the more effective the classification in mapping natural hazards.

The reliability or accuracy of remote sensing-estimated natural hazard risk is measured by various forms of accuracy indicators. Some of the commonly used ones include the determination coefficient (R^2), RMSE, the ratio of the performance to the derivation (RPD), and the mean absolute percentage error (MAPE), calculated as

$$R^2 = 1 - \frac{\Sigma_{i=1}^{N}\left(Y_i - \hat{Y}_i\right)^2}{\left(Y_i - \bar{Y}_i\right)^2} \tag{1.20}$$

$$RMSE = \sqrt{\frac{\Sigma_{i=1}^{N}\left(Y_i - \hat{Y}_i\right)^2}{N}} \tag{1.21}$$

$$RPD = \frac{SD_s}{RMSE} \tag{1.22}$$

$$MAPE = \frac{1}{N}\sum_{i=1}^{N}\left|\frac{Y_i - \hat{Y}_i}{Y_i}\right| \tag{1.23}$$

where N = sample size, Y_i = the observed value of the target to be sensed, $\hat{Y}_i$ = the predicted value of the sensed target, $\bar{Y}$ = the average value of all the collected samples, SDs = standard deviation of the observed attribute value of the sensed target. A larger R^2 and RPD value and a smaller RMSE and MAPE are indicative of better model performance.

All the above four accuracy indicators are applicable to numerical attribute values that are continuous, such as soil salt content. For mapping categorical variables, such as the presence/absence of floods that is inherently binary in nature (either flooded or non-flooded), new accuracy indicators are needed. There are two groups of accuracy measures for these variables of a binary attribute. The first is the graphic method known as area under curve (AUC). This diagram illustrates the

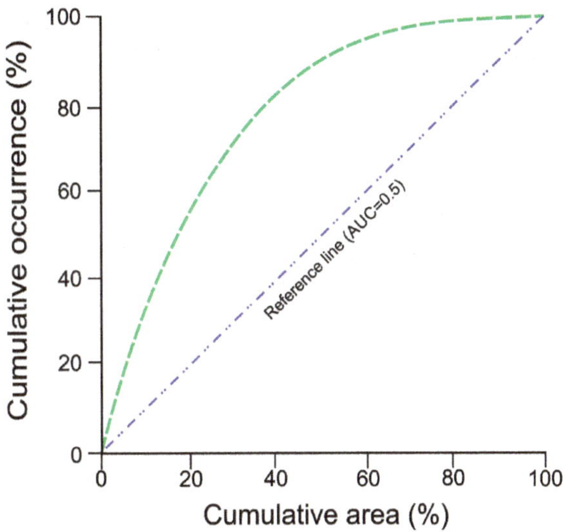

FIGURE 1.12 The AUC concept in indicating the accuracy of remote sensing analysis. Green line: prediction rate (91.5%).

relationship between cumulative occurrence (probability expressed in percentage) vs cumulative area (%), in comparison with the theoretical reference line with the cumulative area (%) or the 1:1 trend line (Figure 1.12). AUC varies between 0.5 and 1, with 1 indicative of perfect predictions and a value of 0.5 random predictions. Along the reference line, the former is strictly proportional to the latter. A higher AUC indicates a higher accuracy (i.e., the curve lies much higher above the reference line), and vice versa.

1.5.3 Prediction Accuracy Measures

In predicting the occurrence of natural hazards, there are four possible outcomes: true positive (TP), true negative (TN), false positive (FP), and false negative (FN) (Table 1.14). TP signifies correct identification of the target, while TN indicates a correctly labeled event that clearly is not the target. Both of them represent correct predictions. In contrast, both FP and FN represent incorrect labeling. The former is known as false alarms as it includes events that are clearly not the target. It is also known as commission errors that identify non-targets as the target, so it inflates the target in the results. FN represents misses in that a true target is not represented in the predicted results. It is identical to omission errors that cause the actual target to be smaller than the results obtained by means of remote sensing. From the four outcomes, six accuracy indicators have been produced.

TABLE 1.14
Contingency Table Showing Four Possible Outcomes of Predicting Natural Hazards that Are Used to Indicate the Prediction Accuracy

	Observed Events		
Predicted Events	**True**	**False**	**Row Sum**
True	a (TP-hits)	b (FP-false alarms)	$a+b$
False	c (FN-misses)	d (TN-correct rejects)	$c+d$
Column sum	$a+c$	$b+d$	$a+b+c+d=n$

They are true-positive rate (TPR), false-positive rate (FPR), probability of detection (POD) or the hit rate (hits/observed events), positive predictive value (PPV), probability of false detection (POFD), and false alarm rate (FAR) calculated using the following equations:

$$TPR = \frac{TN}{TN + FN} = \frac{d}{d + c} \tag{1.24}$$

$$FPR = \frac{TP}{TN + FP} = \frac{a}{d + b} \tag{1.25}$$

$$POD = \frac{TP}{TP + FN} = \frac{a}{a + c} \tag{1.26}$$

$$PPV = \frac{TP}{TP + FP} = \frac{a}{a + b} \tag{1.27}$$

$$POFD = \frac{false\ alarms}{non - events} = \frac{b}{b + d} \tag{1.28}$$

$$FAR = \frac{false\ alarms}{predicted\ events} = \frac{b}{a + b} \tag{1.29}$$

Both FPR and TPR can be calculated based on pixels or patches (polygons), depending on whether the remote sensing imagery is in raster or vector (e.g., segmented) format. POD ($0 \leq$ POD ≤ 1) refers to the detection rate or the probability of labeling a characteristic feature in the presence of such a feature. It indicates the fraction of all observed events correctly predicted without penalty for false alarms. PPV denotes the probability of the existence of a feature given its detection (Brenner and Gefeller, 1997). A high PPV implies that FP outcomes are minimized (Trevethan, 2017). It is the probability of the existence of a hazard event according to the ground truth. A high PPV indicates that the predicted hazards are likely to be "real", while falsely detected hazards are kept to a minimum. POFD and FAR are very similar. They are like the two sides of the same coin. The former is referenced to the non-events, while FAR is calculated in terms of the predicted events.

$$Skill\ score = \frac{score\ for\ the\ forecast - score\ for\ the\ standard\ forecast}{perfect\ score - score\ for\ the\ standard\ forecast} \tag{1.30}$$

The quality of prediction has also been evaluated in terms of skill score calculated using Eq. 1.30. There are two common types of skill score: true skill score (TSS) and Heidke skill score (HSS). The former measures the difference between the probability of correct detections and the probability of unexpected non-events. If the latter probability is high, TSS converges to the probability of correct detections. For the HSS, the "score" is the number correct or the proportion correct. The "standard forecast" is usually the number or the proportion correct by chance. Thus, using the proportion correct, HSS is calculated on the basis of a contingency table (Table 1.14) in the simplified form as

$$HSS = \frac{2(ad - bc)}{[(a + c)(c + d) + (a + b)(b + d)]} \tag{1.31}$$

HSS takes into account correct detections that are not due to chance and is thus more appropriate than other indicators in the case where rare events can occur. It measures the fractional improvement of the forecast over the standard forecast. HSS has a value ranging from $-\infty$ to 1. Negative values indicate that the chance forecast is better, 0 means no skill, and a value of HSS = 1, a perfect forecast. Similar to most skill scores, it is normalized by the total range of possible improvement over the standard, which means HSS can be safely compared on different datasets. HSS is a popular score, partly because it is relatively easy to compute and perhaps also because the standard forecast, chance, is relatively easy to beat. Other standard scores are possible, such as persistence or climatology, but they require additional information to compute, in the form of a separate contingency table.

REFERENCES

Albino, F., B. Smets, N. d'Oreye, and F. Kervyn. 2015. High-resolution TanDEM-X DEM: An accurate method to estimate lava flow volumes at Nyamulagira volcano (D. R. Congo). *J. Geophys. Res. Solid Earth* 120: 4189–4207.

Bagnardi, M., P.J. González, and A. Hooper. 2016. High-resolution digital elevation model from tri-stereo Pléiades-1 satellite imagery for lava flow volume estimates at Fogo Volcano. *Geophys. Res. Lett.* 43: 6267–6275.

Baldi, P., M. Coltelli, M. Fabris, M. Marsella, and P. Tommasi. 2008. High precision photogrammetry for monitoring the evolution of the NW flank of Stromboli volcano during and after the 2002–2003 eruption. *Bull. Volcanol.* 70: 703–715. DOI: 10.1007/s00445-007-0162-1

Behncke, B., A. Fornaciai, M. Neri, M. Favalli, G. Ganci, and F. Mazzarini. 2016. Lidar surveys reveal eruptive volumes and rates at Etna, 2007–2010. *Geophys. Res. Lett.* 43: 4270–4278.

Brenner, H., and O. Gefeller. 1997. Variation of sensitivity, specificity, likelihood ratios and predictive values with disease prevalence. *Stat. Med.* 16: 981–991.

Casciati, F., S. Casciati, C. Fuggini, L. Faravelli, I. Tesfai, and M. Vece. 2016. Framing a satellite based asset tracking (SPARTACUS) within smart city technology. *J. Smart Cities* 2(2): 40–48. DOI: 10.18063/JSC.2016.02.003

Chen, C.S., C.S. Wang, and Y.M. Chang. 2003. A study on precise GPS height using WVR and radiosonde sounding data. *Surv. Land Inf. Sci.* 63: 13–19.

Chien, S., and V. Tanpipat. 2013. Remote sensing of natural disasters. In: *Earth System Monitoring: Selected Entries from the Encyclopedia of Sustainability Science and Technology*, ed. J. Orcutt, 421–438. New York: Springer. DOI: 10.1007/978-1-4614-5684-1_17

Dai, C., and I.M. Howat. 2017. Measuring lava flows with ArcticDEM: Application to the 2012–2013 eruption of Tolbachik, Kamchatka. *Geophys. Res. Lett.* 44(24): 12133–12140. DOI: 10.1002/2017GL075920

Elmahdy, S., T. Ali, and M. Mohamed. 2020. Flash flood susceptibility modeling and magnitude index using machine learning and geohydrological models: A modified hybrid approach. *Remote Sens.* 12: 2695. DOI: 10.3390/rs12172695

Gao, J. 2007. Towards accurate determination of surface height using modern geoinformatic methods: Possibilities and limitations. *Progr. Phys. Geog.: Earth Environ.* 31(6): 591–605. DOI: 10.1177/0309133307087084

Kerle, N., and C. Oppenheimer. 2002. Satellite remote sensing as a tool in lahar disaster management. *Disaster* 26(2): 140–160.

Lewis, S. 2009. Remote sensing for natural disasters: Facts and figures. https://www.scidev.net/global/features/remote-sensing-for-natural-disasters-facts-and-figures/ (accessed July 4, 2022).

Marin, L.E., X. Perez, and E. Rangel. 1998. Comparison of three surveying techniques applied to hydrogeological studies: Level, barometer, and GPS. *Geofisica Int.* 37: 127–129.

Metternicht, G., L. Hurni, and R. Gogu. 2005. Remote sensing of landslides: An analysis of the potential contribution to geo-spatial systems for hazard assessment in mountainous environments. *Remote Sens. Environ.* 98(2–3): 284–303. DOI: 10.1016/j.rse.2005.08.004

Mojaddadi, H., B. Pradhan, H. Nampak, N. Ahmad, and A.H. bin Ghazali. 2017. Ensemble machine-learning-based geospatial approach for flood risk assessment using multi-sensor remote-sensing data and GIS. *Geomatics Nat. Hazards Risk* 8(2): 1080–1102. DOI: 10.1080/19475705.2017.1294113

Okuda, T., F. Kimata, and M. Nakamura 1990. The accuracy of point positioning in differential mode of GPS. *J. Geodetic Soc. Japan* 36: 115–117.

Qi, J., A. Chehbouni, A.R. Huete, Y.H. Kerr, and S. Sorooshian. 1994. A modified soil adjusted vegetation index. *Remote Sens. Environ.* 48: 119–126.

RMA (Resource Management Act). 1991. New Zealand Legislation. https://www.legislation.govt.nz/act/public/1991/0069/211.0/DLM230265.html (accessed July 4, 2022).

Shi, Y. 2022. Aboveground biomass change of the alpine meadow on the Qinghai-Tibet Plateau under climate warming. PhD diss., University of Auckland.

Trevethan, R. 2017. Sensitivity, specificity, and predictive values: Foundations, pliabilities, and pitfalls in research and practice. *Fron. Public Health* 5, 307. DOI: 10.3389/fpubh.2017.00307

Trimble. 2007. *GPS – The First Global Navigation Satellite System*. Sunnyvale, California: Trimble, 144 p.

Van Westen, C. 2000. Remote sensing for natural disaster management. *Int. Arch. Photogramm. Remote Sens.* XXXIII(Part B7): 1609–1617.

2 Earthquakes

2.1 INTRODUCTION

Earthquakes refer to the abrupt and violent shaking of the ground as a result of the release of stress, usually at the interface of two tectonic plates known as the fault line. Earthquakes are commonly triggered by the movement of the two tectonic plates within the Earth's crust or by volcanic activities. The degree and duration of ground shaking are related to the intensity or strength of the earthquake. Some minor aftershocks may follow the major shake, and they last longer than the major shake. Among all the natural hazards, earthquakes are potentially the most catastrophic and destructive, doing enormous damage within minutes, and leaving little time for preparation and evacuation. Earthquakes are highly destructive hazards not only because they can do damage directly to buildings and infrastructure, but also because they can trigger secondary hazards, such as liquefaction, surface rupture, landslides, and even tsunamis. This chapter focuses on the first two types of consequence, while landslides and tsunamis will be covered separately in Chapter 3 and Section 10.5.2 under coastal flooding, respectively.

Earthquake hazard refers to the potential effects of an earthquake on properties and infrastructure, causing injuries and casualties. The most direct and visible damage of an earthquake occurs to buildings and infrastructure. Buildings may be toppled by seismic waves, and building materials may fall during the ground shaking, causing more damage to the occupants and nearby pedestrians or buildings. Ground shaking can take the form of a horizontal shift in opposite directions or just uplift of certain areas above the normal height. In either case, the ground is ruptured as a result of the relative shift of the surface along the weakest fault line. Surface rupture does the most damage to infrastructure and buildings, especially in urban areas. Indirect damage of surface rupture is the disruption to transport that further hampers after-disaster rescue efforts.

Remote sensing plays a vital role in estimating Earth's deformation in general, and earthquake hazards associated with active faults and structure in particular. Ideally, optical satellite images are used for the geomorphological analysis of active faults to fulfill various objectives, such as identification of active faults and repeated earthquake offset to characterize the spatial distribution of slip of a fault in successive earthquakes (Elliott et al., 2016), precise measurement of offsets in geomorphic features and historical fault scarps for earthquakes, in addition to the study of coseismic ruptures and displacement fields from modern and historical earthquakes. Although it is possible to identify lineaments in pre-earthquake images, they are not discussed here because they are not related to earthquakes as a hazard. Besides, the assessment of seismic vulnerability is not covered in this chapter either.

While it is impossible to always predict earthquakes reliably at present, it is possible to detect abnormal activities commonly associated with an imminent earthquake using remote sensing. For instance, a rise in surface and near-surface temperatures by 3–5°C prior to Earth's crust earthquakes can be easily detected from thermal infrared (TIR) imagery to a spatial resolution of 0.5–5 km and with a temperature resolution of 0.12–0.5°C (Tronin, 2006). Such thermal anomalies are typically observed above large faults and their intersections. Thermal anomalies in seismically active regions provide information about the changes in surface temperature associated with an impending earthquake and can serve as earthquake predictors, based on the statistically significant correlation between thermal anomalies and seismic activity. Having a size of $n \cdot 100$ km × $n \cdot 10$ km (n = image spatial resolution), thermal anomalies emerge about 6–24 days before an earthquake and last about a week afterward (Tronin, 2006). They are sensitive to crustal earthquakes with a magnitude higher than 4.5. The thermal anomalies with an amplitude of about 3–6°C are commonly

DOI: 10.1201/9781003354321-2

associated with large faults. Another well-known phenomenon of an imminent earthquake is the release of special gases and aerosols, such as O_3, CH_4, CO_2, CO, H_2S, SO_2, and HCl. The concentrations of these gases in the atmosphere can all be quantified using the methods to be presented in Section 12.5.2. Due to the lack of right sensors that can be deployed near the Earth for long-term monitoring, remote sensing is more useful for the post-event recovery than for the pre-event prevention or hazard zone definition.

Remote sensing is indispensable to earthquake hazard assessment. The area, amount, rate, and type of earthquake damage can all be assessed from remotely sensed data. The remote sensing-yielded information is essential for devising rapid post-earthquake response plans, estimating earthquake-induced loss, and investigating long-term earthquake effects. Remote sensing also serves as the only available information source for emergency response teams and aid agencies, earthquake reconnaissance by engineers, earth scientists, and social scientists, and rapid loss estimates for governmental and insurance agencies. In a word, remote sensing creates a wide range of opportunities for studying earthquakes. This chapter first elaborates on how to assess earthquake damage to buildings from both optical and SAR images using various image processing methods. Discussed at length is the grading of earthquake damage at the individual building and block levels. The pros and cons of optical and radar images in the assessment are also compared and contrasted systematically. The second part of this chapter explores how to assess post-earthquake surface displacement (both direction and rate) using InSAR images and LiDAR data. Finally, this chapter concentrates on how to detect liquefaction from remotely sensed images. Topics covered include the use of suitable indices and mapping of liquefaction-affected areas from very high resolution (VHR) and medium-resolution optical images, as well as radar images. Also explained in the discussion is how to determine the lateral spreading distance of liquefaction.

2.2 ASSESSMENT OF EARTHQUAKE DAMAGE

The enormous damage of powerful earthquakes to buildings and infrastructures is measured by the extent of the affected area, degree, proportion, and type. It is the damage to buildings (Figure 2.1) that has been extensively studied almost exclusively from pre- and post-event remote sensing images, either passive or active. This multi-temporal approach evaluates the ground cover changes between the pre-event and post-event data that must be co-registered with each other. Features such as spectral, textural, edge, spatial relationship, and geometric information may change after an earthquake.

Table 2.1 details the kinds of earthquake damage that have been studied from various image features. Most of them are detectable from images graphically, but not from non-imagery LiDAR data. Although LiDAR data can be quickly acquired over broad areas following an earthquake, post-event data alone cannot detect some specific damages. On the other hand, pre-event LiDAR data are rarely in existence. The possible solution to this absence is to combine LiDAR data with imagery data. Their fusion can improve damage detection but faces numerous obstacles, such as incomplete coverage of all areas due to the restriction to flying over the affected areas and unavailability of pre-event data. So LiDAR-based assessment will not be explored further. The detection is based almost exclusively on imagery data.

2.2.1 Grades of Building Damage

Earthquake damage to buildings and infrastructure can be estimated from remotely sensed images. They produce generalized estimates of damage locations, detailed estimates of damage patterns, and 3D failure geometries. The assessment of earthquake damage may be carried out for individual buildings, which is possible with VHR imagery. According to the European Macroseismic Scale (EMS-98), building damage is graded at five levels: totally collapsed (grade 5), partially collapsed (grade 4), debris generated (grade 3), moderate damage (grade 2), and negligible to slight

FIGURE 2.1 Complete collapse of a school building in Nantou County, Central Taiwan caused by the 21 September 1999 earthquake ($Mw = 7.7$).

TABLE 2.1
Summary of Building Damage Details Detected Based on Consideration of Various Features

Damage Detail	Features Used	References
Collapsed buildings	Shape	Gamba and Casciati (1998)
	Radiometric and morphological	Pesaresi et al. (2007)
	DSM	Turker and Cetinkaya (2005)
	Color, shape, texture, height	Yu et al. (2010)
Damaged and non-damaged	Spectral, texture, spatial relations	Li et al. (2011a)
	Variance and direction of edge	Mitomi et al. (2002)
	Intensity, texture	Cooner et al. (2016)
	Spectral, textural, structural	Rasika et al. (2006)
	Textural and color	Rathje et al. (2005)
	Spectral and textural	Ural et al. (2011)
	Spectral, textural, height geometric	Wang and Jin (2011)
(possibly) Damaged, destroyed, intact	Spectral, intensity, coherence, DEM	Adriano et al. (2019)
Debris, non-debris, no change	Spectral, textural, brightness, shape	Khodaverdizahraee et al. (2020)
(partially) Collapsed, unchanged	Normalized DSM, pixel intensity, segment shape	Rezaeian and Gruen (2007) Rezaeian (2010)
No, slight, low-level damage	SAR texture	Dell'Acqua and Gamba (2012)

Source: Modified from Matin and Pradhan (2021).

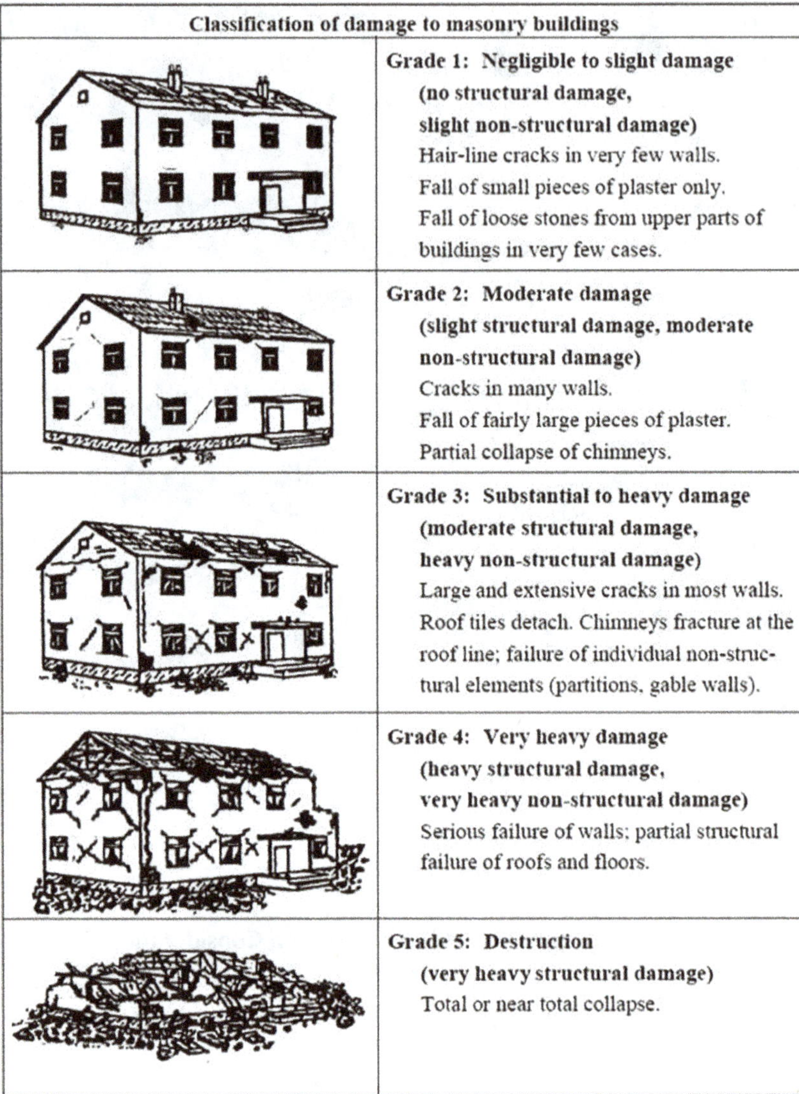

Classification of damage to masonry buildings	
	Grade 1: Negligible to slight damage (no structural damage, slight non-structural damage) Hair-line cracks in very few walls. Fall of small pieces of plaster only. Fall of loose stones from upper parts of buildings in very few cases.
	Grade 2: Moderate damage (slight structural damage, moderate non-structural damage) Cracks in many walls. Fall of fairly large pieces of plaster. Partial collapse of chimneys.
	Grade 3: Substantial to heavy damage (moderate structural damage, heavy non-structural damage) Large and extensive cracks in most walls. Roof tiles detach. Chimneys fracture at the roof line; failure of individual non-structural elements (partitions, gable walls).
	Grade 4: Very heavy damage (heavy structural damage, very heavy non-structural damage) Serious failure of walls; partial structural failure of roofs and floors.
	Grade 5: Destruction (very heavy structural damage) Total or near total collapse.

FIGURE 2.2 Five grades of damage to buildings according to the European Macroseismic Scale (EMS-98). (Source: Grünthal, 1998.)

damage (grade 1) (Figure 2.2). Of the five grades, the easiest to study is collapsed buildings. They include total collapse, partial collapse in excess of 50%, and building structure not distinguishable (the walls have been destroyed or collapsed). Non-collapsed means possible damage and damaged. The former is associated with the presence of possible damage proxies, and the building in question is surrounded by damaged or destroyed buildings, but it is uncertain due to image quality. The latter may refer to minor damage (e.g., the roof remains largely intact but the building suffers partial damage). It includes the partial collapse of the roof and serious failure of walls (Matin and Pradhan, 2021).

The severity of the overall damage is measured by the damage ratio. If individual buildings can be identified from VHR imagery, it is the ratio of the collapsed buildings to all the buildings detected. If the image resolution is too coarse to allow individual buildings to be identified, it is measured by the portion of pixels of collapsed buildings to the total number of building pixels within a district. District-level damage area ratio (DAR) is calculated as

$$DAR_j = \frac{\sum_i D_{i,j} \cdot A^B_{i,j}}{A^P_j} \tag{2.1}$$

where $D_{i,j}$ has a binary value of 0 for damaged buildings and 1 for undamaged buildings; $A^B_{i,j}$ = the area of each building, and A^P_j = the total building area in pixels.

2.2.2 From Optical VHR Imagery

Optical imagery is good at documenting the spatial distribution of damage across a region. Ideally, building damage should be assessed using VHR imagery such as QuickBird and WorldView-3 (Table 1.5). Their sub-meter spatial resolution enables individual buildings to be discerned (Figure 2.3). Post-earthquake air- and space-borne nadir-looking images allow the acquisition of information about the status of the building roofs and the presence/absence of debris abutting the building's lateral walls. They can be interpreted to map the damaged buildings or assess whether they have been toppled. During visual interpretation, the damage to individual buildings is identified from VHR images based on association (e.g., building debris abutting the wall). The level of damage is commonly mapped in accordance with the EMS-98 scales if the image resolution is sufficiently fine. The assessment of damage severity is not always possible with visual interpretation, though. The separation of grade 2 damage from grade 1 is almost impossible even with VHR imagery (Yamazaki et al., 2005). In addition, the identification of grade 3 damage is rather difficult with nadir-viewing images unless a building is abutted by debris. Thus, the more severe the damage, the easier it can be identified via visual interpretation. But such an assessment is feasible only within a small area as it is labor-intensive. More precise building-level damage mapping requires images of meter and sub-meter resolutions. If the satellite imagery has a too coarse spatial resolution (e.g., 10 m or coarser), building damage can only be detected at the block level categorically, such as slight, moderate, and severe (Figure 2.4). With coarser resolution imagery, the damage can be assessed only at the general level, such as the pattern of damage and its severity grade.

FIGURE 2.3 A panchromatic QuickBird band (spatial resolution: 0.6 m) illustrating the damage to buildings caused by the 26 December 2003 Bam earthquake (M_w = 6.6). Left: pre-earthquake image acquired on 30 September 2003; Right: post-earthquake image acquired on 3 January 2004 (Copyright: DigitalGloble).

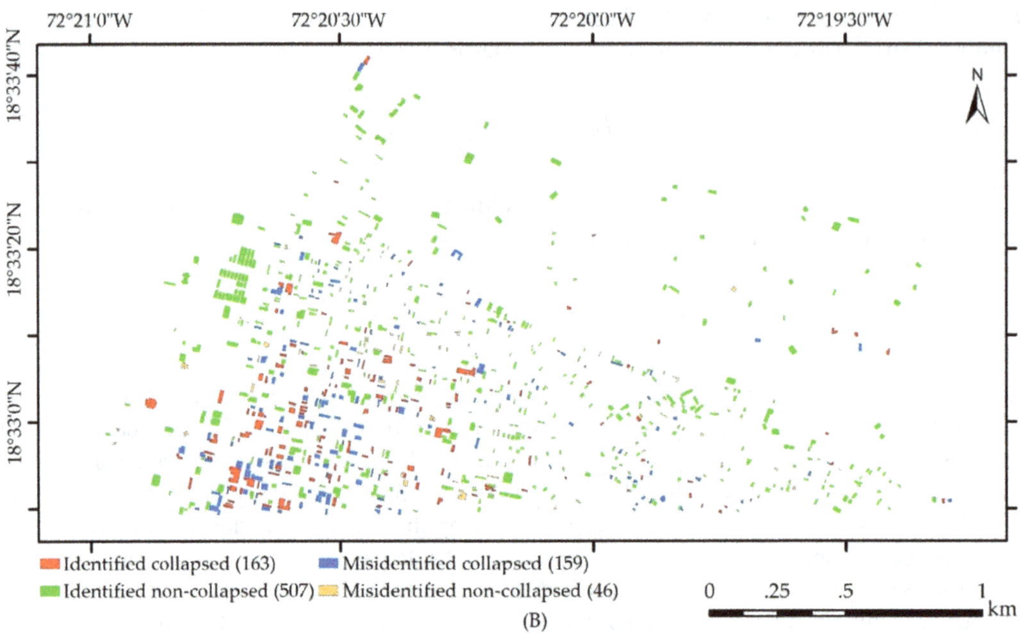

FIGURE 2.4 Distribution of collapsed (red) and non-collapsed (green) buildings after the 2010 Haiti earthquake, and their accuracy of mapping using convolutional neural network classification of post-event QuickBird images. (Source: Ji et al., 2018.)

Building damage is detectable from either multi- or mono-temporal data. The type of data used affects how the damage is detected. Multi-temporal detection is virtually a comparison of the pre-earthquake image with that of the post-earthquake image of the same area, known as change detection. Any changes to buildings in the post-event image represent damage. Some multi-temporal images, such as QuickBird, have multiple viewing angles, obtained via tilting the scanning mirror so as to shorten the revisit period of observing the same target multiple times within a short period. The tilted images of the same area hit by an earthquake may have different off-nadir viewing angles (e.g., 10° pre-event and 24° post-event). Such images cannot be easily superimposed with each other precisely to automatically detect damage to buildings due to the lateral deformation of the ground surface. It is hence imperative that the pre- and post-event images have the same viewing angle. If mono-temporal imagery is used, it must be acquired after the earthquake. It is advantageous to use the post-event image as it avoids additional processing such as image co-registration and normalization (Figure 2.4). From post-event imagery alone, collapsed buildings can be mapped at an average overall accuracy of 78.6% for two test regions using SqueezeNet classification (Ji et al., 2018).

Visual interpretation is slow, tedious, and subjective, all of which can be avoided by image classification. It is fast and can produce digital results on a repetitive basis. Image classification may be implemented as either per-pixel, object-oriented, or knowledge-based. Per-pixel image classification can show earthquake damage at a broad scale, and yield damage intensity at the pixel level, and the percentage of damaged pixels or average damage state within a window of 100 by 100 pixels from VHR QuickBird imagery (Rathje et al., 2005; Rathje and Adams, 2008). This percentage may be used to classify damage into a few severity grades. The accuracy of classification varies with the number of damage grades mapped. The more levels are identified and classified, the lower the classification accuracy, and vice versa. At the binary level of severity (grade 5 vs grades 1–4), the overall accuracy is the highest at 82.6% or 77.9% at grades 1–3 vs grades 4–5. The accuracy drops to 61.5% at three levels of grades 1–2, grade 3, and grades 4–5, and further to 52.8% at four levels in assessing the building damage caused by the Bam, Iran earthquake on 26 December 2003 ($M_w = 6.5$) (Yamazaki et al., 2005).

Since VHR imagery is spectrally poor, spectral information may be supplemented with texture measures, such as contrast (including dissimilarity and homogeneity), orderliness (angular second moment and entropy), and descriptive statistics of the co-occurrence matrix (including mean, variance, and correlation), to enhance the classification accuracy. Another candidate for supplementation is edge density, calculated using the popular and simple Canny edge detector (Rathje et al., 2005). All of these features are treated as the inputs to a supervised (maximum likelihood) classifier of the post-earthquake image. The classifier then produces a few classes that may include damaged areas, undamaged buildings, vegetation, open ground, shadow, and asphalt-paved roads. The severity of damage is determined from the portion of damaged pixels within a window. In the classification, texture features are crucial to distinguishing non-damage classes (all except damaged areas), but the same features are not equally important for the differentiation throughout a study area. It must be noted that this classification yields information on the proportion of damage, not the severity (or grade) of damage to individual buildings, a deficiency that can be overcome with object-based image analysis (OBIA).

In OBIA, if a proper scale is specified for image segmentation, each segment may correspond to a building, forming a 1:1 correspondence between the segmented objects and the building footprint. The segmented pre-event image is then classified into a few discrete classes such as buildings, non-buildings, shadow, vegetation, and unclassified using the usual supervised nearest neighbor method (Gusella et al., 2005). After superimposition with the post-event image, building objects can be further differentiated into collapsed (i.e., piles of debris) and non-damaged based on training samples. Classification of a QuickBird image of the Bam earthquake identified 34% collapsed buildings out of a total of 6,473. Assessed against the ground truth, the collapsed and standing building map has an overall accuracy of 70.5% (Gusella et al., 2005), which is much lower than 91% achieved by Rathje et al. (2005) from the same image using only the post-event data in a semi-automatic supervised classification. This is not surprising given that buildings are classified at a user's accuracy of 80% and a producer's accuracy of 91%. Causes of misclassification are identified as the varied shadow formed by different sensor look angles and image co-registration inaccuracy (an offset on the order of 2–3 pixels).

Instead of space-borne VHR imagery, a post-event aerial photograph of 0.5 m resolution acquired after the Wenchuan earthquake in West China was also classified using an object-oriented method, in which the target area was differentiated into four types of buildings (including collapsed, slightly damaged, and undamaged), roads, shadow, and vegetation (Li et al., 2011b). The evidence of building damage decision-making is based on spectral heterogeneity, shape (compactness and smoothness), texture, and adjacency information. The use of more diverse types of inputs that are spectral, spatial, contextural, and geometric than in other studies achieved a higher than 90% automatic recognition accuracy.

In addition to OBIA, earthquake damage to buildings may be mapped automatically using various artificial intelligence (AI)-based frameworks, including multilayer perceptron (MLP) and neural networks. The use of explainable AI facilitates the interpretation of the outputs of machine learning analysis and the quantification of the relative importance of all the features included in modeling building damage. MLP can differentiate collapsed buildings from non-collapsed buildings on a 0.5 m resolution WorldView-3 image captured soon after the earthquake based on consideration of 15 spectral, textural, and shape features extracted from building polygons, with an overall accuracy of 73.55% after the elimination of redundant features (Matin and Pradhan, 2021). This accuracy is slightly lower than that achieved by Ji et al. (2018), suggesting that mapping of collapsed and standing buildings cannot be more than 80% from VHR imagery regardless of the machine-learning classification algorithm used.

2.2.3 From SAR Imagery

2.2.3.1 SAR Backscattering and Building Damage

SAR imagery is posed as the ideal remote sensing data for assessing areas damaged by earthquakes because of its ability to detect the extent of building damage through clouds and smoke, a feature quite useful and effective for immediate post-disaster damage assessment. Dissimilar to optical imagery that captures reflectance, SAR sensors record both the amplitude and phase of backscattered microwave echoes from the target. A pair of SAR images can be used to produce interferometric SAR (InSAR) that is particularly sensitive to surface relief. Phase-based InSAR analysis can yield quantitative information on relative ground displacement triggered by earthquakes. However, InSAR-generated complex coherence calculated as a complex correlation coefficient is sensitive to satellite geometry, acquisition duration, and wavelength of radar pulses, which is not the case with the backscattering coefficient of the Earth's surface (Matsuoka and Yamazaki, 2004).

SAR imagery is superior to optical images in detecting building damage as it records the backscattered signal, the intensity of which depends on building status (Figure 2.5). Standing buildings have a strong reflection but low backscattering intensity, ending up having a bright tone than the surrounding areas on SAR imagery regardless of the incidence angle (Figure 2.5a). In contrast, a completely collapsed building with a pile of rubble nearby scatters the incident energy in all directions, but a weak reflection, ending up having a dark tone (Figure 2.5b). SAR backscattering coefficient and intensity are correlated significantly in hard-hit areas. These differences enable the differentiation of standing buildings from collapsed buildings based on features derived from the original building footprint.

Dissimilar to optical imagery, the pixel value of which is evaluated directly in detecting building damage, SAR imagery is converted to coherence and correlation between the pre- and post-earthquake images. The complex coherence of these images is expressed as

$$\rho = \frac{E(s_1 s_2^*)}{\sqrt{E(s_1 s_1^*)E(s_2 s_2^*)}} \tag{2.2}$$

where s_1 and s_2 represent the corresponding complex pixel values of the two images (* denotes post-event), and $E(\ldots)$ indicates the expected value.

The intensity correlation is calculated from the two images as

$$\rho_I = \frac{E(I_1 I_2)}{\sqrt{E(I_1^2)E(I_2^2)}} \tag{2.3}$$

where I_1 and I_2 represent the corresponding pixel's intensity.

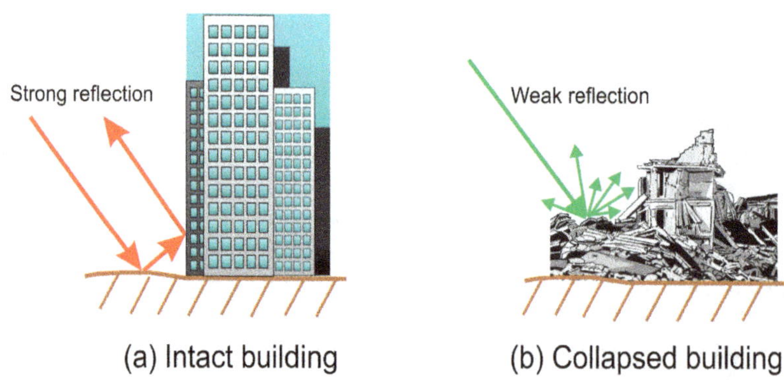

FIGURE 2.5 (a) The intensity of radar back scattering in relation to an intact building, (b) a completely collapsed building.

The complex coherence is affected mostly by the phase difference between radar returns, a distinctive parameter measured by a coherent sensor. It is particularly related to the spatial arrangement of the scatterers within the sensed target, and possibly their displacements. Conversely, the intensity correlation is more related to changes in the magnitude of the radar return (Stramondo et al., 2006).

2.2.3.2 Medium-Resolution SAR

Medium-resolution SAR images usually have a spatial resolution of around 30 m that does not allow the detection of earthquake damage to individual buildings, so damage is assessed at the parcel or district level, at which it is possible to distinguish the level of damage to slight, moderate, high, and completely destroyed, if diverse input features, including intensity, texture, and correlation derived from a local window, are used. It is paramount to specify a proper window size in deriving these parameters. A large window size is desirable for averaging out speckle noise and good for detecting damage to large buildings, but it may weaken the relationship between damage ratio and coherence, leading to a lower accuracy of detection (Matsuoka and Yamazaki, 2004). With medium-resolution images, change detection is the standard method of detecting building damage. Two SAR images, one before and one after the earthquake, are essential. Backscattering intensity and intensity correlation coefficients of the pre- and post-earthquake images are indicative of the surface changes in the earthquake-affected areas and potential damage. More severely damaged areas or areas having a higher damage ratio have a higher and negative difference in the backscattering coefficient (after–before) while the correlation coefficient becomes smaller in less damaged areas, and vice versa because collapsed buildings and debris reduce corner reflection (see Figure 2.6). Thus, hard-hit areas can be potentially detected using these two indices based on discriminant analysis. The discriminant score is a weighted linear combination of these two indices (correlation coefficient and backscattering coefficient difference). A large difference and correlation coefficient suggest a higher discriminant score. The classification of the two index maps into severely damaged and non-damaged areas is as accurate as 78.5% with the training dataset, but the accuracy drops to 71.1% if only the difference image is used or 73.4% if the correlation image is used (Matsuoka and Yamazaki, 2004). Dissimilar to their optical counterparts, the two SAR images must be selected with care and a trade-off if they are going to be interferometrically processed. A long perpendicular separation between the two acquisitions (position) lowers the overall coherence of an image and makes it difficult to identify low coherence (i.e., damaged) areas. Conversely, if the separation is excessively long, the detected change may not relate directly to earthquake damage as the target could have changed after the event. It is worth noting that radar imagery is notoriously prone to speckle noises that can be partially removed via window-based spatial filtering.

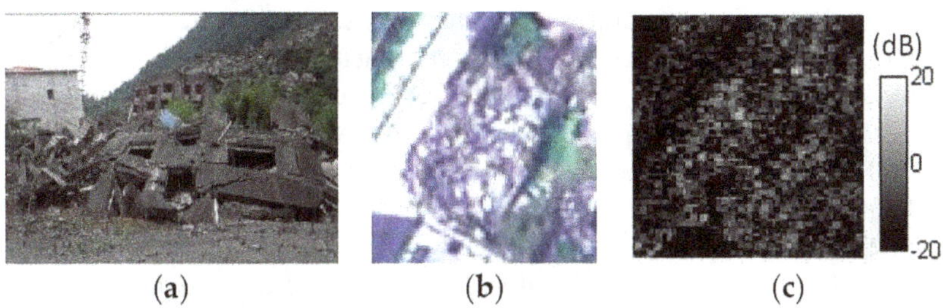

FIGURE 2.6 Comparison of the image properties of a collapsed building (a) on an optical image, (b) with those on an ascending TerraSAR-X staring spotlight image, (c) The scanning direction is across. (Source: Gong et al., 2016.)

2.2.3.3 VHR SAR Imagery

Since 2007, VHR TerraSAR-X, RadarSat-2, COSMO-SkyMed, and TanDEM-X satellites have been launched, yielding multiple-polarized SAR images with a spatial resolution up to 1 m. These VHR SAR images enable the discerning of urban structures or individual buildings in detail. A new TerraSAR-X mode called staring spotlight started to operate in 2013, further improving the images' azimuthal resolution to 0.24 m by widening the azimuth beam steering angle range. Such sub-meter SAR images offer renewed opportunities to assess earthquake-triggered damage to individual buildings even with only the post-event imagery. They are potentially capable of detecting building damage and discriminating basic damage classes, even though one scene of imagery covers a ground area of only about 2.7 km in azimuth × 6 km in range, which makes it unsuitable for large-scale damage assessment.

It is possible to extract standing and collapsed buildings from these images if assisted with a building footprint map available from collaborative projects such as Missing Maps, MapSwipe app, and OpenStreetMap (Gong et al., 2016). The separation of buildings is feasible even with meter-resolution TerraSAR-X and COSMO-SkyMed images if the pre-earthquake imagery is available because collapsed buildings are difficult to identify from post-event SAR images alone. On side-looking SAR images, standing buildings have a cube-like or other regular shape, and possess some unique features, such as a layover area, corner reflection, roof area, and shadow area. In contrast, collapsed buildings have different backscattering properties, such as the absence of strong double-bounce scattering, the destruction of a wall-ground dihedral corner reflector, and hence the layover area (Gong et al., 2016). Shadow is reduced or has completely disappeared (Figure 2.6). Surface roughness decreases after buildings have been flattened, altering brightness and backscattering intensity.

Although SAR imagery is spectrally poor, it can still be effective at separating standing buildings from collapsed ones via visual interpretation, even though this differentiation is hampered by a few factors, such as radar layover, foreshortening effects, and multi-path propagation. So automatic classification is preferred as it is much faster and can produce detailed levels of damage. There is a trade-off between classification accuracy and the level of damage identified. For instance, an overall accuracy of only 55.3% is achieved in assessing the 12 May 2008 Wenchuan earthquake ($M_w = 8.0$) damage to buildings in Old Beichuan County (Wu et al., 2016). This is because the level of damage is rather detailed, including five classes of tilted, outspread multilayer collapse, pancake collapse, piles of debris with some planes, and slight damage, some of which have a rather similar appearance on the image. As expected, the overall accuracy rises to 89.5% if the damage is mapped at binary classes of collapsed and standing buildings. This accuracy is almost identical to the highest accuracy of 90.2% achieved by Gong et al. (2016). Therefore, it is very difficult to map damage grades even from VHR SAR imagery at an accuracy beyond 91%, an accuracy much higher than its counterpart achieved from VHR optical imagery, owing to the use of more textural parameters derivable from SAR images.

Automatic classification allows diverse texture measures to be incorporated into the analysis to compensate for the spectral paucity of SAR imagery, such as first-order statistics (skewness and kurtosis) and second-order image features. The latter refers to texture measures such as gray-level co-occurrence matrix (GLCM) (mean, variance, homogeneity, contrast, dissimilarity, entropy, second movement, and correlation). In addition to two backscattering features, more GLCM-based textural features can be derived from multiple TerraSAR-X staring spotlight images of different polarizations (HH and VV), orbit directions (ascending vs descending), and incidence angles, even though not all of them are equally useful (Wu et al., 2016). Consideration of so many input features is ideally handled by machine learning classifiers such as RF and SVM. The classification of totally collapsed and standing buildings is achieved at an overall accuracy ranging from 87.8 to 90.2%, depending on whether the SAR image is descending or ascending (Gong et al., 2016).

In the classification, not all the considered texture measures are equally important. Of the ten features considered, GLCM second moment is the most effective, while the variance of backscattering and GLCM homogeneity are also useful for classifying damage type (Wu et al., 2016). Spectral features are more important than texture features in distinguishing collapsed and non-collapsed buildings. Some texture measures, such as GLCM-homogeneity, GLCM-dissimilarity, and GLDV-entropy, make only a small contribution towards the separation, and they are better left out of the classification. This exclusion boosts the classification accuracy of collapsed and non-collapsed buildings to 83.68%. The three best features for discriminating the damage type are identified as GLCM second moment, the variance of backscattering, and the GLCM homogeneity among the eight texture measures studied (Wu et al., 2016). These findings are based on RF classification of TerraSAR-X new staring SpotLight mode data of 0.24 m resolution. Whether they still hold with other VHR optical images remains to be determined.

2.2.4 From Integrated Images

Optical imagery and SAR data are so different from each other in their properties that they also affect the operating environment and manner of damage assessment from them. These differences are compared in Table 2.2, together with the main features of optical VHR and medium resolution SAR images in detecting building damage. The level of detection is much more detailed with VHR imagery owing to its fine spatial resolution that enables individual buildings to be discernible, but the assessment must be confined to the block or district level with medium resolution SAR imagery. The rationale of detection or detection cues also differs between them. Both share the commonality of identifying damage at specific levels, with the accuracy decreasing at more specific levels. So only collapsed and standing buildings are commonly extracted from both types of images.

Given that optical and microwave images are complementary in their image properties, naturally, they have been integrated to assess earthquake damage that depends strongly on timeliness (i.e. time delay of the post-seismic images with respect to the destructive event). Naturally, the use of multi-sensor images is conducive to achieving better timeliness. Thus, they have been used jointly to assess earthquake-induced building damage using different ways. One method is to acquire building features from optical images, and the acquired information is then fed to a classifier to extract damaged buildings from SAR images. For instance, built-up areas are automatically extracted from optical Sentinel-2 images using an unsupervised classification of spectral indices and an active learning framework (Hasanlou et al., 2021). Afterward, building damage is detected

TABLE 2.2
Comparison of Optical and SAR Imagery in Detecting Damage to Buildings

Material	Optical VHR	Medium Resolution SAR
Level of detection	Individual buildings	Damaged area or block
Detection cue	Accumulated debris	Changed coherence
Method of detection	Visual interpretation and classification based on spectral and texture features	Change detection of pre- and post-event images
Level of damage detected	A few (three to five), the proportion of damaged buildings possible	Just two grades of totally collapsed and standing buildings
Reliability of detection	Less reliable for lower grade damage, especially at multiple levels	High if using VHR SAR classified using machine learning methods
Operating environment	Clear sky	All-weather, including clouds and smoke
No. of images required	One (post-event only)	Two (one with VHR SAR)
Critical factors to consider		Window size for computing texture
Limitations	Nadir viewing not conducive to revealing damage	Subject to speckle noise

from a Sentinel-1 SAR image using multi-temporal coherence map clustering and similarity measure analysis. After built-up areas have been extracted, attention is then focused on the detection of damaged regions using time-series pre- and post-earthquake coherence data. After they have been masked by the built-up layer, the similarity between the reference signature (damaged pixels) and the target signature is measured using the spectral angle mapper algorithm, from which four levels of damage (no, low, moderate, and complete destruction) are produced via thresholding that minimizes the weight of the variance within each class. The mapping is achieved at an overall accuracy of 70%. In general, highly damaged regions are extracted quite accurately, but some no-damage areas are mistakenly classified as high damage areas (Figure 2.7). Because little human intervention is involved in the automatic classification, the detection can be fast implemented to produce consistent results.

The utility of fusing SAR and optical data in quantifying earthquake damage to buildings has been demonstrated in two case studies, the Izmit, Turkey earthquake of 17 August 1999 (M_w = 7.4) and the 26 December 2003 Bam earthquake of Iran (M_w = 6.6) (Stramondo et al., 2006). In the first case, pre- and post-earthquake single-look ERS-1 SAR images of 4 m × 20 m spatial resolution are integrated with optical IRS1-C imagery of 5.8 m resolution. In the second case, pre- and post-earthquake optical ASTER images of 15 m resolution are fused with EnviSat ASAR images of the same spatial resolution as ERS-1 and -2. The merged data are classified to three levels: slight damage (20–50%), moderate damage (50–80%), and heavy damage (>80%) using the maximum likelihood method based on various input features, including pre-seismic, post-seismic, and co-seismic coherences, plus pre- and post-seismic IRS images. As expected, damaged areas are classified much

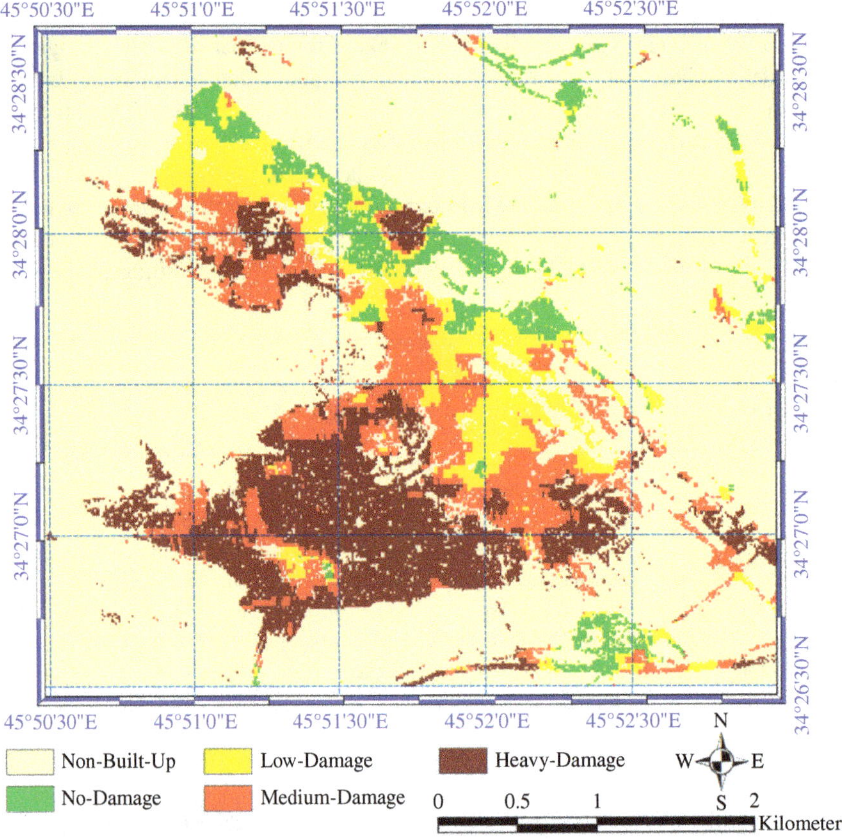

FIGURE 2.7 Damage to built-up areas in Sarpol-Zahab of Western Iran caused by the Iran–Iraq earthquake on 12 October 2017, classified from a Sentinel-1 image. (Source: Hasanlou et al., 2021.)

less accurately from either the optical image or SAR features individually (e.g., 70% from the SAR features, much lower than 81.9% achieved with the 5.8 m panchromatic images) than from the fused data. The fused SAR and optical data achieve the highest accuracy of 89% in a pixel-based classification. The same procedure applied to the Bam test case achieves a correct classification of about 61.5% from SAR data alone, and 70.0% from optical data, both being markedly lower than 76.6% achieved from the fused data (optical with features of pre- and co-seismic coherence, and pre- and co-seismic correlation).

The fusion of optical data and some SAR features significantly improves the damage classification accuracy because side-looking SAR imagery is particularly sensitive to surface changes. Both its backscattering and phase can be changed by debris from partially damaged buildings. The Bam case accuracy is lower than the Izmit case accuracy because the optical imagery used is ASTER, its spatial resolution of 15 m is three times coarser than that of the IRS1-C image. Fine-resolution images yield much detailed damage, such as collapsed buildings, the damage grade of each building, and the number or percentage of collapsed buildings within a district. Besides, the selected post-earthquake SAR images were captured much later after the earthquake event. By then, some of the debris on the ground could have been cleared. In spite of the decrease in the overall accuracy, the omission errors are stubbornly high, suggesting that the remote sensing observations underestimate the number of heavily damaged buildings probably because they are not visible from the nadir-viewing optical images. A heavily damaged building can still stand upright, and its roof can still appear as intact if viewed from above.

Before this section ends, it is crucial to emphasize that remote sensing-based detection of building damage is limited to the exterior appearance or general impression only, no matter which type of imagery is used and how it is processed. Such damage is unable to indicate whether the buildings have been structurally damaged or whether they are still fit for inhabitation or safe to inhabit after restoration. Such an assessment requires the expertise of a structural engineer, usually gained via close inspection of the concerned building on the ground. The value provided by remote sensing is the revelation of those moderately damaged or collapsed buildings that should be excluded from inspection as they are totally un-inhabitable.

2.3 POST-EVENT DISPLACEMENTS

It is significant to document the effects and damages of an earthquake so as to better evaluate infrastructure and engineering design, enhance our understanding of the earthquake process, and identify deficiencies in emergency response and recovery (Rathje and Adams, 2008). The most immediate and obvious effects of an earthquake are surface deformation and rupture, in addition to liquefaction (Figure 2.8). The extent and rate of near-fault surface deformation and plate movement manifest the stresses on faults. Earthquake-induced surface deformation can be easily estimated using GPS. GPS data collected from continuously operating reference stations can reveal crustal motion. DGPS significantly improves the accuracy of horizontal and vertical deformation near the fault over uncorrected data. GPS may yield precise monitoring results, but they are spot observations at most, not spatial distributions. Field distribution of crustal motion is better studied using remotely sensed data. Air- and space-borne panchromatic images (e.g., SPOT), InSAR, and ALS are all able to reveal far-field deformation, especially in near-vertical, strike-slip faults, but not near-field. Pre- and post-event images are the best for determining the spatial distribution of surface deformation and damage. These images are better at detecting horizontal displacement than vertical displacement. Image-based detection requires co-registration of multi-temporal images, preferably after orthorectification and geo-referencing to an identical coordinate system. The correlation of the two images is then calculated to identify the offsets that represent primarily horizontal displacement (Ayoub et al., 2009). In light of significant ground uplift, the horizontal displacement is also affected by the vertical displacement. The two can be differentiated from each other with the use of the third image, or if the two images are stereoscopic, obtained from the same flight path. If the

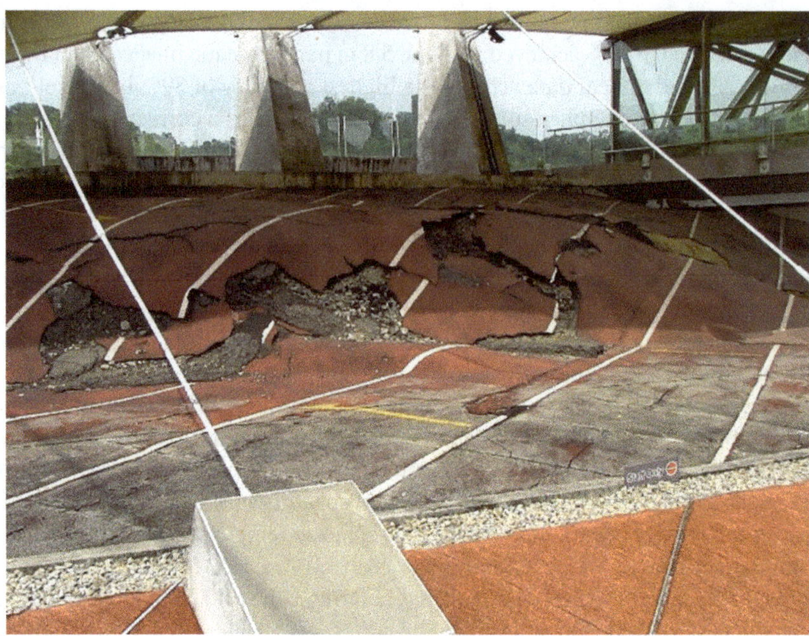

FIGURE 2.8 Uplifting of the sports ground of a high school in Central Taiwan after the 21 September 1999 Nantou earthquake (M_w = 7.7).

images have 1 m resolution, this method is able to identify ruptures with a minimum length of a few km and a minimum displacement of 10s cm, commonly associated with earthquakes breaking the surface with a moment magnitude exceeding about 6.5.

Ideally, the pair of pre- and post-event images used to detect the horizontal displacement should be acquired from the same sensor at the same position, such as optical Sentinel-2 imagery of 10 m resolution. This condition eases their co-registration after orthorectification. In sub-pixel image correlation analysis, the two images are shifted by various pixels in all possible directions, and the spatial cross-correlation between the two images at each shift is calculated. All correlation coefficients are compared with each other to identify the maximum one. And the shift from the former position to the current position of the maximal coefficient represents the displacement. Through sub-pixel level processing, it is possible to decompose the detected displacement into two directions: north–south and east–west, as well as their magnitude (Li et al., 2020). On the basis of the detected horizontal displacement, however, it is difficult to study the geometric features of earthquake rupture if the rupture trace is indistinct. This can be avoided by introducing curl and divergence from the two displacement components. Curl is indicative of rotation, or the direction of strike-slip rupture while divergence shows the magnitude of slip extension.

The cross-correlation between the pre- and post-event images is calculated at the pixel or sub-pixel level. Sub-pixel horizontal displacement may be calculated using COSI-Corr, Medicis, and MicMac (Figure 2.9). COSI-Corr is a module developed using Interactive Data Language and integrated with ENVI, implemented at the Caltech Tectonics Observatory of the US. This tool is the best for accurately orthorectifying and co-registering aerial and push-broom satellite images (Rosu et al., 2015). Medicis is a correlator developed by the Centre National d'Etudes Spatiales (France) for computation of deformation from image pairs. It offers both a spatial and a frequency correlator. The frequency–domain correlation method comprises a few steps. First, the pixel-wise disparities in lines and columns are estimated using the peak correlation method and then the sub-pixel displacement is computed only for couples (line, column) with a correlation score above the threshold of pixel-level correlation. Sub-pixel displacements are measured by applying fractional shifts to the

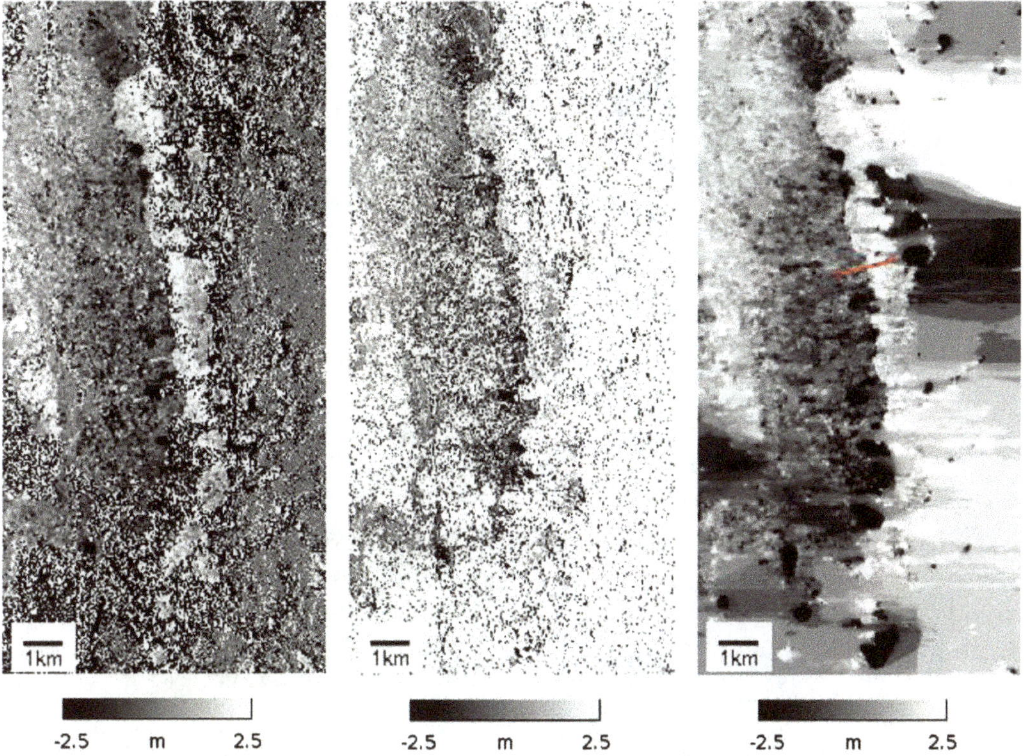

FIGURE 2.9 East–west component of the co-seismic offset field from the 21 March 2008 Yutian earthquake (M_w = 7.1) detected from a pair of orthogonal SPOT images of 25 August 2002 (pre-event) and 26 June 2008 (post-event) using three correlators. Left: COSI-Corr (window size = 32 × 32 pixels); Middle: Medicis (9 × 9 pixels); Right: MicMac (9 × 9 pixels). (Source: Rosu et al., 2015. Used with permission.)

sliding window via image interpolation. MicMac (Multi Images Correspondances par Méthodes Automatiques de Corrélation, http://logiciels.ign.fr/?Micmac) is an open-source software implemented at the Institut National de l'Information Géographique et Forestière, France. It computes the sub-pixel correlation of multi-images in the spatial domain. Of the three correlators, MicMac is the only one using regularization, providing very good results with the use of a small correlation window, which is well suited to acquiring high spatial resolution results near the fault line.

The calculated sub-pixel 2D deformation has a sub-meter to meter-level accuracy with aerial photographs, and meter to 10s m accuracy with satellite images, depending on the image spatial resolution (Figure 2.9). Regardless of their resolution, both aerial photographs and satellite images can detect only horizontal displacement. In comparison, InSAR and differential InSAR (DInSAR) are better options as they are not subject to atmospheric conditions. Pre-earthquake cm-level deformation can be monitored using mm-accuracy InSAR datasets at a high confidence level. InSAR has been successfully used to measure aseismic and coseismic slips across faults (Rathje and Adams, 2008). It offers a window onto interseismic fault processes and large-scale crustal deformation on an unprecedented spatial scale. The principle of InSAR detection of vertical motion will be covered in Section 13.3.1 (ground subsidence), so is not repeated here.

It must be noted that the determined slip distribution from InSAR is color-coded, and a color cycle in the interferograms corresponds to the defined deformation pace in the line-of-sight (LOS) direction only (Figure 2.10). It can be converted to horizontal or vertical displacement only if the radar incidence angle at the time of sensing is known. If overlaid with a DEM, they can reveal downsample LOS displacement. Sensing of coseismic deformation has to rely on InSAR, generated

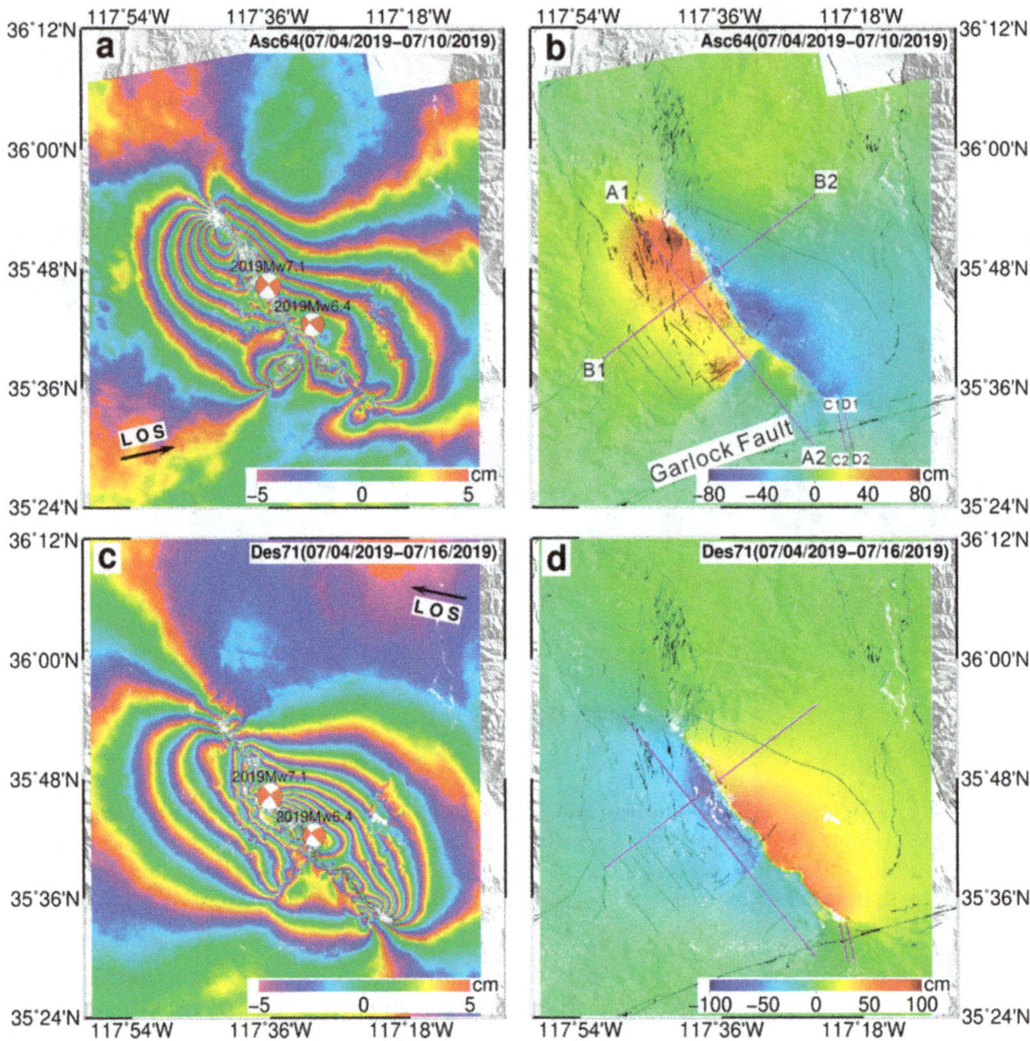

FIGURE 2.10 Coseismic interferograms (a – ascending; c – descending) and the deformation fields (b, d) unwrapped from them of the 2019 Ridgecrest earthquake. A color cycle in the interferograms corresponds to a 10 cm deformation in the line-of-sight (LOS) direction. Negative LOS displacement values in the deformation field suggest a movement away from the satellite. (Source: Li et al., 2020.)

from precisely co-registered radar images. The interferometric phase of the generated interferograms may be filtered using an adaptive spectral filter to remove the phase noise, followed by unwrapping. However, coherent phase unwrapping is not possible in densely vegetated areas, especially with short-wavelength bands, such as X- and C-bands. The decorrelation and unwrapping problems worsen with large displacements, which can be overcome with multi-source data, such as the ALOS-2 ScanSAR L-band interferometric phases, and the Sentinel-1 C-band range offset, ALOS-2 StripMap mode azimuth offsets, and the Sentinel-2 optical image north–south offsets. Offsets derived from multi-source data can detect surface rupture in the forms of co-seismic deformation and slip distribution. The detected results may be made more reliable by using L-band InSAR, the range, and azimuth offset tracking of SAR intensities, and the offset tracking of optical images (Song et al., 2019).

When multi-source data are in use, each individual dataset must be weighted to jointly inverse the fault geometry. One way of determining the weights is to inverse the residual RMS in fitting

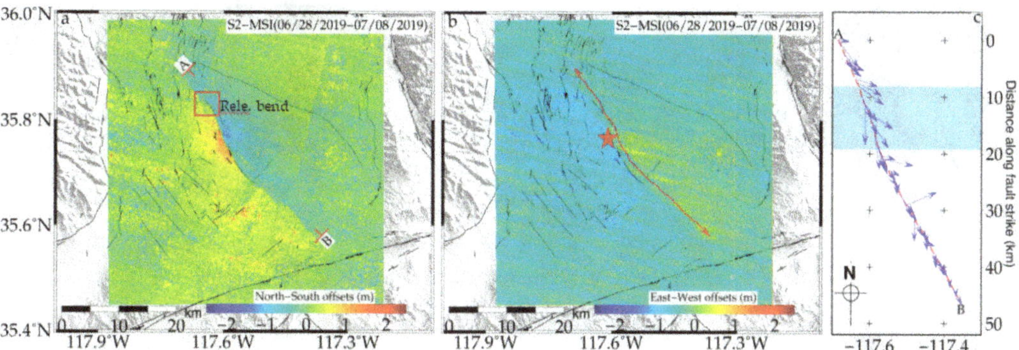

FIGURE 2.11 The horizontal displacement field of the 2019 Ridgecrest earthquake detected from pre- and post-event Sentinel-2 images (28 June–8 July 2019). (a) North–south (NS) displacement, (b) East–west (EW) displacement, (c) Displacement vectors: from the profile AB of the NS and EW horizontal displacements. (Source: Li et al., 2020.)

the refined dip angle and the best slip distribution. The joint inversion method is useful for areas where observations are limited by the presence of dense vegetation as in tropical rainforests or ice covers in Polar Regions. Prior to the inversion, each dataset may be subsampled or downsampled using the quadtree method (e.g., the sampling interval is 1 from 0 to 5 km of the fault trace, 2 from 5 to 20 km, and 4 beyond 20 km) (Song et al., 2019). The initial weight ratio of all the used datasets based on the residuals of individual inversion may be iteratively updated using the residual RMS of each dataset, leading to the final ratio. With the best fitting slip model, it is possible to successively determine the fault dip angle to an accuracy of 0.1 pixels. SAR pixel offset provides unambiguous surface displacement in both the LOS and azimuthal directions by cross-correlating intensities.

Compared with horizontal displacement (Figure 2.11), vertical displacement is much harder to detect reliably. Although it is possible to detect vertical displacement with stereoscopic aerial photographs, the detection is demanding and slow. In comparison, it is much faster using ALS data, from which accurate DEMs are created. This detection involves the subtraction of two DEMs of the same area and the same spatial resolution. Differencing of pre- and post-earthquake ALS point clouds-yielded DEMs reveals the vertical movement, but this differencing requires precise registration of the pre- and post-event data, a task not easily achievable, especially at high accuracy. Earthquake-triggered 3D surface displacement may be detected within a sub-decimeter agreement with field measurements with the use of the moving window iterative closest point method. With moderate spatial coverage, high spatial resolution, high accuracy, and precision, ALS has the potential to map 3-D surface deformation (Zhang, 2016).

2.4 DETECTION OF LIQUEFACTION

Liquefaction refers to the seepage of non-compacted, moisture-laden sandy soils or silts to the ground surface following violent ground shaking in an earthquake, after the highly saturated materials suddenly lose their strength and stiffness under the increased pore pressure. The soils or silts behave more like a liquid than a solid. The seepage occurs along the weakest fissure under an enormous pressure (e.g., the crack in surface sealing in urban areas). In addition to urban areas, the vast alluvial Indo–Gangetic Plains near the border between Nepal and Northern India are also prone to repeated liquefaction in the past under the influence of the Himalayan earthquakes (Gupta et al., 1998). The ideal conditions for liquefaction to take place are highly saturated soils that are loose, sandy, or silty, as commonly found in deltas and coastal lowlands. The likely sites for liquefaction in urban areas are reclaimed land over former river channels where the deposited moist soil layer is rich in silts and lacks cohesion. These potential areas are turned into disasters during strong earthquakes (Figure 2.12). Liquefaction does enormous damage mostly to roads, buildings,

FIGURE 2.12 Liquefaction ejecta on Kilmore Street next to the Avon River in Central Christchurch of New Zealand, triggered by a 5.8-magnitude earthquake in 2011. (Source: Asher Trafford.)

and tanks, which can be studied utilizing remote sensing. On the standard false-color composite of Landsat TM imagery, the limit of the vast liquefaction-prone zone near the border of Nepal and India is discernible owing to the higher soil moisture and unique local soil properties, such as clay content, vegetation, and land use (Gupta et al., 1998). Multispectral remote sensing is crucial to mapping liquefaction-prone areas. With remote sensing, it is possible to detect areas already affected by liquefaction and determine the distance of debris horizontal displacement from both optical and radar images using various methods.

2.4.1 Indices-Based Detection

Since areas affected by liquefaction tend to have a higher moisture content than the surrounding areas, naturally, these areas are easily detectable from NIR bands and spectral indices derived from them, such as the Liquefaction Sensitivity Index (LSeI). It is calculated from NIR and SWIR bands of both the pre- and post-event images using Eq. 2.4 and has been used to identify areas that have a higher level of surface moisture following a seismic event (Ramakrishnan et al., 2006):

$$LSeI = \frac{(B_4 - B_1)\cdot(B_3' - B_2') + (B_3 - B_2)\cdot(B_4' - B_1')}{(B_4' - B_1')\cdot(B_3' - B_2')} \tag{2.4}$$

where B_1, B_2, B_3, and B_4 represent spectral bands, the wavelengths of which are 0.52–59, 0.62–0.68, 0.77–0.86, and 1.55–1.70 µm of pre-event LISS-III (linear-independent self-scanning) imagery; B_1', B_2', B_3', and B_4' denote the post-event bands of the same wavelengths. This index is the most closely correlated with the field-observed liquefaction criteria among all the band ratios attempted. Understandably, the numbering of the four bands will change slightly if they are acquired from a sensor different from LISS-III. This index is considerably simplified with the Wide Field Sensors imagery that lacks B_1 and B_4, so LSeI is simply calculated as

$$LSeI = 1 + \frac{B3 - B2}{B3' - B2'} \tag{2.5}$$

The calculated LSeI value can be sliced into two levels of <2.10 (change) and >2.10 (no change) based on the histogram of the index image. From LSeI, the liquefaction susceptibility index (LSI) is calculated as

$$LSI = 0.852 \cdot LSeI^2 - 10.301 \cdot LSeI + 24.076 \left(R^2 = 0.97\right) \tag{2.6}$$

This index may be grouped to four classes to indicate the level of susceptibility, such as LSI > 40: highly susceptible; LSI 20–40: susceptible; LSI 5–20: moderately susceptible; and LSI < 5: hardly susceptible. With proper calibration with LSI, LSeI can be used as a tool to quickly map liquefied areas.

2.4.2 From VHR Imagery

As shown in Table 1.5, VHR imagery is characterized by a super-fine resolution (e.g., <5 m) but a lack of spectral richness. Such a fine resolution allows even minute liquefied areas to be detected. VHR imagery is best processed using OBIA. Since the pre-event image is not affected by liquefaction, it is processed differently from the post-event image. For instance, PCA can be produced from the pre-event image while image indexing, such as the usual NDVI intended to identify trees and the non-homogeneous feature difference (NHFD) index, is applied to the post-event image, together with the brightness of individual bands (Zhu et al., 2016). Calculated from the ratio of the red edge and coastal bands, NHFD is designed to differentiate bare soil and impervious surfaces (e.g., roads, building roofs, and parking lots). Although none of these features are related to liquefaction, they are crucial to achieving high accuracy in mapping liquefaction-affected areas and are discriminated based on index thresholds. The change induced by liquefaction is then detected using the usual procedure of post-classification comparison. Overlay of the affected areas with the raw image allows further differentiation between bare soil and anthropogenic surfaces. It is also possible to distinguish liquefaction into wet and dry based on moisture content (Figure 2.13). VHR imagery-produced

FIGURE 2.13 Distribution of liquefaction-affected areas by the 2011 Christchurch earthquake, mapped from high-resolution WorldView-2 imagery using maximum likelihood classification. (Source: Rashidian et al., 2020.)

maps can show liquefaction-affected areas in detail but are also plagued by incompetence in areas of dense, tall buildings in urban areas and forests due to the effects of shadow.

The use of space-borne images is limited in that the satellite pass may not coincide with the liquefaction event. The alternative is to make use of airborne images that can be taken days after an earthquake, such as the 22 February 2011 Christchurch earthquake of New Zealand (Morgenroth et al., 2016). In comparison with their space-borne counterparts, airborne images have a much finer spatial resolution or ground sampling distance (GSD) of up to 10 cm, but their spectral sensitivity is confined to only the RGB (red, green, blue) and near-infrared (NIR) spectrum (Table 1.5). If combined with LiDAR data, such fine resolution optical imagery allows liquefaction ejecta to be mapped via OBIA. In this synergy, optical imagery is used to produce an NDVI layer to separate trees from buildings. LiDAR data are converted to a DEM, from which slope is derived. Elevation and slope are jointly used to separate tall buildings from trees. Short objects may be further segmented at multi-scales to map the surfaces into dry liquefaction ejecta on pavement, wet liquefaction ejecta on pavement, wet liquefaction ejecta on grass, and pavement-no liquefaction ejecta (Figure 2.14). In the analysis, the original 81.8% overall accuracy in identifying the presence or

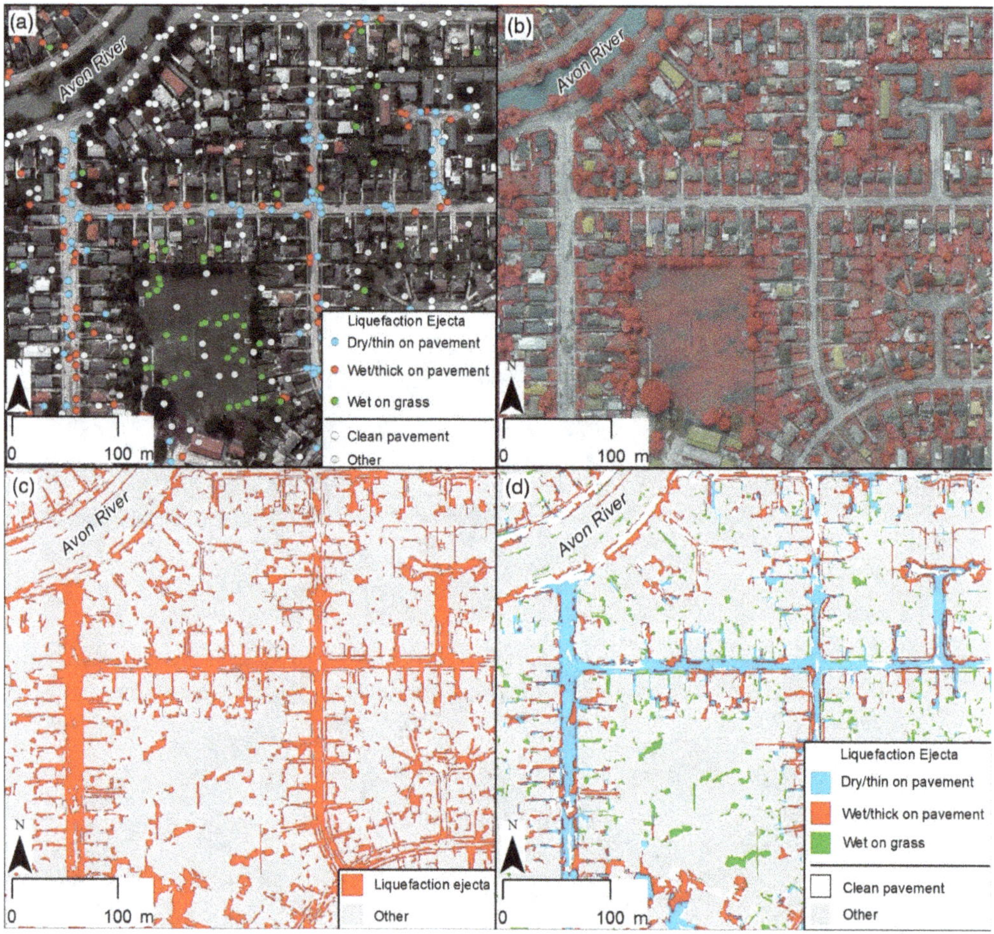

FIGURE 2.14 (a) True color, (b) false-color image of Central Christchurch, New Zealand taken two days after the 22 February 2011 earthquake (GSD: 10 cm), (c) liquefaction ejecta identified from LiDAR data via OBIA, (d) Liquefaction ejecta by surface cover: Thin dry liquefaction ejecta on pavements (blue), wet liquefaction ejecta on pavements (red), wet liquefaction ejecta on grass (green), pavements with no liquefaction ejecta (white), other (grey). (Source: Morgenroth et al., 2016. Used with permission.)

absence of liquefaction ejecta drops to 74.8% if the mapped liquefaction ejecta deposit is further differentiated into four aforementioned types of surface covers due to the spectral and textural variation amongst them being insufficiently unique. It is impossible to achieve higher accuracy because of the aerial photographs' limited spectral resolution that fails to discriminate dark asphaltic pavements of roads from water-saturated fine and medium sand particles dominating liquefaction ejecta. The integrated use of aerial photographs with LiDAR data, if processed using OBIA, can efficiently map the spatial extent of post-earthquake liquefaction ejecta in urban environments shortly after the hazard strikes. The mapped results can support emergency response and clean-up operations, as well as assess the long-term effects of the hazard on infrastructure.

2.4.3 From Medium Resolution Imagery

Liquefaction-induced surface changes are characterized by a high soil moisture content, so their detection from medium-resolution remotely sensed data is possible by quantifying soil moisture using moisture-sensitive green and SWIR bands of optical images such as Landsat 8 OLI and Sentinel-2. The quantification is based on NIR (0.86 μm) and SWIR (1.6 μm) bands, and expressed as the normalized difference water index (NDWI) by Gao (1996). Depending on the wavebands used, NDWI has two more versions involving a green (0.56 μm) band proposed by McFeeters (1996) and Xu (2006). They are calculated as:

$$NDWI_{Gao} = \frac{\rho_{0.86} - \rho_{1.6}}{\rho_{0.86} + \rho_{1.6}} \tag{2.7}$$

$$NDWI_{McFeeters} = \frac{\rho_{0.56} - \rho_{0.86}}{\rho_{0.56} + \rho_{0.86}} \tag{2.8}$$

$$mNDWI_{Xu} = \frac{\rho_{0.56} - \rho_{1.6}}{\rho_{0.56} + \rho_{1.6}} \tag{2.9}$$

NDWI enhances the signal of water but suppresses the influences of soil and vegetation. After NDWI is derived from the pre- and post-event images separately, the two NDWI layers are then differenced. A positive difference suggests an increased surface water content and vice versa. This index successfully detected liquefaction near the epicenter of the 5.4 magnitude earthquake of Pohang in Southeast Korea but failed to detect the same phenomenon far away from the epicenter (Baik et al., 2019). Thus, whether liquefaction can be successfully detected using this simple index depends on the ground moisture level.

Apart from the 1.6 μm SWIR band, a longer SWIR band of 2.2 μm from Sentinel-2 has also been used to detect liquefaction using a new index called temporal difference liquefaction index (TDLI). It is calculated using the same pre- and post-event band as:

$$TDLI = \frac{\rho_{2.2-pre} - \rho_{2.2-post}}{\rho_{2.2-pre} + \rho_{2.2-post}} \tag{2.10}$$

A positive TDLI indicates an increased water content, and thus the occurrence of liquefaction. Moreover, the larger the TDLI value, the higher the possibility of this occurrence. A TDLI value above 0 is considered to represent liquefaction (Figure 2.15). Naturally, the accuracy of detection varies with the thresholds of the index and the image used (Table 2.3). At the highest threshold of 0.150, no indices achieve acceptable accuracy. At a given low threshold, the accuracy varies slightly with imagery. With Landsat 8, TDLI is the second most accurate index after $NDWI_{Gao}$. Sentinel-2 is the most accurate due to its finer spatial (10 m) and temporal resolutions than Landsat's 30 m. A shorter revisit period is conducive to detecting short-lived liquefaction and minimizing the impact

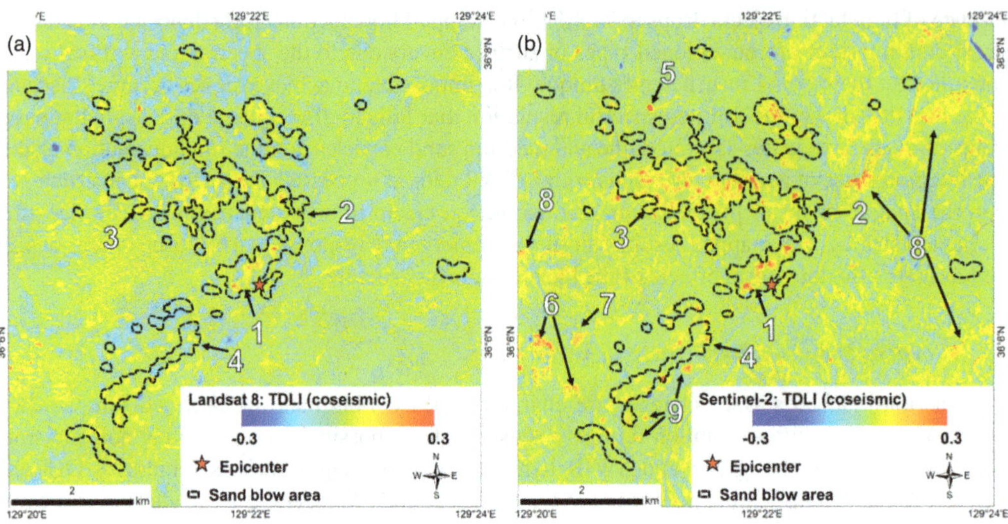

FIGURE 2.15 (a) Comparison of detected soil moisture content based on TDLI from Landsat 8, (b) Sentinel-2, images of Pohang, South Korea. A TDLI value above 0 indicates liquefaction. Star – epicenter; black boundary – outline of the sand blow associated with the 2017 earthquake. (Source: Baik et al., 2019.)

of rain after an earthquake. This index achieves much higher accuracy than all three versions of NDWI derived from both Landsat 8 and Sentinel-2 images. False detection of liquefaction occurs in places near high-rise buildings or areas of topographic shadow.

Instead of detecting soil moisture directly, land surface temperature (LST) has also been used to detect liquefaction indirectly because surface temperature is inversely related to soil moisture (Oommen et al., 2013). Liquefied soils laden with moisture tend to have a lower surface temperature than the unaffected surroundings. Surface temperature is commonly estimated from a single thermal band of a coarser spatial resolution than its optical counterparts. Besides, surface temperature is affected by seasonality. Even if the pre- and post-event images are obtained from the same sensor, seasonal variations in image acquisition cause the same ground to have different temperatures, making the temperature differencing image rather noisy, and the detected results unreliable. If the

TABLE 2.3
Comparison of Accuracy (%) in Detecting Liquefaction Based on Three Types of NDWI and TDLI at Seven Thresholds

	Landsat 8				Sentinel-2			
	NDWI Difference				NDWI Difference			
Threshold	Gao	McFeeters	mNDWI	TDLI	Gao	McFeeters	mNDWI	TDLI
0.000	21.881	99.387	63.190	73.211	58.282	42.536	50.716	95.501
0.025	8.793	82.822	32.106	56.237	26.380	15.542	26.789	75.665
0.050	2.658	36.401	14.928	35.583	10.634	3.885	8.793	53.579
0.075	0.613	7.975	6.544	19.018	4.090	0.204	3.681	37.832
0.100	0.000	0.818	3.272	10.634	2.658	0.000	1.227	26.789
0.125	0.000	0.204	1.227	5.521	0.818	0.000	0.409	18.200
0.150	0.000	0.000	0.613	2.249	0.409	0.000	0.000	11.247

Source: Baik et al. (2019).

optical imagery used has several spectral bands, as with the six bands of Landsat ETM+ (band 6 excluded), it can be transformed into new components using the Tasseled cap (Kauth-Thomas) method. One of the output components is wetness that can be used to detect changes in surface moisture via image differencing. Prior to the transformation, the pre- and post-event images may be standardized to remove any inherent variability between them. The differencing image can be further classed to produce positive and negative changes. This method generates promising results in mapping the surficial expression of earthquake-induced liquefaction, but the accuracy of mapping remains unknown.

2.4.4 From Radar Imagery

Apart from optical imagery, liquefaction can also be detected from actively sensed images. Although radar imagery is not sensitive to changes in soil moisture, it is still possible to identify areas affected by soil liquefaction from pre- and co-seismic pairs. This detection is based mostly on changes in image coherence before and after the earthquake. Coherence can be caused by changes in surface cover and height (relief). Both can trigger decorrelation after the ground surface becomes wetter and structures may be tilted. Decorrelation may stem from scatterer change between observations, causing the sum of scattering returns to vary. Liquefied areas are identified via analysis of surface changes. Changes in surface scattering properties are detectable based on phase-corrected coherence, calculated from the amplitude of SAR data (Ishitsuka et al., 2012). The coherence is influenced by phase variation from noise-induced changes in the signal, and systematic phase variation caused by topographic, atmospheric, or deformation gradients. In addition to liquefaction, coherence can also be altered by surface deformation induced by building collapse during an earthquake. To isolate changes in surface properties attributable to liquefaction, phase-corrected coherence must be calculated from amplitude representing the backscatter intensity of the target. The intensity is mostly affected by surface moisture content and roughness if the incidence angle, polarization, and radar wavelength remain unchanged. Other factors, such as noise in the radar system, perpendicular baseline length of the image pair, and temporal backscatter change in the interim (e.g. surface cover change or vegetation growth), also affect the coherence, but the effects of these factors can be eliminated mostly through coherence differencing or differential interferogram. A coherence threshold must be set to define the change and differentiate genuine liquefaction-induced changes from other ordinary surface cover changes. The threshold value may be based on the average and standard deviation of ordinary change in temporal coherence. Coherence differencing is able to detect liquefaction at an accuracy ranging from 61% to 85% (Figure 2.16).

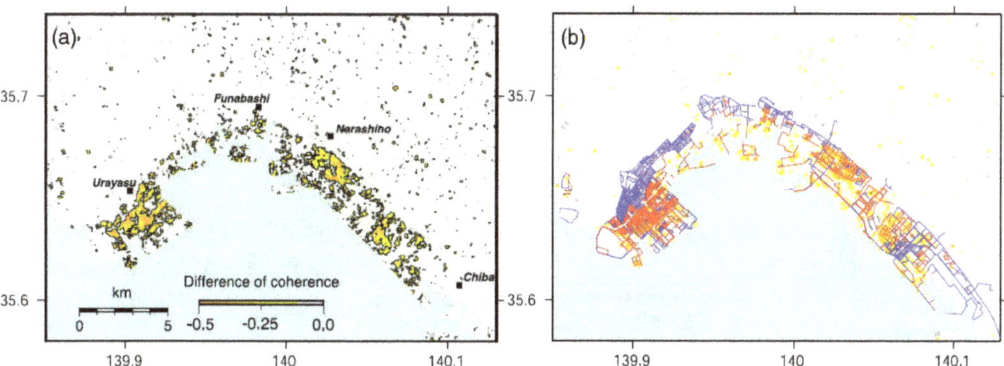

FIGURE 2.16 Areas in Tokyo Bay affected by liquefaction detected using coherence differencing of SAR imagery (a) in comparison with field survey results, (b) Red – liquefaction affected area; blue – not affected area. The level of agreement between the two ranges from 61% to 85%. (Source: Ishitsuka et al., 2012.)

2.4.5 Lateral Spreading Distance

After liquefaction has been successfully detected, it is possible to quantify liquefaction-induced lateral spreading distance by means of remote sensing. It is important to predict the lateral spreading of liquefaction as it can do heavy damage to infrastructure and properties. The spread can be determined by a number of means, one of which is GPS survey. This method is able to reveal the displacement, but is not image-based, so is inefficient. Compared with GPS, image-based detection is efficient, especially over large areas. It is not subject to site accessibility as liquefaction-affected areas are not easy to navigate, let alone survey its position and extent on the ground. There are two broad image-based approaches to the determination: use of a pair of images and LiDAR data. Preferably, the pair of images, one before the event, and another after the earthquake, should have the same spatial resolution, cover an identical ground area, and are recorded at the same local time so that shadows are consistent between them. Both images must be co-registered with each other precisely after they have been orthorectified to minimize geometric inconsistencies between them. Then, the two images are analyzed for their spatial cross-correlation in all possible shifting positions and magnitude. The correlation at every possible location is recorded as a signal-to-noise ratio (SNR). A perfect match is considered to occur if the SNR equals 1. The same correlation analysis can be repeated for each window after the entire scene is partitioned into multiple windows. The detected displacement of liquefaction can be further separated into the two cardinal directions of east–west and south–north (Figure 2.17). Horizontal displacements can be measured accurately to a pace as small as 0.2 to 0.3 m via optical image correlation, even though the images used may have a spatial resolution coarser than 20 m (Rathje et al., 2017).

LiDAR-based detection of lateral spreading distance follows the same correlation analysis except pixel value being replaced by elevation in calculating the cross-correlation. If the LiDAR data use an absolute reference frame for location, the horizontal tectonic displacement caused by fault rupture must be removed from the detected shift to generate the local liquefaction displacements. Lateral tectonic movement can be predicted via a model. Since it always involves a degree of uncertainty, it drags down the accuracy of LiDAR-based detection, and causes the detected displacements to be larger than those from GeoEye imagery of 0.5 m resolution in places further away from a river channel (Rathje et al., 2017). Optical image correlation of satellite imagery pairs provides more accurate and detailed measurements of lateral spreading than non-imagery LiDAR data that also suffer from systematic errors due to the way of data acquisition. So imagery data are preferable to non-imagery LiDAR data in the determination.

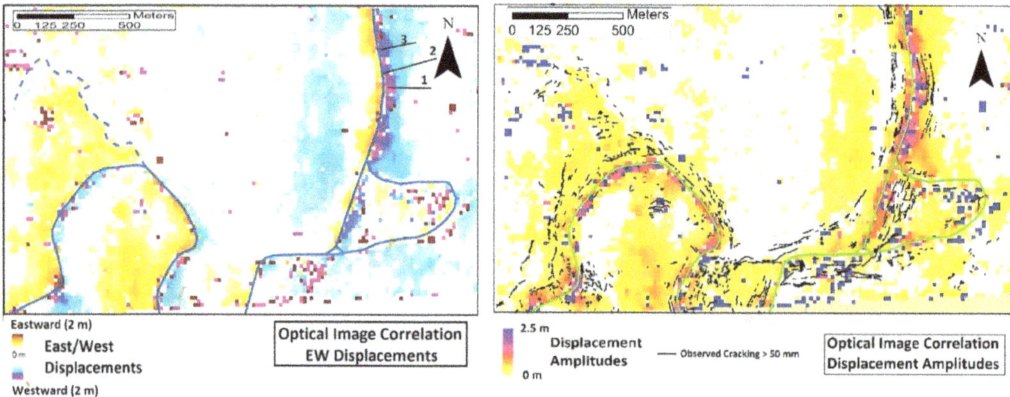

FIGURE 2.17 Horizontal displacements of liquefaction caused by the 2011 Christchurch earthquake, detected via image cross-correlation. Left: east–west displacements; Right: total displacement amplitudes. (Source: Rathje and Franke, 2017. Used with permission.)

REFERENCES

Adriano, B., J. Xia, G. Baier, N. Yokoya, and S. Koshimura. 2019. Multi-source data fusion based on ensemble learning for rapid building damage mapping during the 2018 Sulawesi earthquake and tsunami in Palu, Indonesia. *Remote Sens.* 11(7): 886. DOI: 10.3390/rs11070886.

Ayoub, F., S. Leprince, and J.P. Avouac. 2009. Co-registration and correlation of aerial photographs for ground deformation measurements. *ISPRS J. Photogramm. Remote Sens.* 64: 551–560. DOI: 10.1016/j.isprsjprs.2009.03.005.

Baik, H., Y.-S. Son, and K.-E. Kim. 2019. Detection of liquefaction phenomena from the 2017 Pohang (Korea) earthquake using remote sensing data. *Remote Sens.* 11(18): 2184. DOI: 10.3390/rs11182184.

Cooner, A.J., Y. Shao, and J.B. Campbell. 2016. Detection of urban damage using remote sensing and machine learning algorithms: Revisiting the 2010 Haiti earthquake. *Remote Sens.* 8(10): 868. DOI: 10.3390/rs8100868.

Dell'Acqua, F., and P. Gamba. 2012. Remote sensing and earthquake damage assessment: Experiences, limits, and perspectives. *Proc. IEEE* 100: 2876–2890.

Elliott, J.R., R.J. Walters, and T.J. Wright. 2016. The role of space-based observation in understanding and responding to active tectonics and earthquakes. *Nat. Commun.* 7: 13844. DOI: 10.1038/ncomms13844.

Gamba, P., and F. Casciati. 1998. GIS and image understanding for near-real-time earthquake damage assessment. *Photogramm. Eng. Remote Sens.* 64: 987–994.

Gao, B.C. 1996. NDWI—A normalized difference water index for remote sensing of vegetation liquid water from space. *Remote Sens. Environ.* 58: 257–266.

Gong, L., C. Wang, F. Wu, J. Zhang, H. Zhang, and Q. Li. 2016. Earthquake-induced building damage detection with post-event sub-meter VHR TerraSAR-X staring spotlight imagery. *Remote Sens.* 8: 887. DOI: 10.3390/rs8110887.

Grünthal, G. 1998. *European Macroseismic Scale 1998.* Luxembourg: EU, 99 p.

Gupta, R., A. Saraf, and R. Chander. 1998. Discrimination of areas susceptible to earthquake-induced liquefaction from Landsat data. *Int. J. Remote Sens.* 19(4): 569–572. DOI: 10.1080/014311698215856.

Gusella, L., B. Adams, G. Bitelli, C. Huyck, and A. Mognol, 2005. Object-oriented understanding and post-earthquake damage assessment for the 2003 Bam, Iran earthquake. *Earthq. Spectra* 21(S1): 225–238. DOI: 10.1193/1.2098629.

Hasanlou, M., R. Shah-Hosseini, S.T. Seydi, S. Karimzadeh, and M. Matsuoka. 2021. Earthquake damage region detection by multitemporal coherence map analysis of radar and multispectral imagery. *Remote Sens.* 13: 1195. DOI: 10.3390/rs13061195.

Ishitsuka, K., T. Tsuji, and T. Matsuoka. 2012. Detection and mapping of soil liquefaction in the 2011 Tohoku earthquake using SAR interferometry. *Earth Planets Space* 64: 22. DOI: 10.5047/eps.2012.11.002.

Ji, M., L. Liu, and M. Buchroithner. 2018. Identifying collapsed buildings using post-earthquake satellite imagery and convolutional neural networks: A case study of the 2010 Haiti earthquake. *Remote Sens.* 10(11): 1689. DOI: 10.3390/rs10111689.

Khodaverdizahraee, N., H. Rastiveis, and A. Jouybari. 2020. Segment-by-segment comparison technique for earthquake-induced building damage map generation using satellite imagery. *Int. J. Disaster Risk Reduct.* 46: 101505. https://doi.org/10.1016/j.ijdrr.2020.101505.

Li, S., H. Wu, D. Wan, and J. Zhu. 2011a. An effective feature selection method for hyperspectral image classification based on genetic algorithm and support vector machine. *Knowl. Based Syst.* 24: 40–48.

Li, X., W. Yang, T. Ao, H. Li, and W. Chen. 2011b. An improved approach of information extraction for earthquake-damaged buildings using high-resolution imagery. *J. Earthq. Tsunami* 5(4): 389–399. DOI: 10.1142/S1793431111001157.

Li, C., G. Zhang, X. Shan, D. Zhao, Y. Li, Z. Huang, R. Jia, J. Li, and J. Nie. 2020. Surface rupture kinematics and coseismic slip distribution during the 2019 Mw7.1 Ridgecrest, California earthquake sequence revealed by SAR and optical images. *Remote Sens.* 12(23): 3883. DOI: 10.3390/rs12233883.

Matin, S.S., and B. Pradhan. 2021. Earthquake-induced building-damage mapping using explainable AI (XAI). *Sensors* 21(13): 4489. DOI: 10.3390/s21134489.

Matsuoka, M., and F. Yamazaki. 2004. Use of satellite SAR intensity imagery for detecting building areas damaged due to earthquakes. *Earthq. Spectra* 20(3): 975–994. DOI: 10.1193/1.1774182.

McFeeters, S.K. 1996. The use of the normalized difference water index (NDWI) in the delineation of open water features. *Int. J. Remote Sens.* 17: 1425–1432.

Mitomi, H., M. Matsuoka, and F. Yamazaki. 2002. Application of automated damage detection of buildings due to earthquakes by panchromatic television images. In *Proc. of the 7th US National Conference on Earthquake Engineering*, Boston, MA, USA, 21–25 July 2002.

Morgenroth, J., M.W. Hughes, and M. Cubrinovski. 2016. Object-based image analysis for mapping earthquake-induced liquefaction ejecta in Christchurch, New Zealand. *Nat. Hazards* 82(2): 763–775. DOI: 10.1007/s11069-016-2217-0.

Oommen, T., L.G. Baise, R. Gens, A. Prakash, and R.P. Gupta 2013. Documenting earthquake-induced liquefaction using satellite remote sensing image transformations. *Environ. Eng. Geosci.* 19(4): 303–318.

Pesaresi, M., A. Gerhardinger, and F. Haag. 2007. Rapid damage assessment of built-up structures using VHR satellite data in tsunamiaffected areas. *Int. J. Remote Sens.* 28: 3013–3036.

Ramakrishnan, D., K. Mohanty, S. Nayak, and R.V. Chandran. 2006. Mapping the liquefaction induced soil moisture changes using remote sensing technique: An attempt to map the earthquake induced liquefaction around Bhuj, Gujarat, India. *Geotech. Geol. Eng.* 24: 1581–1602. DOI 10.1007/s10706-005-3811-1.

Rashidian, V., L.G. Baise, and M. Koch. 2020. Using high resolution optical imagery to detect earthquake-induced liquefaction: The 2011 Christchurch Earthquake. *Remote Sens.* 12(3): 377. DOI: 10.3390/rs12030377.

Rasika, A., N. Kerle, and S. Heuel. 2006. Multi-scale texture and color segmentation of oblique airborne video data for damage classification. In *Proc. of the ISPRS Midterm Symposium 2006 Remote Sensing: From Pixels to Processes*, Enschede, The Netherlands, 8–11 May 2006, pp. 8–11.

Rathje, E.M., and B.J. Adams. 2008. The role of remote sensing in earthquake science and engineering: Opportunities and challenges. *Earthq. Spectra* 24(2): 471–492. DOI: 10.1193/1.2923922.

Rathje, E.M., M. Crawford, K. Woo, and A. Neuenschwander 2005. Damage patterns from satellite images from the 2003 Bam, Iran earthquake. *Earthq. Spectra* 21: S295–S307.

Rathje, E.M., and K. Franke 2017. Remote sensing for geotechnical earthquake reconnaissance. *Soil Dyn. Earthq. Eng.* 91: 304–316.

Rathje, E.M., S. Secara, J. Martin, S. Ballegooy, and J. Russell. 2017. Liquefaction-induced horizontal displacements from the Canterbury earthquake sequence in New Zealand measured from remote sensing techniques. *Earthq. Spectra* 33. DOI: 10.1193/080816EQS127M.

Rezaeian, M. 2010. *Assessment of Earthquake Damages by Image-Based Techniques*. ETH Zurich: Zurich, Switzerland, Volume 107.

Rezaeian, M., and A. Gruen. 2007. Automatic classification of collapsed buildings using object and image space features. In *Geomatics Solutions for Disaster Management*, 135–148. Berlin/Heidelberg, Germany: Springer.

Rosu, A.M., M. Pierrot-Deseilligny, A. Delorme, R. Binet, and Y. Klinger. 2015. Measurement of ground displacement from optical satellite image correlation using the free open-source software MicMac. *ISPRS J. Photogramm. Remote Sens.* 100: 48–59. DOI: 10.1016/j.isprsjprs.2014.03.002.

Song, C., C. Yu, Z. Li, Y. Li, and R. Xiao. 2019. Coseismic slip distribution of the 2019 Mw 7.5 New Ireland earthquake from the integration of multiple remote sensing techniques. *Remote Sens.* 11: 2767. DOI: 10.3390/rs11232767.

Stramondo, S., C. Bignami, M. Chini, N. Pierdicca, and A. Tertulliani. 2006. Satellite radar and optical remote sensing for earthquake damage detection: Results from different case studies. *Int. J. Remote Sens.* 27(20): 4433–4447. DOI: 10.1080/01431160600675895.

Tronin, A.A. 2006. Remote sensing and earthquakes: A review. *Phys. Chem. Earth, Parts A/B/C* 31(4–9): 138–142.

Turker, M., and B. Cetinkaya. 2005. Automatic detection of earthquake-damaged buildings using DEMs created from pre-and postearthquake stereo aerial photographs. *Int. J. Remote Sens.* 26: 823–832.

Ural, S., E. Hussain, K. Kim, C.-S. Fu, and J. Shan. 2011. Building extraction and rubble mapping for city port-au-prince post-2010 earthquake with GeoEye-1 imagery and lidar data. *Photogramm. Eng. Remote Sens.* 77: 1011–1023.

Wang, T.-L., and Y.-Q. Jin. 2011. Postearthquake building damage assessment using multi-mutual information from pre-event optical image and postevent SAR image. *IEEE Geosci. Remote Sens. Lett.* 9: 452–456.

Wu, F., L.X. Gong, C. Wang, H. Zhang, B. Zhang, and L. Xie. 2016. Signature analysis of building damage with TerraSAR-X new staring SpotLight mode data. *IEEE Geosci. Remote Sens. Lett.* 13(11): 1696–1700. DOI:10.1109/lgrs.2016.2604841.

Xu, H. 2006. Modification of normalised difference water index (NDWI) to enhance open water features in remotely sensed imagery. *Int. J. Remote Sens.* 27: 3025–3033.

Yamazaki, F., Y. Yano, and M. Matsuoka. 2005. Visual damage interpretation of buildings in Bam City using QuickBird images following the 2003 Bam, Iran earthquake. *Earthq. Spectra* 21(S1): S329–S336. DOI: 10.1193/1.2101807.

Yu, H., G. Cheng, and X. Ge. 2010. Earthquake-collapsed building extraction from LiDAR and aerophotograph based on OBIA. In *Proc. of the 2nd International Conference on Information Science and Engineering*, Hangzhou, China, 4–6 December 2010, pp. 2034–2037.

Zhang, X. 2016. LiDAR-based Change Detection for Earthquake Surface Ruptures. PhD thesis, University of Houston, 144 p.

Zhu, J., L.G. Baise, and M. Koch. 2016. Mapping earthquake induced liquefaction surface effects from the 2011 Tohoku earthquake using satellite imagery. In *Proc. of the IEEE International Geoscience and Remote Sensing Symposium* (IGARSS), Beijing, China, 10–15, 2328–2331.

3 Landslides

3.1 INTRODUCTION

Landslides are defined as the downslope movement of earth materials under the influence of gravity either at a snail's pace (e.g., creeps), ephemerally, or instantly as in a rockfall. The mobilized debris may move at a uniform speed translaterally as a whole or certain parts of it move faster than others non-uniformly (e.g., rotationally), causing deformation. In either case, the distance of downslope travel varies with the debris momentum and the underlying slope, which is indirectly related to the local topography and the landslide triggers. There is no universal consensus on the velocity of debris movement and the debris materials involved. They may refer to mass movements on a slope, including rockfalls, topples, lateral spread, and debris flows that involve little or no sliding (Figure 3.1). Landslides fall into various types according to different criteria, such as the nature of displaced materials, the type of movement/causes, or a combination of both. The materials involved in a landslide can be sediment, earth, unconsolidated debris, rocks, and even snow. They move in the fashion of slump, slide, flow, fall, or avalanche, of which snow avalanche will be covered in Chapter 13.

The exact physical characteristics of a landslide vary widely. A representative landslide encompasses three components: the head zone, the path or chute, and the toe (Figure 3.2). The head or source zone is defined as the upper part of a landslide path from which the mobilized debris originates. It is the zone where the initial slide or slump takes place. The upper border of this zone is termed the initial scarp or head. The scarped surface left behind by the displaced materials lies below the original surface. The middle section of a landslide chute is called the tongue, along which the debris from the source zone is transported downslope. This section is also known as the "body" as it is the largest component of a landslide. The toe represents the position where the landslide path terminates. It is also the place where the mobilized debris is deposited or accumulates. The accumulated debris lifts the surface of the original ground. It must be emphasized that not all the three zones are present or distinct in all landslides, subject to the amount of mobilized debris and the age of landslides. For instance, the transport zone may be missing if the displaced materials do not experience much movement as in a slump due to the underlying surface having a gentle topography. With rain-triggered landslides, the mobilized materials are likely to have been washed downslope by the rainwater to the nearby channels instead of accumulating at the toe. Shaking-induced liquefaction causes the lateral spread of nearly flat sediment plains without any distinction between the source and destination zones. The toe may be indistinct if the debris has been eroded many years after the landslide event. From the perspective of debris transportation, all landslides fall into two general zones of depletion and accumulation (Figure 3.2).

Depending on the manner of debris movement, some landslides are rather hazardous. Landslide hazard is defined as the potential damage of a landslide to the natural and built environment and the threat to human safety over a certain time. For instance, a small landslide can be very tame and innocuous under normal circumstances but can pose a severe hazard during heavy rain. The impact of landslide hazards is highly local, affecting mostly the downslope areas in the landslide path. Landslide hazard is commonly studied via land susceptibility. All areas highly susceptible to landslides are said to possess a high level of hazard. Landslides are a serious natural hazard in many mountainous parts of the world such as the European Alps and the Himalayas. They are especially common in those regions prone to heavy rain storms, such as New Zealand and Taiwan. As a kind of destructive natural hazard, landslides can damage properties and infrastructure and cause loss of lives. Globally, landslides caused 2,312 casualties and injured/affected another 200,000 people in

DOI: 10.1201/9781003354321-3

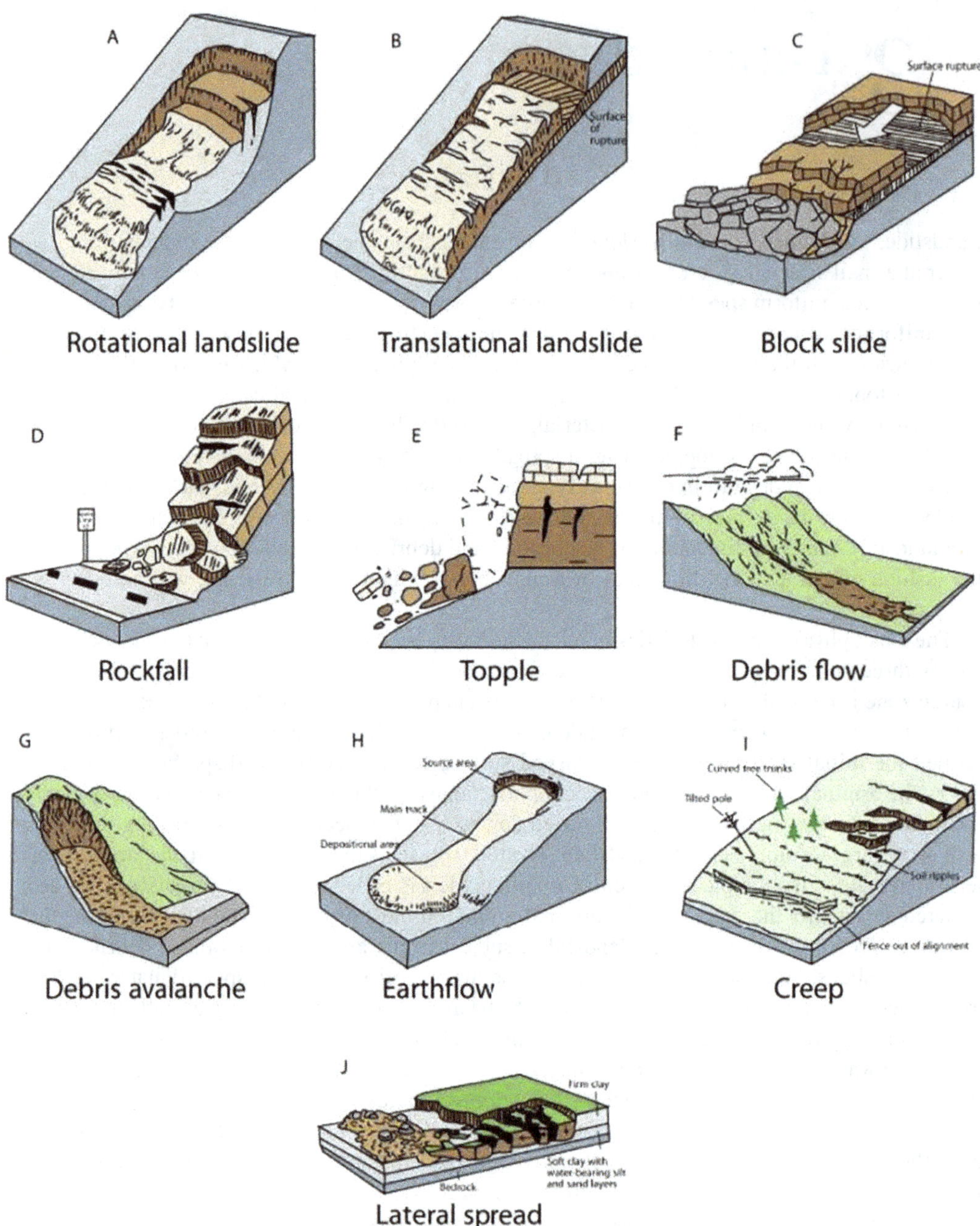

FIGURE 3.1 Major types of landslides by the manner of debris displacement. (A)–(J): manner of debris motion. (Source: U.S. Geological Survey, 2004.)

2017 (CRED [Centre for Research on Epidemiology of Disasters], 2018). Therefore, it is important to study landslides as a major natural hazard. Among all the methods of study, remote sensing offers the most cost-effective means of mapping landslides and assessing their potential hazard.

This chapter expounds how to study landslide hazards from various remotely sensed data that are analyzed using sophisticated methods. It starts with a discussion on how to identify landslide triggers and recognize landslides visually on remote sensing images. Afterward, the focus shifts to how to map landslides from images, using mostly digital image analysis based on per-pixel and

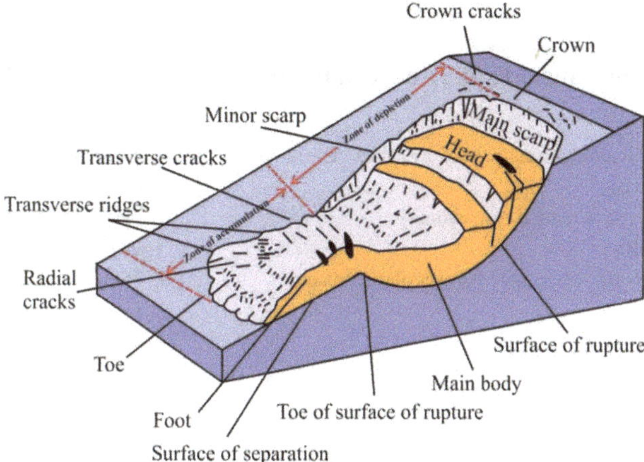

FIGURE 3.2 Nomenclature commonly used to describe the various components of a landslide. (Source: U.S. Geological Survey, 2009.)

object-oriented image classification. The pros and cons of each method are critically evaluated, and its best use is emphasized. Also discussed extensively is how to study landslides from non-imagery LiDAR data, including recognition and delineation of landslide components, estimation of landslide debris volume, and quantification of downslope creeping rate. The pros and cons of airborne LiDAR vs terrestrial LiDAR sensing of landslide features are also compared and contrasted. Lastly, this chapter elaborates on how to assess landslide vulnerability, zone landslide hazards, and monitor landslides using both optical and radar images. In particular, how to monitor the deformation field of a landslide and quantify its rate using InSAR is explained in detail, together with a critical evaluation of its functionality.

3.2 SENSING OF LANDSLIDES

3.2.1 Identification of Triggering Factors

A landslide results from the complex interactions among climate variables, environmental factors, and the local settings, including morphology, topography, hydrology, geology, and land use. The factors that contribute to landslide events can be geological, morphological (topographic), anthropogenic, and external triggers (Table 3.1). These factors suggest that, strictly speaking, a landslide is not just a natural hazard. To some degree, it can be classified as a mixture of natural and human-made hazards. Essentially, these variables fall into two broad groups of internal and external factors. External factors pertain to the environmental settings that create a high potential for landslides to eventuate or directly activate them. This group of factors may include land use, topographic relief, surface gradient, slope configuration, and orientation, all of which are crucial to terrain stability and the redistribution of rainwater over a slope. Inevitably, changes in land cover exert an impact on landslide susceptibility, especially if the terrain is destabilized (Alcantara-Ayala, 2004). In addition, slope stability is also affected by the antecedent soil moisture conditions, soil cohesion and viscosity, land cover, the underlying bedrock/pedology and its dipping, and the size and distribution of internal structures (i.e., faults, joints, bedding), all of which are included in the second class of factors. It is these factors, especially those that exert a significant impact on landslides, that are studied via remote sensing.

The accurate assessment of landslide hazards requires the separation of landslide potential from landslide triggers. The former refers to the propensity of slope materials to slide and is jointly dictated by the environmental and geological settings. A slope having a high propensity of slide may

TABLE 3.1

Checklist of Landside Causal Factors according to Cruden and Varnes (1996)

Geological Causes

- Material properties, including weak/sensitive/weathered/sheared/jointed/fissured materials
- Adversely oriented mass/structural discontinuity
- Contrast in permeability
- Contrast in stiffness (stiff, dense material over plastic material)

Morphological Causes

- Tectonic/volcanic uplift
- Glacial rebound
- Fluvial/wave/glacial erosion of slope toe
- Erosion of lateral margins
- Subterranean erosion
- Deposition loading slope or its crest
- Vegetation removal via forest fires

Physical Causes

- Intense rainfall
- Rapid snow melt
- Prolonged exceptional precipitation
- Rapid drawdown of floods and tides
- Earthquake
- Volcanic eruption
- Freeze-and-thaw/shrink-and-swell weathering

Human Causes

- Excavation of slope of its toe
- Loading of slope or its crest
- Drawdown of reservoirs
- Deforestation
- Irrigation
- Mining
- Artificial vibration
- Water leakage from utilities

be in a stable state if its equilibrium is not tipped by external events, such as heavy rainfall in a hurricane or violent ground shaking in an earthquake. Apart from the internal condition that affects slope stability, external forces also induce changes to destabilize the ground within a short period. These forces are termed triggering factors, such as downpours, earthquakes, volcanic eruptions, storm waves, and rapid stream erosion. Earthquakes and heavy rainfall events commonly trigger landslides in mountainous areas that may be fully vegetated. Rainfall events trigger landslides in two ways of infiltration and overland flow. First, the rapid infiltration of rainwater saturates the soil, increases slope load, and causes a temporary rise in pore-water pressure, eventually triggering landslides, especially shallow landslides. Second, in loose or weak soils, intense rainfall may increase overland flow and cause rill formation, both of which lead to small debris flows. Landslides can also be triggered by earthquake tremor that increases stress and weakens the cohesion of cemented soils. Such landslides are commonly located in loosely compacted regolith of a steep terrain. In addition, landslides can also be activated as a result of liquefaction indirectly, which is triggered by earthquakes. During an earthquake, ground shaking raises pore-water pressure and reduces soil strength, which eventually causes the soil to liquefy. These triggers activate landslides by weakening the support underneath the surface, reducing the strength of slope materials, increasing the slope load or stress during rainfall, or a combination of all three. In other words, they are all characterized by a weakened sheer strength and an increased shear stress that tips the former equilibrium.

All the external forces cause a near-immediate response in the form of a landslide by rapidly increasing the stress or by reducing the strength. However, not all rain events cause slope failures. Rather, there is a certain threshold of intensity and duration beyond which landslides are initiated. As with rainfall, earthquakes also have a landslide-triggering threshold. This relationship is known as the intensity-duration threshold. The rule of thumb for rainfall-triggered landslides is intense-and-short vs moderate-and-long events. With earthquakes, more factors are at play. Whether an earthquake can trigger landslides is affected by a number of factors, such as the magnitude of shaking and depth of the tremor epicenter. Earthquake-triggered (shaking) landslides are localized failures of steep slopes, quite different from rainfall-triggered ones in that the displaced earth materials may just collapse, leaving a deep scarp behind. The displaced materials are unlikely to move downslope immediately (Table 3.2). It is important to understand the morphological differences

TABLE 3.2
Comparison of the Characteristics of Earthquake- and Rainfall-Triggered Landslides

Landslide/Rockfall	Earthquake-Triggered	Rainfall-Triggered
Materials	Rocks, debris	Debris
Main movement	Slump	Slide
Components	Only source zone distinct	Gentle body with transport zone clearly identifiable
Movement	Local	Extended until channel
Depth	Deep-seated	Shallow
Surface gradient	Steep, discontinuous	Gentle, continuous

between landslides triggered by these two causes as they affect how they are best detected and monitored using different types of remote sensing data and different analytical methods.

3.2.2 Recognition and Identification

Landslides are mapped almost exclusively from remote sensing images to produce landslide inventory maps that illustrate the spatial distribution and extent of landslide chutes. Aerial photographs and satellite images allow many landslides in an area to be mapped quickly regardless of their location. The bird's-eye view of remote sensing images enables all landslide paths to be recognized and delineated indoors easily so long as they are not obstructed by overgrown vegetation. These images vary widely in their spatial resolution, including VHR images (see Table 1.5). While large-scale photographs allow detailed mapping of landslides, small-scale photographs are suitable for mapping tens of landslide paths scattered over a large area. Coarse-resolution images may be used to map landslide-affected areas, but not individual landslides. For this reason, if possible, fine-resolution images should be used for the mapping.

Remote sensing of landslides involves a few typical steps, including recognition, identification, detection, and classification, all referring to the inference and reasoning of landslide presence and ascertaining their properties and meanings (Figure 3.3). Recognition is the act of ascertaining the presence of landslides and differentiating them from the surroundings. It is not concerned with their nature and spatial extent. The recognition of a landslide from images can be potentially hampered by two issues: vegetative cover and image spatial resolution. In order to facilitate the recognition, the images should be captured in a season when the landslides or landslide-affected areas are the most distinct from the intact areas in the vicinity, such as when the vegetation is vigorously growing.

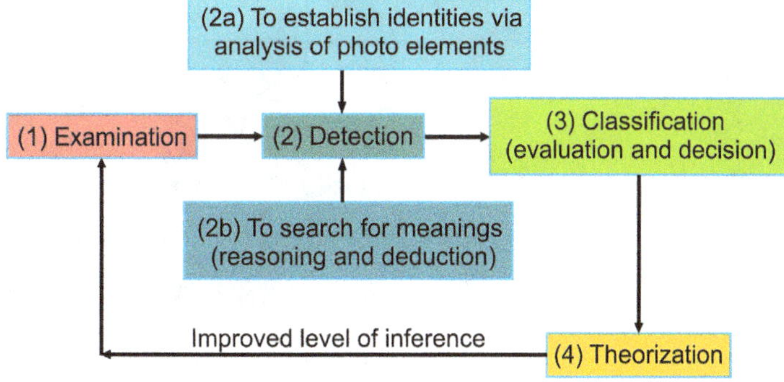

FIGURE 3.3 Procedure of and main steps in mapping landslides from remote sensing images.

However, overgrown vegetation may disguise the landslides and make them difficult to discern on nadir-viewing images.

The second issue is the minimum detectable size of landslides. The ability to detect landslides on imagery depends primarily on their physical size relative to the image spatial resolution and the distinctiveness of the landslide-affected area from their immediate environs. The minimum detectable landslides vary among Landsat, SPOT, and aerial photos at scales of 1:50,000, 1:25,000, and 1:10,000 (Mantovani et al., 1996). There is an inverse relationship between image spatial resolution and the physical size of a detectable landslide. In general, a landslide is not recognizable on an image unless it is several times larger than the image spatial resolution. Besides, the width of the landslide also exerts an influence on its detectability. If a landslide path is extensive to exceed the minimum GRD but is rather skinny, it may still not be successfully recognized. It is relatively easy to recognize a landslide from a fine-resolution image than on a coarse-resolution image. If the landslide is rather small, it may not even register on medium-resolution space-borne images. The ability of an image to allow a landslide to be recognized can be enhanced by merging the fine-resolution PAN band with coarse-resolution multispectral bands from the same sensor, an image processing technique commonly known as PAN-sharpening. This fusion takes advantage of the strength of both types of bands in landslide identification because a fine spatial resolution is critical to recognizing small landslides, while the composites of multispectral images facilitate the recognition of landslides in vegetated areas.

On aerial photographs, landslides are characterized by a sudden slump of the head. Large-scale deep-seated landslides in vegetated areas are easily recognizable from their crown scarp that has been displaced (Figure 3.4). They are spots where topographic discontinuity is the maximal and corresponds to convex slopes. Earthquake-triggered landslides are unmistakenly identifiable from satellite images based on four unique features: (i) deposit of fresh debris at the foot of a steep and

FIGURE 3.4 An abrupt change in slope curvature at the crown of the Mud Creek landslide, with a denudated surface. Such a landslide is best studied using a terrestrial laser scanner. (Source: USGS air photo taken on 27 May 2017.)

tall slope over other structures or extending into river channels; (ii) the circular scarp left behind by the mobilized debris is denudated of vegetation; (iii) the freshly stripped surface has a bright tone/color different from other intact bare grounds; and (iv) the debris movement has left clearly observable paths behind (Sato and Harp, 2009).

Identification goes one step further by classifying the recognized landslides based on their geometry, spatial arrangement, or content (Philipson, 1997). Similar to recognition, identification aims to confirm the recognized landslides. The landslide identified from a one-time image is easily confused with other non-vegetated patches in agricultural areas. For instance, the bare ground may represent a newly plowed field instead of landslides. Such confusion can be resolved from multitemporal images recorded before and after the landslides, or over a long period. Once denudated of vegetation, a landslide is not covered with vegetation immediately. However, after the vegetation has regenerated naturally, the landslide becomes less distinctive and harder to identify. Thus, such seasonal changes in surface cover help to differentiate genuine landslides from other vegetation-to-non-vegetation changes caused by tillage.

Landslide detection has been indiscriminately treated as recognition. Subtly, detection is one step further than identification and involves both recognition and classification. It always focuses on the spatial extent or the outline of a landslide if the spatial resolution permits, while its internal variation is ignored. After a landslide is detected, it is narrowed down to its type. So detection involves the abstraction of the directly observed tonal or color disparity between landslides and intact areas. The reliability of detection is commonly enhanced by making use of at least two-time images, one before the event, and another after, in a process commonly known as change detection. After all, landslides are essentially manifested as a surface cover change from vegetated to non-vegetated (or partially vegetated in certain parts). So the detection is naturally based on the use of NDVI that is rather sensitive to the quantity of biomass. The easiest way of detecting landslides is via manual interpretation that aims to demarcate the boundary of a landslide and attach meaning to the identified landslide such as its material type and relationship with the surrounding environment, usually based on the knowledge gained from remote sensing images or via field visits. This topic is so broad that it will be covered in Section 3.3.1.

3.3 LANDSLIDE MAPPING

In order to study the distribution, type, pattern, and recurrence of landslides, it is important to map them. Landslides are commonly mapped using three methods: field visits, from remote sensing images, and using LiDAR data, of which LiDAR-based mapping will be discussed in Section 3.4. In the field, the observed landslides (head) are logged using a GPS and represented as a point on the map (Figure 3.5c). This mapping method treats a landslide as a single point and ignores its spatial extent. This treatment is permissible if the logged point data are to be rasterized at a coarse resolution to identify the topographic features of landslides from a medium-resolution DEM. If the DEM resolution is fine enough to allow the determination of more detailed landslide traits, however, the landslide needs to be mapped more precisely as a polygon by traversing along its perimeter. The logged outline is more informative than the point data.

The field method is advantageous in that all landslides can be identified regardless of their physical size. However, it is ineffective, physically demanding, and time-consuming. The mapping becomes a daunting challenge for those landslides located in rugged and remote mountainous forests. To navigate to such sites is not a mean task in the field, let alone traversing the landslide perimeter, so this method is restrictive. For these reasons, field-based mapping is not practiced widely. Nevertheless, field visits still play an integral part in remote sensing of landslides for two reasons. First, ground truth data about landslides are needed in order to supply samples to train the computer to map landslides automatically from remote sensing images digitally; second, ground samples are needed to validate the quality and assess the accuracy of the landslide inventory map produced using other methods. In comparison, remote sensing is much more effective and efficient at

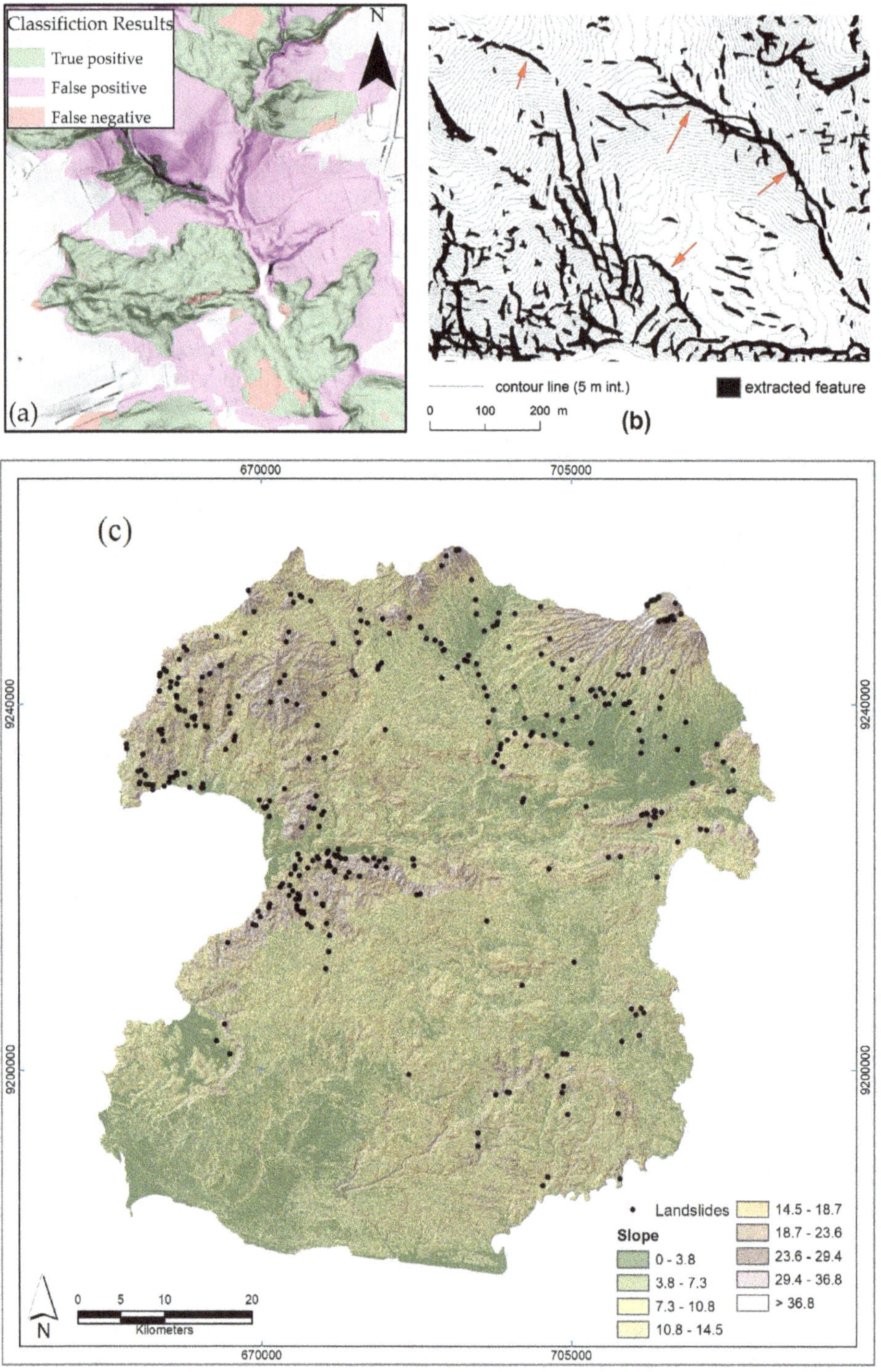

FIGURE 3.5 Three forms of representing landslides mapped using three methods. (a) Landslide-affected areas are mapped as polygons using OBIA of LiDAR data, (b) lines/polygons of large, deep-seated landslide scarps detected automatically from 1.5 m DEM using surface curvature (red arrows: position of the scarps as mapped in the field), and (c) landslides represented as dots at the scarp using GPS and manual interpretation of coarseresolution images against the backdrop of slope. (Sources: [a] Pawluszek-Filipiak and Borkowski, 2020; [b] Lin et al., 2013. Used with permission; [c] Wiguna, 2019.)

mapping landslides in remote and inaccessible areas. Remote sensing-based mapping of landslides may be implemented as manual interpretation or automatic classification. The former is covered in Section 3.3.1. The latter, to be covered in Section 3.3.2, is fast but does not produce the most accurate results all the time.

3.3.1 Manual Mapping

Manual mapping produces a landslide inventory map by delineating landslide outlines, usually on large-scale aerial photographs or VHR satellite images. Manual detection of landslides is based on head scarps and denuded slopes associated with individual landslide paths and may be aided by field surveys. Different types of landslides have different manifestations, all being crucial clues for their identification. The heavily relied clues are vegetative coverage and topographic discontinuity, especially the change in local morphology (Table 3.3). Vegetation cover and its changes in historical images (e.g., presence of bare land/absence of vegetation in the recent image but the presence of vegetation in the earlier image) are the foremost and sure indicators of landslides. Any areas that are

TABLE 3.3
Visual Characteristics of Mass Movement Types Based on Morphological, Vegetation, and Drainage Characteristics Visible in Stereoscopic Images

Type of Movement	Features	Characteristics
Rockfall	Morphology	Slope is mostly >45° where the bedrock is directly exposed. Distinct rock wall or free face in association with scree slopes (20–30°) and dejection cones
	Vegetation	Linear scars along frequent rockfall paths; low density on active scree slopes
	Drainage	No specific characteristics
Translational slide	Morphology	Joint-controlled crown in rock slides; smooth planar slip surface; relatively shallow surface material over bedrock; *D*/*L* ratio <0.1 and large width; hummocky runout and rather chaotic relief with block size decreasing with distance
	Vegetation	Source area and transportation path denuded, often with lineations in the transportation direction; differential vegetation on body in rock slide; no land use on body
	Drainage	Absence of bonding below the crown; disordered or absent surface drainage on body; streams deflected or blocked by the frontal lobe
Rotational slide	Morphology	Abrupt changes in morphology characterized by concave and convex slopes; semilunar crown and lobate frontal part; back-tilting slope facets, scarps, and hummocky morphology on depositional part; *D*/*L* ratio of 0.3–0.1; a slope of 20–40°
	Vegetation	Clear vegetation contrast with surroundings; absence of land use indicative of activity; differential vegetation according to drainage conditions
	Drainage	Contrast with non-failed slopes; bad surface drainage or ponding in niches or back-tilting areas; seepage in the frontal part of the runout lobe
Compound slide	Morphology	Concave and convex slope morphology; concavity often associated with linear grabenlink depression; no clear runout but gentle or bulging frontal part; back-tilting facets associated with (small) antithetic faults; *D*/*L* ratio of 0.3–0.1, relatively broad in size
	Vegetation	Same as rotational slides, although slide mass will be less disturbed
	Drainage	Imperfect or disturbed surface-drainage ponding in depressions and in the rear part of slide

Source: Modified from Soeters and van Westen (1996).

TABLE 3.4
Criteria for Manually Recognizing Rainfall-Induced Landslides

Photographic Clue	Description	Discrimination Rule
Size	Must be sufficiently large to register	Size smaller than image resolution is not recognized, if too small, represented as lines instead of polygons
Tone/color	Light, gray light brightness	Brightness value > threshold
Location/association	Near ridges, cut-off slopes, road-sides	Trigger events and buffer zone of the feature
Shape	Spoon-shaped, elongated-oval, dendritic, rectangular, triangular	Location-specific and topography-specific, area-to-perimeter ratio is small for elongated landslides
Movement direction	The drop direction of the landslide is the gravitational vector on the ground surface	Roughly perpendicular to the streams and topography-specific
Slope	Varying with landslide type, (i) shallow-seated landslides >45%; (ii) deep-seated landslides ~40%; (iii) debris flows ~10–20%	Slope > threshold
Shadow	Dependent on whether the landslides are in the shadow-side or sunny-side	Solar azimuth is related to the slope aspect
Texture	Smooth texture if debris is fine sediment or coarser if made up of pebbles or rocks	Texture < threshold (as vegetation has an even coarser texture)

Source: Modified from Liu et al. (2012a).

not covered by the original vegetation or not by any vegetation at all potentially represent landslides. Apart from vegetation, morphology and drainage are also useful, even though they are less commonly used. The key to correctly interpreting landslides on stereoscopic images is the distinctive topographic and morphologic form shaped by landslides (Metternicht et al., 2005). Topographic discontinuity, especially the abrupt change in surface elevation at the scarp of the crown, is also indicative of landslides. However, topographic discontinuity does not allow the toe of the landslides to be demarcated reliably if its surface has a gentle topography, or its abrupt changes in surface gradient have been eroded years after the landslide event. Thus, this clue may enable the recognition of landslides but may not allow the delineation of the precise outline of a landslide chute in some cases.

On optical images, individual landslide chutes are recognizable from a number of image clues that are used more universally and frequently than morphologic clues. These image properties encompass size, tone/color, shape, location, texture, and slope that are all reliable indicators of landslide presence (Table 3.4). Size is important and governs whether a landslide can be mapped or not. Its importance has been discussed in Section 3.2.2 already and is not repeated here. Landslides have an elongated, irregular shape, even though the toe zone can be circular in shape. In agricultural areas, especially silvicultural areas, landslides can still be distinguished from vegetation removal by farming activities based on the shape and color of denudated patches. Dissimilar to the elongated shape of a landslide, cultivated cropland usually has a rectangular or square shape. In addition, vegetation removed through slash-and-burn tends to have a darkened color.

Location of landsides is also a useful clue to reinforce the interpreted landslides based on other cues. Geographically, landslides are commonly located in mountainous areas of a steep terrain distant from tilled fields, where the land is unlikely to be cultivated. However, the utility of the same cues varies with the method of identification. For instance, location is useful in visually recognizing landslides. If automatically detected, image tone becomes the most useful since landslides are commonly associated with denudated surfaces, in huge contrast to the intact vegetative cover of non-landslide areas. Its tone tends to be lighter or even brown on a color photograph (Figure 3.6).

FIGURE 3.6 Appearance of landslides amid paddocks of hilly New Zealand on a color aerial photograph. The landslides are characterized by a light tone/color formed by stripping of surface vegetative cover, in stark contrast to vegetated slopes of green color.

In comparison, location is not so useful (Table 3.5). Instead, the shape of a landslide, especially its perimeter-to-area ratio, is a useful clue.

After a landslide has been interpreted, it may be desirable to demarcate the three zones of head, path, and toe based on shape and surface features (Table 3.6). Morphology is important to the demarcation of these zones. The source area or the head zone has a hummocky surface in the past

TABLE 3.5
Comparison of Useful Clues in Landslide Recognition

Clue	Pros	Cons
Tone	Bright, direct, intuitive	Easy confusion with non-vegetated bare ground
Shape	Elongated, skinny, a low perimeter-to-area ratio	Useable mainly in OBIA, but not manual interpretation
Location	Uplands in mountainous areas with a steep topography	Not easy to consider in automatic recognition
Topographic continuity	Indicative of scarp only	Unable to indicate the spatial extent of landslide path, 3D view essential
Change	The temporal change in surface cover before and after the landslide event can confirm recognition from a one-time image	Precise delineation of changed extent requires co-registration of the image pair used

TABLE 3.6

Remote Sensing Interpretation Characteristics of Various Landslides

Landform	Feature
Landslide scarp	a. Horseshoe-shaped scarp or cliff
	b. Fresh, open cracks that are evidence of recent movement and indicate slipping or tipping of the landslide body
	c. Cracks often parallel to the edge of a scar
Landslide mass	a. Shovel-shaped depression
	b. Discordant natural vegetation on the landslide boundary
	c. A hollow landform in the upper part and a topography characteristic of a gentle slope or tableland in the lower part
Lateral boundary	a. Hydrological characteristics of two gullies generated in the same source area
	b. Echelon-type shear failure
Slope toe	a. Bulging landform at the toe
	b. Toe extending into river channel or narrowing (curving) of the original channel
	c. Formation of landslide dam upstream of the slope toe
	d. Radial cracking

Source: Modified from Central Geological Survey (2006).

and is distributed with actual landslides. This zone is the most recognizable. However, the differentiation of a landslide into different components is not always feasible if the photo is not viewed stereoscopically or has a small scale.

During manual interpretation, the observed landslides (scarp location) are digitized in vector format as polygons on the computer screen if the interpreted aerial photographs have a large scale (Figure 3.5b). The produced landslide inventory map shows the location, type, and spatial extent of landslides, as well as their physical properties, such as length, width, and shape, but not their 3D information, such as slope gradient. Visual interpretation of landslides is simple and straightforward but tends to be slow, tedious, and laborious. The reliability of photo-interpreted landslide paths is subject to photo scale, spatial resolution, the season of photography, the distinctiveness of the landslide chutes, as well as personal bias. It is error-prone, especially when the image resolution is coarse and the study area is extensive containing hundreds of landslides, some of which are rather small. Aerial photographs and field inspections can result in incomplete information, especially in densely vegetated areas. For these reasons, the manual method is not practiced widely in reality. A better alternative is computer-based automatic mapping.

3.3.2 Automatic Mapping

Automatic mapping is a process of producing landslide inventory maps from digital analysis of image pixel values in the multispectral domain using computer software, commonly known as image classification. Multispectral images are suitable for automatic analysis in the digital environment, in which landslides may be mapped as a separate category of non-vegetated ground. If the satellite image used has a moderate resolution (e.g., 20-m Sentinel-2 imagery), the landslides may be represented as a cluster of pixels, indicating the area that has been affected by landslides, but not individual paths (Figure 3.5a). This classification can be implemented as either per-pixel or object-oriented.

3.3.2.1 Per-Pixel Classification and Vegetation Indexing

Per-pixel image classification is implemented as unsupervised and supervised. Unsupervised classification is synonymous with clustering analysis in which all pixels in the input image are grouped into a pre-determined number of spectral clusters specified by the analyst, without any prior

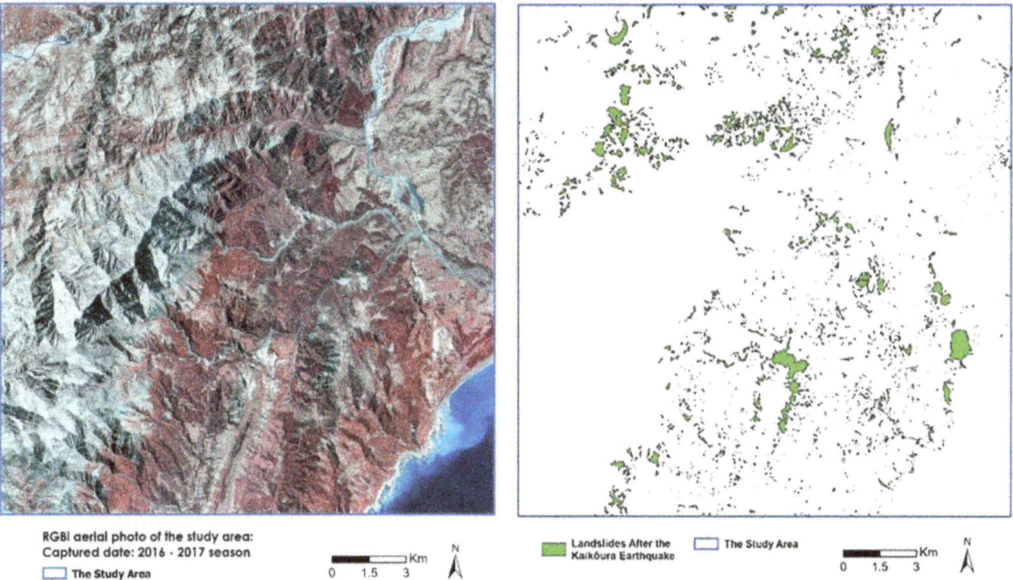

FIGURE 3.7 Landslides triggered by the 14 November 2016 Kaikōura earthquake of South Island, New Zealand. Left: RGB aerial photo captured after the earthquake; Right: spatial distribution of landslides mapped from a post-event Sentinel-2 image using unsupervised image classification after non-landslide areas above (e.g., bare mountain tops); and below (e.g., river channels) certain elevation are masked out. (Source: Chen, 2022.)

knowledge of their identity. Some clusters likely correspond to landslides (Figure 3.7). The exact cluster identity is ascertained in a post-classification session via visual comparison with the original color composite image or via field inspection.

In order to minimize the commission errors that represent false alarms (the classified landslide pixels are not genuine landslides), the classified results may be further refined via intersection with other layers, such as NDVI and elevation to remove those landslides in areas where they are unlikely to occur such as river channels and roads with no vegetation cover at the foothill or bare ground above the treeline. The generation of perfect landslide maps may require painstaking post-classification processing to filter incorrectly classified pixels in the shadow or exposed river channels. The analysis of pre- and post-landslide images with a spatial resolution of 8 m (multispectral) and 2 m (panchromatic) mapped thousands of landslides with a combined area of tens of thousands of ha at 9% omission errors or 16% commission errors (Tsai et al., 2010). With an average size of 2.4 ha, the majority of the detected landslides are smaller than 10 ha. Such a small size means that most of them will be missed out if the imagery's spatial resolution is coarser than 50 m × 50 m, or 0.25 ha.

The performance of per-pixel image classification in accurately mapping landslides is compromised by topographic shadow. The mapping accuracy is also degraded by the overreliance on only spectral information in the decision-making, while other useful spatial and geometric clues of landslides, such as shape, texture, and location (or spatial context), are completely ignored. Understandably, this paradigm of image classification does not produce satisfactory results, especially when the image is noisy, and the assumption of image classification is frequently violated. For instance, if some of the original vegetation is still distributed on the surface of translaterally displaced debris, it will not be classified as landslides. Besides, classification errors can also arise from the classification procedure and the coarse spatial resolution image used. If a pixel of 10-m resolution is classified as a landslide, the landslide will have a minimum area of 100 m^2, even though its genuine area on the ground may be only half of this size (e.g., 50 m^2). This source of error can be minimized by using fine-resolution images.

Landslides change the ground area in three ways: disrupt surface topography to make its gradient discontinuous, change its land cover from vegetated to non-vegetated, and reduce its biomass. Thus, to accurately map landslides, these three changes must be considered simultaneously. Per-pixel image classifiers make use of the clues related to land cover only. No consideration is given to the other two changes in the decision-making. This deficiency can be partially compensated for by integrating multi-temporal images and orthophotos good at detecting land cover with topographic data that allows surface height change to be detected. In the automatic method of detection involving pre- and post-event images, they must be co-registered, and possibly PAN-sharpened, orthorectified, and corrected for atmospheric disturbance (Mondini et al., 2011). Even after such processing, automated change detection of SPOT multispectral images identified only 70% of the landslides, with the missing ones caused by their sub-pixel width smaller than 10 m (Nichol and Wong, 2005).

3.3.2.2 Object-Oriented Image Analysis

Some of the limitations inherent in per-pixel image classification are overcome with OBIA that takes into account the spatial context of landslide pixels and their geometric properties (e.g., shape measured by the area-to-perimeter ratio) in its decision-making. Besides, OBIA, if implemented using machine learning methods, allows the incorporation of non-spectral variables in the decision-making, such as spatial properties of landslides that may include shape, size, orientation, elongation, and compactness. Naturally, it tends to produce more reliable classification results. OBIA can increase the success rate in mapping main scarps and landslide bodies. This mapping may be facilitated by cracked bedrock outcrops, coarse colluvial deposits, and gently undulating surfaces. This method produces objective results much faster and more reliably for deep-seated landslides with distinctive characteristics than for shallow landslides. The accuracy of mapping is assessed by overlaying the training samples over the classified landslides shown as polygons (Figure 3.8). There are three possibilities: hit (true positive), miss, and false positive or false negative, quantitatively measured as POD and POFD.

The relative performance of the pixel-based and object-oriented image analysis has been comparatively assessed by Pawluszek-Filipiak and Borkowski (2020) in mapping the most landslide-affected area in Poland from Sentinel-2A and LiDAR data (Figure 3.9). As shown in Table 3.7, the accuracy is affected by how the samples are split for training and testing. At a high split ratio, OBIA is slightly less accurate than its pixel-based counterpart. However, the relativity is reversed after the ratio drops to less than 0.25. Thus, the achievement of a highly accurate outcome is not guaranteed even with OBIA unless the required inputs are properly parameterized. One possible way of improving the accuracy is to combine both methods. Nevertheless, the accuracy of intersecting pixel-based analysis with OBIA is not markedly improved unless the intersected results are further processed using either median filtering or via removal of small elongated objects (Figure 3.9). Both post-classification processing methods helped to improve the accuracy markedly.

Both per-pixel classification and OBIA are constrained by their inability to incorporate non-imagery features in the decision-making, such as elevation and distance to roads. This deficiency is easily overcome with machine learning-based image classification.

3.3.2.3 Machine Learning-Based Mapping

With OBIA, many parameters about a landslide patch become available. More metrics can be produced from their statistics, such as mean, standard deviation, percentile, and even co-variance. In addition, environmental factors that may dictate where landslides can occur, such as topographic position index, terrain parameters, and surface relief and curvature, may need to be considered in the mapping. The input variables to a classifier become more diverse and numerous if non-image data, such as DEM, roads, and rivers are considered. From a DEM, more variables can be derived, such as slope aspect, elevation, and slope gradient (Table 3.8), as well as morphological parameters (e.g., curvature, hillshade, roughness, and flow direction). The consideration of so many variables, together with spectral bands and vegetation indices derived from them, is ideally handled

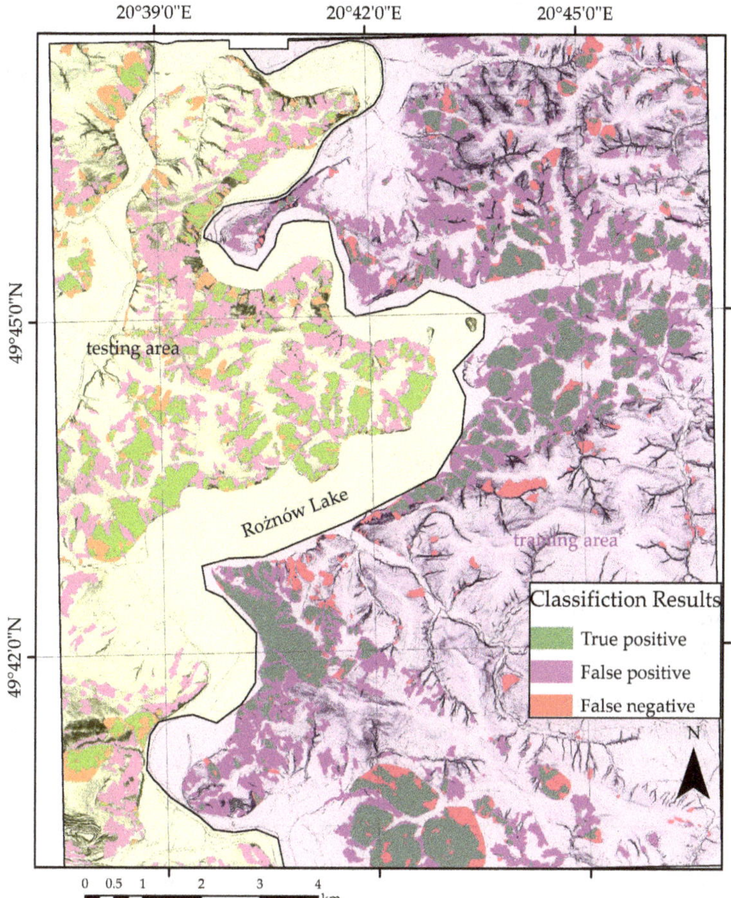

FIGURE 3.8 The quality of object-oriented classification of landslides determined via superimposing training and testing area on the OBIA results. (Source: Pawluszek-Filipiak and Borkowski, 2020.)

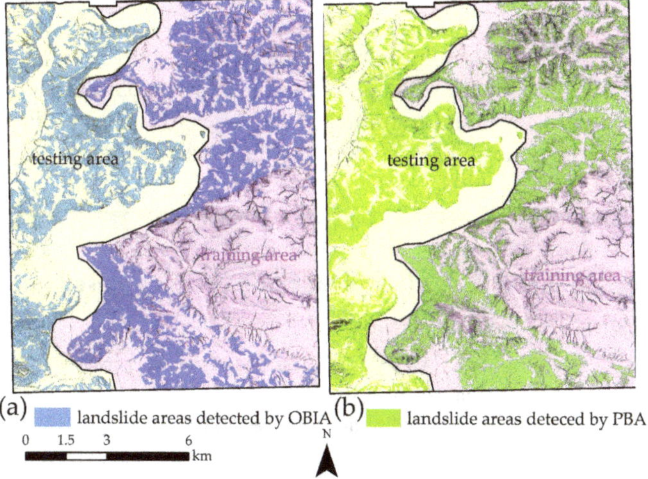

FIGURE 3.9 Comparison of (a) landslide-affected areas mapped using OBIA, (b) landslide-affected areas mapped using PBA. (Source: Pawluszek-Filipiak and Borkowski, 2020.)

TABLE 3.7

Comparison of Pixel-Based Analysis (PBA) with OBIA in Mapping Landslides from Sentinel-2A Data in Six Testing Areas

Testing Area	Method	TTR[a]	*F*1 Score	POD	POFD	Overall Accuracy (%)
1	PBA-RF	1	0.57	0.83	0.29	74
	OBIA-RF		0.58	0.88	0.31	73
2	PBA-RF	0.4	0.53	0.85	0.31	72
	OBIA-RF		0.53	0.88	0.33	71
3	PBA-RF	0.35	0.44	0.83	0.34	69
	OBIA-RF		0.46	0.87	0.34	69
4	PBA-RF	0.25	0.42	0.80	0.34	68
	OBIA-RF		0.46	0.86	0.33	70
5	PBA-RF	0.19	0.42	0.79	0.33	68
	OBIA-RF		0.45	0.85	0.32	70
6	PBA-RF	0.15	0.40	0.78	0.33	68
	OBIA-RF		0.43	0.84	0.32	70
	PBA intersection with OBIA		0.48	0.73	0.23	76
	PBA&OBIA refined 1[b]		0.50	0.62	0.15	81
	PBA&OBIA refined 2[c]		0.50	0.71	0.19	80

Source: Pawluszek-Filipiak and Borkowski (2020).

[a] Training–testing sample split ratio; POD – the probability of detection; POFD – the probability of false detection

[b] Small elongated objects removed

[c] Median filtering

by machine learning-based image classifiers (Chen et al., 2017). No matter how many variables are included in the inputs, machine learning algorithms can quantify their importance in decision-making. Thus, the number of input variables is drastically reduced according to the specified importance threshold. Any inputs that fall below this threshold contribute little to the mapping outcome and can be safely excluded from consideration. In this way, only those factors critical to the accurate mapping of landslides are chosen, thereby significantly reducing the complexity of decision-making in the classification.

The construction of machine learning models such as RF requires training samples to be fed to the computer. The available training samples are split into two parts, usually at a ratio of 80% vs 20%. The former is used for initial training and the latter for model validation or calibration. In training the computer, a delicate balance between the prediction error and overfitting must be

TABLE 3.8

Inputs Commonly Used in Object-Oriented Machine Learning-Based Image Classification

Object-Feature Domains (No.)	Features (No.)
Layer features (57)	max (11), min (11), standard deviation (11), mean (11), ratio (11), brightness (1), max different (1)
Texture features (60)	GLCMall dir. [ent (12), mean (12), Cor (12), Con (12), Stdv (12)]
Geometry features (7)	Shape index, density, main direction, roundness, length–width ratio, area, number of pixels

Source: Chen et al. (2017).

TABLE 3.9
Quantitative Evaluation of Six Versions of RF in Mapping Landslides from VHR Imagery, Together with the Estimated Standard Deviations, All in Percentage (%)

Method	Completeness	Correctness	Quality	UA	PA	OA	Kappa
RF	71.9 ± 0.13	91.5 ± 0.07	67.1 ± 0.45	89.7 ± 0.44	72.9 ± 0.72	95.0 ± 0.08	77.6 ± 0.43
RFC	**84.1 ± 0.36**	90.9 ± 0.11	77.0 ± 0.48	89.2 ± 0.52	**84.2 ± 0.61**	96.3 ± 0.09	84.4 ± 0.38
RFO	63.8 ± 0.88	**97.6 ± 0.13**	61.2 ± 0.56	**96.9 ± 0.42**	64.1 ± 0.97	94.6 ± 0.13	74.3 ± 0.76
RFOC	67.8 ± 0.89	97.4 ± 0.12	65.9 ± 0.69	**96.9 ± 0.44**	67.6 ± 0.97	95.1 ± 0.13	77.0 ± 0.74
RFCO	81.5 ± 0.47	93.5 ± 0.08	**77.6 ± 0.39**	92.9 ± 0.53	83.0 ± 0.66	**96.7 ± 0.10**	**85.7 ± 0.44**

Source: Chen et al. (2017).
Abbreviations: RFC: RF + Closing operation; RFO: RF + Opening operation; RFOC: RF + Opening followed by Closing operation; RFCO: RF + Closing followed by Opening operation
Note: Bolded figures: the highest accuracy in a row

struck. If the computer is highly trained, it can achieve high accuracy with the training dataset but loses its ability to generalize with unseen samples, and vice versa. Validated against 20% of the available samples, multiple landslide paths can be mapped at an accuracy ranging from 73% to 87% among four study sites (Stumpf and Kerle, 2011).

Different machine learning algorithms have different utilities in mapping landslides. If combined with mathematical morphology operations, RF is the best for landslide mapping. Landslide mapping based on the RF method can be implemented in a few forms, such as RF and Closing operation (RFC), RF and Opening operation (RFO), RF and Opening followed by Closing operation (RFOC), and RF and Closing followed by Opening operation (RFCO), in addition to the classic RF landslide mapping (Table 3.9). In terms of correctness and user's accuracy (UA), RF slightly outperforms RFC but both of them are not as good as the other three models. Both RFO and RFOC have a similar performance that is the best among all the five models. Of these two variants, RFCO is better at mapping landslides among all the variants in terms of mapping quality (77.6%) and Kappa (85.7), achieving the highest overall accuracy of 96.7% (Chen et al., 2017). Clearly, the consideration of morphological parameters improves the mapping accuracy.

3.3.3 A Comparison

Both manual interpretation and computer-assisted digital analysis have their strengths and limitations. The manual method works best with high-resolution stereoscopic images that clearly show landslide tracks and their extent. Manual detection focuses on abnormal or irregular contour shapes, and areas devoid of vegetation. In contrast, digital mapping is fast, objective, and requires little human intervention except supplying training samples to the computer (Table 3.10). It is best applied to satellite images of a medium resolution rather than aerial photographs due to more spectral bands available than the three primary bands of blue, green, and red commonly associated with aerial photographs. The manual method may be slow and tedious, but it is still necessary as it performs a few functions intricately associated with the automatic method. One function of manual interpretation is to supply ground truth to validate the landslide inventory map produced digitally. In addition, manual interpretation also supplies training samples for computer classification. Nevertheless, it can make use of only three spectral bands at most and does not allow ancillary information to be incorporated into the decision-making process. Besides, the delineated landslide outline is only 2D. In comparison, digital analysis is not subject to such restrictions. It can produce mostly objective and repeatable classification outcomes. Digital analysis is best performed for repeated and ongoing

TABLE 3.10
Comparison of Manual Interpretation with Digital Analysis in Mapping Landslides

Item	Manual Interpretation	Digital Image Analysis (Per-Pixel)
Number of clues used	Many and diverse	Limited and narrow
Speed	Slow, tedious	Fast, automatic
No. of bands used	Three	Unlimited
Consideration of ancillary knowledge	Possible but difficult	Possible if presented in the digital format
Nature	Subjective, human bias unavoidable	Objective, and repeatable
Quality	Specific, toe, path and head distinguishable	Generic, only landslide pixels
Accuracy	Generally high	Variable, depending on the inputs
Output format	Vector showing landslide outlines	Raster of landslide pixels unless object-oriented analysis
Best use	Large-scale images, to validate digital analysis results	Repeated and ongoing analysis using the same routine

mapping of landslides as the same method can be applied to newly acquired data without any modifications.

Per-pixel image analysis results are likely to be in raster format that can be converted to vector format via vectorization, if necessary. With object-oriented methods, landslides are always mapped as polygons. However, it may not be feasible to differentiate a landslide into different zones, which is essential in assessing landslide potential. This limitation can be overcome with visual interpretation, or with the use of an additional DEM layer derived from LiDAR data.

3.4 NON-IMAGERY SENSING

How landslides should be detected automatically depends on the nature of the remote sensing data used, such as whether they are graphic images or non-graphic LiDAR data. The increasing availability of high-resolution LiDAR data provides an alternative for automatically mapping landslides efficiently. These dense, high-precision data are processed to construct the surface of the Earth, from which landslide-related information is derivable. Since its inception, LiDAR has been widely used to study natural hazards, including landslides, such as detection and characterization of landslides, landslide hazard assessment and susceptibility mapping, monitoring, and modeling (Table 3.11). Landslides of various scales and ages can be studied from LiDAR data. LiDAR data are advantageous to aerial photographs in forested areas where the landslides are likely to be obscured by tree canopy. In particular, LiDAR-based detection is rather strengthy in remote, hilly, and inaccessible regions, especially for large-scale, deep-seated landslides with an evident crown that is impossible to interpret and map from aerial photographs accurately (Van Den Eeckhaut et al., 2012). High-resolution LiDAR data are especially good at detecting old, previously unknown landslides (Van Den Eeckhaut et al., 2012). Airborne LiDAR data enable the identification of morphological structures along the sliding surface. Large-scale LiDAR-derived maps are especially significant in improving field survey-generated inventories of landslides, even if they have a subdued morphology in hilly regions.

LiDAR data can be used to derive a wide variety of information on landslides, including statistics of slope failures, landslide susceptibility, and hazard potential, either directly or indirectly. The features that can be retrieved from LiDAR data include landslide path (morphologic properties), landslide components, the velocity of movement (via repeat flights), and volume of the displaced debris (Table 3.12). Each of them is acquired using a unique method, based on different identification cues or ancillary data, and has unique requirements.

TABLE 3.11

Summary of LiDAR Capability in Four Areas of Studying Landslides

Applications	Landslide	Debris Flow	Rockfall
Detection and characterization	Mapping of geomorphic features	Detection of mobilizable debris volume Hydromorphic characterization	Rock face imaging and characterization Calculation of discontinuity orientation
Hazard assessment and susceptibility mapping	Support for mapping	Input for mapping hazard based on geomorphologic approach	Some attempts for susceptibility and hazard mapping
Monitoring	Surface displacements Volume budget	Sediment budget Morphologic changes in channel	Surface displacement Detection of pre-failure displacements Quantification of rockfall activity (volume)
Modeling	HRDEM allows accurate modeling by improving geometrical characterization	Input for spreading modeling	High-resolution DEM for trajectory modeling

Source: Jaboyedoff et al. (2012).

While the derivation of certain information, such as the spatial extent of landslides and differentiation of a landslide path into various components, is feasible from analysis of one-time LiDAR data, the generation of information on the velocity of displaced materials, debris volume, and landslide susceptibility modeling have to rely on multi-temporal LiDAR data or data from other sources.

TABLE 3.12

Landslide Features That Can Be Extracted from LiDAR Data

Landslide Features	Implementation	Ancillary Data/ Cues	Affecting Factors	Requirements
Landslide path (morphologic properties)	Visual interpretation, topographic analysis	Elevation change; contrast in roughness and texture	Tree factors	Fine-resolution DEM
Components (scar, body, toe)	Zones of uplift and subsidence	GPS; change in curvature	Forest cover	Surface roughness, suitable for large-scale, deep-seated landslides
Susceptibility to landsliding	Multivariate modeling based on past events; regression analysis	Topographical, lithological, and soil maps	Various factors	Data in raster format
Velocity and displacement (stability)	Tracking of mobilized materials	GPS, theodolite, and radar data	Timing of multi-source data; non-uniform velocity	Reference points (e.g., trees) essential
Volume of displaced materials and sediment yield	DEM differencing	(ortho)Photographs, topographic maps	Leaf-on or leaf-off in forested areas	Accurate co-registration essential

3.4.1 Recognition of Landslides

The recognition of landslides from LiDAR data differs drastically from aerial photographs because primarily LiDAR data record only 3D coordinates of dense points, from which the height of the ground is computed. Non-graphic LiDAR data are thus unable to indicate landslides or allow landslides to be identified directly. Instead, the identification of landslides has to rely on their topographic signatures from LiDAR data. Topographic cues vital for recognizing landslides include the abrupt variation in surface elevation and slope gradient. If mapped from single-pulse LiDAR derivatives, landslides are recognized via landslide-induced geomorphometric changes in slope gradient, roughness, and curvature that can be reliably derived from LiDAR data. However, it is much simpler and easier to recognize landslides from the height difference between multi-temporal LiDAR data recorded before and after the landslides automatically. Detailed landslide features are recognizable from LiDAR data owing to their capability of detecting bare ground elevation. The recognition is based on pre- and post-event DEMs generated from irregularly spaced LiDAR point cloud data. Their differencing is effective at detecting changes in surface elevation and morphology caused by landslides. In addition to height difference, the disparity in roughness between bedrock landslide and its adjacent intact terrain, as captured by some statistical parameters, enables the automatic detection and mapping of the overall landslide complex. DEM analysis may be used to objectively delineate landslide features, and generate mechanical inferences about landslide behavior. The success of detection, nevertheless, depends on the DEM grid size. A small grid size is critical to inventorying small landslides and indispensable in estimating landslide depth, minimum sediment transport, and denudation rates. The improved topographic information from small grid-size DEMs is conducive to identifying recent rainfall-induced landslides. Conversely, coarser grid (>10 m) DEMs do not allow the identification of small landslides. Potential landslide sites can be extracted from DEM-derived filter values on the basis of characteristic eigenvalues and slope filter values (Glenn et al., 2006).

LiDAR-based detection is simple conceptually, but the detected results can be very noisy if it relies on only elevation change that can also be caused by vegetation growth. Provisions must be made to accommodate vegetation growth in vegetated areas. For instance, changes less than ±0.20 m in areas of high grass cover are excluded from consideration to capture genuine surface cover change (Stumvoll et al., 2021). Besides, more morphologic clues should be used, including surface roughness, slope, semi-variance, and even fractal dimension, all of which can be calculated from the variation in grid cell height (Glenn et al., 2006). The use of more parameters, such as the contrast in median roughness, texture, and continuity of roughness elements, enables a landslide complex to be differentiated from the adjacent earth flows. If assisted by expert knowledge, more landslides can be identified from LiDAR-derived hill shade, slope, and contour maps than from aerial photographs.

The detected results can be made more accurate by basing the automatic detection on the landslide scarp, or by approximating the surface with a bivariate quadratic equation fitted in a local neighborhood (Lin et al., 2013). The generation of satisfactory results, however, is contingent upon how to set up two critical factors: the window size at which curvature is calculated and its threshold. The former is jointly governed by the actual size of the landslide crown and the spatial resolution of the DEM used for the recognition. The latter is determined by experimenting with a number of thresholds and then comparing the results to see which one produces the best outcome. So the process is not as straightforward as it seems. If based on such parameters as slope, curvature, object height calculated from the difference between DSM and DEM, and topographic wetness index, landslides are recognized at an overall accuracy of only 65.8% (Liu et al., 2012a). The main causes of inaccuracy are identified as the inclusion of river banks, and upstream erosion. This accuracy rose to 72.5–76.6% after the mapped landslides were filtered using a threshold of 50 m^2.

Although LiDAR data are effective at mapping landslides in hilly and densely vegetated areas, they are ineffective at mapping surface cover that can be accomplished using optical images. Naturally, LiDAR data have been combined with optical image-derived vegetation index layers to detect landslides after they have been converted into regular grid DEMs. Owing to the integration of both types of data, the mapping accuracy of an active landslide increased from 65% to 91% using LiDAR data, or 76% using only orthophotos (Kamps et al., 2017). It must be noted that this improvement is for deep-seated landslides. If the landslides are shallow, the synergetic use may not achieve such a large pace of improvement in accuracy.

3.4.2 Identification of Landslide Components

In reality, it may be necessary to differentiate the three sections (the source zone, the transport zone, and the toe) of a representative landslide path in order to assess the potential damage of a landslide. After a landslide path has been identified, it is possible to differentiate it into the three zones based on surface elevation and its variability from LiDAR data (Glenn et al., 2006). They provide a complete displacement field for the whole landslide body, not limited to a single-point measurement (Scaioni et al., 2014), on a 3D model. This trait facilitates the identification and separation of one component from another based on surface topographic characteristics. In particular, differencing of multi-temporal DEMs of a fine resolution quantifies the changed height and is crucial to the separation. Each zone can be more precisely delineated on the basis of local relief. For instance, if an area has a lowered elevation in the post-event DEM, it must represent the source zone. Conversely, the toe zone is characterized by an increased surface height. Besides, it also has a relatively high surface roughness, and high vertical and lateral movement, while the body has a relatively low surface roughness (Ventura et al., 2011). Another topographic clue is surface gradient. The erosional zone has a steep surface, in contrast to the depositional zone that has a gentle surface. The transport zone has a transitional slope between that of the toe (T), the body (B), and the upper block (UB) (Figure 3.10). The successful differentiation of morphological components of a landslide requires fine-resolution data that can also reveal the material type and activity of the slide, which is unable to achieve otherwise (Lin et al., 2013). Not surprisingly, high-quality LiDAR-derived maps show much more landslides than those mapped during field surveys. However, the demarcation of the

FIGURE 3.10 Separation of a large landslide into three components of toe (T), body (B), and upper body (UB) from LiDAR data. (Source: Glenn et al., 2006. Used with permission.)

precise boundary of a landslide has to rely on the zones of uplift and subsidence, the volume of removed and/or accumulated debris, and the average rate of vertical and horizontal displacements, all of which will be covered in the respective sections below.

3.4.3 Detection of Landslide Deformation and Movement

Once a landslide is initiated, it will move downhill during which the topographic surface may be deformed. This deformation can be detected using multi-temporal LiDAR data alone or in conjunction with other non-LiDAR data. In either case, the two types of data must be recorded at different times. Dissimilar to their mono-temporal counterpart, multi-temporal LiDAR data can reveal the magnitude of the debris movement during the interim, and the nature of landslides or landslide stability (Figure 3.11). Since LiDAR data capture the landslide state at a single time only (probably the current state), they have to be coupled with other remotely sensed data to detect the movement of landslides, especially in characterizing large, slow-moving landslides (Roering et al., 2009). Common candidates for combination are GPS data, laser theodolite (total station) data, field observations for currently active landslides, and even historical movement data. This combination is strong at detecting vertical and horizontal displacements of the mobilized debris, estimating the velocity of debris movement, and mapping deformation structures (scarps, cracks, folds) affecting landslides at different times. If assisted by digital image classification, it is possible to study the landslide process, such as the influence of underlying topography on earthflow mobility. If combined with orthogonal historical aerial photographs, LiDAR data allow objective mapping of earthflow movement, from which the rate of movement is calculated. The combination of displacement orientations with stacked radar data enables the estimation of downslope velocities of the central

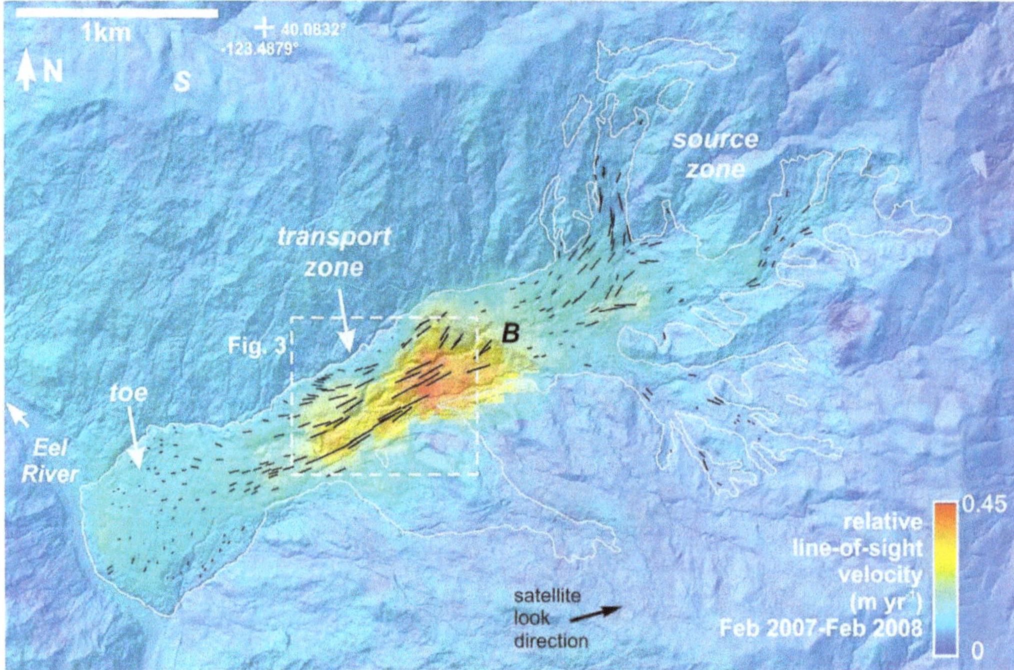

FIGURE 3.11 Earthflow velocity in the LOS direction atop a shaded relief map derived from airborne LiDAR of 1 m spacing. The thin white line denotes the active landslide boundary. Black lines mark the starting and ending location of individual trees mapped on historic aerial photos and a recent unfiltered LiDAR DEM. (Source: Roering et al., 2009.)

transport zone of landslides (Roering et al., 2009). The detection of landslide movement, however, becomes more challenging or even impossible in forested areas. Thus, the displacement of trees growing on earthflow surfaces has to be used as the surrogate target of detection. Their tracking allows the determination of the area that has experienced movement. It is hence more advantageous to capture LiDAR data during the leaf-off season so as to make the optimal use of multi-temporal LiDAR data in forested areas.

The velocity of landslide creeping or downslope movement is calculated from the position of the tracked features, such as a tree or the landslide toe automatically using Eq. 3.1. The two planimetric positions of (x_e, y_e) and (x_s, y_s) are used to calculate the distance of downslope travel. If the surface is steep and the velocity is high, it may be desirable to decompose the distance into the actual distance instead of the projected 2D horizontal distance with the assistance of a DEM. The actual 3D distance is then divided by the time interval between the two images, and the results are expressed in m·year^{-1} (Figure 3.11). The same detection can be repeated for every pixel in the two images, resulting in a field of movement velocity. However, if detected using LiDAR data, this velocity applies to the line-of-sight rate as LiDAR is mostly side-viewing sensing instead of nadir-viewing. This velocity can be translated into the horizontal rate if the scanning angle of each LiDAR point is known.

$$Velocity = \frac{distance}{time} = \frac{\sqrt{(x_e - x_s)^2 + (y_e - y_s)^2}}{t_e - t_s} \tag{3.1}$$

where t_s = time of start monitoring; t_e = time of end monitoring.

3.4.4 Estimation of Debris Volume

The volume of landslide debris is 3D in nature. Thus, its estimation requires elevational data in the form of a regular grid DEM created from LiDAR data. LiDAR-based estimation of the volume of the displaced debris is a step further than the recognition and characterization of profound morphologic changes caused by landslides. The estimation of landslide-triggered sediments requires multi-temporal LiDAR data, usually through DEM differencing. Two DEMs of the same area should be used, one before the event and another after. The disparity in height at the same grid cell (Δh) is then calculated by subtracting the post-event DEM from the pre-event one. This elevation disparity indicates only the change in height at the given grid cell and the mass transfer from the erosional zone to the tracks and the depositional zone. If the difference between the current and historic DEMs is negative, it indicates erosion, otherwise, deposition of the mobilized debris. The total quantity of displaced debris is calculated by summing Δh multiplied by the DEM grid area, or

$$Volume = \sum_{i=1}^{n} \Delta h \cdot R^2 \tag{3.2}$$

where R = spatial resolution or grid size of the DEM used and n = the total number of landslide grid cells in the DEM. The above equation just shows the net change in the volume of the displaced debris. This estimation can be further differentiated between the depletion ($\Delta h < 0$) and depositional ($\Delta h > 0$) zones. The volume of eroded debris is estimated using the same equation for those grid cells, the Δh of which is >0 for the depositional zone, or <0 for the erosional zone (Figure 3.12).

Although multi-temporal LiDAR surveys can quantify the volume of debris transferred out of a slope (Baldo et al., 2009), the subtraction of multi-temporal DEMs is unable to indicate the direction of displacement directly or the pace of displacement in longitudinal studies if the landslide debris under study is in slow motion. The generation of such information is still possible; however, if the debris object is recognized based on 2D transects in the main direction of displacements.

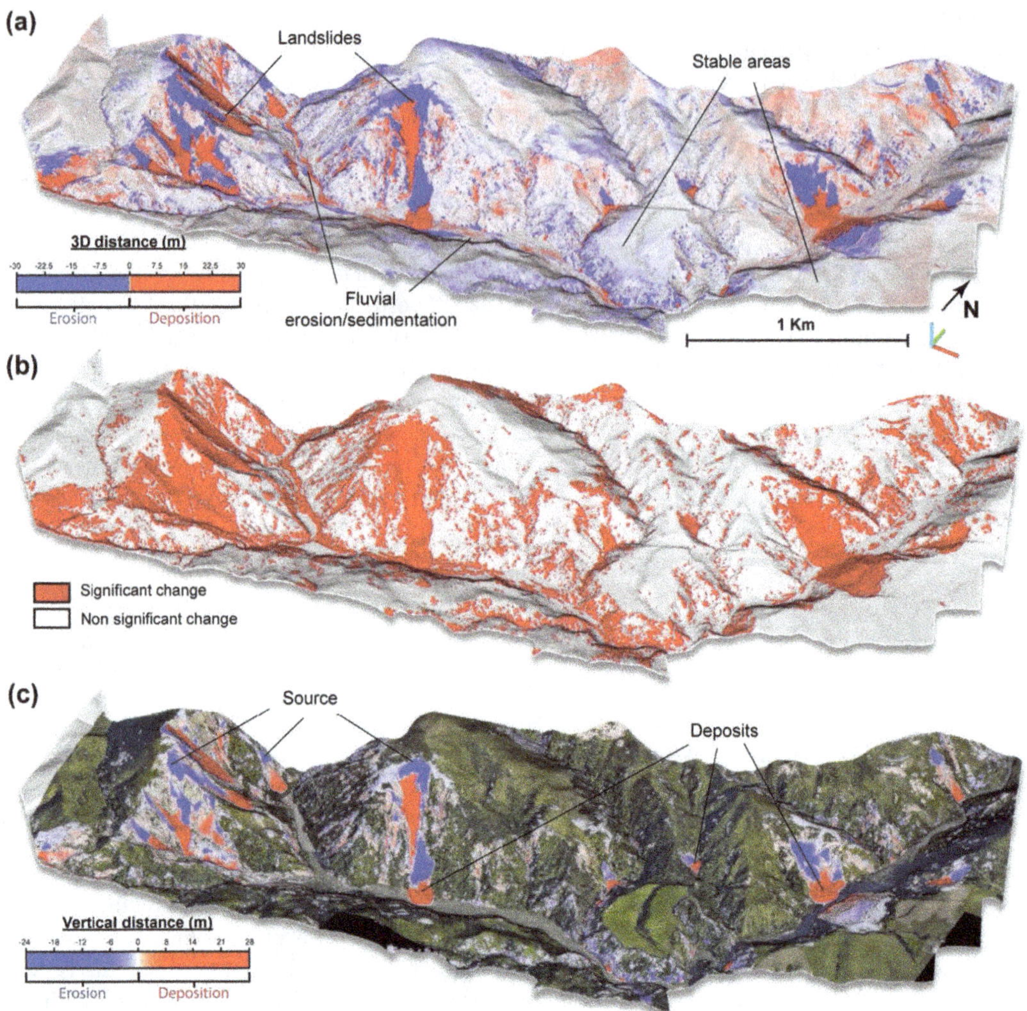

FIGURE 3.12 Perspective maps showing the spatial distribution of erosional and depositional zones of earthquake-trigged landslides in Kaikōura, New Zealand using LiDAR data. (Source: Bernard et al., 2020.)

A comparison of such objects enables the displacements to be quantified at a precision of ±30 mm in all directions. In order for this method to work, the two DEMs must have the same spatial resolution. If not, the DEM grid cell size must be unified by specifying an identical grid cell in a spatial interpolation. After the change in surface relief has been determined via DEM subtraction, it may be visualized graphically by color-coding the magnitude of change. The continuous change in elevation can be categorized into more than ten classes to illustrate the spatial pattern of elevation change (Figure 3.13), with erosion shown in one color of different shades and deposition in another color. The changed part may be superimposed with aerial photographs or satellite images to illustrate the relationship between the magnitude of change and surface land cover.

Apart from debris volume, differential DEMs created from LiDAR datasets recorded at different times can also be used to estimate sediment yields from landslides. Subtraction of the two topographic surfaces constructed from multi-temporal LiDAR data quantifies the amount of sediment originating from landslides and the total amount of post-event sediment discharge (Matsuoka et al., 2008). This quantification is identical to the determination of debris volume within the landslide path in operation. The only additional requirement of this analysis is the outline of the

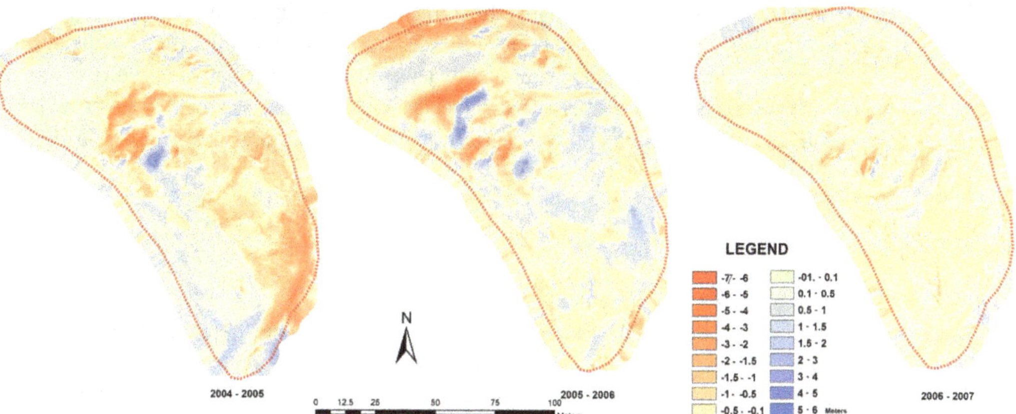

FIGURE 3.13 Change in surface elevation caused by landslides in three periods, detected from multi-temporal LiDAR data-derived DEMs. Red – erosion; blue – deposition. (Source: Baldo et al., 2009. Used with permission.)

landslides. The net change in all pixels' height represents the amount of sediment lost to the catchment (Figure 3.13). However, it remains mostly unknown how much of it remains in the catchment, and the portion that has been discharged to the downstream channel unless the channel bed is dry and exposed. Prior to DEM differencing, it is essential to accurately geo-reference the two datasets separately before they are superimposed with each other. Alternatively, the two datasets may be precisely co-registered with each other to avoid artificial changes. Any inaccuracy in image co-registration degrades the accuracy of the estimated sediment volume. Once such requirements are fulfilled, slope gradient maps from LiDAR surveys can be produced to quantify and reconstruct sediment movement patterns over various sectors of a landslide and reveal the erosion rate for the source zone (Liu et al., 2012b). In addition, deformation trends and characteristics of landslides can be detected and accurate deformation is quantified.

Due to the short history of LiDAR data existence, two-time LiDAR data of the same area may not be available all the time. In this case, the estimation of the displaced debris volume is commonly accomplished by combining LiDAR data with data gathered using other traditional 3D methods. This integration is able to identify the vertical and horizontal ground displacements associated with landslide movement. Common candidates for synergy are GPS data, digital photogrammetric maps, orthorectified historical aerial photographs, topographic maps, and radar interferograms. These data are complementary to LiDAR data in their resolution, accuracy, coverage, and time of acquisition. Ground-collected GPS data are useful in refining the horizontal accuracy of LiDAR data, and in mosaicking multiple LiDAR scans to fully cover the area under study in one huge tile, even though it is less accurate than a single scan tile. If combined with GPS, LiDAR can potentially determine the volumetric evolution of a landslide complex (e.g., translational landslide evolving into mudflows) and numerically evaluate landslide morphological evolution (Baldo et al., 2009). More research is needed to produce a detailed budget of errors for LiDAR-estimated debris volume so as to broaden the application areas in which LiDAR data can be competently used to detect small landslide features and to yield accurate information on landslide movement.

The challenge facing LiDAR sensing of landslides is its limited accuracy achievable because LiDAR-generated DEM elevation inevitably involves a degree of uncertainty. The rate of success, defined as the area ratio of landslide areas to non-landslide areas included in the detected elevation change, is 79% for large-scale landslides and debris flows (e.g., over 10 m deep and 100 m wide) in areas of a moderate slope, but drops to 65% in steep-sloped areas. The accuracy of elevation difference as measured by RMSE is affected by slope gradient. The RMSE lies between 4 and 5 m on slopes <30° but rises to 5–6 m on slopes >30° in moderate topography and further to 5–9 m in steep terrain. One way of improving the quality of the detected change in elevation is to compare the elevations in areas where no known landslides have taken place. Theoretically, the two elevations should be identical. If not, their discrepancy represents a systematic offset between the two sets of height and can be used to calibrate all other differences. In the presence of shadow, shadow areas and those suffering from spatial mismatch due to image co-registration inaccuracy can be excluded from estimation to increase accuracy (Tsutsui et al., 2007).

3.4.5 Airborne or Terrestrial LiDAR?

LiDAR sensing of landslides is implemented either airborne or terrestrially, each having its strengths and limitations (Table 3.13). The former is mobile, while the latter is stationary. The mobile LiDAR scanner is carried aboard a flying platform such as a helicopter or a light plane. It is in constant motion during the scanning of the area under study (Figure 1.1). The scanner transmits dense signals to the target below perpendicularly to the motion of the plane. The entire study area is scanned as the plane flies forward, acquiring dense LiDAR point cloud data quickly. In this way, the entire area is scanned at an almost constant range, resulting in the LiDAR point cloud data having a mostly uniform density and accuracy across the whole area of study. Airborne LiDAR covers a large ground area per scan and allows landslides over extensive areas to be studied quickly, which is difficult to accomplish via field surveys or even from aerial photographs. If the ground area to be covered is too extensive, however, it may be necessary to scan it in multiple strips to cover it fully. Multiple scans must be stitched together using ground control points. Although the number of scan strips may be reduced through a higher flight altitude, the reduced LiDAR data density and resolution can cause the omission of small landslides from the acquired data. This method of sensing is expensive but flexible.

Dissimilar to airborne LiDAR, terrestrial LiDAR scans the targeted area from a stationary scanner mounted atop a tripod near the front of landslide paths (Figure 3.14). The area of interest is scanned through the horizontal rotation of the scanner, usually less than 180°. The surface displacement field of landslides is calculated from a few fixed positions emplaced in the front of the landslides at high accuracy. Terrestrial LiDAR is good at studying individual landslides and enjoys the advantage of providing highly accurate results owing to the use of the immobile scanner and hence more accurate ground control. It is inexpensive and rather versatile in monitoring landslides, suitable for emergencies that require rapid topographic information for assessing landslide hazards

TABLE 3.13

Comparison of the Pros and Cons of Airborne vs Terrestrial LiDAR Sensing of Landslides

Mode of Sensing	ALS	Terrestrial
Area of sensing	Large	Small
Speed of sensing	Fast	Slow
Point density (pt·m^{-2})	0.5–100	50–10,000
Data uniformity	Spatially uniform	Spatially variable
Range of sensing	Highly uniform	Highly variable
Best use	Many landslides on gentle slopes	Single landslides on cliffs, slow-moving landslides
Pros	Large area of coverage, uniform point density	Inexpensive, flexible, good for repeated sensing from the same position, long-term monitoring
Cons	Expensive, accuracy subject to GPS, not so accurate (e.g., >cm), some areas next to cliffs could be hidden	Point density variable, hidden area rife; small area of sensing, multi-scans need to be stitched

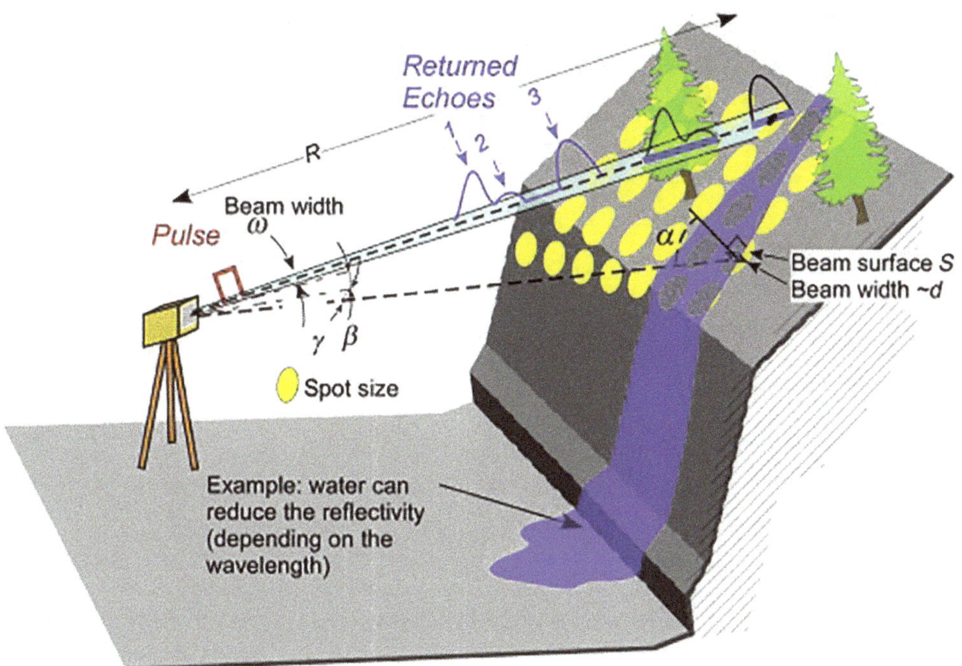

FIGURE 3.14 Sensing of landslide features on a steep slope using terrestrial LiDAR. (Source: Jaboyedoff and Derron, 2020.)

at a short notice (Jaboyedoff et al., 2012). The main limitation of terrestrial LiDAR is that the differencing of LiDAR-derived DEMs yields information about altitude changes only (Kasperski et al., 2010).

Airborne LiDAR and terrestrial LiDAR have their own best uses. The former has found wide applications in studying landslides, especially in remote, hilly, and inaccessible regions. They range from recognition of landslide paths and identification of landslide components, determination of the motion type and stability of landslides, to the modeling of landslide susceptibility. Airborne LiDAR is good at detecting a large number of landslides on a gentle topography and can be used to produce highly accurate and quantitative results about landslide volume and create slope profiles (Jaboyedoff et al., 2012). The less rigid ground control during the flight means that the vertical accuracy of airborne LiDAR is lower than its terrestrial counterpart. This accuracy is affected by a number of factors, including the accuracy of source (ground control), accuracy in stitching multiple scans, and the accuracy of LiDAR point data themselves. At present, the achievable LiDAR accuracy level is reportedly 0.50 m with a systematic error of 0.09 m. Higher accuracy is needed for generating more precise estimates of landslide debris volume. Because of the limited accuracy achievable, airborne LiDAR still faces some challenges in certain applications, such as determining landslide age and estimating landslide depth, minimum sediment transport, and denudation rates despite the considerable potential of LiDAR in deriving a variety of landslide-related information successfully.

Although airborne LiDAR is able to detect landslide paths in densely vegetated areas, the accuracy of identification is lowered in forested areas or mountainous terrains of a relatively large relief. During scanning over inhospitable areas, several problems may arise, such as weak or even absent GPS signals, unfavorable atmospheric conditions, and gaps behind steep terrains. In steep terrains, LiDAR radiation cannot reach the "hidden" ground, causing "gaps" in the data coverage known as LiDAR shadow (Figure 3.15), reminiscent of topographic shadow on aerial photographs. For this reason, airborne LiDAR is suited to sensing landslides on gently rolling slopes. For slow-moving

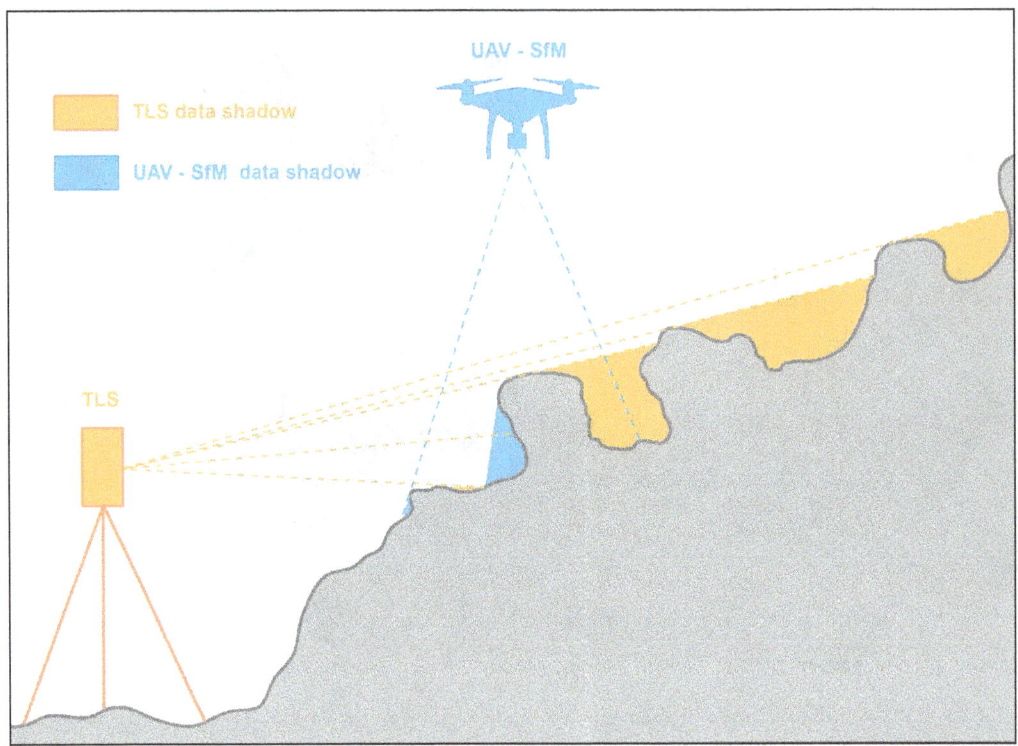

FIGURE 3.15 Viewing geometry and data coverage of terrestrial laser scanning (TLS) that cannot reach areas below the line of sight, in comparison with airborne UAV scanning in a rugged terrain. (Source: Šašak et al., 2019.)

landslides (e.g., a couple of centimeters a year), airborne LiDAR is incompetent due to the pace of change falling within the uncertainty level of measurement. Thus, it must rely on terrestrial LiDAR.

Terrestrial LiDAR has been used to study massive landslides that pose a significant geohazard, such as damming a channel to form a lake. Usually, such landslides are hanging on a steep slope and extend some distance downslope, visible mostly from a single point in front of them, such as the active Séchilienne landslide in Isère, France (Figure 3.16). Monitoring of its 3D displacements is achievable by scanning it repeatedly from the same position, and the LiDAR point clouds acquired in different years are used to create surfaces of the terrain, from which the volume of registered collapses is calculated, and the associated subsidence and toppling are monitored. From these data, the landslide movement along the main geological structures is inferred, together with the manner of landslide evolution and its relationship with structural settings (Kasperski et al., 2010).

On a local scale (e.g., 0.05 m raster), terrestrial LiDAR is able to detect the real surface change to an error range of ±0.05 m with 0.05 m steps under optimal conditions (e.g., areas devoid of trees). The accuracy drops to a range of ±0.20 m with 0.20 m steps in areas of less optimal conditions where genuine elevational change is disguised by changes in vegetation cover (Stumvoll et al., 2021). Terrestrial LiDAR data must be processed. If not, raw terrestrial LiDAR data with a displacement range of 7.4 mm $< \sigma <$ 11.2 mm cannot be used to detect mm-level rockfall displacement because the maximum precursory displacement is estimated at 45 mm, about 2.5 times the standard deviation of the model comparison (Abellán et al., 2009). Thus, the actual displacement is rather similar to instrumental error in magnitude (e.g., 7.2 mm at a distance of 50 m). However, after applying the 5 × 5 nearest neighbor filtering, the precursory displacement became clearly detectable at the mm-level from terrestrial LiDAR data.

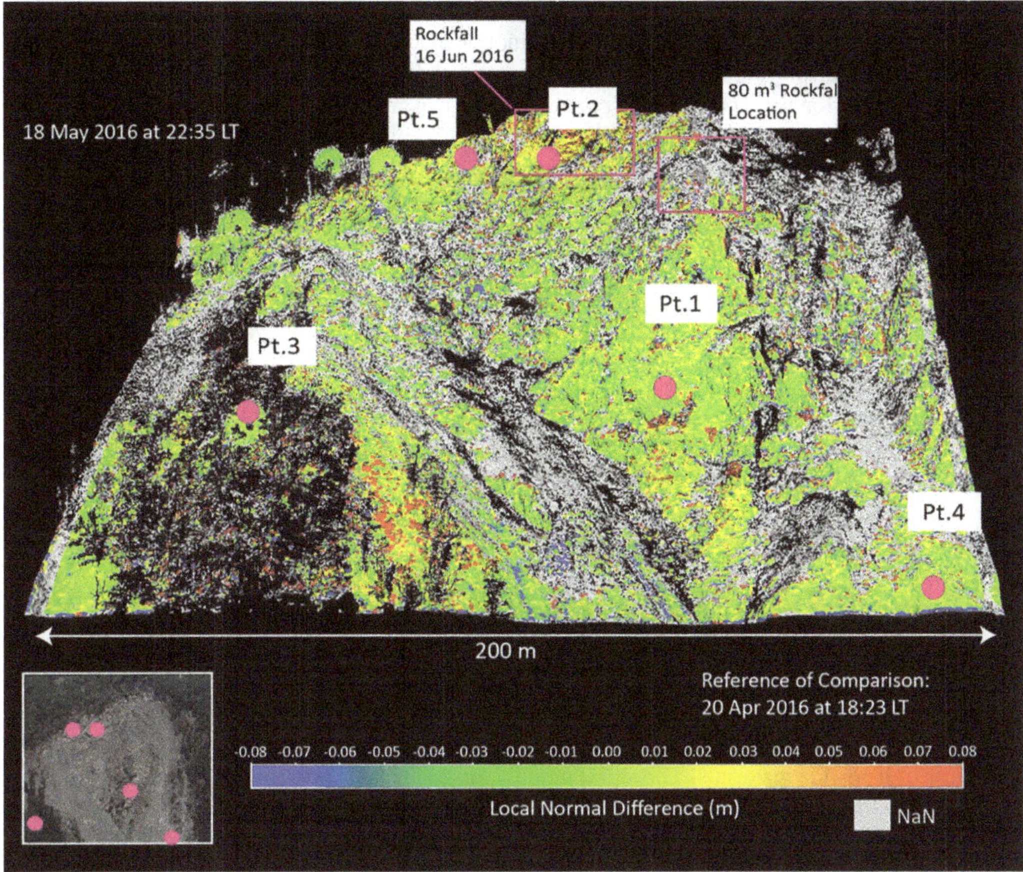

FIGURE 3.16 Differencing image of two terrestrial laser scans of the Séchilienne Landslide in France, the more recent 18 May 2016 scan at 22:35 local time relative to a reference scan on 20 April 2016 at 18:23 local time. (Source: Kromer et al., 2017.)

There is a relatively larger variation in the range of terrestrial laser scanning, so point density varies more widely among the acquired point cloud data than with airborne scanning. The achievement of a more spatially consistent point density requires restricting the area of sensing to a narrow range as accuracy is noticeably lower at a longer range. Terrestrial laser scanning is severely disadvantaged in having a limited ground cover, even though it is affected by the scanning range available. To prevent low accuracy at a far range, each scan must be confined to a small area, and multi-scans are needed to cover a large ground area, but digital mosaicking of multi-scans is laborious and time-consuming. Because of this limitation, terrestrial LiDAR is suitable for small areas or for studying a few landslides hanging over a cliff where all of them are visible from one fixed perspective and have a more or less uniform distance to the fixed scanner to ensure a consistent sensing range. Another limitation of this implementation is the required access to the study site. Because of these limitations, terrestrial LiDAR has not found wide applications in studying landslides despite the accurate results attainable.

3.5 LANDSLIDE VULNERABILITY AND HAZARD ZONING

Landslide vulnerability refers to the propensity of earth materials to move downhill under the influence of gravity. It is synonymous with terrain instability. While landslides can be mapped directly from remote sensing images using various image processing methods, landslide susceptibility, defined as the likelihood of a landslide occurring in an area under similar topographic and hydrological conditions as has occurred in the past, cannot be mapped from images. Despite this inability, remote sensing still plays an indispensable role in modeling landslide susceptibility and zoning landslide hazards for five reasons: (i) remote sensing can map the distribution of existing landslides. Their concentration in certain areas indicates the likelihood of future landslide occurrence; (ii) topographic settings that are so crucial to terrain stability can be studied from DEMs derived from LiDAR data; (iii) factors that affect landslides, such as land cover, can be mapped from multispectral images; (iv) hydrological conditions such as soil moisture content can be gathered from remote sensing images; and (v) crucial geographic information such as distance to fault lines and likely places of heavy rainfall can be generated from remotely sensed data. Remote sensing can predispose all the factors (geomorphology, topography, land cover, hydrology, and geology), although the accuracy of these maps varies with their scale.

Apart from landslide susceptibility, remote sensing is also indispensable to landslide hazard zoning. Spatially, landslides are not evenly or randomly distributed in the landscape and over different parts of a slope. Some locations are more prone to landslides than others. In order to mitigate landslide hazards, it is essential to zone them. Landslide hazard zoning refers to the process of partitioning a landscape or a region into smaller, homogeneous, and irregularly shaped subareas, each having a unique level of landslide susceptibility or risk that is expressed either verbally or quantitatively. Several approaches are available for zoning landslide hazards, such as the discrete element method (Lo et al., 2018), simulation, or modeling. The former allows the exploration of the landslide movement processes and assessment of the scope of influence with regard to debris deposition. Numerical simulation based on the PFC3D (Particle Flow Code in 3 Dimensions) model can determine landslide movement and deposit ranges, but is unable to reveal landslide susceptibility, even though this model can accurately simulate the falling, colliding, shattering, sliding, and piling behaviors of landslide materials. The simulation is impossible to run without the input of a number of parameters that must be determined on-site.

In comparison, remote sensing-based modeling is much simpler. It is based on the simultaneous consideration of all the factors that contribute to terrain instability or have played a role in the initiation of landslides, such as elevation, gradient, slope orientation, flow direction, flow length, flow accumulation, distance to channels, distance to highways, distance to faults, and slope loading. Some of them are directly derivable from LiDAR-generated DEMs, such as slope gradient, aspect, surface curvature, and soil drainage. Historic landslides can be mapped from aerial photographs or satellite images captured decades ago or periodically; the spatial concentration and pattern of historic and current landslides reveal the spatial distribution of topographic vulnerability. Understandably, it is impossible to model terrain susceptibility to landslides and assess and zone landslide hazards without remote sensing.

The utility of remote sensing in landslide hazard zoning varies with the scale of study and the nature of analysis (Table 3.14). Remote sensing is the best at mapping, but less useful for deterministic analysis and frequency analysis, especially on the regional scale, at which more non-imagery data are needed. For instance, micro-topography (e.g., landslide scarp and gullies) may be detected from LiDAR data, while current landslides are mapped from images. The production of an accurate and reliable hazard zoning map requires in-depth knowledge about the triggering mechanism of landslides in an area, its geo-environment, and the historic behavior of landslides.

In the zoning, the simultaneously considered relevant factors may be topographic, geologic, hydrologic, pedologic, and even land use (Table 3.15). More sub-variables may be found under a general category, such as (i) geomorphology: terrain mapping units and watershed shape; (ii) topography (or morphometry): slope gradient, slope aspect, elevation, and curvature profile; (iii) geology: lithology, fractures, joints, lineaments, and tectonic structures; (iv) land cover: land use, vegetation cover, vegetation indices; (v) hydrology: soil moisture and drainage system; and (vi) anthropogenic influencing factors: transportation infrastructure, quarrying and mining, dams and reservoirs. It is worth noting that not all of them are relevant in a given area, nor are these

TABLE 3.14

The Feasibility (F) and Utility (U) of Remote Sensing in Landslide Hazard Analysis at Three Scales

Types of Analyses	Main Characteristics	Regional Scale	Medium Scale	Large Scale
Distribution	Direct mapping of mass movement features yields information for those sites where landslides have occurred in the last	F = 2 U = 3	F = 3 U = 3	F = 3 U = 3
Qualitative	(Semi)Direct methods for renumbering the geomorphological map to a hazard map or merging several maps into one using subjective rules based on the analyst's experience	F = 3 U = 3	F = 3 U = 2	F = 3 U = 1
Statistical	Indirect methods in which statistical analysis is used to predict the likelihood of mass movement from a number of parameter layers	F = 1 U = 1	F = 3 U = 3	F = 3 U = 2
Deterministic	Indirect method in which parameters are combined to calculate slope stability	F = 1 U = 1	F = 1 U = 2	F = 2 U = 3
Frequency	Indirect methods in which earthquakes and/or rainfall data are correlated with known landslides dates to obtain threshold values with a certain frequency	F = 2 U = 2	F = 3 U = 3	F = 3 U = 2

Source: Mantovani et al. (1996). Used with permission.

Note: 1 = low: it would take too much time and money to gather sufficient information in relation to the expected output; 2 = moderate: a considerable investment would be needed, which only moderately justifies the output; 3 = good: the necessary input data can be gathered with a reasonable investment related to the expected output

TABLE 3.15

Geo-Environmental Factors Commonly Considered in Landslide Analysis

Variables	Sub-Variables	Relevance for Landslide Analysis
Topography	Slope gradient	Most important factor in mass movement
	Slope aspect	Reflects soil moisture and vegetation
	Slope length/curvature	Reflects slope hydrology
	Flow direction	Used in hydrological modeling
	Flow accumulation	Used in hydrological modeling
	Internal relief	Indicator for the type of terrain in small-scale assessment
	Drainage density	Indicator for the type of terrain in small-scale assessment
Geology	Lithology	Engineering characteristics of a rock type
	Weathering	Determines the depth profile
	Discontinuities (fracture and joints)	Relevance for rock slides
	Structural aspects	Predictor for rock slides
	Lineaments/faults	Used in predictive mapping
Soil	Soil types	Engineering properties of soil types
	Soil depths	Useful for stability analysis
	Geotechnical properties	Soil characteristics for stability analysis: grain size, cohesion, friction angle, bulk density
	Hydrological properties	Parameters for groundwater modeling

(*Continued*)

TABLE 3.15 (*Continued*)

Geo-Environmental Factors Commonly Considered in Landslide Analysis

Variables	Sub-Variables	Relevance for Landslide Analysis
Hydrology	Water table	Reflects spatial and temporal variation of groundwater table
	Soil moisture	Main component of stability assessment
	Hydrological components	Interception, evapotranspiration, through fall, overland flow, infiltration, and percolation
	Stream network	Stream-induced erosion
Geomorphology	Physiographic units	Subdivision of terrain for small-scale mapping
	Terrain mapping units	Units for lithology, morphography, and processes
	Geomorphological units	Genetic classification of main landform building processes
	Geomorphological subunits	Geomorphological subdivision of the terrain in the smallest units
Land use	Land use map	Land cover is the main component in stability analysis
	Land use changes	Main components in stability analysis
	Vegetation characteristics	Evapotranspiration and root systems
	Roads	Related to slope cuts during construction
	Buildings	Related to slope cuts during construction

Source: modified from Van Westen et al. (2008). Used with permission.

variables equally important to landslide hazards. Their exact relationship with the landslide mechanism is not always clearly understood. Thus, their importance in zoning has to be established empirically or via machine learning methods because of the compounding effects of multi-factors in the eventuation of a landslide. In practice, the number of variables considered falls far shorter than the theoretical number due to the difficulty of generating their geospatial data as whatever variables are taken into consideration; their spatial layer at a proper scale must be available.

The simultaneous consideration of such a large number of diverse factors is possible via the modeling approach. It is underpinned by an assumption that those areas suffering from a higher spatial concentration of landslides than elsewhere in the past are more hazardous and susceptible to landslides in the future. Essential in the zoning is a distribution map of current landslides that is used to establish the association between the considered variables and landslide occurrence, which is used as the evidence to weigh them. The relationship between current landslides and all the considered factors is then analyzed statistically. For instance, if more landslides are associated with steep terrain, it is assumed that all other terrains with a similar slope gradient will be similarly susceptible to landslides. A larger weight is assigned to this variable. If implemented in raster format, the association is built on the basis of the number of pixels. The susceptibility is expressed as the ratio of frequency, which can be used as the weight to indicate the relative importance of this factor (or a specific attribute value, e.g., 30–35° slope gradient) in the zoning.

In the zoning, the distribution of current landslides is divided into two parts, one for model construction and another for model validation. The first part (training dataset) is used to compare current landslides with their affecting factors and this comparison helps to establish their statistical relationship automatically by the computer. The established model is then applied to the second part of the dataset to check model accuracy in a process known as model validation, during which a number of accuracy measures are produced, such as the relative operative characteristic and the percentage of correctly observed landslide pixels.

The end product of zoning is a map of topographic instability created via analyzing the diverse data using the multivariate statistical method or one of the machine learning algorithms, such as binary logistic regression (BLR), bivariate statistical analysis, and multivariate adaptive regression spline models (MARSplines). They can identify potentially hazardous areas related to landslides, in addition to mapping landslide susceptibility through a spatial comparison of the past landslide events with all the considered conditioning factors (Wang et al., 2015b). BLR is a multivariate statistical method that allows the consideration of both categorical (e.g., land cover) and continuous predictor variables (e.g., distance to fault lines and channels) in a regression analysis. Other analytical methods include conditional probability (CP), ANN, and SVM (Yilmaz, 2010). They have been compared in zoning landslide susceptibility graphically (Figure 3.17), in which the

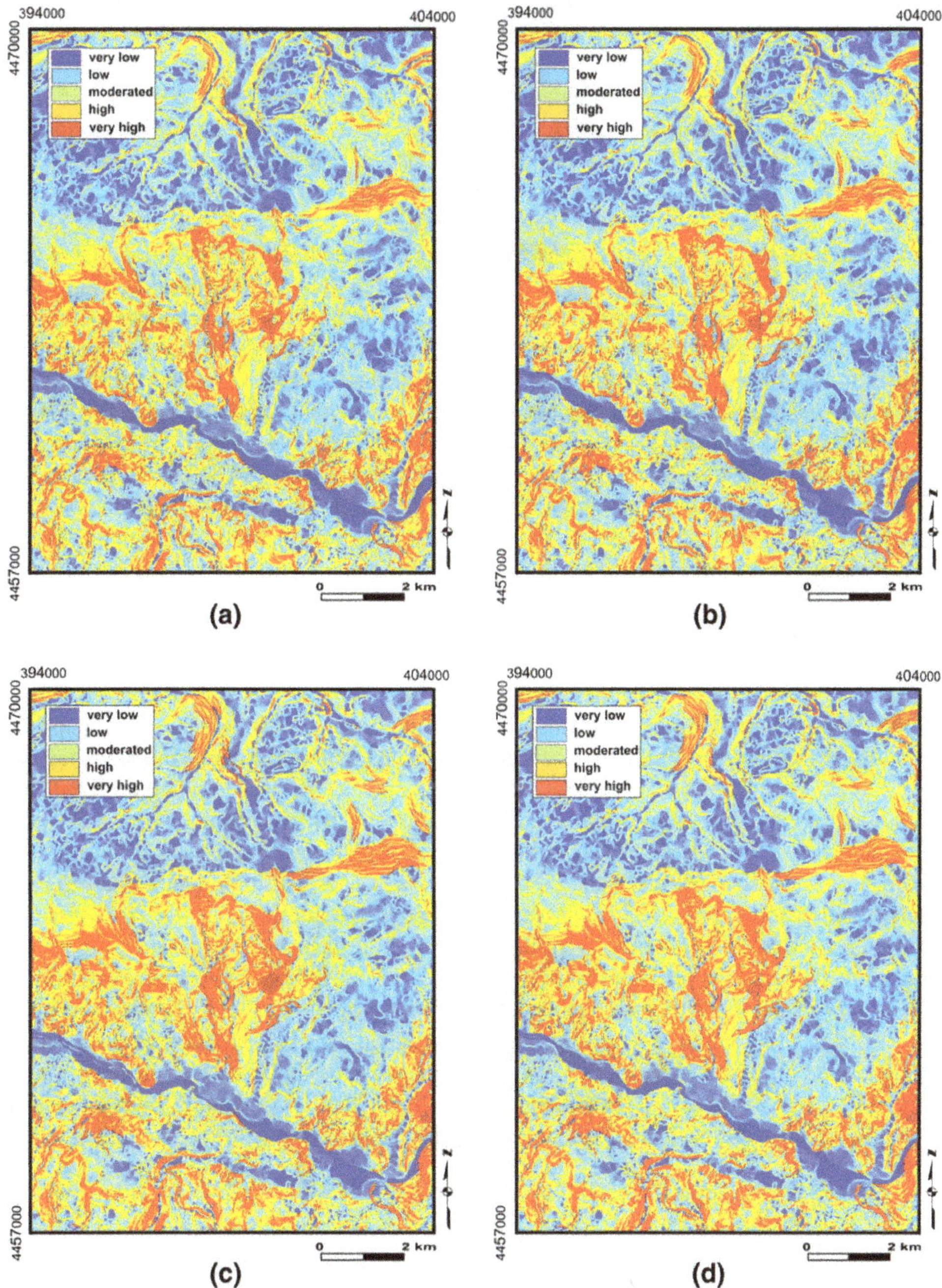

FIGURE 3.17 Comparison of zoned landslide susceptibility (hazard) of Koyulhisar, Turkey, based on consideration of nine variables using four methods. (a) Conditional probability (CP), (b) logistic regression (LR), (c) ANN, and (d) SVM. (Source: Yilmaz, 2010. Used with permission.)

original index value is categorized into four or five classes to indicate the level of susceptibility. MARSplines is found to be a superior model with a higher predictive power to BLR and bivariate statistical analysis (Wang et al., 2015b). It offers the best predictions in areas with a medium level of susceptibility and higher.

With the use of BLR analysis based on 12 variables, areas prone to landslides (presence) and non-landslides (absence) are zoned at an overall accuracy of 86.8% (AUC = 0.91) in part of the tectonically active Garhwal Lesser Himalaya, India (Mathew et al., 2009). One possible explanation for this high accuracy is the large landslide size (600 m^2 from Indian RS P6 LISS IV satellite imagery data, and a minimum of 100 m^2 obtained in the field). Thus, it is impossible to conclude that BLR is better than machine learning methods in the zoning. Judged by AUC, ANN (0.846) is very similar to SVM (0.841) in their accuracy, followed by LR (0.831). CP is the least accurate among the four, achieving an AUC of 0.827, but it is a simple method in landslide susceptibility mapping and highly compatible with GIS spatial operation of multiple features because feature input, calculation, and output are rather simple with this model (Yilmaz, 2010).

The production of a hazard zoning map involves sophisticated data manipulation that is ideally accomplished in a GIS. For instance, how far the mobilized debris is displaced becomes known only after a landslide. Prior to the actual event, the flow length can be modeled using GIS. In this way, it is possible to generate a landslide hazard map from each of the considered variables (e.g., geological instability, rock mass rating or geomechanical classification of rock outcroppings in the field, current landslides, and microtopography) and then derive the overall hazard or the integrated landslide hazard that reflects the intrinsic state of slope land from the past toward the future (Lee et al., 2018). This assessment method may yield a rather high accuracy of 91% in forecasting landslides and rockfalls along roads following intense rainfall events and typhoons, but it requires intensive fieldwork and does not consider other factors. So it is applicable to the prediction of regional, heavy rainfall-triggered landslides only.

3.6 LANDSLIDE MONITORING

Landslide monitoring aims to observe the activity of the same landslide(s) over a period of time to see how it evolves by comparing multi-temporal observations. Implicitly, monitoring is synonymous with time-series analysis or multi-temporal remote sensing. It involves observing landslide conditions (e.g., areal extent, speed of movement, surface topography, and soil humidity) repeatedly using the same data and method (Mantovani et al., 1996). Essentially, monitoring is a kind of change detection during which the change to landslide activity (e.g., deformation) is tracked. The monitored change may refer to landslide deformation or movement. Landslide deformation is the change to the morphometry of the landslide body. Monitoring landslide deformation is similar to monitoring landslides except that the results are quantitative and multi-dimensional. It can reveal the speed of change or creep in a particular direction over time and indicate the rate at which the landslide path evolves or how its components are displaced. Monitoring differs from detection in that it involves at least two states, the initial and the terminal states. Monitoring can produce time-series results of change. Based on what has been observed in the past, it is possible to predict the behavior of the monitored landslide in the future. Thus, monitoring is important for predicting when a landslide is likely to eventuate so that warnings can be issued prior to the disaster striking to mitigate damage and losses.

3.6.1 Monitoring Methods

Landslides are monitored using four methods: GPS, theodolite, LiDAR, and images, each having its unique strengths and limitations (Table 3.16). As with landslide mapping, landslide deformation can be monitored using GPS. A time-lapse GPS survey reveals how the landslides have changed. This monitoring may be based on a number of GPS points strategically distributed along the deforming slope.

TABLE 3.16
Comparison of Four Means of Monitoring Landslides

Monitoring Method	Theodolite	GPS	Imagery	LiDAR
Philosophy of monitoring	Strategic position of a landslide surveyed multiple times	Positions of permanent or semi-permanent stations logged	At least two-time images are compared	DEMs of the same area produced from multi-temporal LiDAR data are differenced
Main features	Point-based results, short-term	Point-based results, long-term, repetitive	Areal-based results, can be decades-long	Areal-based, mostly height change
Strength	Highly precise	Repeatable	Extent of change known	Ability to monitor the volume of change
Limitations	Slow, sparse	Not so accurate	Moderate accuracy	Moderate accuracy
Best use	Deformation monitoring	Creeping/deformation monitoring	Spatial extent change, velocity of movement	Change in debris volume and sediment yield

GPS-based repetitive and periodic surveys at these points, all referenced with respect to a fixed station, can reveal the deformation (Bayer et al., 2017). Differential, dual-frequency global navigation satellite system (GNSS) receivers should be used to achieve the highest accuracy possible. Long-term GPS monitoring is usually achieved using monitoring stations, either permanent or semi-permanent. Semi-permanent GPS surveys have proven capable of detecting landslides caused by tectonic and geological movements. These sites must not creep, and be temporally stable, especially for the landslides covered by snow and ice. In this kind of frigid environment, local and seasonal changes in elevation are not related to genuine landslide movement and thus have to be excluded from the measurements to reveal landslide movement. Such artificial changes can be modeled by annual cyclical sinusoids (Wang et al., 2015a), and then subtracted from the monitored changes.

A challenge facing GPS monitoring is the establishment of a stable local reference frame and assessing its accuracy. This reference frame can be established from thousands of permanent continuous GPS stations installed around the world, such as the GEONET in Japan (http://www.gsi.go.jp) and the CORS network in the US (http://geodesy.noaa.gov/CORS) for studying local ground deformations. Wang et al. (2014) established a stable Puerto Rico and Virgin Islands Reference Frame (SPRVIRF) from the International GNSS Service Reference Frame of 2008. With the joint use of the precise post-positioning processing and SPRVIRF, it is feasible and practical to monitor mm-level landslide creeping and deformation using a single GPS data logger to about 2-mm horizontal and 6-mm vertical RMSE. At these accuracy levels, slow-moving landslides at a rate of 1–2 mm per annum can be monitored over the long term. Nevertheless, GPS monitoring can yield results only at the monitored stations. The generation of the displacement field requires additional processing of spatial interpolation if there are sufficient monitoring stations. GPS can reveal deformation at the logged points only, which may not be spatially representative (e.g., the logged points are concentrated heavily at easy-to-access locations). Besides, it is difficult to log dense points all over the landslide surface. For this reason, GPS is not used as frequently as remote sensing images.

Theodolite monitoring is almost identical to GPS monitoring in principle and implementation, even though it produces more reliable results than GPS but is not repeatable. Besides, theodolite-based monitoring is slow, the monitored area is usually small, and the duration of monitoring is rather short. This method of monitoring shares the same limitations as GPS in that it is unable to provide a spatial distribution of changes to all landslides. In comparison, LiDAR is much more effective at monitoring the change in surface elevation, which may reveal landslide activities. However, LiDAR data are not good at revealing the spatial extent of change unless sophisticated algorithms are used to process them. Besides, LiDAR data are not multi-temporal by default.

These limitations are effectively overcome by image-based monitoring, a topic so broad that it will be covered in a separate section below.

3.6.2 InSAR-Based Monitoring

Images allow a huge ground area to be monitored. If acquired from satellites orbiting the Earth, they have a long history of existence and are available periodically. These characteristics enable landslides to be monitored from multi-temporal images spanning over decades. If the images are recorded at a short temporal interval from each other (e.g., daily), it may be possible to monitor the process of landslides via an animation created from multiple landslide maps. Image-based monitoring, however, is complex and challenging as the pre- and post-event image pairs must be spatially co-registered with each other to a high geometric accuracy to render image geometric inaccuracy to be negligible or much smaller than the change itself. Such surface change is ideally monitored from VHR optical and microwave images (Metternicht et al., 2005). Of all the existing images, only those possessing the 3D-viewing capacity can serve this purpose well as deformation is inherently 3D in nature and has a temporal component. If optical images are used, they must be time-series and repeatable, and accurately geo-referenced. The processing of time-series images increases the workload of monitoring, but the task can be automated by using sophisticated software. Most optical images are acquired in the nadir-viewing direction, which makes them insensitive to surface height change. Besides, optical images have a temporal resolution on the order of days and are subject to cloud contamination. They are not suited to monitoring fast-changing landslides, which is better accomplished from radar images that capture the backscattered radiation from the target. Since radar images are sensitive to surface relief, they have been used much more widely than optical images in monitoring landslides.

Landslide deformation can be effectively monitored from SAR images that are inherently side-looking, and thus sensitive to topographic relief. The most commonly used type of radar imagery is multi-temporal InSAR that is good at retrieving the spatiotemporal rate of slow deformations and monitoring land displacement processes related to landslides (Figure 3.18). The displacement in multiple directions increases the difficulty of phase unwrapping of interferograms, precipitating the use of more complex 2D and 3D phase unwrapping algorithms. They enable the identification of the movement of landslides in two directions (Liu et al., 2013) but increase the unwrapping complexity. The wrapped and unwrapped phase values enable the inference of different movement types (rotational, translational, and complex sliding) if integrated with information on the location of specific morphological (scarps, grabens, and lobes) or topographic (steep slope, shape of slope) background knowledge (Schlögel et al., 2015). Based on the spatial distribution of phase values, the detected deformation zone can be further differentiated into specific units of subsidence, transit, and stability. Instead of interpreting the phrase, the interferograms can be used to construct DEMs, from which time-series DEMs can be produced, and their comparison is able to reveal the morphological changes associated with landslides (Du et al., 2017).

For space-born InSAR monitoring to be successful, the displacements must take place to a relatively large spatial extent, and there must be houses located inside the study area so that a large number of persistent scatter (PS) candidate pixels can be selected (Bayer et al., 2017). The lack of PS candidates in rural areas, where landslides are likely located, may cause the density of results spatially sparse, especially in forested areas. So artificial corner reflectors have to be installed in the field of sensing. These corner reflectors are needed in areas where no existing buildings can serve as candidate PS pixels. A 12-day interferogram constructed from consecutive Sentinel-1 data using a simplified PS-InSAR method detects the landslide deformation pattern after 2D phase unwrapping. It shows the accumulated deformation clearly distinguishable from the surrounding areas. The deformation time series of the landslides illustrates the mean displacements of the entire landslide (Crosetto et al., 2016). InSAR is more sensitive to vertical displacement than horizontal creep, but

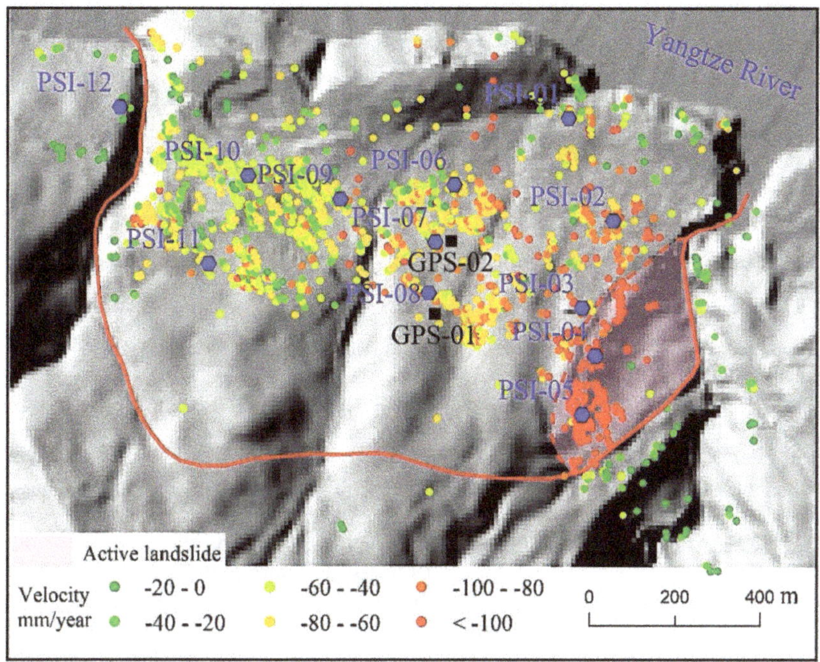

FIGURE 3.18 Spatial distribution of landslide deformation velocity detected from multi-temporal InSAR generated from multi-temporal Sentinel-1 images. GPS – global positioning system; PSI – persistent scatterer InSAR feature points. (Source: Zhou et al., 2020.)

the opposite is true with images. Horizontal displacement is better monitored from images rather than from non-imagery LiDAR point cloud data.

One method of overcoming the low PS candidate pixels in rural areas is to make use of coherent scatterer InSAR (CSI) to map landslide surface displacements by combining PS with distributed scatterers (DS). Through the use of the generalized likelihood ratio test, statistically homogeneous pixels are identified and a phase linking algorithm is used to estimate the optimal phase for each DS pixel. The joint exploitation of both PS and DS targets dramatically increases the spatial density of measurement points, which enhances the reliability of phase unwrapping (Dong et al., 2018). Validated against in situ GPS measurements, the CSI results from time-series InSAR data retain an RMSE of about 10.5 mm·year^{-1} in monitoring landslide surface displacements. Although CSI produced a spatial distribution pattern of surface displacement rates similar to those obtained using traditional PS InSAR feature points, it detected more than ten times more measurement points than the other two methods in vegetated mountainous areas, owing obviously to the use of DS targets.

Space-borne InSAR data-derived deformations agree well with ground measurements obtained via inclinometer and GPS, and small differences exist between time-series deformation determined using a ground-based InSAR instrument. InSAR can be further processed to derive differential interferometric data or DInSAR. The coupling of radar data with DInSAR data enables landslide activities to be monitored at the mm accuracy level, even for slow-moving landslides (Scaioni et al., 2014). Nevertheless, DInSAR-based monitoring faces several challenges in obtaining the true phase of a corner reflector, the difficulty of co-registering corner reflector pixels in light of extremely low coherence of their surrounding area, and computing the interferometric phase of two co-registered corner reflectors without the flat earth term and the corners' height contribution (Ye et al., 2004).

No matter which kind of remote sensing images is used, it is not feasible to systematically monitor relatively rapid movements (e.g., mudflows, rockfalls, and debris slides) confined to small areas and on steep slopes or in narrow valleys using remote sensing images because of their long revisit period. Instead, these images are good at detecting the net change from the pre- and post-event images. Monitoring of the landslide process is possible only when the debris is displaced at an extremely slow pace (e.g., less than a few centimeters per month), affecting large areas of a sparse vegetation cover. The temporal interval may be shortened by combining multiple images from different sensors. However, this may still be further complicated by cloud covers on each image.

REFERENCES

Abellán, A., M. Jaboyedoff, T. Oppikofer, and J.M. Vilaplana. 2009. Detection of millimetric deformation using a terrestrial laser scanner: Experiment and application to a rockfall event. *Nat. Hazards Earth Syst. Sci.* 9: 365–372. DOI: 10.5194/nhess-9-365-2009

Alcantara-Ayala, I. 2004. Hazard assessment of rainfall-induced landsliding in Mexico. *Geomorphology* 61(1–2): 19–40. DOI: 10.1016/j.geomorph.2003.11.004

Baldo, M., C. Bicocchi, U. Chiocchini, D. Giordan, and G. Lollino. 2009. LiDAR monitoring of mass wasting processes: The Radicofani landslide, Province of Siena, Central Italy. *Geomorphology* 105(3–4): 193–201.

Bayer, B., A. Simoni, D.A. Schmidt, and L. Bertello. 2017. Using advanced InSAR techniques to monitor landslide deformations induced by tunneling in the Northern Apennines. *Italy. Eng. Geol.* 226: 20–32.

Bernard, T.G., D. Lague, and P. Steer. 2020. Beyond 2D inventories: Synoptic 3D landslide volume calculation from repeat LiDAR data. *Earth Surf. Dynam.* DOI: 10.5194/esurf-2020-73

Central Geological Survey. 2006. Annual Report – Establishment of Environmental Geology Database for Urban and Suburban Slopelands Area. New Taipei (in Chinese).

Chen, R. 2022. GIS-based Assessment of Geomorphological Influences on Earthquake-triggered Landslides in Kaikōura. MSc thesis, University of Auckland, New Zealand, 78 p.

Chen, T., J.C. Trinder, and R. Niu. 2017. Object-oriented landslide mapping using ZY-3 satellite imagery, random forest and mathematical morphology, for the Three-Gorges reservoir, China. *Remote Sens.* 9: 333. DOI: 10.3390/rs9040333

CRED (Centre for Research on Epidemiology of Disasters), 2018. *Natural Disasters 2017.* Brussels: CRED. https://cred.be/sites/default/files/adsr_2017.pdf

Crosetto, M., O. Monserrat, N. Devanthéry, M. Cuevas-González, A. Barra, and B. Crippaet. 2016. Persistent scatterer interferometry using Sentinel-1 data. *Int. Arch. Photogramm. Remote Sens. Spatial Inf. Sci.* XLI-B7. DOI: 10.5194/isprsarchives-XLI-B7-835-2016

Cruden, D.M., and D.J. Varnes. 1996. Chapter 3: Landslide types and processes. *Landslides: Investigation and Mitigation*, Special Report 247 National Research Council (U.S.) Transportation Research Board, pp. 36–75.

Dong, J., L. Zhang, M. Tang, M. Liao, Q. Xu, J. Gong, and M. Ao. 2018. Mapping landslide surface displacements with time series SAR interferometry by combining persistent and distributed scatterers: A case study of Jiaju landslide in Danba, China. *Remote Sens. Environ.* 205: 180–198.

Du, Y., Q. Xu, L. Zhang, G. Feng, Z. Li, R. Chen, and C. Lin. 2017. Recent landslide movement in Tsaoling, Taiwan tracked by TerraSAR-X/TanDEM-X DEM time series. *Remote Sens.* 9: 353.

Glenn, N., D.R. Streutker, D.J. Chadwick, G.D. Thackray, and S.J. Dorsch. 2006. Analysis of LiDAR-derived topographic information for characterizing and differentiating landslide morphology and activity. *Geomorphology* 73(1–2): 131–148.

Jaboyedoff, M., and M.-H. Derron. 2020. Chapter 7 – Landslide analysis using laser scanners. In *Developments in Earth Surface Processes*, ed. P. Tarolli and S.M. Mudd, vol 23, 207–230. DOI: 10.1016/B978-0-444-64177-9.00007-2

Jaboyedoff, M., T. Oppikofer, A. Abellán, M.H. Derron, R. Loye, R. Metzger, and A. Pedrazzini. 2012. Use of LiDAR in landslide investigations: A review. *Nat. Hazards* 61(1): 5–28.

Kamps, M., W. Bouten, and A. Seijmonsbergen. 2017. LiDAR and orthophoto synergy to optimize object-based landscape change: Analysis of an active landslide. *Remote Sens.* 9: 805.

Kasperski, J., C. Delacourt, P. Allemand, P. Potherat, M. Jaud, and E. Varrel. 2010. Application of a terrestrial laser scanner (TLS) to the study of the Séchilienne landslide (Isère, France). *Remote Sens.* 2(12): 2785–2802. DOI: 10.3390/rs122785

Kromer, R.A., A. Abellan, D.J. Hutchinson, M. Lato, M.-A. Chanut, L. Dubois, and M. Jaboyedoff. 2017. Automated terrestrial laser scanning with near real-time change detection – Monitoring of the Séchilienne landslide. *Earth Surf. Dynam.* 5: 293–310. DOI: 10.5194/esurf-5-293-2017

Lee, C.-F., W.-K. Huang, Y.-L. Chang, S.-Y. Chi, and W.-C. Liao. 2018. Regional landslide susceptibility assessment using multi-stage remote sensing data along the coastal range highway in northeastern Taiwan. *Geomorphology* 300: 113–127.

Lin, C.-W., C.-M. Tseng, Y.-H. Tseng, L.-Y. Fei, Y.-C. Hsieh, and P. Tarolli. 2013. Recognition of large scale deep-seated landslides in forest areas of Taiwan using high resolution topography. *J. Asian Earth Sci.* 62: 389–400. DOI: 10.1016/j.jseaes.2012.10.022

Liu, J.K., K.H. Hsiao, and P.T.Y. Shih. 2012a. A geomorphological model for landslide detection using airborne LIDAR data. *J. Mar. Sci. Technol.* 20(6): 629–638.

Liu, S.-W., D.-H. Guo, W.-T. Chen, X.-W. Zheng, S.-Y. Wang, and X.-J. Li. 2012b. The application of airborne LiDAR technology in landslide investigation and monitoring of three gorges reservoir area. *Geol. Chin.* 39(2): 507–517.

Liu, P., Z. Li, T. Hoey, C. Kincal, J. Zhang, and J.-P. Muller. 2013. Using advanced InSAR time series techniques to monitor landslide movements in Badong of the three gorges region, China. *Int. J. Appl. Earth Observ. Geoinform.* 21: 253–264.

Lo, C.M., Z.-Y. Feng, and K.-T. Chang. 2018. Landslide hazard zoning based on numerical simulation and hazard assessment. *Geomatics Nat. Hazards Risk* 9(1): 368–388. DOI: 10.1080/19475705.2018.1445662

Mantovani, F., R. Soeters, and C.J. van Westen. 1996. Remote sensing techniques for landslide studies and hazard zonation in Europe. *Geomorphology* 15: 213–225.

Mathew, J., V.K. Jha, and G.S. Rawat. 2009. Landslide susceptibility zonation mapping and its validation in part of Garhwal Lesser Himalaya, India, using binary logistic regression analysis and receiver operating characteristic curve method. *Landslides* 6: 17–26. DOI: 10.1007/s10346-008-0138-z

Matsuoka, A., T. Yamakoshi, K. Tamura, J. Maruyama, and K. Ogawa. 2008. Sediment yield from seismically-disturbed mountainous watersheds revealed by multi-temporal aerial LiDAR surveys. In *Sediment Dynamics in Changing Environments*, Proc. of a symposium held in Christchurch, New Zealand, December 2008, 208-216. I IAHS-AISH Publication 325.

Metternicht, G., L. Hurni, and R. Gogu. 2005. Remote sensing of landslides: An analysis of the potential contribution to geo-spatial systems for hazard assessment in mountainous environments. *Remote Sens. Environ.* 98(2–3): 284–303. DOI: 10.1016/j.rse.2005.08.004

Mondini, A.C., F. Guzzetti, P. Reichenbach, M. Rossi, M. Cardinali, and F. Ardizzone. 2011. Semi-automatic recognition and mapping of rainfall induced shallow landslides using optical satellite images. *Remote Sens. Environ.* 115: 1743–1757.

Nichol, J., and M.S. Wong. 2005. Satellite remote sensing for detailed landslide inventories using change detection and image fusion. *Int. J. Remote Sens.* 26(9): 1913–1926. DOI: 10.1080/01431160512331314047

Pawluszek-Filipiak, K., and A. Borkowski. 2020. On the importance of train–test split ratio of datasets in automatic landslide detection by supervised classification. *Remote Sens.* 12, 3054. DOI: 10.3390/rs12183054

Philipson, W.R. 1997. *Manual of Photographic Interpretation* (2nd ed.). American Society of Photogrammetry and Remote Sensing Science and Engineering Series, Washington, DC.

Roering, J.J., L.L. Stimely, B.H. Mackey, and D.A. Schmidt. 2009. Using DInSAR, airborne LiDAR, and archival air photos to quantify landsliding and sediment transport. *Geophy. Res. Lett.* 36: L19402. DOI: 10.1029/2009GL040374

Šašak, J., M. Gallay, J. Kaňuk, J. Hofierka, and J. Minár. 2019. Combined use of terrestrial laser scanning and UAV photogrammetry in mapping alpine terrain. *Remote Sens.* 11(18): 2154. DOI: 10.3390/rs11182154

Sato, H.P., and E.L. Harp. 2009. Interpretation of earthquake-induced landslides triggered by the 12 May 2008, M7.9 Wenchuan Earthquake in the Beichuan area, Sichuan Province, China using satellite imagery and Google Earth. *Landslides* 6(2): 153–159. DOI: 10.1007/s10346-009-0147-6

Scaioni, M., L. Longoni, V. Melillo, and M. Papini. 2014. Remote sensing for landslide investigations: An overview of recent achievements and perspectives. *Remote Sens.* 6. DOI: 10.3390/rs60x000x

Schlögel, R., C. Doubre, J.P. Malet, and F. Masson. 2015. Landslide deformation monitoring with ALOS/PALSAR imagery: A D-InSAR geomorphological interpretation method. *Geomorphology* 231: 314–330.

Soeters, R., and C.J. van Westen. 1996. Chapter 8: Slope instability recognition, analysis, and zonation. In *Landslides: Investigation and Mitigation. Transp. Res. Board Spec. Rep.* 247: 129–177.

Stumpf, A., and N. Kerle. 2011. Object-oriented mapping of landslides using random forests. *Remote Sens. Environ.* 115: 2564–2577. DOI: 10.1016/j.rse.2011.05.013

Stumvoll, M.J., E.M. Schmaltz, and T. Glade. 2021. Dynamic characterization of a slow-moving landslide system – Assessing the challenges of small process scales utilizing multi-temporal TLS data. *Geomorphology* 389, 107803. DOI: 10.1016/j.geomorph.2021.107803

Tsai, F., J.-H. Hwang, L.-C. Chen, and T.-H. Lin. 2010. Post-disaster assessment of landslides in southern Taiwan after 2009 Typhoon Morakot using remote sensing and spatial analysis. *Nat. Hazards Earth Syst. Sci.* 10, 2179–2190. DOI: 10.5194/nhess-10-2179-2010

Tsutsui, K., S. Rokugawa, H. Nakagawa, S. Miyazaki, C.T. Cheng, T. Shiraishi, and S.D. Yang. 2007. Detection and volume estimation of large-scale landslides based on elevation-change analysis using DEMs extracted from high-resolution satellite stereo imagery. *IEEE Trans. Geosci. Remote Sens.* 45: 1681–1696.

U.S. Geological Survey. 2004. Landslide Types and Processes, Fact Sheet 2004-3072.

U.S. Geological Survey. 2009. USGS Science for a changing world – Landslide Hazards. https://www.usgs.gov/programs/landslide-hazards (accessed on July 5, 2022).

Van Den Eeckhaut, M., J. Poesen, F. Gullentops, L. Vandekerckhove, and J. Hervás. 2011. Regional mapping and characterisation of old landslides in hilly regions using LiDAR-based imagery in Southern Flanders. *Quat. Res.* 75(3): 721–733.

Van Den Eeckhaut, M., Hervas, J., Jaedicke, C., et al. (2012). Statistical modeling of Europe-wide landslide susceptibility using limited landslide inventory data. *Landslides*, 9, 357–369. DOI: 10.1007/s10346-011-0299-z

Van Westen, C.J., E. Castellanos, and S.L. Kuriakose. 2008. Spatial data for landslide susceptibility, hazard, and vulnerability assessment: An overview. *Eng. Geol.* 102: 112–131. 10.1016/j.enggeo.2008.03.010.

Ventura, G., G. Vilardo, C. Terranova, and E.B. Sessa. 2011. Tracking and evolution of complex active landslides by multi-temporal airborne LiDAR data: The Montaguto landslide (Southern Italy). *Remote Sens. Environ.* 115(12): 3237–3248.

Wang, G., T.J. Kearns, J. Yu, and G. Saenz. 2014. A stable reference frame for landslide monitoring using GPS in the Puerto Rico and Virgin Islands region. *Landslides* 11: 119–129. DOI: 10.1007/s10346-013-0428-y

Wang, G., Y. Bao, Y. Cuddus, X. Jia, J. Serna, and Q. Jing. 2015a. A methodology to derive precise landslide displacement time series from continuous GPS observations in tectonically active and cold regions: A case study in Alaska. *Nat. Hazards* 77: 1939–1961. DOI: 10.1007/s11069-015-1684-z

Wang, L.J., M. Guo, K. Sawada, J. Lin, and J. Zhang. 2015b. Landslide susceptibility mapping in Mizunami City, Japan: A comparison between logistic regression, bivariate statistical analysis and multivariate adaptive regression spline models. *Catena* 135: 271–282.

Wiguna, S. 2019. Modelling of Future Landslide Exposure in Sukabumi, Indonesia in Two Scenarios of Land Cover Changes. Master thesis, University of Auckland, New Zealand, 150 p.

Ye, X., H. Kaufmann, and X.F. Guo. 2004. Landslide monitoring in the three Gorges area using D-INSAR and corner reflectors. *Photogramm. Eng. Remote Sens.* 70: 1167–1172.

Yilmaz, I. 2010. Comparison of landslide susceptibility mapping methodologies for Koyulhisar, Turkey: Conditional probability, logistic regression, artificial neural networks, and support vector machine. *Environ. Earth Sci.* 61: 821–836.

Zhou, C., Y. Cao, K. Yin, Y. Wang, X. Shi, F. Catani, and B. Ahmed. 2020. Landslide characterization applying Sentinel-1 images and InSAR technique: The Muyubao landslide in the Three Gorges reservoir area, China. *Remote Sen.* 12, 3385. DOI: 10.3390/rs12203385

4 Land Degradation

4.1 INTRODUCTION

Land degradation alludes to the lowering of land quality, loss of soil fertility, or an increase in soil salinity. After farmland has been degraded, its biological and economic productivity is reduced or completely lost. This set of definitions, tailored for the evaluation of land degradation at the global and continental scales, is too vague for evaluating land degradation at the national or local scales (Costantini and Lorenzetti, 2013). Thus, it is modified to mean the impairment of soil qualities and is differentiated from land desertification, a type of severe land degradation, in this chapter. Here land degradation is defined as the physical, biological, and chemical deterioration of soil quality to such a level to adversely impact the land production value (Table 4.1). Physical degradation refers to the compaction and coarsening of soil texture after fine silt has been eroded. In the worst case, only coarse sand and pebbles are left behind. Chemical degradation is defined as the change in the soil's chemical properties and composition, as exemplified by soil salinization and alkalization. Biological degradation pertains to the loss of soil nutrients and reduced soil fertility stemming from soil erosion. In the extreme case (e.g., desertification), it can completely obliterate surface vegetation. Thus, land degradation occurs in a number of forms, including soil erosion, salinization, and waterlogging.

Land degradation stems from both natural processes and improper tillage, such as a high evaporation combined with a high soil salt content. This process is known as land salinization. The accumulation of salt near the surface reaches a high level to have exceeded the tolerance threshold of vegetation for an extended period. The net effect is a reduced biomass or crop yield. Waterlogging is a phenomenon in which the land is submerged underneath water for an extended period following a heavy rainfall or flood event. Such water is not discharged or has not dissipated timely due to the lack of topographic gradient. The accumulation of the stagnant water in shallow depressions stifles the submerged vegetation, stuns its growth, and results ultimately in a landscape resembling that of severe soil salinization. The only difference between the two is the shape of the waterlogging-affected areas that tend to be large circular polygons completely devoid of vegetation.

Land degradation is a major environmental hazard in many parts of the world. Degradation can occur to irrigated and non-irrigated farmland, rangeland, woodland, and forest in arid, semi-arid, and dry sub-humid areas. It can cause grave environmental, ecological, and economic consequences, such as reduced biodiversity, loss of habitat and arable land, and threatened sustainable agriculture. Globally, land degradation has affected 2 billion ha (22.5%) of farmland, pasture, forest, and woodland (Gibbs and Salmon, 2015) and resulted in the disappearance of around 5–10 million ha of agricultural land per annum. The lost productivity in drylands caused by land degradation is valued at between US$13 and 28 billion a year. It is thus very important to study land degradation, especially using remote sensing.

Remote sensing is the ideal and cost-effective means of studying land degradation, especially on a global scale. It plays an indispensable role in mapping, assessing, and monitoring degraded lands at multiple spatial and temporal scales. Remote sensing offers long-term data useful for land degradation assessment at a spatial scale ranging from regional, national, continental to global. Satellite images are available from multi-platforms and multi-sensors at multi-scales over a broad spectrum for this purpose. These data are characterized by widely ranging temporal, spatial, and spectral resolutions. Data of different spatial resolutions and ground cover are suitable for studying degradation at multiple scales. Space-borne imagery such as MODIS is an efficient data source for monitoring degradation at the global scale. For regional monitoring, images of a finer resolution such as ETM+, ASTER, SPOT-5, and Sentinel-2 (Table 1.6) are useful. In general, images of a finer spatial resolution are suitable for studying land degradation on a smaller scale. Irrespective of the

DOI: 10.1201/9781003354321-4

TABLE 4.1

Four Criteria Used to Group the Severity of Chemical and Physical Land Degradation to Four Levels Based on Soil Properties

Chemical Degradation	Salinization Increase in EC ($dS \cdot m^{-1} \cdot year^{-1}$)	Alkalization Increase in ESP ($EPS \cdot year^{-1}$)
None to slight	<0.5	<5
Moderate	0.5–3	0.5–3
High	3–5	3–7
Very high	>5	>7
Physical Degradation	**Compaction Increase in Bulk Density ($g \cdot cm^{-3} \cdot year^{-1}$)**	**Waterlogging Increase in Water Table ($cm \cdot year^{-1}$)**
None to slight	<0.1	<1
Moderate	0.1–0.2	1–3
High	0.2–0.3	3–5
Very high	>0.3	>5

Source: El-Baroudy (2011). Used with permission.

satellite images used, they enable degradation to be studied cost-effectively as the same detection procedure can always be adhered to without modifications.

This chapter first introduces how to map degraded land automatically using image classification and purposely built indices. The discussion then progresses to how to detect and assess degradation severity using post-classification comparison, decision-tree analysis, and complex statistical analysis. The second half of this chapter explains how to map salinized land. Included in the discussion are the spectral behavior of soil salt and the most sensitive wavelengths for sensing it. In particular, the emphasis is placed on the utility of multispectral and hyperspectral images in retrieving soil salt content quantitatively. This chapter then details how to use machine learning algorithms to quantify the spatial distribution of soil salt content based on the most important predictor variables, and quantifying its spatial distribution using machine learning algorithms. Finally, this chapter explores how to sense a special type of land degradation, waterlogging, using a simple index. As mentioned previously, land desertification will be covered separately in Chapter 5 because the topic requires extensive coverage that cannot be conveniently accommodated in a single chapter.

4.2 MAPPING AND MONITORING

4.2.1 Image Classification

Although land degradation is manifested as physical changes in soil, it is the biological manifestation that is the most revealing of degradation and reliably detected from remotely sensed data. Satellite imagery is sensitive to the disparity between healthy vegetation and the absence of vegetation in land degradation-affected areas. Land degradation is detected on the assumption that healthy vegetation is not subject to land degradation, while degraded land is distributed with less biomass. So the best time of studying land degradation is the peak growing season when vegetation is growing most vigorously. The vegetation that is not so vigorous is construed as being affected by land degradation indirectly unless the impact is so severe that it has disappeared from the affected area.

On a local scale, degraded land is relatively easy to map from multispectral images of a moderate spatial resolution if it is not covered by healthy vegetation in an agricultural area. The mapping can be achieved using supervised classification to group land covers. Image classification may be per-pixel-based or object-oriented (see Section 1.4.1), in which all land covers are classified into various categories, some of which are related to degraded land (Figure 4.1). This method works well only

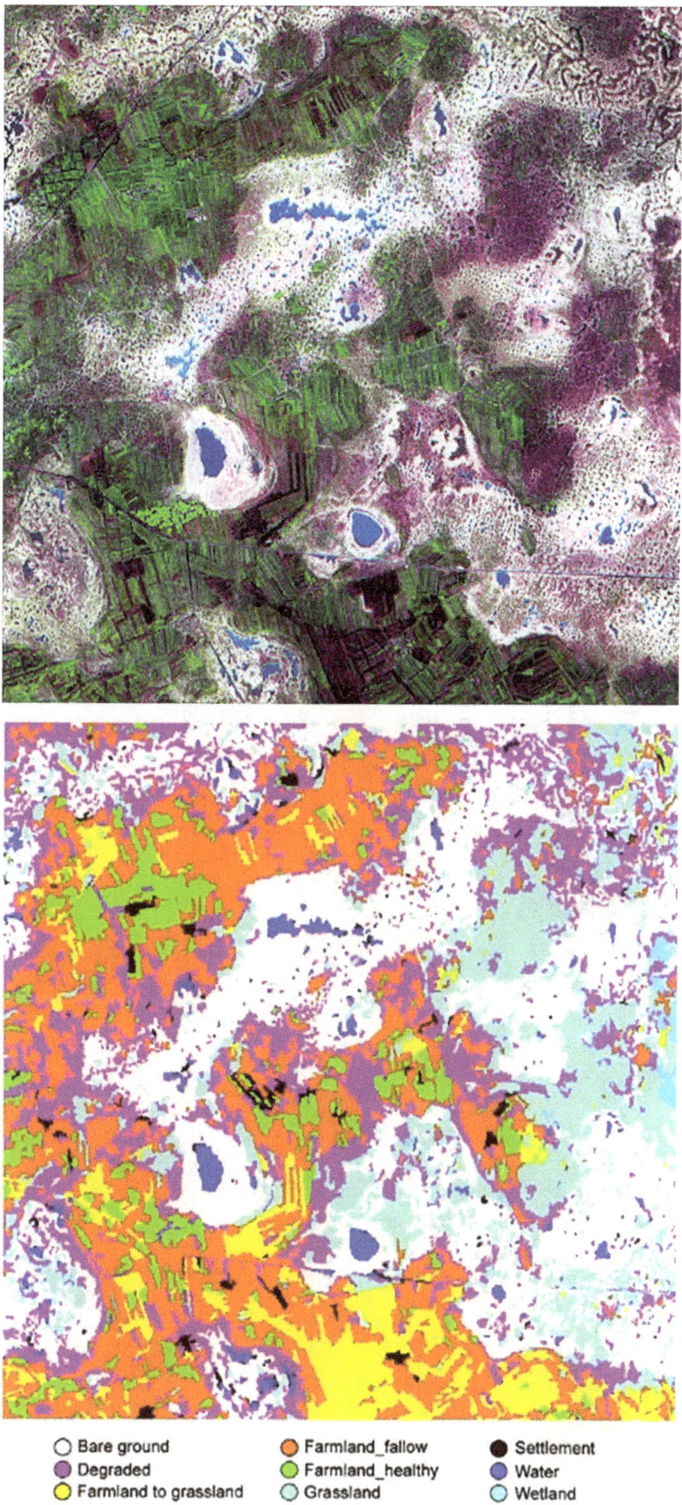

FIGURE 4.1 Distribution of degraded land mapped from a 15-m resolution ASTER image using object-oriented image classification. Top – false-color composite of bands 2 (blue), 3 (green), and 4 (red); Bottom: classified results of nine covers, including degraded land. (Source: Gao, 2008.)

when non-vegetated areas are all caused by land degradation as in grazing land, but not so in agricultural and urban areas. In agricultural areas, if some plots are recently tilled or the planted crops have not fully grown to cover the surface densely, the reliability of the mapped non-vegetated areas will be reduced unless the cause of the reduction in biomass is attributed solely to land degradation (e.g., it cannot be attributed to the varying sowing time). In urban fringe areas, transitional areas may be temporally bare, but they have nothing to do with land degradation.

In addition to maximum likelihood classification, knowledge-based approaches have also been used to produce a land cover map of nine types, one of which represents degraded land in the arid rangelands of North Africa (Mahyou et al., 2016). The joint consideration of satellite images, bioclimate data (e.g., aridity index), and lithology enables the degraded land to be mapped into four levels of severity: slight (e.g., wooded steppe), moderate (e.g., alfa grass steppe), severe (e.g., desert steppe), and extremely severe (bare soil and sand) (Figure 4.2). These four groups are formed by amalgamating the original ten land covers produced in a knowledge-based classification. After the re-grouping, the mapping is deemed rather accurate with an overall accuracy of 93% (overall Kappa statistics = 91%).

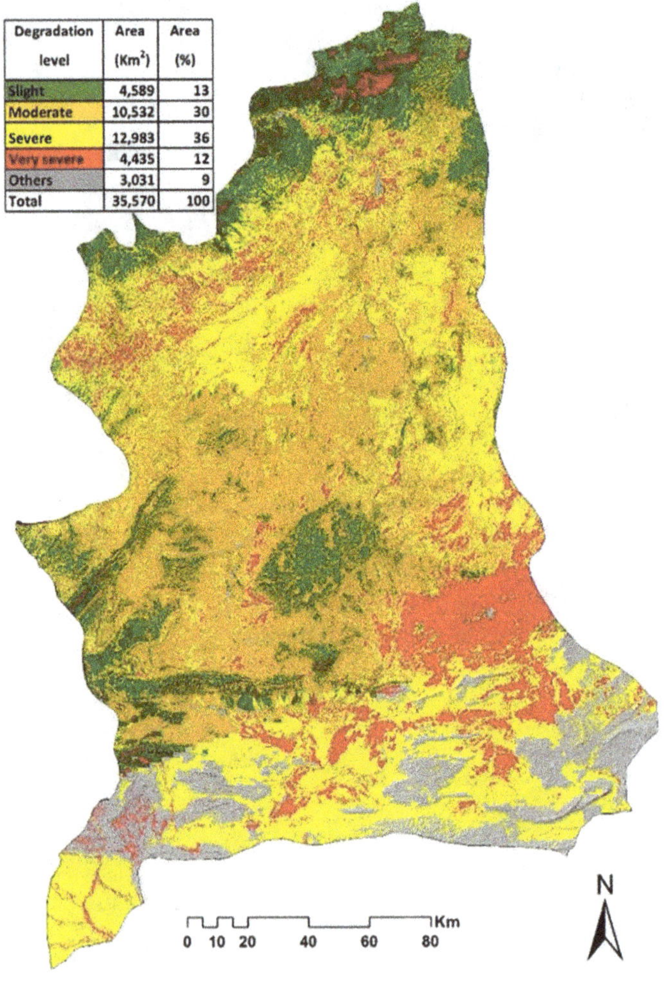

Degradation level	Area (Km^2)	Area (%)
Slight	4,589	13
Moderate	10,532	30
Severe	12,983	36
Very severe	4,435	12
Others	3,031	9
Total	35,570	100

FIGURE 4.2 Severity of rangeland degradation on the high plateaus of Eastern Morocco mapped from Landsat-5, climate and lithological data using knowledge-based classification. (Source: Mahyou et al., 2016. Used with permission.)

4.2.2 Indicator Differencing and Thresholding

Surface changes in vegetative cover, especially vegetation dynamics, have been used to study degradation because they are relatively easy to quantify from remote sensing data. Vegetation dynamics have been widely adopted as the proxy for mapping land degradation on regional and global scales. Of all multispectral bands, NIR bands combined with red bands are particularly effective at revealing degradation via vegetation indices derived from them. These indices enable easy and objective quantification of the condition of degradation-impacted vegetation and are treated as reliable indicators of land degradation. So far a wide range of proven indices have been developed, each having its strengths and limitations, but some are especially effective at revealing the influence of bare ground on the derived index value. The most commonly used indices are NDVI, EVI, and SAVI, all being the major indicators of vegetation condition and productivity (see Section 1.4.3).

Hyperspectral images with more than 100 spectral bands offer more choices in deriving vegetation indices involving 2 or 3 bands than multispectral images. Two-band indices include degradation index (DI), normalized DI (NDI), and ratio index (RI), calculated as follows:

$$DI = \rho_i - \rho_j \tag{4.1}$$

$$NDI = \frac{\rho_i - \rho_j}{\rho_i + \rho_j} \tag{4.2}$$

$$RI = \frac{\rho_i}{\rho_j} \tag{4.3}$$

where ρ_i and ρ_j = reflectance of two randomly selected bands i and j with a wavelength between 0.35 and 2.50 μm. Three-band indices (TBI) are calculated in a similar manner to subtracting two bands, followed by ratioing by the third band. Theoretically, the use of more spectral bands better captures the spectral signature of salinized land. The performance of these indices is evaluated through their correlation with soil salinity. Two-band indices of RI, DI, and NDI are significantly correlated with soil salinity. In particular, NDI calculated from $[(\rho_{12} - \rho_7)/(\rho_{12} + \rho_7)]$ of Sentinel-2 imagery has a maximum absolute Pearson correlation coefficient (PCC) of 0.52. Of all the two-band indices, SWIR band 12 and two other bands usually yield higher PCCs than others. Among all TBI, TBI4 $[(\rho_{12} - \rho_3)/(\rho_3 - \rho_{11})]$ is the best, achieving a maximum PCC of 0.54 (Wang et al., 2019a). Individually, the blue band (band 2) is the best.

Different VIs derived from different satellite images have different utilities in studying land degradation. A Landsat data-derived map of EVI change over South Africa at 30-m resolution exhibits an overall trend corresponding closely to that of VI derived from an 8-km resolution AVHRR EVI product (Venter et al., 2020). It is able to reveal patterns of degradation (e.g., woody plant encroachment, desertification), and restoration (e.g., increased rangeland productivity, alien clearing) over selected landscapes. The produced EVI trend layer is able to distinguish climatic from anthropogenic drivers of vegetative change, which is of tremendous value in monitoring ecosystem changes. Seasonally integrated NDVI or cumulative NDVI is more closely correlated with total phytomass than either maximum or maximum–minimum NDVI (Holm et al., 2003). It enables total phytomass in the heterogeneous arid shrubland of Western Australia to be estimated at a reasonable accuracy. Theoretically, an increase in VI value means lessened degradation and vice versa. However, thresholds are essential to grouping the continuous range of NDVI differences into a few categorical classes to show the progression of degradation severity.

The use of NDVI-based thresholds for long-term studies of degradation is limited in that inevitably, its temporal variation is subject to large inter-decadal variability of rainfall and persistent droughts. This variation does not necessarily indicate degradation but can cause a temporary decline in natural vegetation growth. Detection of land degradation on the sub-continent scale in the presence of underlying environmental trends (e.g., increased rainfall) is problematic with trend analysis of NDVI (Wessels et al., 2012). Even residual trends analysis has a limited range of applicability within which it can serve as a reliable indicator of land degradation, outside which it is an unreliable indicator of land degradation. The change in the vegetative cover is ideally monitored from time-series vegetation indices derived from multi-temporal images.

Apart from VIs, other indicators of land degradation such as rain use efficiency (RUE) and net primary productivity (NPP) have also been used. RUE is defined as the ratio of (annual) aboveground NPP to (annual) precipitation and has been widely used to measure land degradation. RUE is useful in identifying trends of degradation patterns, and can potentially serve as a competent indicator of landscape degradation on a broad scale, but is unable to quantify degradation severity (Holm et al., 2003). RUE is not as a sound indicator of phytomass as NDVI, even though it is essentially another form of NPP derived from cumulative NDVI.

4.2.3 Decision-tree Analysis

The mapping of land degradation using the decision-tree method is based normally on a number of degradation indicators and their critical statistical parameters. The joint use of a set of indices derived from NDVI, such as the mean of the annual amplitude (difference between annual maximum and minimum NDVI) ($NDVI_{AMP}$) over a period, annual average NDVI ($NDVI_{AVE}$), annual maximum NDVI ($NDVI_{MAX}$), annual minimum NDVI ($NDVI_{MIN}$), cumulative change of the annual NDVI amplitude (CC_{AMP}), average NDVI-CC_{AVE}, maximum NDVI (CC_{MAX}), and minimum NDVI (CC_{MIN}), aims to increase the reliability of detection. Cumulative changes of these indices are calculated by summing up the differences between the annual values minus the value of the first year of analysis until the current year. The diverse inputs are typically analyzed based on various threshold criteria in a decision-tree analysis to identify degradation. In its simplest form, a tree is binary with only two branches at each node established through a cyclic analysis of the training dataset comprising the predictor and target variables (e.g., degraded vs non-degraded) (Figure 4.3).

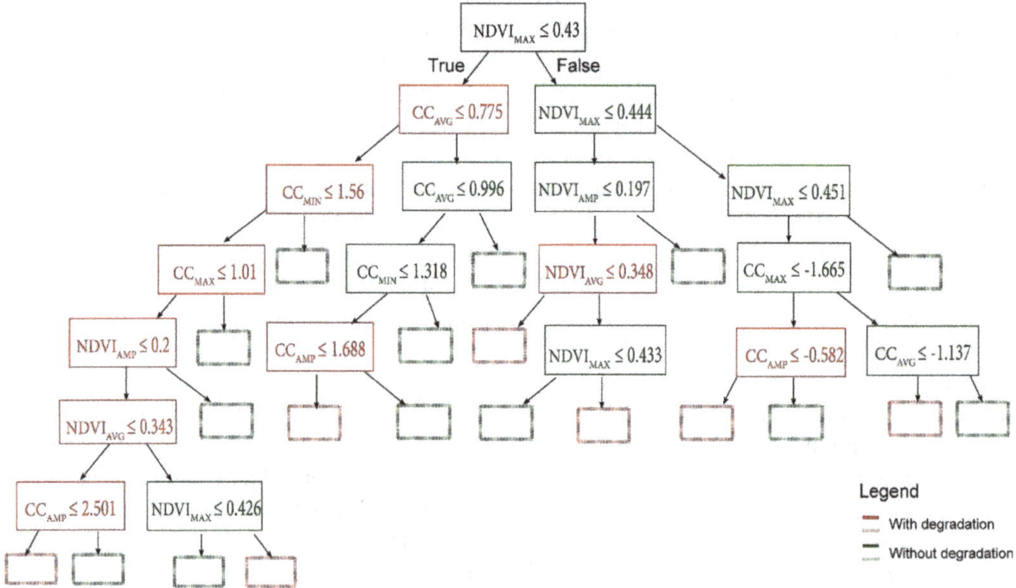

FIGURE 4.3 A CART-based binary decision-tree classifier for detecting degraded areas based on a combination of various degradation indicators and their thresholds. Left arrow – condition met; right arrow – condition not met (e.g., false); red box – degraded; green box – non-degraded. (Source: Vieira et al., 2021. Used with permission.)

The algorithm searches for multiple combinations of the independent variables, starting from the root node that encompasses the entire dataset. At the next level of nodes, the dataset is split into two groups, corresponding to the two branches in a binary partitioning. This partitioning process is terminated once the split subsets are deemed sufficiently homogeneous. In order to gauge the accuracy of the established tree, the available dataset is usually divided into two parts, training and validation. The training dataset is used to construct the tree, and the validation dataset is used to check its predictive power with unseen samples. The outcome of validation is expressed as the producer's accuracy, the user's accuracy, and the overall accuracy. This simple approach is capable of providing fine, detailed, and accurate information on degradation and underpins the effects of persistent droughts, which in some cases cause a false indication of degradation (Vieira et al., 2021).

4.2.4 Degradation Monitoring

Since the United Nations Conference on Desertification in 1977, several international organizations (e.g., United Nations Environment Programme, International Soil Reference and Information Centre, and Millennium Ecosystem Assessment) have commissioned and/or executed programs for monitoring and assessing land degradation on regional/global scales. These programs include but are not limited to Global Assessment of Human-Induced Soil Degradation (GLASOD), Land Degradation Assessment in Drylands (LADA) (http://www.fao.org/nr/lada), and Global Assessment of Land Degradation and Improvement (GLADA). GLASOD is based on expert opinion and hard remotely sensed evidence of soil degradation only, neglecting vegetation and water. As the follow-up projects of GLASOD, LADA and GLADA are carried out at three spatial scales, global, national, and sub-national, all aim at identifying the status and trends of land degradation in drylands, hotspots (i.e., areas with the most severe land constraints, their actual degradation, and areas at risk of degradation, drought or flood), and bright spots where policies and actions have slowed or reversed land degradation. LADA adopts a participatory approach that makes the assessment subject to human bias. Only GLADA makes use of remote sensing and climate data to produce maps of changes in NDVI and RUE over the past decades. As an ongoing assessment program sponsored by the Food and Agriculture Organization, GLADA aims to quantify degradation events during 1981–2003 with the use of NDVI. The first quantitative GLADA used yearly averaged linear trends of NDVI that was translated to NPP as a proxy measure. GLADA seems to pick up land degradation in moist areas better than in arid areas (Vogt et al., 2011), but the 8-km spatial resolution of the underlying data is too coarse to show any details that can be compared with ground truth.

4.3 DETECTION AND ASSESSMENT

The change in land degradation is detectable using several methods, such as mapping the areas that have been severely degraded using the aforementioned methods and then spatially comparing the time-series maps of land degradation in a post-classification session.

4.3.1 Post-classification Comparison

Post-classification comparison works only when the land has been so severely degraded that it has resulted in the disappearance of vegetation from the worst-affected areas. The change to the degraded land is detected from multiple land cover maps of different years. All of the maps are produced using the aforementioned per-pixel or object-oriented image classification methods. Initially, a number of clusters may be produced in an unsupervised classification and, subsequently, they are merged to form only two categories of sand (or degraded) and non-sand in a post-classification session. Apparently, the same standards must be applied when grouping all the identically classified maps. Otherwise, artificial changes may arise, which can be minimized by producing the maps from images of the same spatial resolution. Spatial comparison of these maps is possible only after they have been co-registered with each other. This method is easy to implement and understand, but whether the multiple classification results are directly comparable to each other requires exercise of caution. Since these results are highly dependent on the training samples, a degree of uncertainty is introduced during selecting training samples representative of different degrees of degradation. This method can yield objective results only if degraded land is treated as a single class without differentiating the degree of degradation. Understandably, the reliability drops if degradation is further differentiated into multiple levels of severity since not all levels are based on the same severity thresholds or criteria.

4.3.2 Complex Statistical Analysis

Instead of comparing degraded land directly, land degradation in arid and semi-arid regions has been assessed via the analysis of degradation indicators, such as remote sensing-generated time-series RUE data and residual trend analysis (RESTREND). Both can be used to quantitatively assess land degradation. However, the use of RUE is limited in that it lacks a baseline reference with which current degradation can be compared. This deficiency is overcome with RESTREND that examines the trend of pixel-wise NDVI residuals, which is the difference between the observed NDVI and the theoretical NDVI predicted from rainfall, or the NDVI residuals. For a pixel to be eligible for processing using RESTREND, it must meet three criteria:

i. The vegetation–precipitation relationship must be significant and positive (e.g., slope > 0). Ideally, the regression coefficient R^2 should be larger than 0.3 (alpha = 0.05) as recommended by Wessels et al. (2012);
ii. If the residuals exhibit a trend, it must be gradual and monotonic over the study period; and
iii. The vegetation–precipitation relationship must be consistent temporally (e.g., no structural changes within the vegetation under study).

In most cases, RESTREND is effective at estimating the extent of human-induced land degradation (Radda et al., 2021). However, standard RESTREND tests of criteria (i) and (iii) fail to identify a trend in pixels if the rate and direction of change vary with time in case of rapid degradation or intensive restoration. RESTREND is effective at estimating the extent of human-induced land degradation (Radda et al., 2021). RESTREND results are meaningful only when NDVI is strongly correlated with precipitation. If the correlation between them is broken down by severe or rapid degradation, this method is unlikely to identify the areas experiencing severe degradation.

This deficiency can be circumvented by examining the long-term trends from time-series data to detect where degradation causes the relationship between NDVI and precipitation to break down, known as the breakpoint. Breakpoints are commonly identified using the Breaks For Additive Seasonal and Trend (BFAST) method. It works by decomposing time-series NDVI data into seasonal, trend, and remaining components so as to detect the significant structural changes in land use. The performance of this method depends on the annual phonological signal of vegetation. Without it, it detects lots of pseudo-breakpoints, as in drylands where drought-tolerant vegetation is highly responsive to flood events. This limitation can be overcome using time-series segmentation (TSS)-RESTREND (Burrell et al., 2017). It reliably distinguishes the effects of climate and degradation process in dryland by predicting NDVI for every pixel via regression analysis. The intercept of the linear regression equation represents the theoretical NDVI in light of no rainfall. The predicted NDVI highlights the climatic effects on NDVI. The trend of the residual NDVI is determined via regression, with a negative slope indicating degradation, and a positive slope showing improved vegetation conditions not related to precipitation.

This analytical method involves a number of steps, including (i) identification of breakpoints in the complete time series using BFAST, (ii) a Chow test for breakpoint significance in the

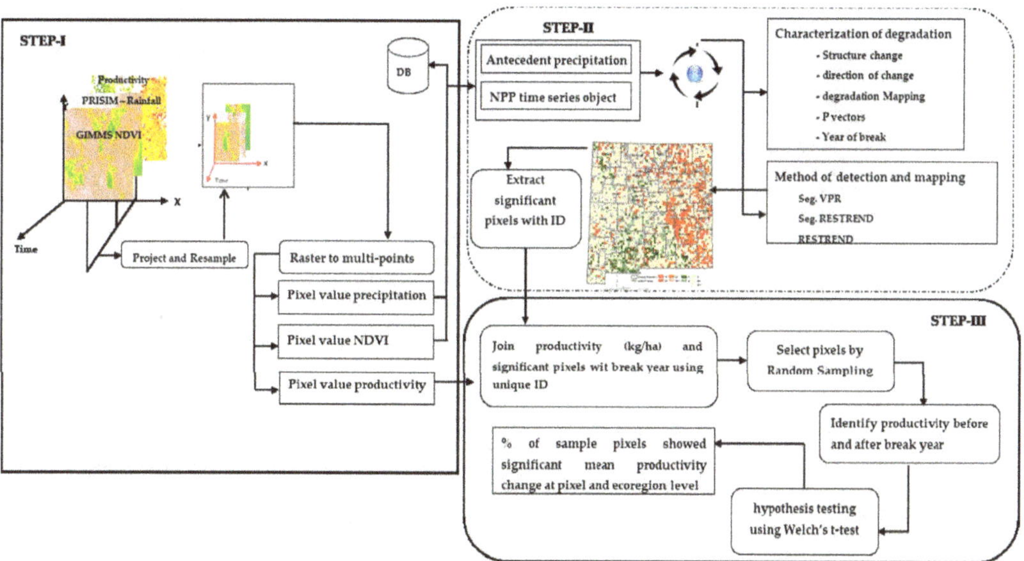

FIGURE 4.4 Procedure of implementing the TSS-RESTREND method to identify the trend of land degradation. (Source: Gedefaw et al., 2021.)

residuals, and (iii) test of the impact of significant breakpoints on the vegetation–precipitation relationship (Figure 4.4). Included in step (i) are data preparation, such as the acquisition of NPP data, precipitation, and productivity data, data reprojection, resampling, and extraction of pixel-level values. Included in step (ii) (TSS-RESTREND) are residual fitting, determination of p-vector, detection of the trend of productivity change (i.e., decline, rise, non-significance change, and indeterminant), and ecosystem structure change. Step (iii) is the evaluation that includes stratified random sampling for validation, and Welch's t-test on randomly selected pixels to test the significant change in productivity before and after the break years. If a breakpoint does exert a significant influence on the vegetation–precipitation relationship, this relationship is recalculated dependently on either side of the breakpoint after the $NDVI_{max}$ has been divided so as to detect the existence of any structural changes. Owing to these processing steps, TSS-RESTREND is able to detect many more degraded pixels than the standard RESTREND (Burrell et al., 2017).

4.3.3 Assessment of Degradation Cause, Severity, and Risk

Land degradation can be caused by various factors, both natural and anthropogenic. Natural factors are related to climate change, such as environmental desiccation. A warmer temperature regime accelerates evapotranspiration and is conducive to frequent droughts. Human activities are related to exploitive land uses, such as improper land reclamation for agriculture, overgrazing of grassland/rangeland, improper irrigation of farmland in arid and semi-arid regions, and excessive fuelwood collection in shrubland. After land degradation has been detected, it is possible to further differentiate it into human-induced and naturally occurring. This differentiation requires sensitive analytical procedures instead of residual trend analysis. A very simple method of qualitative assessment is to examine the locations of degraded land to see whether they are related to natural and environmental factors such as rainfall and soils. If the correlation is loose, it is concluded that degradation is attributed mostly to human factors, such as improper land use practices (Prince et al., 2009). This method is simple and fast, but crude and imprecise. More precise and definitive attribution of degradation causes requires more sophisticated and stringent methods of analysis, such as the aforementioned RESTREND.

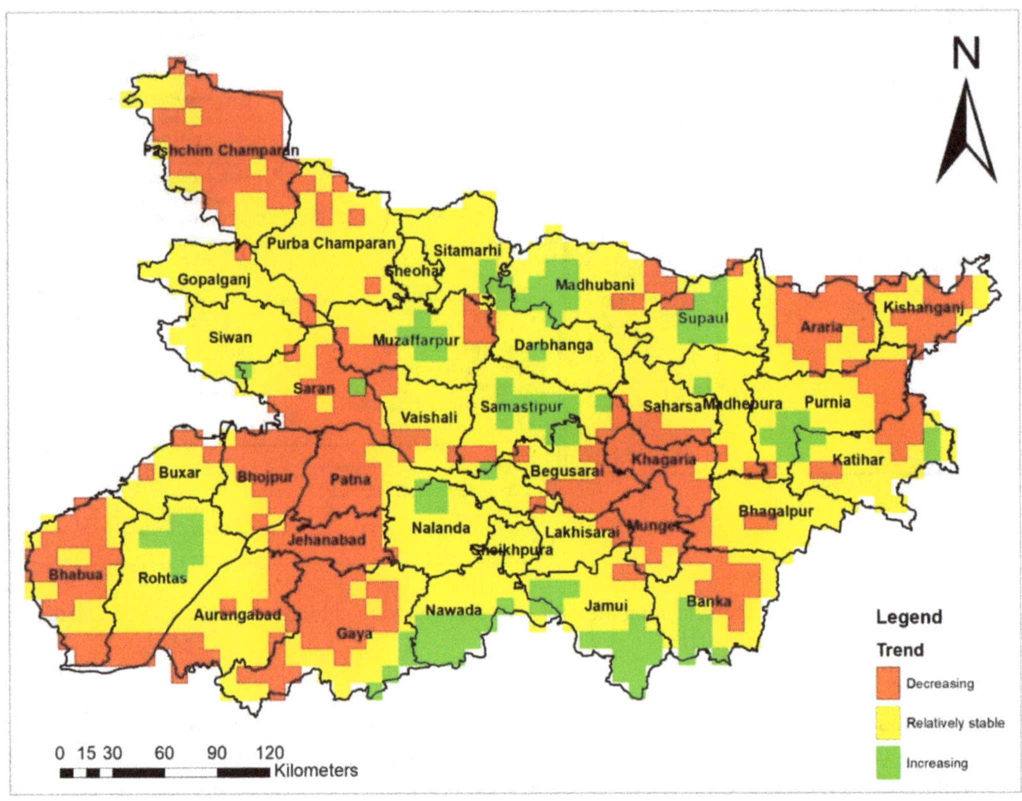

FIGURE 4.5 Spatial distribution of NDVI residual (RESTREND) in Bihar, India. The decreasing trend indicates land degradation from 1995 to 2011 as a consequence of human activities. (Source: Radda et al., 2021. Used with permission.)

If the environment has not changed, a decreasing trend of NDVI indicates land degradation caused by human activities (Figure 4.5).

The severity and spatial extent of rangeland degradation are judged from multiple criteria that include physical, chemical, and water table information (El-Baroudy, 2011). Severity is commonly expressed in a few discrete categories. More precise grading of severity requires the use of more inputs or consideration of more criteria in addition to the usual vegetation parameters, such as grazing level and cultivation intensity (Table 4.2). The joint consideration of these inputs, together with lithology, bioclimatic data, and field measurements, enables degraded rangeland in arid regions to be mapped and monitored at high accuracy (Mahyou et al., 2016). These criteria must be adapted for an area with a different geographic setting or land use practice from those of the original area of study. Some of the criteria may have to be removed (e.g., overgrazing) for agricultural land but new criteria of assessment may be added. No matter how many criteria are considered, realistic and objective assessment of degradation severity or hazard is not possible if the required spatial data cannot be generated remotely (e.g., soil compaction). For instance, if soil salinity is quantified from satellite images remotely (see Section 5.4), the accuracy of retrieval may be insufficiently high to warrant a credible grading. A criterion can be considered only if its spatial data can be obtained at a reasonable accuracy level.

Degradation severity is assessed using a comprehensive index or via trend analysis of the considered factors. Both methods lack localized benchmarks, causing the assessment outcome to be subjective. An objective and reliable means of grading degradation severity is to subtract the potential productivity from the actual productivity. The former is estimated using the local NPP scaling

TABLE 4.2
Criteria for Assessing Rangeland Degradation Severity

Criteria	Slight	Moderate	Severe	Very Severe
Steppe composition	Wooded steppes	Alfa grass steppes 2	Mixed steppes	Sand dune associated with Thymelaea microphylla, Lygeum spartum
	Alfa grass steppes	Chamaephytic steppes	Desert steppes	Aristida pungens, Stipa tenacissima, Artemisia herba alba, bare soil and desert rock
Cover (%)	>20	11–20	6–10	<5
Biomass productivity (ton·ha^{-1}·year^{-1})	>1.5	1–1.5	0.3–1	<0.3
Level of cultivation	Low Cultivated areas not present or negligible	Slightly cultivated Patches of cultivated areas apparent and scattered on	Widespread cultivation Rangeland appears covered by areas already cultivated	Widespread cultivation Rangeland appears covered by areas already cultivated
Overgrazing	Low Appearance of very light grazing, key species show no evidence of grazing or negligible use	Medium Forage species completely utilized. Half the available forage of key species has been utilized	High Entirely covered as uniformly as natural features and facilities will allow. Most desirable forage plants are sparse or missing	Very high Completely utilized. Low value species dominate and have been utilized. Key species appear to have been completely utilized. Dominated by species with greater adaptation to disturbance pressure

Source: Mahyou et al. (2016). Used with permission.

(LNS) method (Eq. 4.4) in homogeneous land capability classes, and the actual productivity is modeled using remote sensing data (e.g., NPP is calculated from the integral NDVI or ΣNDVI of the growing season) (Prince et al., 2009). The difference represents the impact of degradation or the yield reduced by degradation.

$$LNS = \left| NPP_{act} - NPP_{ref} \right| \tag{4.4}$$

$$LNS(\%) = \frac{LNS}{NPP_{ref} - NPP_{min}} \tag{4.5}$$

One way of determining the reference NPP or NPP_{ref} is to make use of the percentile of the frequency distribution of NPP (e.g., 85th). Actual NPP (NPP_{act}) is obtainable directly by means of remote sensing, such as the MODIS NPP data product (MOD173A). This method is able to show the location and severity (e.g., the value of the difference) of degradation over a region. The LNS method may be further improved by dividing land capability classes (LCC) based on NDVI or environmental variables (e.g., climate, soil, and vegetation) using remote sensing data, and the reference NPP is calculated for each LCC to represent the productivity under the optimal environmental conditions (Li et al., 2020). The LNS values (%), defined as the ratio of the current NPP to the reference NPP, are used to assess the extent, severity, and trend of land degradation. For instance, a

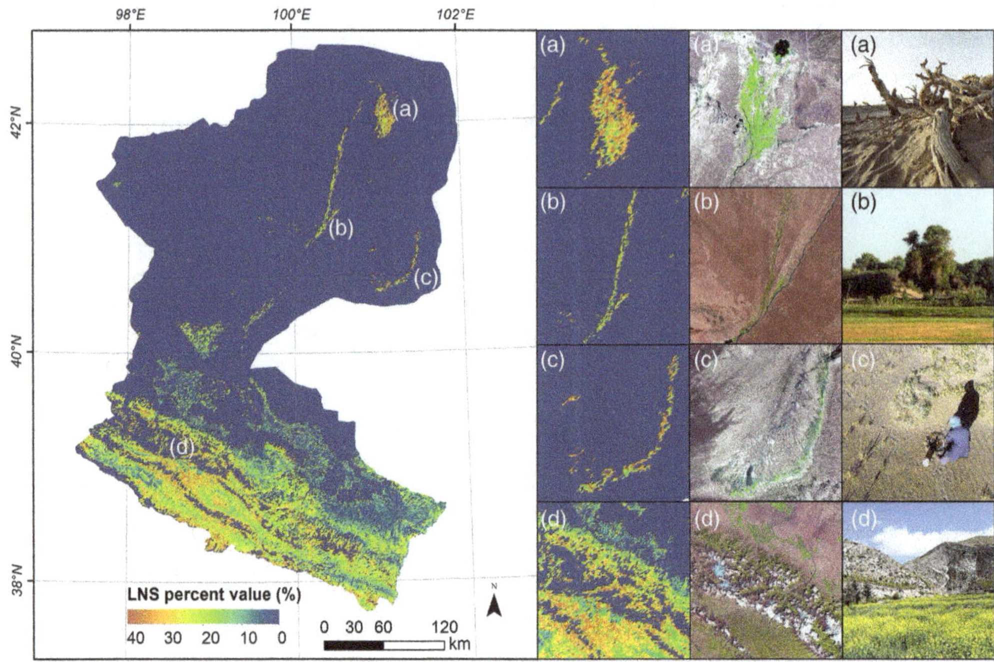

FIGURE 4.6 Severity of typical land degradation measured by average local NPP scaling (LNS) value (%) in the Heihe River Basin of East China during 2000–2014. (Source: Li et al., 2020. Used with permission.)

value of 0–10% represents slight degradation, 10–30% moderate degradation, and over 30% severe degradation (Figure 4.6).

Satellite-based assessment and quantification of degradation severity yield spatially consistent and quantitative results that are readily repeatable. But if the assessment is based solely on NPP, it measures actual rather than potential changes. This method neglects degradation in soil quality. NPP captures the degradation taking place after the acquisition of the earliest data, rather than the complete status of land (e.g., it cannot detect degradation that may have occurred prior to the acquisition of the remotely sensed data used). Besides, NPP can be compounded by other biophysical changes, such as fertilizer use at a variable rate.

Similar to land degradation severity, land degradation hazard has also been assessed from various indicators. Since it can occur in various forms, the indicators used for the assessment also vary with the type of degradation. As shown in Table 4.3, land salinization hazard is assessed from an

TABLE 4.3
Criteria for Grading the Level of Degradation Hazard Based on Four Indicators

			Hazard Level			
Critical/Hazard Type	**Indicator**	**Unit**	**Low**	**Moderate**	**High**	**Very High**
Salinization	EC	dS·m^{-1}	≤S	4-8	8–16	>16
Alkalinization	ESP	%	≤SP	10–15	15–30	>30
Soil Compaction	Bulk density	g·cm^{-3}	≤mlk	1.2–1.4	1.4–1.6	>1.6
Waterlogging	Water table level	cm	≥mte	150–100	100–50	<50

Source: El-Baroudy (2011). Used with permission.
Note: EC – electrical conductivity; ESP – exchangeable sodium percentage

indicator different from that of waterlogging. For the four types of hazards, only four indicators are considered. The value of each indicator is graded into four ranges, corresponding to four levels of hazard. Since these indicators are not so easily quantifiable from remotely sensed data, they will not be explored further except that the grading of land degradation status and hazard requires manipulation of the mapped land degradation with data from other sources, usually carried out in a GIS.

4.4 LAND SALINIZATION

Land salinization refers to the accumulation of salt in the soil to an excessively high level to adversely impact normal vegetation growth and health. Soil salt originates from two sources: weathered parent materials and primary minerals, and irrigation water of an inferior quality. The former is known as primary salinity that occurs naturally. The latter is anthropogenic and prevalent in irrigated farmland in arid and semi-arid regions around the world. The use of water of an inferior quality for irrigation, together with poorly drained soil, eventually elevates soil salinity and even results in waterlogging. An elevated concentration level of soluble salts (e.g., sodium, carbonates, and bicarbonates) in the surface soil leads to land salinization mostly in arid and semi-arid areas, where salt gradually accumulates in the soil as a result of strong evapotranspiration. The excessive level of salt concentration in the soil blocks the pores of the soil, hinders aeration, and interferes with plant metabolism. Salinization adversely affects crop growth by increasing the osmotic potential of soil solution and preventing plants from absorbing moisture and nutrients, exerting a hazardous impact on crop yield (Singh, 2015). Thus, land salinization is also considered an environmental and natural hazard exacerbated by anthropogenic activities.

Soil salt is commonly measured against electrical conductivity (EC). This proxy is reliable and much easier to measure than salt ions. It is commonly graded into five severity levels (Table 4.4).

TABLE 4.4

Five Levels of Soil Salinity and Their Corresponding EC Range Based on Richards (1954) "soil Salinity"

Salinity Class	EC Range (dS m^{-1})	Description
Extremely saline	>16	Only highly salt-tolerant plants survive and the community will be dominated by two or three species. Extensive bare saline areas occur with salt stains or crystals evident (on some soils a dark organic stain may be visible). Topsoil may be flowery or puffy with some plants surviving on small pedestals
Highly saline	8–16	Salt-tolerant species like Sea Barley-grass may dominate large areas and only salt-tolerant plants remain unaffected. Large, bare saline areas may occur showing salt stains or crystals. At the upper end of the range, halophytic plants may dominate the plant community
Moderately saline	4–8	Salt-tolerant species begin to dominate the vegetation community and all salt-sensitive plants are markedly affected by soil salinity levels. Small bare areas up to 1 m^2 may be present and salt stain/crystals may sometimes be visible on bare soil at the upper end of the range
Slightly saline	2–4	Salt-tolerant species such as Sea Barley-grass are often abundant. Salt-sensitive plants in general show a reduction in number. There are no bare saline patches and no salt stain/crystals are evident on the bare ground
Non-saline	<2	No vegetation appears affected by salinity and a wide range of plants present

Source: Taghadosi et al. (2019a).

They are non-saline (0–2 dS·m^{-1}), slightly saline (2–4 dS·m^{-1}), moderately saline (4–8 dS·m^{-1}), highly saline (8–16 dS·m^{-1}), and extremely saline. The last class is broken down further into three sub-levels: 16–32, 32–64, and > 64 dS·m^{-1} by Gorji et al. (2020). At the extremely high level, only highly tolerant plants of salt can survive. For this reason, some mapping is carried out at the categorical level using these EC range thresholds. However, EC is unable to distinguish specific types of salt ions in the soil.

4.4.1 Multispectral Sensing of Salt

The presence of salt in the soil changes its texture, moisture content, organic matter, and surface roughness, in turn, all of which affect the spectral behavior of the soil. There is a positive correlation between bare soil salt content and the in situ measured reflectance over the spectral range of 0.50–2.50 μm (Figure 4.7). At the shortest wavelengths, the reflectance is rather low but it rises quickly until 0.60 μm, beyond which it increases slowly with wavelength until 2.20 μm where it declines sharply. The higher the salt concentration, the higher the reflectance. This general pattern of the reflectance curve remains largely unchanged at all concentration levels except the extremely low level. In spite of the existence of spectral disparity at different levels of salt concentration, it remains uncertain whether the exact level of salt content can be quantified from space-borne images accurately for four reasons: (i) the distance of sensing is much further from space than from the ground. The signal from the salt can be degraded during its transmission from the ground to the sensor, during which the atmosphere inevitably exerts an influence on the received signal; (ii) the sensor may not be sensitive enough to discern the subtle variation in the reflected energy; (iii) the recorded image may not have the proper spectral and spatial resolutions to enable the subtle reflectance disparity to be retrieved at a sufficiently high accuracy; and (iv) the ground can be covered by vegetation or surrounded by water, both of which interfere with the spectral signal of soil salt.

So far soil salinity has been quantified and monitored from multispectral satellite images, largely because of their low cost (mostly free these days) (Table 4.5). Their improved spatial and spectral resolutions over the recent decade have renewed the opportunities to map salinity of an extremely low level. Remote sensing images that can potentially retrieve soil salt have evolved with the launch of new satellites. In the past, soil salt was estimated from Landsat TM data that have a coarse spatial

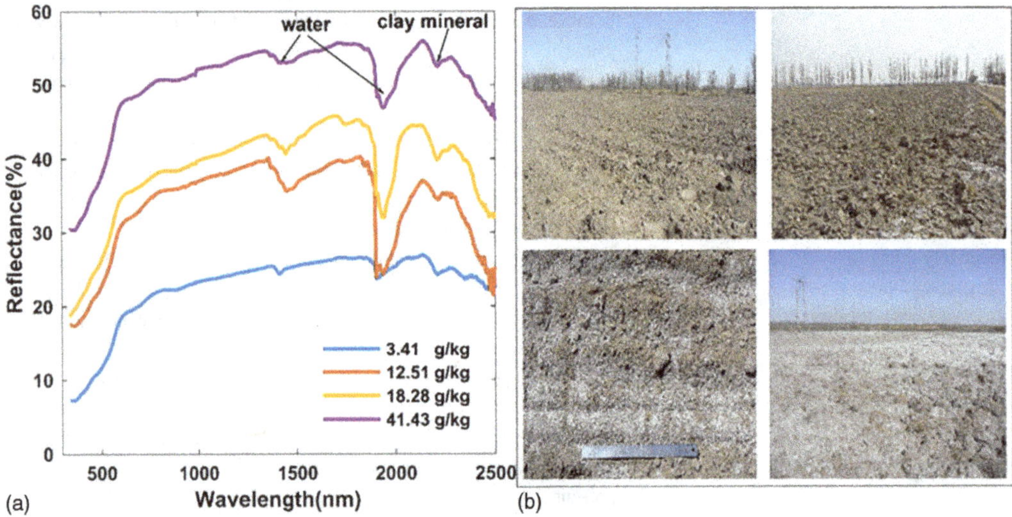

FIGURE 4.7 In situ (a) measured spectral response of bare soil at four levels of salt content, (b) the soil appearance on the ground. (Source: Wang et al., 2019b.)

TABLE 4.5
Comparison of Spectral Bands and Spatial Resolution (pixel size) among Sentinel-2 Multispectral Instrument (MSI), Landsat-8 OLI, SPOT-6/7, and HJ-1 Images that can Be Used to Map Soil Salinity.

	Sentinel-2				
Bands	**Wavelength (nm)**	**Pixel Size (m)**	**Landsat-8 OLI**	**SPOT-6/7**	**HJ-1**
Coastal-aerosol	430–450	60	433–453		
Blue	460–520	10	450–515	450–520	400–520
Green	540–580	10	525–600	530–590	520–600
Red	650–680	10	630–680	625–695	630–690
Red-edge 1	700–710	20			
Red-edge 2	730–750	20			
Red-edge 3	770–790	20			
NIR-1	780–900	10	845–885	760–890	760–900
NIR-2	850–870	20			
Water–vapor	930–950	60			
Cirrus	1360–1390	60	1360–1380		
SWIR-1	1560–1650	20	1560–1660		
SWIR-2	2100–2280	20	2100–2300		

Note: Spatial resolution (m): OLI = 30; SPOT-6 = 6 (multispectral bands); HJ-1 = 30

(30 m) and spectral resolution (only seven bands). They have been superseded by Sentinel-2 data (Table 4.5). Not only has the number of spectral bands increased but also the bandwidth narrowed and the spatial resolution refined, all of which make Sentinel-2 data potentially the ideal choice in mapping soil salt content (Figure 4.8). The spectral response of low EC levels does not vary noticeably with Sentinel-2 spectral bands except for a slight decrease in band 12. In general, the longer

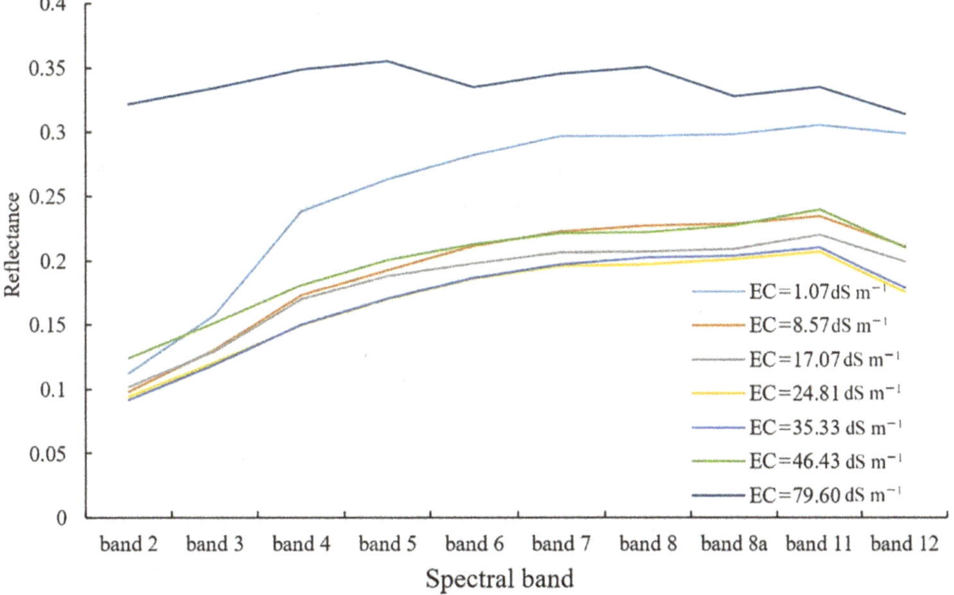

FIGURE 4.8 Spectral behavior of soil with seven levels of salinity in ten Sentinel-2 spectral bands. (Source: Wang et al., 2021.)

TABLE 4.6

Correlation Coefficient between EC Values and Landsat TM Spectral Bands, Indices, and Environmental Variables

Variables	Elevation	Slope	Band1	Band2	Band3	Band4	Band5	Band7	NDVI	NDSI	SAVI
r	*−0.1805*	−0.1378	**0.2498**	**0.2902**	**0.2562**	**0.3837**	0.1283	0.0581	**−0.1445**	0.1445	**−0.1328**

Source: Elnaggar and Noller, 2010.
Note: *Italic*: significant at p-value <0.05; **boldface**: significant at p-value <0.001

the wavelength, the higher the reflectance except at the extremely high EC level when it fluctuates mildly with wavebands.

Different wavebands have different utilities in predicting soil EC levels (Table 4.6). Of the seven Landsat TM bands, only band 4 achieves the highest correlation of 0.384 while bands 5 and 7 are much less effective. NDVI derived from 20-m resolution SPOT multispectral imagery proves to be a poor predictor of soil salinity. Although the salinity index (SI) from the image is closely correlated with field-measured salinity, it significantly underestimates salinity in areas of high levels of salt exposure (Abdelkader et al., 2006).

The above findings have been confirmed by Wang et al. (2018). All individual spectral bands over the wavelength range of blue to SWIR (0.40–2.30 μm) have a limited capability in estimating soil salt content, irrespective of the sensor used to acquire the images (Table 4.7). The blue and green bands are marginally better than the other bands. The highest R^2 value is just 0.24, while the RMSE error is stubbornly high at over 14.18 g·kg^{-1}. Thus, single-band models are unable to predict soil salt adequately. More spectral bands are needed. For instance, the use of two bands from Landsat OLI and Sentinel-2 MSI (multispectral instrument) improved the estimation accuracy (R^2) to over 0.5 in estimating EC (Eqs. 4.6 and 4.7). MSI bands 2 and 4 achieved higher accuracy than their OLI counterparts (Davis et al., 2019) because their 10-m spatial resolution is three-time finer than 30 m of OLI bands. Overall, MSI has superior spatial and temporal resolutions to OLI. The adjusted R^2 (0.04–0.54) and RMSE of 1.15 for OLI are highly similar to 0.05–0.67 and 1.17 for MSI, suggesting that these two sensors have a similar salinity modeling capacity. Thus, given the same spectral

TABLE 4.7

Capability of Individual HJ-B and Landsat OLI Bands in Predicting Soil Salt Content in the Ebinur Lake Wetland of Northwest China

Imagery	Band	Wavelength	Prediction Model	R^2	RMSE	RPD
	Blue	0.400–0.520	$177.6x - 4.04$	0.17	14.77	1.08
HJ-B CCD	Green	0.520–0.600	$266.9x - 12.78$	0.22	14.41	1.11
	Red	0.630–0.690	$176.5x - 7.3$	0.13	14.94	1.07
	NIR	0.760–0.900	$119.2x - 7.6$	0.13	15.13	1.06
	Blue	0.450–0.515	$120.3x - 8.19$	0.24	14.18	1.13
	Green	0.525–0.600	$105.6x - 9.14$	0.23	14.27	1.12
	Red	0.630–0.680	$85.4x - 5.9$	0.18	14.76	1.08
Landsat OLI	NIR	0.845–0.885	$73.7x - 6.35$	0.16	14.98	1.07
	SWIR-1	1.560–1.660	$65.45x - 4.51$	0.15	15.05	1.06
	SWIR-2	2.100–2.300	$74.94x - 5.2$	0.15	15.01	1.06

Source: Modified from Wang et al. (2018). Used with permission.

resolution, a finer spatial resolution is conducive to achieving a higher prediction accuracy probably because it can capture the spatial variability of soil salt better than coarse-resolution bands.

$$EC_{OLI} = 2.0080 + 0.0698B_2 - 0.0156B_4 \left(adjusted\ R^2 = 0.51\right) \quad (4.6)$$

$$EC_{MSI} = 2.0065 + 0.0666B_2 - 0.0156B_4 \left(adjusted\ R^2 = 0.67\right) \quad (4.7)$$

In addition to individual spectral bands, a large number of indices have been developed and used to quantify soil salinity (Table 4.8), derived from either two or three bands. The performances of three two-band and seven TBI in mapping soil salinity have been comparatively assessed by Wang et al. (2019a), but their findings are not conclusive. In general, TBI are marginally better than their two-band counterparts, but some indices such as NDI produce a higher r than a few TBI (Table 4.9).

As for the best spectral range of sensing, the VNIR and SWIR bands of Sentinel-2 MSI images have a limited capability in discriminating soil salinity classes because the spectral signal of salt is seriously confused with the soil optical properties (i.e., color and brightness) in the VNIR bands (Bannari et al., 2018). The SWIR bands are more closely correlated (R^2 of 0.50 and 0.64, for SWIR-1 and SWIR-2, respectively) with EC than the VNIR bands ($R^2 < 9\%$), including the red-edge and NIR bands. Thus, Sentinel-2 MSI SWIR-1 and SWIR-2 bands are the most promising in potentially detecting and discriminating saline soils of various concentration levels competently in an arid landscape. For low and moderate salinity levels, the SWIR bands are more sensitive than other bandwidths (Bannari et al., 2008). It must be noted that these findings are obtained from individual bands. In reality, it is a common practice to derive salinity indices from multiple bands that may include both VNIR and SWIR bands. So the performance of the same band may vary if combined with other bands.

TABLE 4.8

Vegetation and Soil Salinity Indices that Have Been Used to Map Soil Salinity

Index	Formula	Sources
NDVI	$(NIR - R)/(NIR + R)$	Rouse et al. (1974)
ENDVI	$(NIR + SWIR_2 - R)/(NIR + SWIR_2 + R)$	Weng et al. (2008)
EVI	$2.5(NIR - R)/(NIR + 6R - 7.5B + 1)$	Liu and Huete (1995)
SAVI	$(NIR - R)(1 + L)/(NIR + R + L)$	Huete (1988)
MSAVI	$NIR + 0.5 - 0.5\sqrt{(2NIR + 1)^2 - 8(NIR - R)}$	Peng et al. (2018)
DVI	$NIR - R$	Clevers (1988)
Ratio VI	NIR/R	Jordan (1969)
Normalized difference salinity index	$(R - NIR)/(R + NIR)$	Khan et al. (2001)
Brightness index (BI)	$\sqrt{R^2 + NIR^2}$	Khan et al. (2001)
Salinity index (SI)	$\sqrt{B \times R}$	Khan et al. (2001)
Salinity index (SI2)	$\sqrt{G^2 + R^2 + NIR^2}$	Yahiaoui et al. (2015)
Salinity index (SI – 1)	ALI9/ALI10	Bannari et al. (2008)
Salinity index (SI – 2)	(ALI6 – ALI9)/(ALI6 + ALI9)	Bannari et al. (2008)
Salinity index (SI – 3)	(ALI9 – ALI10)/(ALI9 + ALI10)	Bannari et al. (2008)
Salinity index (S1)	B/R	Abbas and Khan (2007)
Salinity index (S2)	$(B - R)/(B + R)$	Abbas and Khan (2007)
Salinity index (S3)	$G \times R/B$	Abbas and Khan (2007)
Salinity index (S5)	$B \times R/G$	Abbas and Khan (2007)

Source: Allbed and Kumar (2013).

TABLE 4.9

Correlation *(r)* of Two-Band and Three-Band Indices with Soil Salinity

No. of Bands	Index Abbreviation	Optimal Band Combinations	*r*
	RI	B_4/B_{12}	0.41
Two-band indices	DI[a]	$B_3 - B_7$	–0.16
	NDI[b]	$(B_{12} - B_7)/ (B_{12} + B_7)$	0.52
	TBI_1	$B_{12}/(B_{8a} \times B_{8a})$	–0.43
Three-band indices	TBI_2	$B_7/(B_7 + B_{12})$	–0.51
	TBI_3	$(B_6 - B_{12})/(B_{12} + B_{11})$	0.50
	TBI_4	$(B_{12} - B_3)/(B_3 - B_{11})$	0.54
	TBI_5	$(B_{12} + B_{11})/B_3$	0.53
	TBI_6	$(B_{11} - B_3)/(B_{11} - 2B_3 + B_{12})$	0.44
	TBI_7	$B_{11} - 2B_{12} + B_5$	–0.49

[a] DI = difference index

[b] NDI: normalized DI

Source: Wang et al. (2019a). Used with permission.

With the assistance of the constructed models in Eqs. 4.6 and 4.7, the original image pixel value of multispectral bands and their ratio can be transformed into maps of soil salt concentration (Figure 4.9). These maps may be further classified into a number of groups to show the spatial distribution of soil salt. The spatial pattern of EC is mostly similar between OLI (Figure 4.9a) and Sentinel-2 MSI (Figure 4.9b), with minor variations in high EC level classes.

4.4.2 Hyperspectral Sensing

In assessing the feasibility of remotely sensing soil salinity, field-measured spectra are used as the ground truth, from which soil salinity parameters are derived. The level of soil salinity

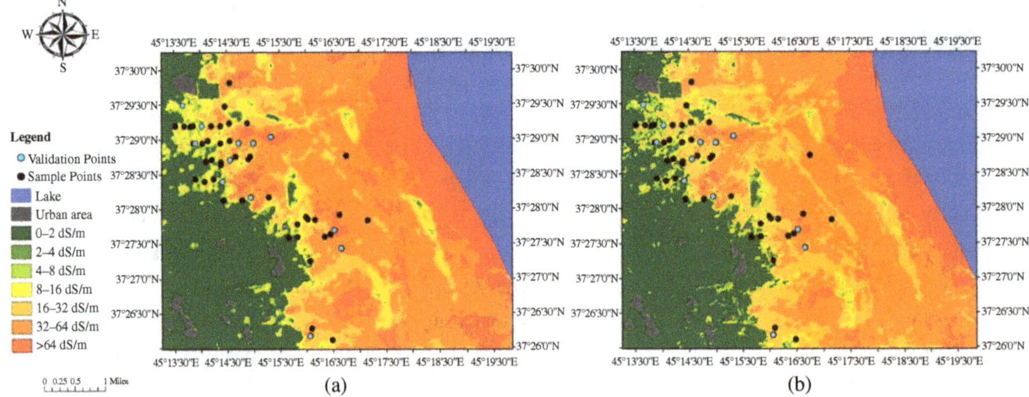

FIGURE 4.9 Distribution of soil EC in the Urmia Lake basin, Northwestern Iran, mapped at seven classes from two data sources using multiple regression. (a) OLI; (b) Sentinel-2B MSI data. (Source: Gorji et al., 2020. Used with permission.)

is quantified from hyperspectral data based on the relationship between the in situ measured hyperspectral data and the measured parameters related to soil salinity indirectly. Compared to multispectral images, hyperspectral bands are better suited to the quantification of soil salinity because of their ability to discern the subtle spectral signal of soil salts. Hyperspectral bands are acquired using a high-resolution spectroradiometer that can have a 1.4-nm sampling interval from 350 to 1,000 nm or a 2-nm interval from 1,000 to 2,500 nm. Space-borne hyperspectral data are commonly acquired by the Hyperion sensor. Hyperspectral imagery, with its fine spatial and spectral resolutions, enables soil salinity to be mapped in greater detail than multispectral bands. It represents another alternative data source for mapping soil salinity. Those hyperspectral bands of Hyperion imagery with a wavelength in the regions of 1,487–1,527, 1,971–1,991, 2,032–2,092, and 2,163–2,355 nm are especially sensitive to soil salt content and should be used for its retrieval (Weng et al., 2008). In spite of the superiority of hyperspectral bands over multispectral bands, the highest accuracy of estimation based on a linear model involving only one spectral band is achieved at lower than R^2=0.4, an accuracy considered unacceptable (Table 4.10). Soil salt content is more accurately retrieved from the optimal hyperspectral bands of HJ-CCD images and Landsat OLI data with the assistance of various spectral indices, such as difference index (DI), RI, and NDI (Wang et al., 2018). Individually, each hyperspectral band (including its first and second derivatives) or the indexing of multiple bands still has a limited performance with the best being R_{1600}/R_{820}. Even so, its R^2 value is merely 0.45 (Table 4.10).

A comparison of Tables 4.9 and 4.10 reveals that neither multispectral nor hyperspectral bands produced an acceptable accuracy, even though hyperspectral bands yielded marginally better results than multispectral bands. These results suggest that individual bands alone have a limited capacity in mapping the concentration of soil salt quantitatively. More sophisticated mapping methods are required to improve the estimation accuracy by making use of multiple bands and their derivatives, a topic to be covered in Section 4.4.5.

TABLE 4.10

Effectiveness of Optimal Hyperspectral Bands (Parameters) in Predicting Salt Content in the Ebinur Lake Wetland of Northwest China

Spectral Parameter (x)	Prediction Model	R^2	RMSE	RPD
R_{873}	$70.18x - 9.8$	0.34	13.18	1.21
R_{961}	$56.77x - 6.08$	0.22	14.39	1.11
R_{1300}	$50.83x - 5.49$	0.23	14.31	1.12
R_{1397}	$58.97x - 7.88$	0.36	12.54	1.27
R_{1940}	$60.15x - 4.91$	0.39	12.78	1.25
R'_{2098}	$21{,}864x + 10.97$	0.26	14.81	1.08
R''_{459}	$23{,}539x + 16.21$	0.25	14.08	1.13
$1/R_{615}$	$4.9x - 3.73$	0.31	14.40	1.11
$1/R_{2132}$	$4.81x - 2.98$	0.32	13.37	1.19
R_{1600}/R_{820}	$21.56x - 10.18$	0.45	10.85	1.47
$(R_{860} - R_{1240})/(R_{860} + R_{1240})$	$213x - 17.8$	0.30	14.56	1.10
$CR_{572} - CR_{2222}$	$13.63x - 12.04$	0.25	12.05	1.33
R_{900}/R_{970}	$11.85x - 11.74$	0.43	10.03	1.59

Source: Wang et al. (2018). Used with permission.

Note: ′ – first derivatives; ″ – second derivatives. RPD – residual predictive deviation.

4.4.3 Importance of Spatial vs Spectral Resolution

The best way of assessing the relative importance of spatial and spectral resolutions in quantifying soil salt content is to compare the results obtained from medium-resolution images with those of VHR images. The former includes Landsat-8 OLI and Sentinel-2 that are freely available to the public (Table 4.5). VHR images have not been widely used to study soil salinization except for IKONOS and QuickBird. IKONOS imagery has four multispectral bands (blue, 0.40–0.52 μm; green, 0.52–0.60 μm; red, 0.63–0.69 μm; and near-infrared, 0.76–0.90 μm) with a spatial resolution of 4 m (Table 1.5). Linear regression of EC against these bands and the *R*/NIR ratio yielded a maximal R^2 of 0.65 (Allbed et al., 2014). The use of the three visible bands makes little difference to the R^2 value, even though band 3 combined with SI achieves an R^2 value of 0.65. Unsupervised classification of PAN-sharpened QuickBird multispectral bands of blue (0.45–0.52 μm), green (0.52–0.60 μm), red (0.63–0.69 μm), and NIR (0.76–0.90 μm) of 0.6-m spatial resolution produced six covers, three of which represented salt-affected areas at three levels of salinity of low, medium, and high (Setia et al., 2013). They have a PCC of ≤0.051 with EC. The correlation hardly varies from the blue band (0.048) to the red band (0.051). Neither of the SI indices (NDSI, S1–S5) achieved a correlation higher than 0.052. These results suggest that a fine spatial resolution is not so critical to the successful and accurate estimation of soil salinity.

Whether spectral resolution is critical to accurate salinity retrieval is best evaluated using radar data because they are unexceptionally mono-spectral. The lack of spectral richness of SAR imagery is partially compensated for by its multiple VV and VH polarized images, from which texture can be derived. Radar data have been rarely used to study soil salinity due to the lack of a suitable theoretical model for simulating radar backscatter of soil based on its salt content. Thus, radar return intensity is relied on and is found to be related to in situ measured soil EC (Taghadosi et al., 2019b). Some of the 31 feature images derived from the VV and VH polarized Sentinel-1 SAR data are ill-suited because of the small range of pixel values. Others are noisy, leaving only 19 features useable. They are used to create 15 parameters in three categories of radar intensity (3), histogram-based first-order texture (6), and second-order GLCM texture (6), all of which are used as the explanatory variables to predict soil salt. Texture features are derived from histograms using first- and second-order statistics by taking into account the relative position of pixels with respect to each other in the image. With the assistance of SVM regression analysis, a maximum R^2 value of 0.98 is achieved, confirming that spectral information is not critical to the achievement of a very high accuracy from radar data. It is the physical manifestation of soil salt in soil texture (e.g., its influence on vegetation and soil structure) that is critical to the accurate quantification of soil EC levels.

4.4.4 Predictor Variables

The concentration level of soil salt has been quantitatively predicted from remotely sensed data using a large number of predictor variables to achieve the highest accuracy possible. They fall into a number of types, including individual spectral bands and their derivatives such as texture, indices from multiple bands, and even environmental variables. Apart from salinity indices (SI), various vegetation indices have also been used for the quantification, including NDVI, EVI, SAVI, and MSAVI, in addition to individual bands (Table 4.8). Salinity-specific indices include SI, normalized difference SI (NDSI), and soil salinity sodicity index (SSSI1 and SSSI2) calculated from visible light bands as follows:

$$SI1 = \frac{\rho_b \times \rho_r}{\rho_g} \tag{4.8}$$

$$SI2 = \frac{\rho_g + \rho_r}{2} \tag{4.9}$$

$$SI3 = \sqrt{\rho_b \times \rho_r} \tag{4.10}$$

$$SSI1 = ALI9 - ALI10 \tag{4.11}$$

$$SSI2 = \frac{ALI9 \times ALI10 - ALI10 \times ALI10}{ALI9} \tag{4.12}$$

where ALI9 and ALI10 = simulated Advanced Land Imaging (EO-1) bands 9 and 10. b = blue band, g = green band, r = red band.

SSS1 and SSS2 are designed to monitor low and moderate levels of soil salinity sodicity, respectively. Two other indices for studying salinity in vegetated areas are vegetation soil SI (VSSI) proposed by Seifi et al. (2020) and canopy response SI (CRSI) developed by Scudiero et al. (2016). They are calculated as follows:

$$VSSI = 2\rho_g - 5(\rho_{NIR} + \rho_r) \tag{4.13}$$

$$CRSI = \sqrt{\frac{NIR \cdot R - G \cdot B}{NIR \cdot R + G \cdot B}} \tag{4.14}$$

Two new indices, ARVI (atmospherically resistant vegetation index) and SARVI (soil-adjusted and atmospherically resistant vegetation index), have also been proposed to map soil salinity (Kaufman and Tanré, 1992). The most recent addition to the existing VIs is the generalized difference vegetation index (GDVI) developed by Wu et al. (2014) for salinity analysis. It is calculated as follows:

$$GDVI = \frac{\rho_{NIR}^n - \rho_r^n}{\rho_{NIR}^n + \rho_r^n} \tag{4.15}$$

Where ρ^n_{NIR} and ρ^n_r = reflectance of the NIR and red bands, respectively; n = the power number from 1 to n. When $n = 1$, GDVI = DNVI. When $n = 2$, GDVI is better correlated with leaf area index (LAI) in all biomes than other VIs (Wu et al., 2014). It is particularly sensitive to low vegetation biomass. But $n = 3$ or higher causes GDVI to be saturated and insensitive to dense vegetation and is not recommended to use.

In order to assess Sentinel-2 MSI utility, Wang et al. (2019a) made use of 36 indices from its spectral bands, some of the unique ones are red-edge bands (Table 4.11). Most of them are similar to their counterparts in Table 4.8 except that the red-edge band is used in the calculation. In this narrow wavelength range, the reflectance of vegetation rises sharply and abruptly, so it is more sensitive to vegetation conditions (e.g., under stress caused by an excessively high soil salt content) than the ordinary red band. Naturally, the red-edge-based salinity indices require hyperspectral bands to derive.

In spite of the large number of indices proposed to quantify salinity, their effectiveness or relative performance to other existing indices has not been comprehensively evaluated except that the NIR band and VSSI are reasonably correlated with $EC_{1:5}$ ($R^2 = 0.8$ and 0.7, respectively). They have an adequate performance and are better than other indices. Surprisingly, the NIR band alone outperforms other indices such as NDVI, SI4, and NDSI, all of which achieve a similarly low accuracy

TABLE 4.11

Newly Proposed Spectral Indices Based on Red-Edge for Mapping Soil Salinization

Index	Formula	Equation
Intensity index 1 red-edge	$(G + \text{Red-edge}^{a})/2$	$(B3 + B5^{a})/2$
Intensity index 2 red-edge	$(G + \text{Red-edge} + \text{NIR})/2$	$(B3 + B5 + B8a)/2$
Normalized difference salinity index red-edge	(Red-edge – NIR)/(Red-edge + NIR)	(B5 – B8a)/(B5 + B8a)
Salinity index I red-edge	B/Red-edge	B2/B5
Salinity index II red-edge	(B – Red-edge)/(B + Red-edge)	(B3 – B5)/(B3 + B5)
Salinity index III red-edge	$G \times$ Red-edge/B	B3 × B5/B2
Salinity index IV red-edge	$G \times$ Red-edge/G	B2 × B5/B3
Salinity index V red-edge	Red-edge × NIR/G	B5 × B8a/B3
Salinity index 1 red-edge	$\sqrt{B + Red\_edge}$	$\sqrt{B2 + B5}$
Salinity index 2 red-edge	$\sqrt{G \times Red\_edge}$	$\sqrt{B3 \times B5}$
Salinity index 3 red-edge	$\sqrt{G^2 + Red\_edge^2 + NIR^2}$	$\sqrt{B3^2 + B5^2 + B8a^2}$
Salinity index 4 red-edge	$\sqrt{G^2 + Red\_edge^2}$	$\sqrt{B3^2 + B5^2}$

Source: Simplified and modified from Wang et al., 2019a. Used with permission.

[a] The same equation can be repeated two more times using B6 and B7 that correspond to red edges 2 and 3, respectively, for all the remaining 11 indices

(Table 4.12). It is the most effective among the compared indices (Nguyen et al., 2020). All five salinity indices except SI4 are weakly correlated with salinity with an EC range of 0–9.2 dS·m^{-1} (Nguyen et al., 2020). In comparison, NDSI, NDVI, SAVI, and VSSI bear a closer correlation with salinity ($R^2 > 0.41$). However, the NIR band alone achieves an R^2 value of 0.80 in predicting EC in the exponential format, the highest among all the indices. These findings demonstrate that Landsat-8 OLI images have a promising potential for spatiotemporally monitoring the level of soil salinity in the topsoil layer without sophisticated computation. It is not the band but the estimation model that is critical to achieving high accuracy in estimating soil salts.

TABLE 4.12

Indices Commonly Used to Assess Soil Salinity and Their Effectiveness in Estimating Soil EC from Landsat OLI Bands

Indices	Formula	Source	R^2
Salinity index – SI1	$\sqrt{G^2 + R^2}$	Abdelkader et al. (2006)	N/A
Salinity index – SI2	$\sqrt{G \times R}$	Abdelkader et al. (2006)	N/A
Salinity index – SI3	$\sqrt{B \times R}$	Khan et al. (2001)	N/A
Salinity index – SI4	$R \times$ NIR/G	Abbas and Khan (2007)	0.48
SI5	B/R	Abbas and Khan (2007)	N/A
Normalized difference salinity index (NDSI)	(R – NIR)/(R + NIR)	Khan et al. (2001)	0.46
NDVI	(NIR – R)/(NIR + R)	Khan et al. (2001)	0.41
Soil-adjusted VI ($L = 0.5$)	$(1 + L)\times\text{NIR} - R/L + \text{NIR} + R$	Alhammadi and Glenn (2008)	0.62
Vegetation soil salinity index	2G – 5(R + NIR)	Dehni and Lounis (2012)	0.71
NIR band			0.80

Source: Modified from Nguyen et al. (2020).

TABLE 4.13

Integrated Pearson Correlation Coefficient between Salinity Measured with an Electromagnetic Instrument Called EM38 and Predictor Variables or Their Transformations

Predictor Variables	EC (20)	EM_V (59)	EM_H (59)
SAVI	−0.799	−0.824	−0.791
SARVI	−0.811	−0.820	−0.781
NDVI	−0.829	−0.837	−0.806
GDVI	−0.867	−0.858	−0.838
EVI	−0.761	−0.798	−0.759
NDII	−0.807	−0.779	−0.740
Ln(SAVI)	−0.869	−0.836	−0.820
Ln(SARVI)	−0.881	−0.786	−0.757
Ln(NDVI)	−0.891	−0.847	−0.831
Ln(GDVI)	−0.899	−0.837	−0.829
Ln(EVI)	−0.869	−0.825	−0.798
Ln(NDII[a])	−0.891	−0.783	−0.743
Exp(GDVI)	−0.840	−0.854	−0.828
Exp(NDVI)	−0.802	−0.818	−0.783
Exp(EVI)	−0.678	−0.733	−0.689
Exp(SAVI)	−0.777	−0.809	−0.773
LST	0.711	0.633	0.651
Ln(LST)	0.712	0.633	0.650
PC1	0.495	0.638	0.591
PC2	0.403	0.565	0.556
TCB[b]	0.470	0.594	0.592

Source: Wu et al., 2014. Used with permission.

[a] Normalized difference infrared index

[b] Tasseled Cap Brightness

For instance, Wu et al. (2014) obtained a much higher PCC in retrieving soil salinity measured with an electromagnetic instrument, using six vegetation indices, their logarithmic and exponential transformations, together with two principal components (PC) and LST (Table 4.13). A rather high negative correlation (>0.678) exists between salinity and all the predictor variables except LST, the correlation coefficient of which is positive. Only the principal components and brightness temperature have a coefficient below 0.5. These findings are different from those in Tables 4.9 and 4.12 because only 20 samples are used. Another cause of the higher correlation coefficient could be the level of salt concentration.

Instead of relying on the PCC, Wang et al. (2019a) evaluated the effectiveness of nine Sentinel-2 spectral bands in predicting soil EC based on a linear model using both validation and cross-validation R^2 values (Table 4.14). Individually, all the bands achieved a similar accuracy based on cross-validation R^2. The validation R^2 value is higher than cross-validation R^2 for most bands except SWIR-1 and SWIR-2. Such findings suggest that Sentinel-2 imagery is competent in predicting soil EC. The different conclusions between this study and others lie likely in the salinity level of sensing. A higher level can be quantified more accurately than a lower level.

TABLE 4.14

Effectiveness of Sentinel-2 Spectral Bands in Predicting Soil EC (dS·m^{-1}) Based on a Linear Model

Predictor Variable	Model						
	Slope	Intercept	R^2_{CV}	$RMSE_{CV}$	R^2_V	$RMSE_V$	RPD
Blue (B2)	143.94	−7.37	0.52	16.97	0.70	13.86	1.30
Green (B3)	144.41	−10.56	0.54	16.72	0.70	13.89	1.29
Red (B4)	130.659	−11.19	0.56	16.55	0.67	14.39	1.25
Red-edge 1 (B5)	130.070	−13.40	0.57	16.41	0.66	14.53	1.24
Red-edge 2 (B6)	133.06	−15.30	0.57	16.36	0.65	14.68	1.22
Red-edge 3 (B7)	136.37	−17.25	0.59	16.10	0.66	14.65	1.23
NIR narrow (B8a)	131.32	−17.06	0.60	16.02	0.65	14.81	1.21
SWIR-1 (B11)	108.99	−15.84	0.56	16.51	0.58	15.82	1.14
SWIR-2 (B12)	88.75	−7.15	0.52	17.07	0.51	16.74	1.07

Source: Wang et al. (2019a). Used with permission.
Note: CV = cross-validation; V = validation; RPD = residual predictive deviation

The effectiveness of several SIs derived from both Landsat-8 OLI and Sentinel-2A bands has been assessed and compared (Table 4.15). In general, an R^2 value above 0.7 is achieved with either of the three SIs individually, even though the accuracy is lower with the validation dataset. However, the accuracy is the highest from the joint consideration of all the three indices, even with the validation dataset. Furthermore, there is little difference in accuracy between Landsat-8 OLI and Sentinel-2A data. Nevertheless, the high accuracy of predicting EC from these indices must be treated with caution as it is expressed categorically at seven classes, not continuously nor quantitatively, as in the study of Nguyen et al. (2020). Predictably, the accuracy will drop significantly if EC is expressed as a continuous variable.

When judging which predictor is the best, care needs to be taken to check the nature of the model (linear vs exponential), not just the R^2 value. With the use of exponential models, Nguyen et al. (2020) achieved an R^2 value over 0.5 from most indices. The only exception is NDVI ($R^2 = 0.41$), as

TABLE 4.15

Comparison of Landsat-8 OLI and Sentinel-2A Data in Estimating Seven EC Classes Based on Three SI Indices

Satellite data	Index	R^2	RMSE	MAE	Validation R^2
Landsat-8 OLI	SI1	0.72	10.2	8.38	0.63
	SI2	0.73	10.0	8.16	0.65
	SI3	0.72	10.46	7.92	0.61
Multiple regression model	ES=−122.6 + 425.5×SI1 − 120.8×SI2 + 325×SI3				0.77
Sentinel-2A	SI1	0.72	10.25	8.41	0.61
	SI2	0.71	10.38	8.12	0.64
	SI3	0.74	9.95	8.11	0.63
Multiple regression model	EC = −63.71 − 423.38×SI1 + 129.17×SI2 + 750.14 × SI3				0.75

Source: Gorji et al. (2020). Used with permission.
Note: $n = 42$; mean = 14.7 (0.57–78); number of validation samples = 12; mean = 15.8 (1.05–65.12)

TABLE 4.16

Accuracy of Exponential Models Based on Singular Indices for Estimating Soil $EC_{1:5}$ from Landsat-8 OLI Data

No.	Regression Model	R^2	*p*-Value	Standard Error of EC (dS m^{-1})
1	$28.013e^{-13.39NIR}$	0.80	2.71E–08	1.29
2	$26.068e^{-14.73SI4}$	0.48	0.00761	2.44
3	$2.485e^{3.2449NDSI}$	0.46	0.00019	1.88
4	$5.671e^{-4.984NDVI}$	0.41	0.00034	2.15
5	$7.406e^{-9.16SAVI}$	0.62	3.78E–07	1.69
6	$72.86e^{2.8802VSSI}$	0.71	1.39E–05	1.76

Source: Nguyen et al. (2020).

shown in Table 4.16. The highest R^2 of 0.80 is much higher than that in Table 4.9, but these values are all based on non-linear models. It is inappropriate to compare the findings of all these studies for five reasons: (i) the soil salt is defined and mapped differently. In some studies, it is mapped into seven categories (Gorji et al., 2020). In others, the actual salt is expressed continuously in g·kg^{-1}; (ii) the number of observations used in establishing the models also varies widely from as low as 16 (Wu et al., 2014) to 42 (58). In general, a low number of observations can be fitted with a model more accurately; (iii) the salt content level is not the same. A higher content of salts is more spectrally distinctive and easier to discern on the satellite images and mapped at higher accuracy; (iv) some data have been corrected for the atmospheric effects, while others are not. In general, the correction for the atmospheric effects is conducive to achieving higher accuracy with the same data and from the same predictor variables, and (v) the estimation models may involve different predictor variables or have different complexity (e.g., linear vs exponential) (Table 4.16). Non-linear models tend to enable the distribution of scatter points to be fitted more accurately, but their meaning in retrieving salt content is ambiguous.

4.4.5 Analytical Methods

Soil salt content is mapped directly via visual interpretation and image analysis. Visual interpretation is good at detecting the nature, extent, and spatial distribution of salt-affected soils from false-color composites (RGB 432) of multispectral bands. It allows salt-affected soils to be easily and visually identified if combined with wetness indices (Elnaggar and Noller, 2010). However, visual interpretation is imprecise and cannot show the spatial variability of salinity quantitatively, which can be overcome using image classification. In image classification, all pixels in the input image are grouped into a discrete number of covers, some of which may be related to salinity level, such as slightly salinized, moderately salinized, and severely salinized. Image classification can be supervised or unsupervised, depending on whether training samples are involved in the classification.

Unsupervised classification of remote sensing imagery allows differentiation of severity levels of salt-affected soils, but these levels may not match those determined from in situ measured EC and sodium adsorption ratio, primarily because the expression of salinity is strongly influenced by paddock-level variations in crop type, growth, and prior land management (Setia et al., 2013). This method is simple and quick as it does not require in situ sampling or laboratory analysis. Requiring training samples, supervised classification is slower but the classification results tend to be more informative. Supervised classification (e.g., minimum distance to mean) of 20-m SPOT images produced 11 classes, including bare soils, 3 cultivated classes, and 2 uncultivated classes (Abdelkader et al., 2006). Although these land cover categories discriminate soil EC in highly saline areas, they do not lend sufficient accuracy. More critically, image classification at most generates only imprecise information on soil salt, and the results are subject to the training sample quality.

More accurate and objective results are still obtainable using regression analysis to construct prediction models such as those shown in Eqs. 4.6 and 4.7. The construction of such models requires in situ samples of soils and analysis of their salt content in the lab. The analysis of salts by type is a lengthy and costly process. Furthermore, it is unlikely that a particular type of salt can be detected from remotely sensed data to any creditable accuracy level. In practice, all types of salts are commonly lumped into one general group. Alternatively, the salinity level may be expressed as the easily measured EC regardless of the chemicals in the soil.

Before remotely sensed data are used to estimate soil salinity, they may undergo transformation that aims to produce non-redundant and informative datasets (features). Data transformation also facilitates subsequent model construction using machine learning. With hyperspectral data, there are over 100 spectral bands to choose from. Their dimension can be reduced via PCA. PCA converts data linearly into a low-dimensional sub-space by identifying the new orthogonal directions in which the raw data have maximal variance, to which the raw data are projected. Whatever variance that cannot be accounted for in this direction is then projected to another orthogonal direction (axis). In this way, the variance of the newly created components is always unique (e.g., it does not overlap with other components at all) and declines exponentially from the first component to the last. Therefore, most of the variance in the raw bands is retained in a handful of components effectively and efficiently.

4.4.5.1 Regression Modeling

Regression models play an essential role in attaching a quantitative value to a pixel to indicate its salt content level on the ground. Thus, it requires in situ sampling that must be synchronized with the recording of the satellite data if the salinity level changes quickly with time. To build a reliable relationship between pixel values or their transformation such as ratio or index, at least 30 samples should be collected. A large sample size is highly desirable and advantageous in that some of them can be used to validate the constructed model. In order for the samples to be linked to their spectral properties on the image, they must be precisely positioned using a GPS unit in the field. Naturally, the images have to be geo-referenced to the same coordinate system used to geo-reference the in situ samples.

The estimation model is established via regression analysis during which the overall residuals (differences between the observed and the estimated soil salt) among all the observations are minimized, while residuals between adjacent observations are independent of each other (e.g., absence of trend). In regression analysis, the dependent variable is predicted from the explanatory variables selected by the analyst as accurately as possible (e.g., the smallest residuals). Accurate models can be established between salinity indices and the EC classes (Table 4.17). In all regression analyses, the relationship between the dependent and independent variables is assumed to be linear. If not, data may be transformed logarithmically or exponentially to allow the construction of a linear model. So far simple (multiple) linear regression is the most common in expressing soil salt concentration from pixel values. As shown in Tables 4.9 and 4.10, linear models, while simple to construct and easy to understand, have limited predictive power. Non-linear models seem to be more accurate than linear ones (Table 4.17).

$$SSC = 4.343 + 29.469\rho_{426} + 185.743\rho_{1971} - 294.310\rho_{1991} + 97.417\rho_{2213} \tag{4.16}$$

If more than one predictor variable is considered for inclusion in a model, the classic linear regression must be extended to multivariate linear regression (MLR), a stepwise linear regression in which all the considered features are sequentially and incrementally (i.e., stepwise) added to the initial model to check whether their inclusion in the model improves its predictive power. If the inclusion of an extra variable in the model does not improve its prediction power or the improvement fails to meet the predefined criterion, it is excluded from the model. When multiple variables are included in the same model (Eq. 4.16), inevitably, they will be correlated with each other, an

TABLE 4.17

Formats of Linear and Multivariate Linear Estimation Models of Soil Salinity by Ground Cover Type and Their Accuracy

Type of Model	No. of obs	Model	Multiple R^2	Error Scope	*F*-Ratio
Integrated for vegetated	19	EC = −2.87 − 23.27Ln(NDVI)	0.874	±5.240	111.137
and non-vegetated areas	58	EM_V =535.403 − 487.905GDVI	0.729	±64.168	142.839
Integrated for non-vegetated area	58	EM_V = -2,725.05 + 10.018LST − 509.404NDII	0.650	±73.230	47.360
	58	EM_H = 1,627,956.14 + 1148.84LST − 345,815.62Ln(LST) − 245.198NDII	0.649	±58.240	30.869
Vegetated area specific	43	EM_V = 64.359 − 319.306Ln(GDVI)	0.372	±54.680	24.331
Non-vegetated area	16	EM_V = 1,502.43 − 1,166.35Exp(PC1)	0.444	±90.811	11.200
specific	16	EM_H = -223.22 − 1,043.11Ln(TCB)	0.557	±74.549	17.585

Source: Wu et al. (2014). Used with permission.
Note: H – horizontal; V – vertical

issue known as collinearity. Their overlapping effect must be discounted so as to understand how exactly an individual factor affects soil salt and to produce the most accurate prediction possible. This problem can be overcome with partial least-squares regression (PLSR). It overcomes the multicollinearity problem by merging PCA with canonical correlation analysis. PLSR is more suited to quantitatively estimate soil salt content than stepwise regression. MLR is almost equally capable as the PLSR model that achieves an RMSE of 0.74 (R^2 = 0.897) in producing a soil salt content map (Table 4.18). The salinity map produced from the MLR model shows the predicted value to be higher than the actual content, suggesting that PLSR is more suited than stepwise regression for quantitatively estimating soil salt content over a large area (Weng et al., 2008).

The predictor variables in a regression model are selected based on a number of criteria, such as PCC and importance in projection (VIP), or using gray relational analysis (GRA) and RF. The correlation coefficient simply indicates the association between two variables. There is no guarantee that the relationship is caustic. VIP calculates a score that accounts for the amount of variation of the dependent variable by an independent variable. GRA is a systematic analytical approach to understanding the complex interrelationships among multiple variables. The effect of each predictor

TABLE 4.18

Comparison of Stepwise Regression with PLSR in Estimating Soil Salt ($g{\cdot}kg^{-1}$) from Hyperion Data Using Calibration (*n* = 61) and Validation (*n* = 30) Datasets

		Calibration (*n* = 61)		Validation (*n* = 30)	
Model Type	Model/Accuracy Parameter	Model I	Model II	Model I	Model II
PLSR	No. spectral bands	158	43		
	No. of PCs	5	3		
	R^2	0.897	0.893	0.937	0.930
	RMSEc	0.740	0.753	0.542	
	RMSEcv	0.838	0.823	0.574	
MLR	R^2	0.892		0.951	
	RMSEc	0.786		0.470	

Source: Modified from Weng et al. (2008).

TABLE 4.19

Effects of Feature Selection Strategies on the Accuracy of Modeling Soil Salinity (dS·m^{-1}) Using PLSR from 67 Spectral Covariates (9 Sentinel-2 MSI Spectral Bands, 12 Soil Salinity Indices, 36 Red-Edge Indices, 3 Two-Band Indices, and 7 Three-Band Indices).

Variable Selection Strategy	LV_s	R^2_V	$RMSE_V$	R^2_{CV}	$RMSE_{CV}$	RPD
All	4	0.78	12.12	0.85	10.75	1.48
PCC	2	0.82	11.18	0.87	9.97	1.61
VIP	2	0.82	11.13	0.85	10.43	1.61
GRA	3	0.88	9.08	0.89	9.25	1.98
RF	4	0.92	7.58	0.93	8.35	2.36

Source: Wang et al. (2019a). Used with permission.

Abbreviations: LV – the optimal number of latent variables; CV – cross-validation; V – validation; RPD – residual predictive deviation (ratio of standard deviation).

Note: The number of parameters selected by all strategies is 20 except all (67)

variable is judged quantitatively by the gray correlation degree. The use of all variables in PLSR results in the least accurate model in terms of the validation R^2 (0.78) (Table 4.19). The accuracy remains the same with both PCC and VIP but is slightly higher with GRA. The highest accuracy is achieved by RF. It is the best in terms of the cross-validation accuracy indicators, so should be adopted for all PLSR analyses in light of a large number of input features to consider (e.g., >60).

4.4.5.2 SVM and SVR

As shown in Table 4.20, SVR with the RBF kernel is much more accurate than other variants in modeling soil salinity from Sentinel-2 multispectral data based on 22 predictor variables (R^2 = 87.42% and RMSE = 5.196) (Taghadosi et al., 2019a). Although the polynomial kernel with a degree of 1 achieves the same R^2, its RMSE is larger. Whether all features or a select few are used does not affect SVR with the RBF kernel, but the use of fewer but more selective variables reduces the prediction RMSE. The sequential selection of predictor variables is critical to improving MLR performance. In comparison, other implementations of SVR have a subdued and only marginal improvement. The accuracy does vary with the SVR kernel function. In general, the linear kernel is the worst performer. The polynomial kernel is better, but its performance also varies with the degree. The third degree is much worse than the first and second degrees, but the difference between the two is negligible.

TABLE 4.20

Comparison of Accuracy in Retrieving Soil EC from Sentinel-2 Data Using MLR and SVR in Which the Number of Predictor Variables Is Selected Sequentially with All 22 Variables

Model	Kernel Type	R^2	RMSE	No. of Features Selected	R^{2a}	RMSE[a]
MLR	N/A	0.83	4.31	14	0.63	6.38
	Linear	0.91	5.56	16	0.85	6.18
SVR	Polynomial (degree = 1)	0.94	4.76	15	0.87	5.96
	Polynomial (degree = 2)	0.75	5.59	5	0.64	7.02
	Polynomial (degree = 3)	0.72	5.62	9	0.64	6.86
	Radial basis function	0.86	5.05	13	0.87	5.19

Source: Taghadosi et al. (2019a).

[a] Based on all 22 predictor variables

4.4.5.3 ANN

Based on bootstrap optimization, the BP neural network can be used to automatically weight the predictive variables using a non-linear differential function with a small sample size. It is particularly suited to retrieve soil salt content from hyperspectral data in which the use of a small sample size is the norm (Wang et al., 2018). Table 4.21 compares the performance of the same indices derived from the derivatives of spectral bands from three sensors in estimating soil salt using the bootstrap neural network model. In particular, the difference index (R1377–R2134) based on the first derivative of the hyperspectral data yields an R^2 value of 0.83. This value improves to 0.95 with the bootstrap-BP neural network model developed using 1,000 bootstrap samples comprising optimal remote sensing indices (DI, RI, and NDI) and soil salt content. The optimal model of Landsat OLI images is the first derivative (Wang et al., 2018). The optimal model based on hyperspectral data yields an RMSE of 4.38 $g{\cdot}kg^{-1}$, and an RPD of 3.36. In terms of RPD, surprisingly, the optimal model is the first derivative based on Landsat OLI, not the hyperspectral data. It yields an R^2 value of 0.91, RMSE of 4.82 $g{\cdot}kg^{-1}$, and RPD of 3.32, but the differences in accuracy between hyperspectral and multispectral data are rather small. The soil salt is quantified much more accurately using ANN than the accuracy levels shown in Tables 4.9 and 4.10. These accuracy levels are also higher than those obtained with the raw bands, suggesting that ANN is crucial to accurate retrieval of soil salt from remote sensing images, irrespective of whether they are hyperspectral or multispectral.

A comparison of Table 4.21 with Table 4.14 reveals that individual bands, no matter whether they are multispectral or hyperspectral, are not correlated with soil salt ($g{\cdot}kg^{-1}$) in simple linear regression models. However, the use of RI, DI, and NDI derived from these bands achieves a much higher R^2 with the bootstrap-BP neural network regardless of the data source or their spatial resolution, even though raw hyperspectral data have a lower R^2 value of 0.76 than raw HJ-B CCD data (R^2 = 0.86). Landsat OLI has more spectral bands of the same spatial resolution than HJ-B CCD (30 m), the same three indices derived from Landsat OLI raw data achieve the lowest R^2 value of only 0.65. However, OLI data benefit more from the first derivative than from the second derivative that achieves an R^2 value of 0.84, lower than 0.91 of the first derivative (Table 4.21). It is the hyperspectral data that benefit the most from the continuum removal processing. It uplifts the model's R^2 to 0.95, the highest among all the inputs and data transformations.

TABLE 4.21
The Impact of Data Transformation on the Performance of the Bootstrap-BP Neural Network Model

Data Source	Data Transformation	Parameters	R^2	RMSE	RPD	Epoch
Hyperspectral	Previous studies	WI, MSI, R1940	0.53	9.03	1.77	9
	Raw bands	RI, DI, NDI	0.76	6.97	2.30	7
	Continuum removed	RI, DI, NDI	0.95	4.38	3.66	9
	1st derivative	RI, DI, NDI	0.55	9.41	1.70	8
	2nd derivative	RI, DI, NDI	0.67	8.21	1.95	8
HJ-B CCD	Raw bands	RI, DI, NDI	0.86	5.72	2.80	7
	1st derivative	RI, NDI	0.69	9.88	1.62	9
	2nd derivative	RI, DI, NDI	0.85	6.21	2.58	7
Landsat OLI	Raw bands	RI, DI, NDI	0.65	5.42	2.95	9
	1st derivative	RI, DI, NDI	0.91	4.82	3.32	12
	2nd derivative	RI, DI, NDI	0.84	5.83	2.75	7

Source: Wang et al. (2018). Used with permission.

4.4.5.4 Decision Trees and Random Forest

As a non-parametric statistical method, decision-tree analysis can handle data enumerated on both the ordinal and categorical scales. Without making any assumptions about the data distribution, the decision rules are interpretable and can be extrapolated to other similar areas. Using a decision tree with 10 trial boosting, Elnaggar and Noller (2010) obtained an overall accuracy of 98.8% in mapping soil EC into five classes (<2, 2–4, 4–8, 8–16, >16) from 15 environmental variables (Table 4.22). Some of them are vegetation indices, such as NDVI and SAVI (plus 6 TM bands). Image-derived variables are not so useful in mapping soil salinity over large areas. Compared to decision trees, RF is advantageous in that it can quantify the importance of the input variables based on two considerations: the size of the input variables subset and the number of trees in the forest. If predicted from multiple salinity indices, the prediction can be achieved at a similar accuracy level of R^2=0.76 from both Landsat-8 OLI and Sentinel-2 MSI data (Table 4.15). It must be noted that the mapped salinity is categorical and the threshold of each class is pre-determined.

4.4.5.5 A Comparison

The performance of various analytical methods in constructing the model for estimating soil salt has been comprehensively evaluated using 60 samples, 50 of which are used for model construction, and 10 for testing and comparison. They were collected in Shizuishan City of Ningxia, Northwest China, where the soil salt ranges from 2.92 to 290.86 g·kg^{-1} (Wang et al., 2019b). However, most of the samples have a low level, so the data are logarithmically transformed and expressed as a percentage after the transformation. The RF–PLSR model constructed from a variety of spectral covariates, including individual band reflectance, salinity indices, red-edge indices, newly constructed 2D indices, and TBIs, is the best among the five models (raw, PCC, Variable importance in projection, GRA, RFR) with the R^2 value, RMSE value, and RPD being 0.92, 7.58 dS·m^{-1}, and 2.36, respectively (Figure 4.10). A comprehensive evaluation of five models (RFR, SVR, gradient-boosted regression tree – GBRT, multilayer perceptron regression, least angle regression – LAR) reveals a ranked relative performance of RFR > LAR > SVR > MLPR > GBRT (Wang et al., 2019b). As shown in Figure 4.10, RFR achieves almost the same accuracy as SVR, the difference in accuracy between them lies likely within the range of random variation. These accuracies are highly comparable to those of LAR and multilayer perceptron regression. GBRT is obviously inferior with an accuracy (R^2=0.77) about 13% lower than other models. RFR is the best with a strong performance for high-dimensional data. Similar to MLPR, RFR models are resistant to noise in the input

TABLE 4.22
Accuracy of Mapping Soil EC into Five Classes from Landsat and Environmental Data Using Decision-Tree Analysis with 10 Trials of Boosting

Salinity Class	0–2	2–4	4–8	8–16	>16	Row Sum	User's Accuracy (%)
0–2	6,987	17	3	4	5	7,016	99.59
2–4	25	63				88	71.59
4–8	11		27			38	71.05
4–16	13			380	5	398	95.48
>16	7			1	93	101	92.08
Column sum	7,043	80	30	385	103	7,641	
Producer's accuracy (%)	99.20	78.75	90.00	98.70	90.29		98.81
Kappa							0.92

Source: Elnaggar and Noller (2010).

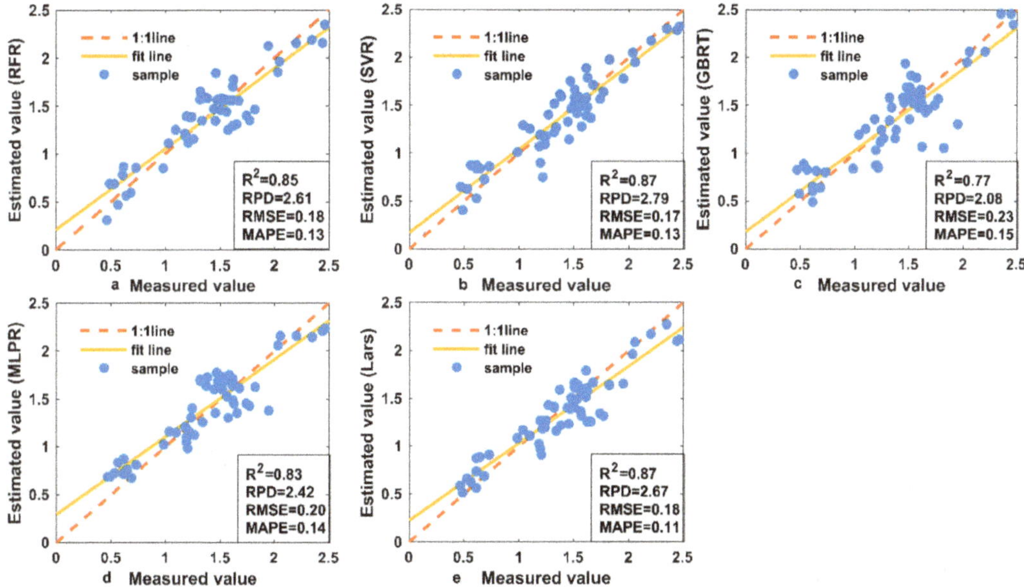

FIGURE 4.10 Comparison of in situ measured with spectra-derived soil salt contents estimated using five methods: (a) RFR, (b) SVR, (c) gradient-boosted regression tree, (d) multilayer perceptron regression, and (e) least angle regression. Accuracy indicators are obtained using leave-one-out cross-validation. (Source: Wang et al., 2019b.)

data. SVR models have a satisfactory stability. The LARS model is the most accurate in estimating soil salinity. Judged by all the accuracy indicators, LARS is the best, followed by RFR and SVR. Their accuracy is highly comparable to that of MLPR. The least reliable is GBRT that is noticeably less accurate than the other four models. These findings are based on field-measured spectra data. Whether they remain valid with remotely sensed spectral bands requires further study to confirm.

The relative performance of SVM is compared to ANN in estimating soil EC from the soil backscattering coefficient (σ^0_{soil}), groundwater depth (GD), SI, and surface evapotranspiration (SET), of which σ^0_{soil} is obtained from Sentinel-1A SAR data, GD and SI are derived from Landsat-8 OLI imagery, and SET from MODIS evapotranspiration product (MOD16). The comparison is based on 80 samples, of which 60 are used for model development and 20 for model validation. The SVM regression algorithm is more accurate than ANN, achieving an R^2 value of 0.82 (RMSE = 2.01) using the training dataset or R^2 = 0.88 (RMSE = 1.36) using the test dataset (Jiang et al., 2019). This accuracy is marginally better than ANN's R^2 = 0.79 and RMSE = 2.20 obtained using the training dataset (R^2 = 0.68 and RMSE = 2.25 with the testing dataset) (Table 4.23). As illustrated in Figure 4.11, ANN overestimates EC with both the training and test dataset in most cases.

TABLE 4.23

Comparison of SMV and ANN Accuracy in Mapping Soil Salt Content Measured by EC (dS·m⁻¹)

Accuracy Measure	Training Dataset (n = 60)		Test Dataset (n = 20)	
Algorithm	**SVM**	**ANN**	**SVM**	**ANN**
R^2	0.82	0.79	0.88	0.68
RMSE	2.01	2.20	1.36	2.25

Source: Jiang et al. (2019).

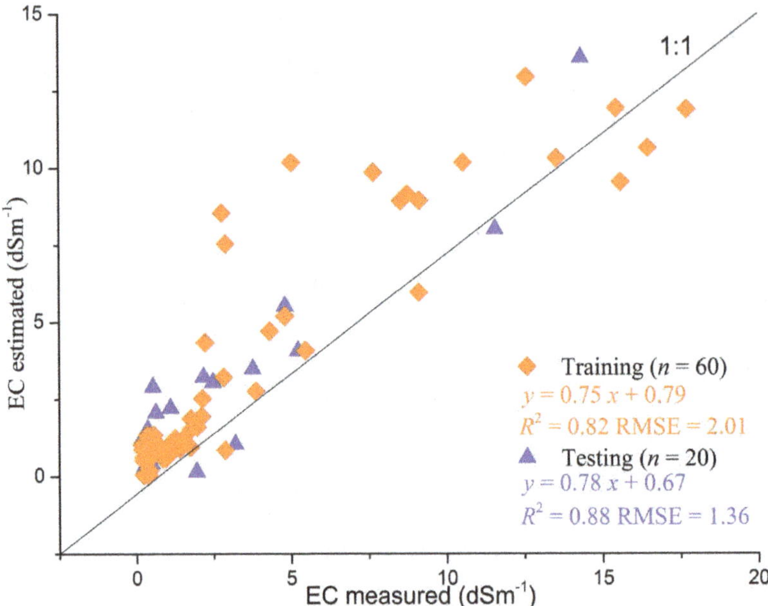

FIGURE 4.11 Scatterplot of the field measured EC and the estimated EC predicted from four predictor variables using the ANN model by training and test samples. (Source: Jiang et al., 2019.)

A comparison of these results with those in Figure 4.10 reveals that SVM achieved an R^2 of 0.87, almost identical to 0.88 achieved with the test dataset using SVM, even though the dependent variable of soil salt is expressed differently in the estimation. Naturally, the space-borne data have a higher RMSE value, suggesting that the estimated EC is less accurate than that modeled from the in situ measured spectral data. Such a close similarity indicates that indeed soil salt as measured by EC can be reliably estimated using either in situ measured spectral data or from space-borne images if the various manifestations of soil salt are considered in the estimation using SVM. The application of the constructed SVM model translates the input satellite image to a map of salt content distribution (Figure 4.12).

4.5 WATERLOGGING

Waterlogging refers to the accumulation of surface water on shallow depressions that are dry under normal circumstances because the soil is saturated with moisture. The water body is usually rather shallow and small, so is not commonly called lakes, just small ponds. Spatially, several ponds may lie in close proximity to each other, depending on the local topography. The duration of waterlogging is ephemeral or perennial. If the water table rises at or above the ground surface, the duration is much longer. The main causes of waterlogging in arid and semi-arid regions are traced to evapotranspiration and poor drainage, coupled with improper water management. Waterlogging differs from flooding in that the water volume involved is small and the water level rises slowly. The water lingers longer than flood water which is caused by heavy rainfall. Once the rainwater is absorbed by the soil or evaporates into the atmosphere, the accumulated water disappears quickly. In comparison, waterlogging occurs in saturated soil and lasts much longer on the surface. The stagnant water deprives plant roots of oxygen, stunts plant growth by reducing soil aeration around the root zone, and, if persisting for an extended period, can cause them to perish eventually. Waterlogging is considered hazardous as it stifles vegetation, retards its growth, and reduces crop yield. Besides, it can raise soil salinity and lead to land degradation or even desertification after the water recedes.

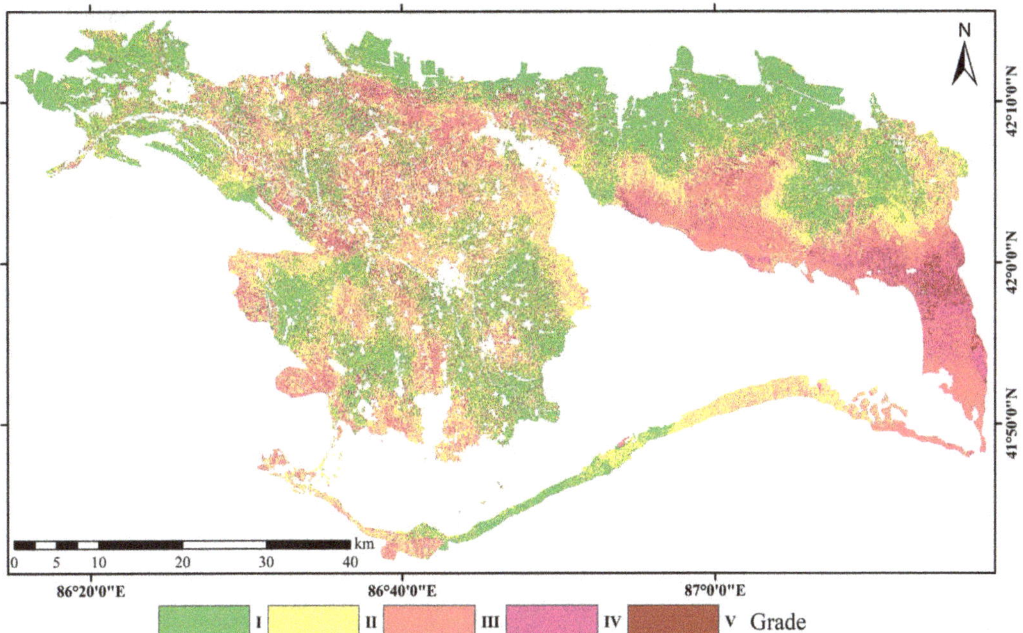

FIGURE 4.12 Distribution of soil salt expressed as EC (dS·m^{-1}) in the Yanqi Basin, the Xinjiang Uyghur Autonomous Region of Northwest China at five levels (I: EC=0–2; II: EC = 2–4; III: EC = 4–6; IV: EC = 6–8; V: EC = 8–10), produced from Landsat-8 OLI imagery using SVM analysis. (Source: Jiang et al., 2019.)

Waterlogged areas can be identified from satellite images based on a few characteristic indicators such as high soil moisture, stagnant water, and perennial vegetation (Choubey, 1997). These areas can be detected, monitored, and mapped from remotely sensed images of a moderate resolution either visually or digitally. Visual interpretation relies on the recognition of water bodies based on image tone, texture, and association. Water has a dark tone as most of the incident energy is absorbed or transmitted, resulting in little energy being reflected back to the sensor. High soil moisture and shallow standing water show up as a bluish color on the standard false-color composite (IR–red, red–green, green–blue) of photographs, against the characteristic pink or red color of vegetation. The water color can change from dark blue to black, depending on water depth, in stark contrast to moist areas that have a gray or light gray color. It exhibits a drastic contrast with the adjacent areas that have a light tone and coarse texture. The contrast between water and non-water features is maximal in the VNIR spectrum. Waterlogged areas can be visually interpreted and separated from cropped/vegetated areas at a rather high overall accuracy of 96.3% owing to the distinct spectral signature of water from that of the intact land surrounding it (Dwivedi and Sreenivas, 2002).

Because of their distinct spectral uniqueness from non-water features, waterlogged areas can be digitally mapped at rather high accuracy even using a single threshold on an IR band, or image classification. It is possible to further differentiate the waterlogged areas as very critical, critical, less critical, and non-critical based on water depth, even though the retrieval of water depth from remotely sensed images is more complex and demanding than the mapping of water bodies (Prajapati et al., 2021). With the use of time-series satellite images, waterlogged areas may be further differentiated into perennial (e.g., presence of water bodies in all time-series images) and seasonal (e.g., presence of water bodies in the rainy season images only) (Figure 4.13). Another method of quickly delineating water logged areas is density slicing based on the minimum and maximum gray values. For instance, low and high moist areas have a pixel value of 15–25 in IRS-A LISS-I infrared band 4 (Choubey, 1997). They are determined from the pixel values of water and peripheral water bodies (wet- and drylands) exclusive of vegetation in high and forest areas. The thresholds

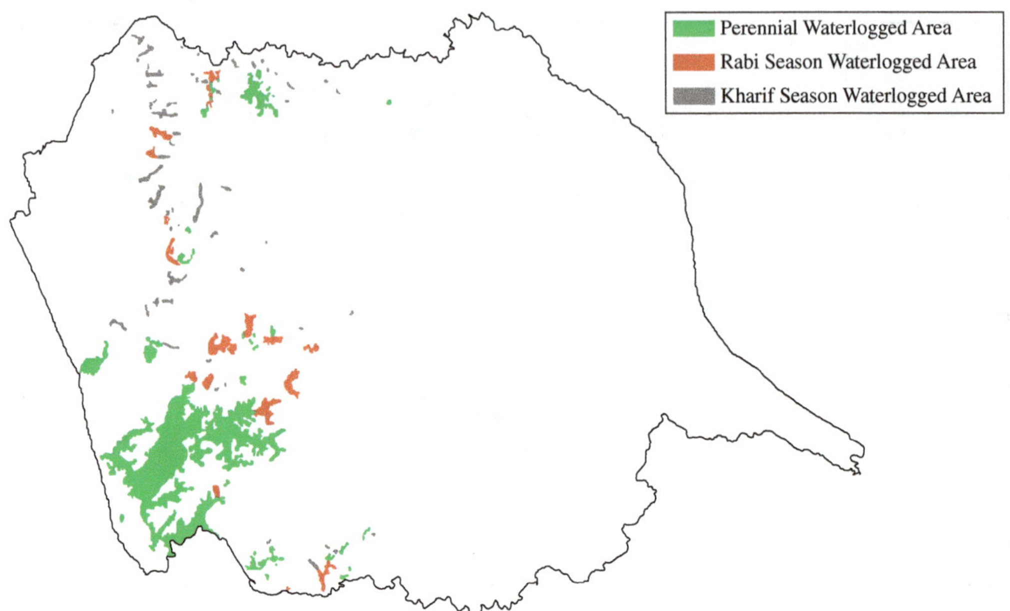

FIGURE 4.13 Perennial and seasonal waterlogged areas associated with the Kallada irrigation project in Kerala, India, mapped using thresholding of two-season IR bands. (Source: Prajapati et al., 2021.)

have to be relaxed slightly to 15–25 for standing water, 15–26 for waterlogged areas, and 15–28 for identifying areas sensitive to waterlogging. If assisted with field observations or DEM data, areas vulnerable to waterlogging can also be identified by modifying these thresholds. It must be borne in mind that the use of a single threshold from a single band is risky and unable to take advantage of the rich spectral information of the image used. A better alternative is to make use of indices, such as normalized difference water index (NDWI) (Eq. 4.17).

Instead of using a single threshold, multiple thresholds derived from various spectral bands may be combined to improve mapping reliability and accuracy, especially in delineating the interface between land and water. For instance, a pixel is considered to represent water if its pixel value or digital number in the NIR band is lower than that of the red and green bands, and its NDWI is no less than 0.32 (Chatterjee et al., 2003). However, the exact threshold may vary with the geographic area under study and the image used. Besides, it is subject to whether the image has been calibrated for the atmospheric effects. Thus, the image-specific, area-specific threshold may need to be modified for other areas.

$$NDWI = \frac{\rho_g - \rho_{NIR}}{\rho_g + \rho_{NIR}} \tag{4.17}$$

NDWI is good at detecting water features that have a positive index value, while soil, terrestrial, and vegetation features have a zero or negative value. A simple NDWI-based threshold enables these covers to be separated. If the same detection is carried out with multi-temporal data, the change in waterlogged areas can be monitored using NDWI, as well (Dwivedi and Sreenivas, 2002).

REFERENCES

Abbas, A., and S. Khan. 2007. Using remote sensing techniques for appraisal of irrigated soil salinity. In: *Inter. Congress on Modelling and Simulation (MODSIM)*, eds. L. Oxley and D. Kulasiri, 2632–2638. Brighton: Modelling and Simulation Society of Australia and New Zealand.

Abdelkader, D., N. Hervé, and W. Christian. 2006. Detecting salinity hazards within a semiarid context by means of combining soil and remote-sensing data. *Geoderma* 134: 217–230. DOI: 10.1016/j.geoderma.2005.10.009.

Alhammadi, M.S., and E.P. Glenn. 2008. Detecting date palm trees health and vegetation greenness change on the eastern coast of the United Arab Emirates using SAVI. *Int. J. Remote Sens.* 29(6): 1745.1765.

Allbed, A., and L. Kumar. 2013. Soil salinity mapping and monitoring in arid and semi-arid regions using remote sensing technology: A review. *Adv. Remote Sens.* 2: 373–385.

Allbed, A., L. Kumar, and P. Sinha. 2014. Mapping and modelling spatial variation in soil salinity in the al Hassa oasis based on remote sensing indicators and regression techniques. *Remote Sens.* 6(1):1137–1157. DOI: 10.3390/rs6021137.

Bannari, A., A. El-Battay, R. Bannari, and H. Rhinane. 2018. Sentinel-MSI VNIR and SWIR bands sensitivity analysis for soil salinity discrimination in an arid landscape. *Remote Sens.* 10, 855. DOI: 10.3390/rs10060855

Bannari, A., A.M. Guedona, A. El-Hartib, F.Z. Cherkaouic, and A. El-Ghmari. 2008. Characterization of slightly and moderately saline and sodic soils in irrigated agricultural land using simulated data of advanced land imaging (EO-1) sensor. *Communi. Soil Sci. Plant Anal.* 39(19–20): 2795–2811. DOI: 10.1080/00103620802432717

Burrell, A.L., J.P. Evans, and Y. Liu. 2017. Detecting dryland degradation using time series segmentation and residual trend analysis (TSS-RESTREND). *Remote Sens. Environ.* 197: 43–57. DOI: 10.1016/j.rse.2017.05.018

Chatterjee, C., R. Kumar, and P. Mani. 2003. Delineation of surface waterlogged areas in parts of Bihar using IRS-1C LISS III data. *J. Indian Soc. Remote Sens.* 31(1): 57–65

Choubey, V.K. 1997. Detection and delineation of waterlogging by remote sensing techniques. *J. Indian. Soc. Remote Sens.* 25: 123–135. DOI: 10.1007/BF03025910.

Clevers, J.G.P.W. 1988. The derivation of a simplified reflectance model for the estimation of leaf area index. *Remote Sens. Environ.* 25(1): 53–69. DOI: 10.1016/0034-4257(88)90041-7.

Costantini, E.A.C., and R. Lorenzetti. 2013. Soil degradation processes in the Italian agricultural and forest ecosystems. *Italian J. Agron.* 8: e28, 233–243.

Davis, E., C. Wang, and K. Dow. 2019. Comparing Sentinel-2 MSI and Landsat 8 OLI in soil salinity detection: A case study of agricultural lands in coastal North Carolina. *Int. J. Remote Sens.* 40(16): 6134–6153. DOI: 10.1080/01431161.2019.1587205.

Dehni, A., and M. Lounis. 2012. Remote sensing techniques for salt affected soil mapping: Application to the Oran region of Algeria. *Procedia Eng.* 33: 188–198.

Dwivedi, R.S., and K. Sreenivas. 2002. The vegetation and waterlogging dynamics as derived from space-borne multispectral and multitemporal data. *Int. J. Remote Sens.* 23(14): 2729–2740. DOI: 10.1080/01431160110076234

El-Baroudy, A.A. 2011. Monitoring land degradation using remote sensing and GIS techniques in an area of the middle Nile Delta, Egypt. *Catena* 87(2): 201–208. DOI: 10.1016/j.catena.2011.05.023.

Elnaggar, A.A., and J.S. Noller. 2010. Application of remote sensing data and decision-tree analysis to mapping salt affected soils over large areas. *Remote Sens.* 2(1): 151.165.

Gao, J. 2008. Mapping of land degradation from ASTER data: A comparison of object-based and pixel-based methods. *GISci. & Remote Sens.* 45(2): 1–18. DOI: 10.2747/1548-1603.45.1.1

Gedefaw, M.G., H.M.E. Geli, and T.A. Abera. 2021. Assessment of rangeland degradation in new Mexico using time series segmentation and residual trend analysis (TSS-RESTREND). *Remote Sens.* 13(9):1618. DOI: 10.3390/rs13091618

Gibbs, H.K., and J.M. Salmon 2015. Mapping the world's degraded lands. *Appl. Geogr.* 57: 12–21.

Gorji, T., A. Yildirim, N. Hamzehpour, A. Tanik, and E. Sertel. 2020. Soil salinity analysis of Urmia Lake basin using Landsat-8 OLI and Sentinel- 2A based spectral indices and electrical conductivity measurements. *Ecol. Indic.* 112: 106173.

Holm, A.M., S.W. Cridland, and M.L. Roderick 2003. The use of time-integrated NOAA NDVI data and rainfall to assess landscape degradation in the arid shrubland of Western Australia. *Remote Sens. Environ.* 85: 145–158.

Huete, A.R. 1988. A soil-adjusted vegetation index (Savi). *Remote Sens. Environ.* 25 (3): 295–309. DOI: 10.1016/0034-4257(88)90106-X.

Jiang, H., Y. Rusuli, T. Amuti, and Q. He. 2019. Quantitative assessment of soil salinity using multi-source remote sensing data based on the support vector machine and artificial neural network. *Int. J. Remote Sens.* 40: 284–306.

Jordan, C.F. 1969. Derivation of leaf-area index from quality of light on the forest floor. *Ecology* 50(4): 663–666. DOI: 10.2307/1936256.

Kaufman, Y.J., and D. Tanré. 1992. Atmospherically resistant vegetation index (ARVI) for EOS-MODIS. *IEEE Trans. Geosci. Remote Sens.* 30(2): 261–270. DOI: 10.1109/36.134076.

Khan, N.M., V.V. Rastoskuev, E.V. Shalina, and Y. Sato. 2001. Mapping salt-affected soils using remote sensing indicators – a simple approach with the use of GIS IDRISI. In *Proc. the 22nd Asian Conf. on Remote Sens.*, 5–9 November 2001, Center for Remote Imaging, National University of Singapore.

Li, F., J. Meng, L. Zhu, and N. You. 2020. Spatial pattern and temporal trend of land degradation in the Heihe River Basin of China using local net primary production scaling. *Land Degrad. Develop.* 31(4): 518–530. DOI: 10.1002/ldr.3468.

Liu, H.Q., and A. Huete. 1995. A feedback based modification of the NDVI to minimize canopy background and atmospheric noise. *IEEE Trans. Geosci. Remote Sens.* 33(2): 457–465. DOI: 10.1109/36.377946.

Mahyou, H., B. Tychon, R. Balaghi, M. Louhaichi, and J. Mimouni. 2016. A knowledge-based approach for mapping land degradation in the arid rangelands of North Africa. *Land Degrad. Develop.* 27(6): 1574–1585. DOI: 10.1002/ldr.2470.

Nguyen, K.A., Y.A. Liou, H.P. Tran, P.P. Hoang, and T.H. Nguyen. 2020. Soil salinity assessment by using near-infrared channel and vegetation soil salinity index derived from Landsat 8 OLI data: A case study in the Tra Vinh Province, Mekong Delta, Vietnam. *Prog. Earth Planet. Sci.* 7(1). DOI: 10.1186/s40645-019-0311-0.

Peng, Y., M. Fan, J. Song, T. Cui, and R. Li. 2018. Assessment of plant species diversity based on hyperspectral indices at a fine scale. *Sci. Rep.* 8, 4776. DOI: 10.1038/s41598-018-23136-5.

Prajapati, G.S., P.K. Rai, V.N. Mishra, P. Singh, and A.P. Shahi. 2021. Remote sensing-based assessment of waterlogging and soil salinity: A case study from Kerala, India. *Results in Geophys. Sci.* 7, 100024. DOI: 10.1016/j.ringps.2021.100024.

Prince, S.D., I. Becker-Reshef, and K. Rishmawi. 2009. Detection and mapping of long-term land degradation using local net production scaling: Application to Zimbabwe. *Remote Sens. Environ.* 113: 1046–1057.

Radda, I.A., B.M. Kumar, and P. Pathak. 2021. Land degradation in Bihar, India: An assessment using rain-use efficiency and residual trend analysis. *Agric. Res.* 10(3): 434–447. DOI: 10.1007/s40003-020-00514-y.

Richards, L.A. 1954. Diagnosis and improvement of saline and alkali soils. *Soil Sci.* 78(2), 154. DOI: 10.1097/00010694-195408000-00012.

Rouse, J.W., R.H. Haas, J.A. Schelle, D.D. Deering, and J.C. Harlan. 1974. Monitoring the Vernal Advancement or Retrogradation of Natural Vegetation. Greenbelt, MD, USA: NASA/GSFC, Type III Final Report.

Scudiero, E., T.H. Skaggs, and D.L. Corwin. 2016. Comparative regional-scale soil salinity assessment with near-ground apparent electrical conductivity and remote sensing canopy reflectance. *Ecol. Indic.* 70: 276–284. DOI: 10.1016/j.ecolind.2016.06.015

Seifi, M., A. Ahmadi, M.-R. Neyshabouri, R. Taghizadeh-Mehrjardi, and H.-A. Bahrami. 2020. Remote and vis-NIR spectra sensing potential for soil salinization estimation in the Eastern coast of Urmia hyper saline lake, Iran. *Remote Sens. Appl.: Soc. Environ.* 20, 100398. DOI: 10.1016/j.rsase.2020.100398

Setia, R., M. Lewis, P. Marschner, R. Raja Segaran, D. Summers, and D. Chittleborough. 2013. Severity of salinity accurately detected and classified on a paddock scale with high resolution multispectral satellite imagery. *Land Degra. Develop.* 24(4): 375–384.

Singh, A. 2015. Soil salinization and waterlogging: A threat to environment and agricultural sustainability. *Ecol. Indic.* 57: 128–130.

Taghadosi, M.M., M. Hasanlou, and K. Eftekhari. 2019a. Retrieval of soil salinity from sentinel-2 multispectral imagery. *Eur. J. Remote Sens.* 52(1): 138–154.

Taghadosi, M.M., M. Hasanlou, and K. Eftekhari 2019b. Soil salinity mapping using dual-polarized SAR Sentinel-1 imagery. *Int. J. Remote Sens.* 40(1): 237–252.

Venter, Z.S., S.L. Scott, P.G.L. Desmet, and M.T. Hoffman. 2020. Application of landsat-derived vegetation trends over South Africa: Potential for monitoring land degradation and restoration. *Ecol. Indic.* 113. DOI: 10.1016/j.ecolind.2020.106206.

Vieira, R.M.S.P., J. Tomasella, A.A. Barbosa, S.P. Polizel, J.P.H.B. Ometto, F.C. Santos, Y. da, C. Ferreira, and P.M. Toledo. 2021. Land degradation mapping in the MATOPIBA region (Brazil) using remote sensing data and decision tree analysis. *Sci. Total Environ.* 782. DOI: 10.1016/j.scitotenv.2021.146900.

Vogt, J.V., U. Safriel, G.V. Maltitz, Y. Sokona, R. Zougmore, G. Bastin, and J. Hill. 2011. Monitoring and assessment of land degradation and desertification: Towards new conceptual and integrated approaches. *Land Degrad. Develop.* 22: 150–165.

Wang, S., Y. Chen, M. Wang, and J. Li. 2019b. Performance comparison of machine learning algorithms for estimating the soil salinity of salt-affected soil using field spectral data. *Remote Sens.* 11(22): 2605. DOI: 10.3390/rs11222605.

Wang, J., J. Ding, D. Yu, X. Ma, Z.P. Zhang, X.Y. Ge, D.X. Teng, X.H. Li, J. Liang, and I. Lizaga. 2019a. Capability of Sentinel-2 MSI data for monitoring and mapping of soil salinity in dry and wet seasons in the Ebinur Lake region, Xinjiang, China. *Geoderma* 353: 172–187.

Wang, J., J. Peng, H. Li, C. Yin, W. Liu, T. Wang, and H. Zhang. 2021. Soil salinity mapping using machine learning algorithms with the Sentinel-2 MSI in arid areas, China. *Remote Sens.* 13, 305. DOI: 10.3390/rs13020305

Wang, X., F. Zhang, J. Ding, H.T. Kung, A. Latif, and V.C. Johnson. 2018. Estimation of soil salt content (SSC) in the Ebinur Lake Wetland National Nature Reserve (ELWNNR), Northwest China, based on a Bootstrap-BP neural network model and optimal spectral indices. *Sci. Total Environ.* 615: 918–930.

Weng, Y., P. Gong, and Z.L. Zhu. 2008. Soil salt content estimation in the yellow River delta with satellite hyperspectral data. *Can. J. Remote Sens.* 34(3): 259.270.

Wessels, K.J., F. van den Bergh, and R.J. Scholes. 2012. Limits to detectability of land degradation by trend analysis of vegetation index data. *Remote Sens. Environ.* 125: 10–22.

Wu, W., A.S. Mhaimeed, W.M. Al-Shafie, F. Ziadat, B. Dhehibi, V. Nangia, and E.D. Pauw. 2014. Mapping soil salinity changes using remote sensing in central Iraq. *Geoderma Reg.* 2(3): 21–31.

Yahiaoui, I., A. Douaoui, Q. Zhang, and A. Ziane. 2015. Soil salinity prediction in the lower Cheliff Plain (Algeria) based on remote sensing and topographic feature analysis. *J. Arid Land* 7(6): 794–805. DOI: 10.1007/s40333-015-0053-9.

5 Land Desertification

5.1 INTRODUCTION

Desertification, also known as desertization, is a natural process in which the original natural-looking, productive land is converted to desert-like dryland. This conversion is usually accompanied by an increase in mobile dunes, a decrease in surface vegetation, or complete elimination of vegetative cover in the worst scenario. As an extreme form of land degradation, desertification signifies the complete loss of land capability for sustainable farming, grazing, or forestry. In most cases, this loss is irreversible or faces a grim prospect of reversibility. Desertification usually takes place in arid, semi-arid, and dry sub-humid regions of the world where the environment is fragile and vulnerable to external changes. Desertification is caused by natural processes (e.g., droughts and climate change) or inappropriate human activities, such as deforestation, exploitative tillage practice, land reclamation, and overgrazing in semi-arid grassland areas (Figure 5.1). The major driving forces, both natural and anthropogenic, of desertification in the Arab region are outlined in Figure 5.2, together with the manifestation of desertification. Most of them are also applicable to other arid and semi-arid regions of the world.

Desertification is considered a semi-natural hazard that threatens the livelihood of farmers and pastoralists, and the environment. This process of change exerts a profound impact not only on the local environment but also on the living conditions in distant areas. During desertification, land fertility is reduced via the deposition of sand blown from elsewhere or the depletion of fine fertile soil particles by wind, leading to deteriorated land productivity. If exacerbated by drought, desertification can result in crop failures. In the worst case, the encroachment of the advancing dunes shrinks arable land and reduces the fertility of the affected farmland to such a level that it is no longer profitable to farm, leading ultimately to farmland abandonment. Moreover, the newly desertified land serves as a sand source to fuel dust storms that degrade air quality and disrupt aviation. The desertification of shrubland reduces biodiversity. Therefore, it is very important to study desertification and monitor its temporal evolution. Accurate assessment of the status, changing trend and rate of desertification is a prerequisite to enacting effective measures to combat the advance of dunes, reverse the degradation process, and rehabilitate the desertified land to productive uses. Since the United Nations Conference on Desertification in 1977, tremendous progress has been made in studying desertification by means of remote sensing.

This chapter elaborates on the sensing of desertification, a particular form of severe land degradation. It starts with a comprehensive introduction to meteorological, physical, and biological desertification indicators and how they are characterized and quantified using remote sensing based on generic vegetation indices and a number of purposely devised spectral indices. The second part of the chapter discusses how to map desertified land and assess desertification severity from various types of remotely sensed data using visual interpretation, image classification, decision-tree analysis, and linear spectral mixing (LSM) analysis. Explained at length is how to monitor long-term desertification, including the detection of changes between healthy and desertified land and the quantitative attribution of desertification causes to human and natural factors. Finally, this chapter explains how to assess, quantify, and grade the severity, vulnerability, and risk of land desertification from diverse variables derived from remotely sensed data and other sources using various analytical methods.

5.2 DESERTIFICATION INDICATORS AND INDICES

5.2.1 Indicators

An indicator is a parameter or its value that supplies qualitative or quantitative information about desertification. Indicators can be socioeconomic or natural. Since socioeconomic indicators are not easily quantifiable via remote sensing, they are not discussed here. Instead, this section concentrates

DOI: 10.1201/9781003354321-5

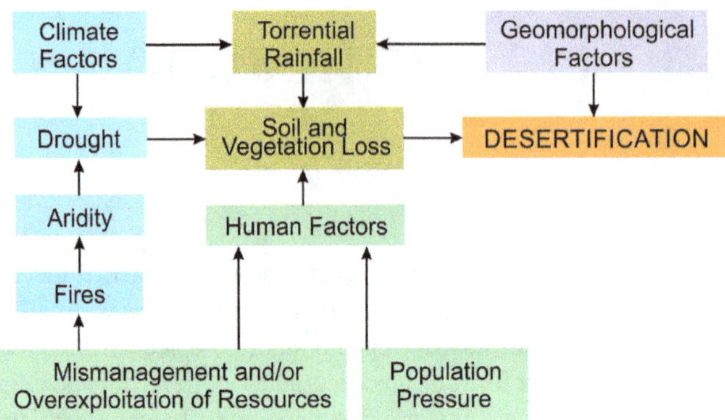

FIGURE 5.1 Factors leading to desertification and their interactions. (Source: Modified from Perez-Trejo, 1994.)

on natural or environmental indicators. Natural indicators of desertification can be hydrological, meteorological, physical, or biological (Table 5.1). Hydrological indicators are related to soil moisture, surface runoff, and rainwater infiltration. The hydrological manifestation of desertification is characterized by a reduced area of water bodies, increased runoff and consequently decreased rainwater infiltration, and deteriorated groundwater resources (Sharma, 1998). These hydrological indicators, together with accelerated soil erosion and sedimentation, can be used to evaluate the magnitude and severity of desertification.

Meteorological indicators are indirectly linked to hydrological indicators involved in water budgeting and RUE. They include rainfall and its variability. Climate factors such as LST and wind

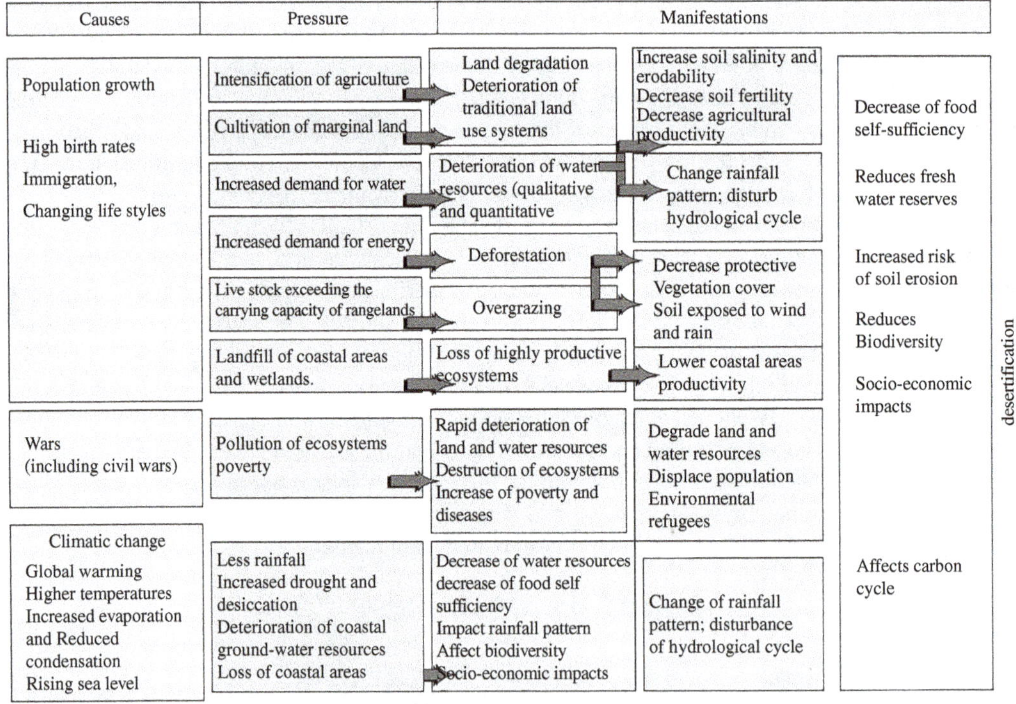

FIGURE 5.2 Major driving forces of desertification, their manifestations, and impacts in the Arab region. (Source: Abahussain et al., 2002. Used with permission.)

TABLE 5.1
Four Groups of Desertification Indicators

Hydrological	Meteorological	Physical	Biological
Soil moisture	Rainfall	Albedo	Vegetation cover
Runoff	Wind speed	Water erosion	Biomass
Water area	Evapotranspiration	Soil salinization and alkalization	Net primary production
Rainwater infiltration	Land surface temperature	Soil erosion (rate)	Species abundance and diversity
	Aridity	Topography	Carrying capacity
	Rain use efficiency (RUE)	Sand sheets and dunes	Desirable and undesirable plant species
		Topsoil gran size	

Source: Modified from Symeonakis and Drake (2004).

mainly affect the evaporation of moisture from the land surface. Desertified areas with a lower vegetative cover tend to be warmer than non-desertified areas under the same environmental settings. A higher temperature promotes stronger evapotranspiration, which in turn reduces precipitation.

Physical indicators refer to the current degradation state and general topography, such as soil erosion, soil salinity, sediment yield, dunes and their mobility, and surface albedo. They also include the dominance or proportion of dunes and their mobility in assessing desertification severity. Aridity, crusting and compaction, soil salinization and alkalization, and even water erosion areas and potential have also been suggested as physical indicators (Krugmann, 1996). Of all the physical indicators, topography is the most unique in that it does not indicate desertification potential at all. Instead, it is useful in revealing the relationship between desertification and the physical environment and affects other indicators, such as evaporation and LST.

Although deemed as a physical indicator, surface albedo is able to indicate biological changes associated with desertification. Desertification is inevitably accompanied by changes in the amount of solar energy reflected from the ground over the 0.28–6.0 μm wavelengths due to reduced vegetation cover and soil moisture, and increased soil erosion. As such, it is a direct function of vegetative cover, biomass, soil moisture, and soil erosion. The increase in albedo detected from satellite data is instrumental in locating the sites of desertification. The processes leading to desertification, such as unsound irrigation and agricultural practices, eventually manifest in the forms of soil salinization, waterlogging, soil erosion, and dune mobility, all of which are related to vegetative cover. This indicator should be treated with caution in areas where agriculture is active, though.

Biological indicators are related to the health, quantity, and species abundance (especially indicator species) of the vegetation community. They may include NPP, biomass, and vegetative cover. These indicators have different utilities in studying desertification. NPP is crucial to understanding the desertification mechanism and how to bring it under control. At a local scale, more indicators are still available, such as soil depth, slope gradient, slope exposure, parent material, rock fragment content, annual rainfall, and aridity (Kosmas et al., 2003), all of which are related to the physical environment, even though some of them straddle between physical and hydrological indicators. As shown in Table 5.1, other indicators can be environmental in general. They facilitate the assessment and monitoring of desertification by providing synthetic information on the status and trends of environmental processes leading to desertification, but they are unable to reveal whether the land has been desertified already. Analysis of environmental indicators can pinpoint areas prone to desertification and define criteria for selecting proper indicators to assess desertification risk. It is impossible to generalize the importance of the four types of indicators as it most likely varies with the geographic area under study.

No matter how effective an indicator is, it is still of limited use if it cannot be derived reliably and easily in the spatial domain. Of the diverse range of indicators in Table 5.1, some may be obtained from field measurements, others are better derived from remotely sensed data either directly or indirectly. In general, those indicators that can be quantified from remotely sensed data have found wide applications in studying desertification, such as detecting desertification and assessing its severity. Changing vegetation and land use, drought, and soil erosion can all be derived readily from remotely sensed data and are useful to study the desertification process via their changes over time. They have been exploited to detect, monitor, and map desertification-affected areas. The degree of successfully detecting desertified land depends largely on the local manifestation of desertification in several environmental components (Figure 5.2).

Since most of the proposed hydrological indicators in Table 5.1 are difficult to derive objectively and accurately from remotely sensed data, they have been seldom taken advantage of in assessing desertification (Harahsheh and Tateishi, 2001). In contrast, other hydrological indicators such as albedo, RUE, surface runoff, LST, soil erosion, and biological evidence (e.g., amount of vegetation) have been widely used to monitor desertification because they are derivable from remotely sensed data quantitatively, objectively, and accurately. Albedo (ρ) is defined as the instantaneous portion of surface-reflected incident radiation flux in relation to the incident radiant flux at the same wavelength. Sensitive to the change in surface cover from vegetation to sand and bare ground, it is a function of wavelength, known as the spectral reflectance pattern. Bare ground has a quite distinctive pattern from that of vegetation, especially in the NIR spectrum. In order to make use of the albedo information in multiple bands, a complex albedo index can be derived from the linear combination of reflectance in multiple bands (Liang, 2001), namely,

$$Albedo = 0.356\rho_b + 0.130\rho_r + 0.373\rho_{NIR} + 0.085\rho_{SWIR1} + 0.072\rho_{SWIR2} - 0.0018 \tag{5.1}$$

where ρ_B, ρ_R, ρ_{NIR}, ρ_{SWIR1}, and ρ_{SWIR2} refer to the reflectance value of TM/ETM+ bands 1, 3, 4, 5, and 7, respectively, or OLI bands 2, 4, 5, 6, and 7.

RUE may appear to be a physical indicator, actually it is a meteorological indicator defined as the ratio of NPP to precipitation (Symeonakis and Drake, 2004). NPP is calculated from NDVI in the form of NPP ($\text{Mg}\cdot\text{ha}^{-1}\cdot\text{a}^{-1}$) = 3.139ΣNDVI – 3.852 (Σ: summation over a period). Although NPP is indeed a biological indicator, it is not used as frequently as vegetation indices because it is unable to reveal whether the land has been desertified. Instead, this indicator plays a key role in assessing the relative role of human activities and climate change in desertification dynamics. For instance, the difference between potential NPP and actual NPP is able to reveal the impacts of climate changes and human activities on desertification.

Compared to biological indicators, physical and meteorological indicators are much harder to derive from remotely sensed data that may involve the use of complex algorithms and non-remote sensing data. For instance, the rate of soil erosion is quantitatively estimated from the empirical universal soil loss equation, and LST is derivable from two thermal bands in a number of steps to convert at-sensor radiance to temperature. Soil moisture requires only a single band to derive, usually the TIR band, in conjunction with in situ samples. Field samples are necessary to convert pixel values in the image/band to the ground-level moisture via regression analysis to establish the statistical relationship between the two sets of variables. The establishment of accurate regression relationships may require remote sensing data to be calibrated for the atmospheric effects. Therefore, physical and biological indicators are used much more commonly than other indicators, especially in assessing global desertification. Of these two types of indicators, it is the biological ones that have found wider use due to their ease of reliable and objective derivation from remote sensing images, and all the derived indicators are directly comparable to one another regardless of seasonality and geographic area.

In reality, multiple indicators of different natures are essential to realistically assess and map desertification. For instance, vegetation indices may be combined with RUE, surface runoff, and soil erosion to monitor desertification, of which soil erosion is the most indicative of the desertification process (Symeonakis and Drake, 2004). The use of four indicators instead of one enables the area most susceptible to desertification to be highlighted. The combination of multiple indices to form a complex indicator system is readily realized with remote sensing and perfectly suited to monitor desertification, such as categorized NDVI values combined with percent vegetative cover mapped into different intensity levels of desertification (Zhang et al., 2008). This complex indicator system is posed ideally for assessing regional desertification over 10,000s km^2. However, it is difficult to envisage that it can yield meaningful information at the local scale.

While it is beneficial to combine multiple indicators to assess desertification severity, two points must be borne in mind. First, they must be treated properly with care as the selected indicators may be correlated with each other. If this is indeed the case, it is important to apply an appropriate weight to each individual indicator in the combination to discount their joint effect. The properly assigned weights to them eradicate double counting of the effect of the same indicator occurring in different forms. Second, all the indicators to be included in the assessment must be quantifiable spatially, preferably from satellite images either alone or in conjunction with field surveys or other spatial data, such as topographic data. All data that are initially collected at certain points (e.g., rainfall) have to be converted into spatial coverage via spatial interpolation before they can be used for the assessment.

5.2.2 Indices

Apart from indicators, desertification is commonly quantified via vegetation indices (see Section 1.4.3). While the relative performance of all the vegetation indices in assessing desertification has not been systematically and comprehensively evaluated and compared, it is generally understood and accepted that MSAVI is superior to other VIs in that it is highly sensitive to change in the ground biomass and vegetative cover (Table 5.2). It also takes into account the influence of the

TABLE 5.2

Comparison of the Strengths and Weaknesses of Five Commonly Used VIs in Detecting Desertification

VI	Strengths	Weaknesses
NDVI	Simple transformation, Able to monitor phenology, quantity, and activity of vegetation, Long time-series data available, No assumptions (land cover classes, soil type, or climatic conditions)	Sensitive to background reflectance and plant canopy (changes in water content), Over-estimated vegetation in semi-arid and arid regions, Over-detection of non-linear (ratio-based index) soil surface light
EVI	Tolerant of background reflectance, high aerosol loads, biomass, biophysical leaf area index, Able to quantify evapotranspiration or water-use efficiency, change over large areas	Requiring surface reflectance in blue, red, and NIR bands, sensitive to plant canopy, Difficult to derive from broadband radiation measurements Non-linear ratio-based index, Narrower ranges over semi-arid areas
SAVI	Suitable for semi-arid and arid regions	Sensitive to atmospheric differences, Less sensitive to changes in vegetation
WDVI	Effective by enhancing canopy reflectance component	Detailed atmospheric correction needed, Distance-based index
MSAVI	Rangeland studies correlated to field data on vegetation cover/desertification	Requiring red and NIR band to calculate, Sensitive to changes in vegetation amount/cover, Sensitive to differences in atmospheric conditions between areas or times

Source: Modified from Albalawi and Kumar (2013).

background soil and can enhance the dynamic range of the vegetative signal in a pixel while minimizing the influence of the background soil. Similar to NDVI, it also requires two spectral bands of red and NIR to derive as it is calculated from NDVI and WDVI (Eq. 1.5).

In addition to vegetation indices, albedo, RUE, and other desert-specific indices have also been used to study desertification, including bare soil index (BSI), normalized difference desertification index (NDDI), and grain size index (GSI). BSI is calculated from four spectral bands of blue, red, NIR, and SWIR wavelengths. NDDI is calculated from MODIS bands 1 and 2 reflectance data product (MODIS09A1) (Feng et al., 2016), and GSI from Landsat visible light bands for characterizing topsoil texture (Xiao et al., 2006). They are calculated as follows:

$$BSI = \frac{(\rho_{SWIR} + \rho_r) - (\rho_{NIR} + \rho_b)}{(\rho_{SWIR} + \rho_r) + (\rho_{SWIR} + \rho_b)} \times 100 + 100 \tag{5.2}$$

$$NDDI = \frac{MODIS1 - MODIS2}{MODIS1 + MODIS2} \tag{5.3}$$

$$GSI = \frac{\rho_r - \rho_b}{\rho_r + \rho_b + \rho_g} \tag{5.4}$$

where ρ_B, ρ_R, ρ_{NIR}, and ρ_{SWIR} = reflectance of blue, red, NIR, and SWIR wavelengths, or Landsat TM/ETM+ bands 1, 3, 4, and 5, respectively, or Landsat OLI bands 2, 4, 5, and 6. r = red band, g = green band, b = blue band. Since desertified areas are characterized by a lack of vegetative cover, BSI can distinguish bare soil from vegetative cover. Able to reveal a whole range of vegetation conditions from full coverage to exposed soil, BSI overcomes the unreliability of common VI in areas sparsely covered by vegetation.

GSI is a proxy for describing the texture of topsoil or grain size. A GSI value around 0.20 indicates high contents of fine sand. GSI > 0 indicates an increase in desertified areas. This index can be used to predict the amount of fine sand component and clay and silt content of topsoil. It is also indicative of the degree of desertification in productive farmland, the fine particles of which have been lost, thus a sign of desertification caused by wind erosion in arid regions. This index has restrictive applicability as it does not function well in vegetated areas.

Finally, NDWI, salinity index calculated as $(\rho_r + \rho_g)/2$, and aeolian mapping index $(\rho_r/\rho_{\text{NIR}})$ have also been used to map desertification status (Othman et al., 2014). If combined with NDVI, these three indices are powerful in characterizing and mapping the desertification process by providing direct measurements. They offer a robust and low-cost alternative for preliminary and large-scale assessments of desertification in arid environments.

5.3 MAPPING OF DESERTIFIED LAND

5.3.1 Visual Interpretation

The mapping of desertification from aerial photographs and satellite images is grounded on its effect on vegetation and its unique image features. The level of mapped details about desertified land and mapping accuracy are both influenced by image spatial resolution that also affects the detectability of different desertification features. So far desertification has been mapped from a variety of remotely sensed materials that fall into two broad categories of airborne and space-borne. In general, large-scale aerial photographs and VHR satellite images are suitable for producing detailed desertification maps. Aerial photographs have a relatively large scale and are capable of mapping desertified areas in fine detail, such as mobile dunes and their pace of advance. Since each photograph covers only a limited ground area, a large number of photographs are necessary to cover a large study area. Manual processing of these photographs is a mammoth and daunting task and a major hurdle in mapping regional desertification. Besides, aerial photographs are also constrained by their limited range of spectral sensitivity as SWIR photography is not available with optical cameras. However, this disadvantage is largely compensated for by their long history of existence, dating back to the 1940s or earlier. Such a long time span potentially extends the monitoring of desertification to several decades. In comparison with aerial photographs, satellite images are the superior data source for mapping regional desertification owing to their synoptic view of a huge ground area and the capability of periodic coverage of the same area. Spanning over a period of nearly four decades, satellite images are the best at detecting desertification in vegetated areas at the regional or global scale. The TIR bands of satellite imagery are especially effective at differentiating vegetative from non-vegetative covers, and hence at identifying areas affected by desertification, or desertified land that has been rehabilitated to productive uses.

Desertification status can be mapped directly from remotely sensed data using a number of methods, the most common being visual interpretation. Visual analysis of satellite images is able to map dunes and saline areas. This method starts with a preliminary analysis of desertification on remote sensing imagery and proceeds to the identification of its characteristic elements and topography (Table 5.3). The established indoor interpretation label system is subsequently verified via intensive in situ investigation and/or modified if necessary. In order to generate an authentic assessment, hundreds of evaluation points representing all unique types or severities of desertification should be included in the field visit. Visual interpretation of remotely sensed data produces land cover maps. Historical land cover maps from aerial photographs are vital to assess the impact of desertification (Elhag, 2006).

TABLE 5.3
Main Characteristics of Sandy Land in the Mu Us Sandy Land of China

Type of Sandy Land	Characteristics
Fixed sand	*Artimisia ordosica* and *Caragana korshinskii* as dominant species with over 40% of vegetation cover; presence of soil crust of >1 cm thick; accumulated humus in surface soil; little sand movement in earth surface.
Semi-fixed sand	*Artimisia ordosica* and *Caragana korshinskii* as dominant species with 20–40% of vegetation cover; soil crust developed under plant associations (0.5–1 cm thick) and poorly developed between plant associations (with thickness <0.5 cm): comparatively steady dunes with little sand shifting
Mobile sand	Little vegetation cover or sparse annual pioneer herbs such as *Agriophyllum arenarium* and *Psammochloa villosa* of <10% coverage, no soil crust; obvious sand shifting.
Complex dunes and salinized and/or damp wetland	Several mosaic forms: mobile dune-meadow (with *Calama-grostis pseudophragmites* and *Carex stenophylla* as dominant species), shifting dune-salinized shrub (with *Nitraria tangutorum* and *Suaeda corniculata* as dominant species) or salinized meadow (with *Achnatherum splendens* as dominant species), fixed or semi-fixed dune-meadow, and fixed or semi-fixed dune-salinized meadow.
Low salinized wetlands	Main vegetation types: meadow, consisting of *Carex stenophylla*, *Achnatherum splendens*, *Puccinellia tenuifolia*, and *Iris ensata*; salinized shrubs, consisting of *Nitraria tangutorum* and *Kalidium cuspidatum*; and marsh, consisting of *Eleocharis palustris* and *Phragmites communis*

Source: Modified from Wu and Ci (2002). Used with permission.

TABLE 5.4

A Classification Scheme Illustrating the Characteristics of Surface Features on the True-Color Composite of Landsat TM Bands 7 (Red), 4 (Green), and 2 (Blue), with an Emphasis on Desertified Land

ID	Surface Features	Characteristics
1	Grassland	Green, fuzzy edge, high degree of vegetation, in the shape of sheet, patch, or strip
2	Forest	Dark green, in the shape of patch, fuzzy edge, stereo (the hue is dark or light because of the effect of topographic shadow)
3	Farmland	Various colors caused by crops planted at different times, in the rectangular shape of strip fabrics
4	Residential area	Magenta color, linearly shaped, fuzzy patch edge
5	Lakes and reservoirs	Black or dark blue, color varying with water depth and quality, in the shape of patch with smooth edges
6	Saline–alkaline land	White or pale, in the shape of patch, mostly associated with alkaline lakes
7	Windbreak trees	Dark green, linearly shaped, located mostly around crop fields
8	Slightly desertified land	Rosy color, in the shape of irregular patches
9	Moderately desertified land	Rosy color, usually located in sparsely forested areas, in the shape of irregular patches
10	Highly desertified land	Slight amaranth, in the shape of irregular patches
11	Extremely desertified land	Pale, in the shape of strips, usually with relief caused by dunes and sand ridges

Source: Fang et al. (2008). Used with permission.

Visual interpretation is suited to map the desertification landscape at a small scale such as 1:500,000 (Wu and Ci, 2002). In order to produce a reasonable map, it is important to come up with a scientific classification scheme (Table 5.4), or a list of all types of land covers present in the study area that can be mapped from the remote sensing material at a reasonable accuracy. These land covers should be logically grouped without overlap between any two covers. The detail levels of the scheme vary with the spatial resolution of the image used. Further descriptions may be supplied to ensure that the same criteria are adhered to and applied consistently throughout the interpretation process or used by different interpreters.

If desertification has taken place at the regional or national level, it is best studied from space-borne imagery. With a bird's eye view, satellite images are particularly effective at revealing the spatial distribution and pattern of desertified land. They allow the precise location and extent of desertification-affected areas to be mapped quickly and efficiently. It is even possible to monitor the evolution of desertified land and study the process of desertification via multi-temporal images. Analysis of time-series satellite images can shed light on the nature and trend of desertification. These images also enable the assessment of the severity and characteristics of desertification, its impacts, and causes.

Visual interpretation is carried out to individual images, so is not subject to the nature of the satellite images, be they multispectral scanner (MSS) available since the 1970s, thematic mapper (TM) available in the 1990s, or enhanced thematic mapper plus (ETM+) available since 2000. These images can all reveal genuine trends of desertification (Duan et al., 2014). After the desertified land has been visually interpreted on a computer screen, it is commonly graded into several severity levels (Figure 5.3). During visual interpretation, the severity is judged from two indicators of vegetative cover (including indicator species) and the portion of

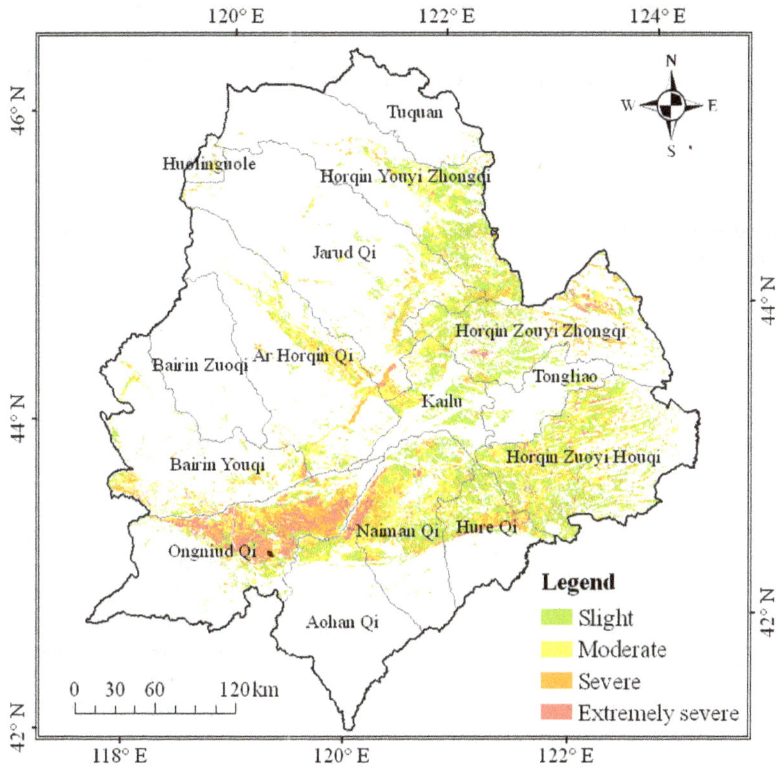

FIGURE 5.3 Aeolian desertified land in the Horqin Sandy Land visually interpreted from a 2010 Landsat ETM+ image. For the definition of severity, refer to Table 5.5. (Source: Duan et al., 2014. Used with permission.)

mobile dunes (Table 5.5). Naturally, it is almost impossible to apply the quantitative thresholds between different severity levels during interpretation. Thus, visually interpreted results are imprecise. Another major limitation of visual analysis is the personal bias in delineating the boundary of desertified areas. Consequently, the interpreted boundaries of desertified areas are likely to vary slightly with the interpreter. Visual interpretation is notoriously slow, and painstakingly demanding, and the results are subjective. A better alternative is to make use of digital image analysis.

5.3.2 Image Classification

Satellite data already in the digital format are better analyzed digitally to map desertified land as digital analysis is fast and can lead to more repeatable and objective results. It also produces more details about desertification than visual interpretation and allows the incorporation of the portion of mobile dunes into the classified results to indicate the level of desertification. Besides, remotely sensed data can be combined with other digital data to derive the desertification susceptibility level and assess the causes and impacts of desertification.

Supervised classification is commonly adopted to classify multispectral Earth resources satellite data, such as Landsat ETM+ images with a spatial resolution of 30 m, into a number of discrete land covers, some of which may be active dunes, fixed dunes, and semi-fixed dunes (Elhadi et al., 2009; Fang et al., 2008). Apart from the raw bands, image classification can also be performed on the aforementioned desertification indicator layers. Supervised classification is based

TABLE 5.5
Grading of Aeolian Desertification in the Horqin Sandy Land of North China by Means of Remote Sensing

			Types of Desertification		
Severity	**Mobile Sand / Wind Eroded Land (%)**	**Vegetation Cover (%)**	**Dune Reactivation or Shifting Sand Intrusion**	**Shrub Encroachment**	**Non-irrigated Farmland**
Slight	<5	>60	Wind erosion pits emerge on the windward slope of dunes, shifting dunes with a punctuated distribution	Predominantly shrub vegetation; sand begins to deposit under the shrub	Slight sand deposit and evident wind erosion emerging on the ridges of farmland in spring
Moderate	5–25	30–60	Obvious erosion and deposit slopes	Sand piles not fully covered; shifting sands emerge on the windward side of shrub, floating sand and sandy gravel among sand dunes	Small pieces of shifting sand emerge on the loess farmland; the thickness of the humus layer was eroded by more than 50%
Severe	25–50	10–30	Sandy land in a state of mobility	Sparse vegetation with little shrub; most shrubs have died	Humus layer nearly eroded out; calcium lamination and parent materials exposed; farmland mostly abandoned
Extremely severe	>50	<10	Mobile sand dunes	Mobile sand and sand ripples	Flat sand or gravel sand

Source: Duan et al. (2014). Used with permission.

on statistical mean and standard deviation of pixel values or machine learning algorithms, such as SVM and decision trees. All supervised classification methods require feeding training samples to the computer. They are selected manually based on their properties in an image (Table 5.5). With properly selected training samples, desertified areas can be classified into four severity levels at an overall accuracy of over 90% (kappa = 0.88). But this method is complex and slow. Both unsupervised and supervised methods are per-pixel classifiers without taking into account spatial variability of pixels. Thus, misclassifications are rife between those spectrally resembling classes. This method suffers from low accuracy in classifying desertification severity. In the presence of heterogeneous covers, automatic image classification may not yield satisfactory results. In this case, spectral mixing analysis rather than image classification facilitates the distinction of the vegetation, water, and sand components on satellite images (Collado et al., 2002).

Better results may be achievable using more image elements or clues other than pixel values, such as texture, VIs, and spatial knowledge, which are difficult to implement with per-pixel classifiers but can be easily accommodated in machine learning methods. They allow additional inputs from non-remote sensing sources to be included in the decision-making, hence leading to more accurate results than those classified solely from spectral data. So far no machine learning algorithms have been used to classify land desertification except decision trees.

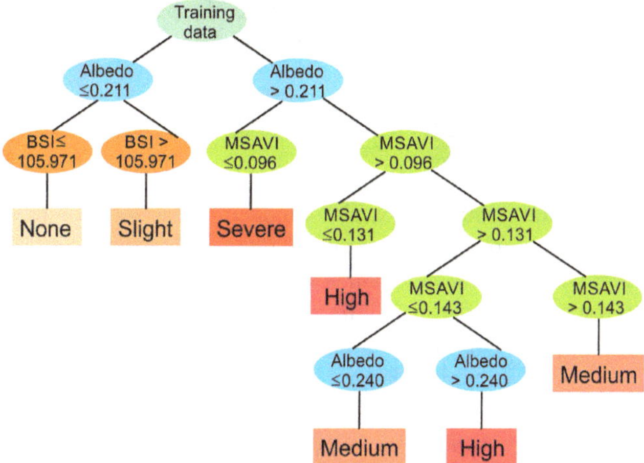

FIGURE 5.4 Two-branch decision trees used to map desertification to four severity levels based on the joint consideration of BSI, MSAVI, and albedo. The decision criteria (thresholds) are determined statistically from the training samples. (Source: Modified from Guo et al., 2017.)

5.3.3 Decision-Tree Analysis

As a binary, hierarchical decision-maker, a decision-tree classifier assigns a pixel in question into one of two possible classes based on the classification rules at each branch. In the decision-making process, an indicator may be evaluated individually in isolation, or two indicators are evaluated simultaneously in tandem. In either case, the decisions are made hierarchically with those at the bottom being more detailed than those at the top of the tree (Figure 5.4). The complexity of a tree or the number of hierarchies varies with the number of classification criteria considered. The more criteria are used, the deeper the tree, and the more branches it has. Multiple rules based on multiple criteria may be combined to form a tree of variable hierarchies. However, the decisions at different hierarchies can be the same, depending on the evaluation criteria adopted.

So far decision-tree analysis has been used to map desertification severity from VIs according to the pre-determined thresholds of NDVI and SAVI, and vegetation cover proportion calculated from $(NDVI - NDVI_{min})/(NDVI_{max} - NDVI_{min})$, all derived from multi-temporal satellite data (Ghebrezgabher et al., 2019). Freely available optical Earth resources satellite images, such as Landsat ETM+ and OLI, are commonly used to produce these indices. MSS images are another source of data if the study period needs to be extended to the 1970s. The severity of desertification is judged from individual indicators over different threshold ranges. Virtually, this classification is a process of converting a given index value into one of the pre-determined severities. In order for this method to function well, it is imperative to set the proper thresholds for different levels of desertification severity. All the criteria used to determine the degree of desertification in a decision-tree analysis must have the same number of grades of severity based on different thresholds. After the desertification state in the respective decade has been mapped, these maps can be overlaid with each other to detect changes in desertified land. The detected changes may be linked to environmental factors to explore their role in the observed desertification process over multiple decades.

5.3.4 Linear Spectral Unmixing Analysis

Although the severity of desertification may be graded based on vegetative cover, no direct relationship between grassland desertification and vegetation cover is easily established, except with spectral mixing analysis. It is strong at identifying the extent or abundance of sandy land at the sub-pixel level. An LSM model is commonly used for this purpose due to its simplicity, reasonable level of effectiveness, and interpretability. The reflectance of a pixel in a given spectral band ρ(*i*) is decomposed as the sum of all endmembers and their abundance, namely,

$$\rho(i) = \sum_{j=1}^{m} Fj\rho j(i) + \Delta\rho(i) \tag{5.5}$$

where $j = 1, 2, \ldots, m$ (m is the number of pixel component or endmember); $i = 1, 2, \ldots, n$ (n is the number of spectral bands). For Eq. 5.5 to be resolvable, $m \leq n + 1$; $\rho j(i)$ = reflectance of endmember *j* in band *i*. It can be obtained from the pixel value after it has been calibrated for observation angles and compensated for atmospheric scattering. The reflectance of pure ground covers is derived either from laboratory measurements, spectral libraries, or via the identification of homogeneous covers in the same image; Fj = the fraction or abundance of endmember *j* in the pixel. Representing the best fit coefficient that minimizes RMSE, Fj is an unknown coefficient to be estimated via the spectral unmixing analysis of the image; $\Delta\rho(i)$ = the disparity between the actual and modeled reflectance of band *i*.

The key to the success of linear spectral unmixing analysis lies in the selection of appropriate and accurate endmembers. Whether an endmember is an effective component that should be selected depends on its representativeness in the pixel and is judged by RMSE in comparison with the field data. RMSE should be as small as possible. In order to maintain the analysis to a manageable level, the number of endmembers should not exceed 4. In practice, three endmembers of vegetation, water, and sand components are commonly selected (Collado et al., 2002). They facilitate the analysis of areas of heterogeneous cover from satellite images. Simple differences between unmixed images of sand or water revealed dune movement, re-vegetation trends, and variations in water bodies, all being parts of the desertification process (Collado et al., 2002). Coupled with temporal trend analysis of fractional images, linear spectral unmixing facilitates the understanding of the desertification process (Kundu et al., 2015).

5.4 MONITORING AND DETECTION

5.4.1 Monitoring

Monitoring aims to track the spatiotemporal change of the same area under study qualitatively to reveal where desertification has worsened and where it has lessened. It can illustrate the spatiotemporal change of desertification graphically but does not reveal the location of change nor the quantity of change. The target of monitoring can be the extent of desertification from one-time images, its trend of change over a period from multi-temporal images, and the spatiotemporal dynamics of desertification. Of these targets, extent monitoring is the simplest and easily fulfilled by classifying the area under study into a binary map of sandy land (desertified) and vegetation (intact), and then calculating the portion of each cover (Kundu et al., 2015). If the images used are recorded at a short interval from each other, time-series analysis of the fractional maps of vegetative cover and bare soil can reveal the desertification process. Analysis of the derived time-series maps can also shed light on the long-term trend of desertification. However, this analysis yields no information on desertification severity if the map just shows categorical land covers, such as active dunes, fixed dunes, semi-fixed dunes, grassland, and unused land (Elhadi et al., 2009). If the monitored area is small, it can be achieved using fine-resolution images. However, if the area under study extends 10,000s km^2, coarse-resolution images such as MODIS have to be used. Such coarse-resolution images may be supplemented with fine-resolution images to explore the change between aeolian desertification and other land covers (Feng et al., 2016).

Monitoring of desertification is implemented as monitoring its indicators, which is commonly accomplished by mapping the same area from the same type of data using the same desertification criteria in multiple years. This monitoring is eased by the free accessibility to most archived medium-resolution images such as the Landsat series of satellite data in Google Earth Engine. These images are effective at monitoring desertification dynamics in arid and semi-arid regions. Remote sensing-based monitoring is fast and cost-effective as the same method can be re-used to analyze data collected from the same sensor repeatedly. If acquired from different sensors, only those satellite images having the same spectral and spatial resolutions should be used, so that the same index (e.g., VI calculated from the NIR and red bands) can be replicated without even modifying the thresholds used in the decision-making. If this proves unrealistic, images of a similar spatial resolution and a similar number of spectral bands should be used. While spatial resolution exerts no impact on the implementation of the mapping method, it does affect the level of mapped details about desertification. The similarity in spatial resolution makes multi-temporal results directly comparable to each other. If desertification is mapped into several severity levels, the same mapping criteria should be applied so that the longitudinal changes of desertification are directly comparable and detectable from all the maps (Guo et al., 2017). Spatial comparison of multiple desertification maps is able to reveal the rate and trend of desertification, and the location of changes. This overlay can also indicate the spatial change of active dunes and the formation of active sand belts (Elhadi et al., 2009). For ease of comparison, the same map legend should be adopted in all maps (Figure 5.5). The desertification maps of different years reveal the change in desertified areas of various severities over time. Those areas that have experienced worsened desertification need to be carefully managed or rehabilitated to prevent further desertification.

Desertification monitoring may be carried out at the global, regional, or even local level, depending on the spatial resolution of the images used. Global monitoring relies on coarse-resolution

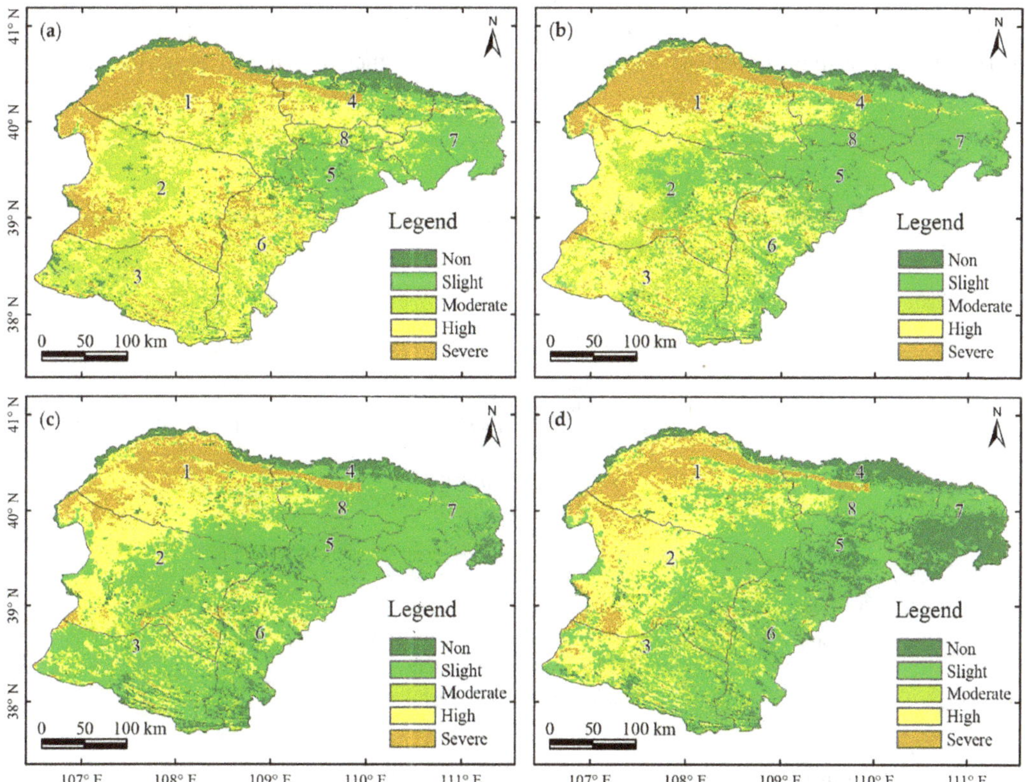

FIGURE 5.5 Evolution of desertification on the Ordos Plateau, China from 2000 to 2015, detected from Landsat TM, ETM+, and OLI images based on MSAVI, BSI, and albedo. (a) 2000; (b) 2006; (c) 2010; and (d) 2015. (Source: Guo et al., 2017.)

images such as MODIS, while regional monitoring is best accomplished using medium-resolution images, such as Landsat series images or SPOT images (Sentinel-2 has a short history of existence, so is not a good choice unless for recent monitoring). If the monitoring period far exceeds the life expectancy of a satellite, multi-sensor data have to be used. For instance, Landsat TM 5 images date back to the 1990s only. A monitoring period of 20 years requires the use of more recent satellite images from other satellites. If the monitoring period is short, caution needs to be taken in interpreting the monitored results as desertification may be affected by the high temporal variability of rainfall. Thus, long-term data are essential to ensuring that the detected trend of desertification is genuine rather than short-term fluctuations caused by the variability of external variables.

Long-term monitoring of desertification is ideally implemented as analysis and spatial comparison of standard time-series data, such as NDVI. Change in desertification is detected via overlay analysis of desertification maps produced from remotely sensed data of different dates or with data from different sources. The derived map indicates desertified areas and those at the risk of desertification, both being essential in assessing desertification severity. At present no universal desertification monitoring systems are in existence except the one developed by Symeonakis and Drake (2004). This system is based on four indicators of vegetation cover, RUE, surface runoff, and soil erosion, all of which are derived from continental scale, publically accessible remotely sensed data. It has the potential to be operational near-real-time for sub-Saharan Africa.

Since it is not so easy to judge whether desertification has worsened due to climate variability, the trend of desertification is assessed from the simultaneous consideration of multiple indices, such as NDVI, WDVI, SAVI, BSI, GSI, aridity index, and human impact index (HII) (Table 5.6). The aridity index is related to climate, calculated by dividing yearly precipitation by mean annual PET. HII is indicative of the non-uniform distribution of human population. The trend of desertification is detected from the slope of the trend of indices used for the assessment. A positive slope of HII suggests a higher population density, and hence a higher pressure of desertification. The

TABLE 5.6

Components and Indices That Have Been Used to Evaluate the Desertification Process

Component	Index	Value Range
Vegetation	$\text{NDVI} = \frac{NIR - R}{NIR + R}$	−1 to 1
	$\text{WDVI} = \text{NIR} - \gamma R$	−1 to 1
	$SAVI = \frac{NIR - R}{NIR + R + L}(1 + L)$	−1 to 1
Soil	$BSI = \frac{(SWIR + R) - (NIR + B)}{(SWIR + R) + (NIR + B)} \times 100 + 100$	0 to 200
	$GSI = \frac{R - B}{R + B + G}$	−1 to 1
Climate	$\text{AI} = P/\text{ET}_p$	0 to >1
Anthropic	HII = spatial distribution of population per pixel (30 m × 30 m) (depending on hydrography, roads, topography, cities, and protected natural areas)	0 to >1

Source: Becerril-Piña et al., 2015. Used with permission.

Note: R = red band, NIR = near-infrared band, G = green band, B = blue band; γ = slope of the soil line; L = soil adjustment factor; P = total annual precipitation (mm); and ET_p: mean annual potential evaporation (mm)

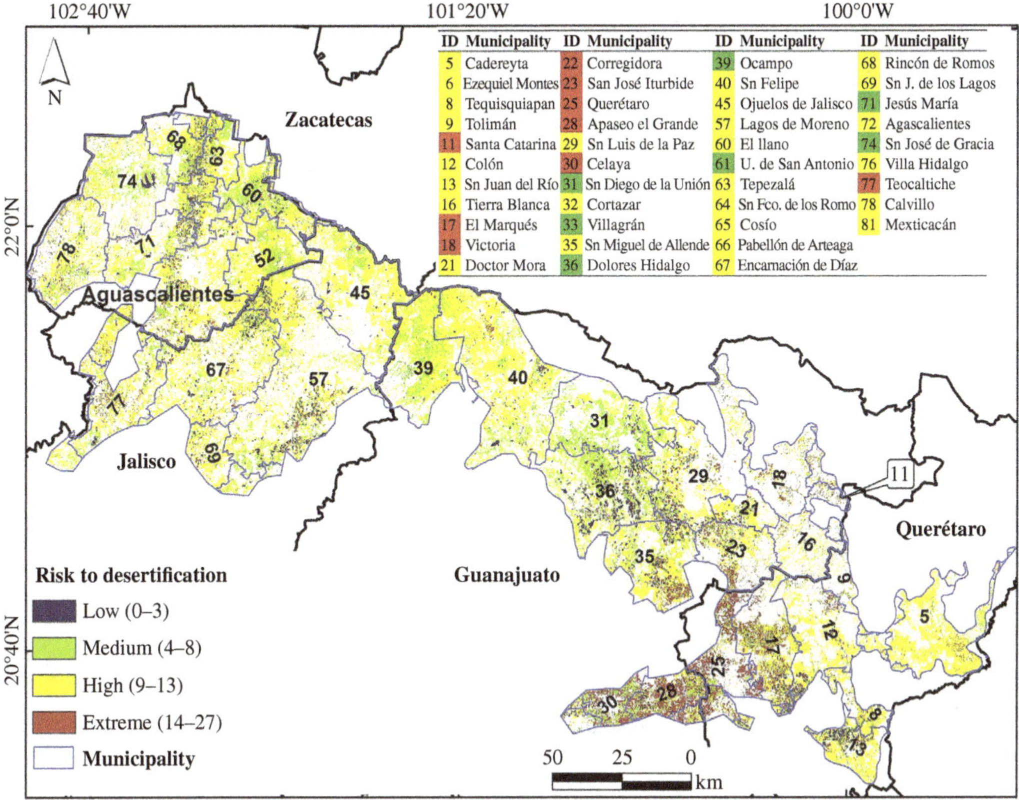

FIGURE 5.6 Desertification risk in the municipalities of the semi-arid highlands, Central Mexico (inset: traffic light alert system). (Source: Becerril-Piña et al., 2015. Used with permission.)

consideration of so many indices can be simplified by deriving the desertification trend risk index (DTRI) (Becerril-Piña et al., 2015). The slope of the Theil–Sen analysis indicates the magnitude of the trend. A lower vegetation cover and higher aridity indicate a heightened risk of desertification. A positive slope of BSI, GSI, and HII indicates vulnerability to desertification.

The spatial dynamics of desertification may be detected from two-time maps showing the spatial distribution of desertified land, and both must contain the same number of desertification classes (Figure 5.6), or via analysis of time-series binary desertification maps of desertified vs non-desertified. More details on this method are provided in the next section.

5.4.2 Change Detection

Detection of desertification aims at yielding quantitative information on its change, both quantitatively and graphically. It involves spatial comparison of time-series desertification maps produced from remotely sensed data. This detection may be implemented as change detection involving two-time images. Change detection can be accomplished via image differencing based on spectral mixing analysis, or post-classification land cover maps (Figure 5.7). If desertification is detected from multi-temporal maps, it can be implemented in two manners, direct subtraction of the classed desertification maps as shown in Figure 5.6, or subtraction of the pixel value indicative of the desertification level. The first type of subtraction leads to change maps directly. The second type of subtraction may be followed by grouping the differencing values to show the spatial distribution of changes. Prior to the subtraction, the two images may be radiometrically corrected if they are

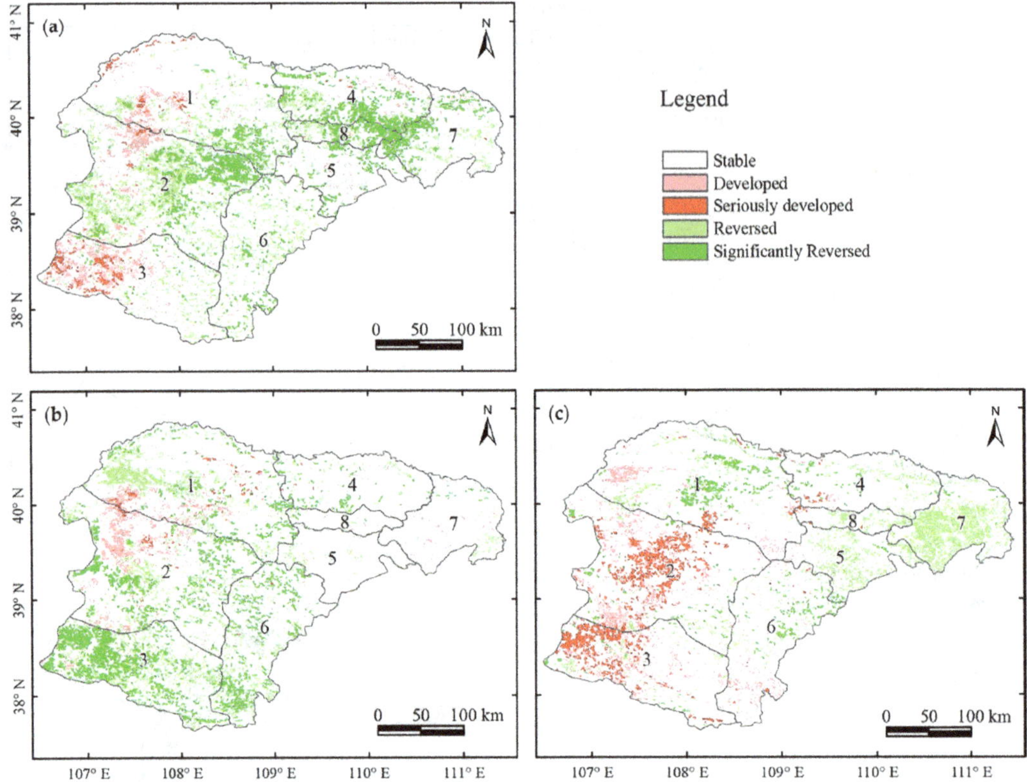

FIGURE 5.7 Spatial dynamics of desertification on the Ordos Plateau of West China over three periods detected by overlaying two-time pairs of desertification maps. (a) 2000–2006; (b) 2006–2010; and (c) 2010–2015. (Source: Guo et al., 2017.)

obtained at different times or in different seasons. Radiometric correction, however, applies mostly to raw remotely sensed data. It is redundant if the satellite data have been classified into discrete land covers prior to the change detection, but essential if the reflectance or albedo of spectral bands is linked directly to the intensity of desertification, as with the second type of subtraction.

Spatial comparison of multiple desertification maps produced using the classification methods described in Section 5.3 is possible only after they have been co-registered with each other spatially. After they are overlaid with each other in a GIS, the changes between desertified land and non-desert covers are explored to quantify the changes between them. Once the long-term changes of desertification have been ascertained, it is also possible to explore where desertification has been reversed and the location of newly desertified areas. The locations of changes may be visualized in the map form. Through overlaying the detected changes with other layers, it is also possible to analyze the interactions between desertification and the ambient environmental settings, such as vegetation, soil, moisture, and human activities (i.e., land-use practices), after all of them have been quantitatively retrieved from remotely sensed data.

Table 5.7 summarizes the recent studies on desertification processes using various remote sensing methods. The study areas vary from regional to national, and the study period is no more than 15 years due to data unavailability. However, the methods used differ widely, so are the indicators of desertification used in the analysis. The more recent studies make use of increasingly sophisticated methods that also enable more information about desertification to be derived, such as quantification of its causes.

TABLE 5.7
Remote Sensing-Based Methods of Detecting Desertification at Different Scales Over the Last Ten Years

Method	Country (Scale)	Data	Study Period	Author
Spectral unmixing analysis, decision-tree analysis of NDVI	Ningxia, China	TM/ETM+	1993–2011	Li et al. (2013)
On-screen visual interpretation	Horqin Sandy Land, Northern China	MSS, TM/ETM+	1975–2010	Duan et al. (2014)
Severity from NDVI, land cover, albedo, soil erosion, drought, wind erosion, population pressure, and aridity	Aurangabad district, central India	LISS I, AWiFS, DEM	1989–2011	Khire and Agarwadkar (2014)
Desertification trend risk index based on four factors, change vector analysis	Semi-arid highlands, central Mexico	TM, population, climate data	1993–2011	Becerril-Piña et al. (2015)
Linear spectral unmixing based on pixel purity index to determine % sand, trend analysis	Western Rajasthan India	TM/ETM+	1990–2009	Kundu et al. (2015)
SVM classification, convergent cross mapping to analyze cause–effect relationship of T, rainfall, wind and human population	Northern China	GF-1, MODIS NDDI product	2001–2015	Feng et al. (2016)
Logistic regression of susceptibility against 16 factors, including TPI and curvature	Djelfa area, Algeria	OLI, SRTM DEM		Djeddaoui et al. (2017)
MSAVI, BSI, and albedo to detect desertification, NPP to assess natural and human impact on it	Ordos Plateau, China	TM, ETM+, OLI	2000–2015	Guo et al. (2017)
Desertification risk based on land cover/land use, NDVI, LST	Northern Nigeria	TM/ETM+/OLI	1984–2015	Joseph et al. (2018)
Decision-tree classification of NDVI, SAVI, and vegetation cover proportion	Eritrea	Landsat data	1970s–2014	Ghebrezgabher et al. (2019)

Note: LISS – Linear Imaging Self-Scanning Sensor; AWiFS – Advanced Wide Field Sensor

5.4.3 Causes of Desertification

A large number of factors play a role in triggering desertification, but it is almost impossible to quantify the influence of individual factors. The easiest and the least convincing way of determining the relationship between land desertification and its causes is to plot out their time-series data in the same chart (Wu and Ci, 2002). If both have a similar trend of variation, it is concluded that desertification is closely related to the variable. Needless to say, the cause–effect evidence based on such qualitative correlation is flimsy, and the conclusion has to be vague and unconvincing, such as "*the material sources and windy, warm and dry climate are the immanent causes of potential land desertification, while the irrational human activities, such as deforestation, reclaiming and grazing in the grassland, are the external causes of potential land desertification*" (Fang et al., 2008). This conclusion is quite understandable as correlation does not necessarily equate to a cause–effect relationship. The quantification of desertification causes may require the use of both remotely sensed data and non-remote sensing data in some instances. For example, simultaneous analysis of high-resolution satellite data and geomorphologic data is critical to identifying the causes of desertification and to deciding whether the identified trend of desertification is long term or ephemeral caused by climate fluctuation.

When multiple factors are at play, it is even more difficult, if not impossible, to isolate the effect of one individual factor from that of other factors due to their joint effects and data paucity. It is, however, possible to separate the influence of natural (e.g., climate change) factors from that of anthropogenic factors (e.g., over-cultivation in marginal land in arid regions). After desertification expansion has been determined, the attribution of its causes to the influence (%) by human activities and the portion caused by climate change is quantifiable via NPP (Zhou et al., 2015). It allows the separation of human influence from natural forces based on the difference between potential and actual annual total NPPs, even in the process of desertification reversion. The effect of climate change on NPP is judged by the temporal trend of potential NPP or PNPP which is the difference between PNPP and annual NPP (PNPP–ANPP). The contribution of human activities to NPP is measured by the trend of human appropriation NPP or HANPP (Guo et al., 2017). The slope of the trend between time t and $t + n$ (interval) is calculated via ordinary least square regression analysis as follows:

$$Slope_{t\to t+n} = \frac{n\sum xy - \sum x \sum y}{n\sum x^2 - \left(\sum x\right)^2} \tag{5.6}$$

where x = time (year); y = PNPP (HANPP), n = time span. The above calculation can be repeated twice, one for climate change and another for human activities. Thus, there are two slopes of S_{cc} and S_{ha}. If both of them are positive, the influence of climate change is 100%; if both are negative, the cause is attributed entirely to human activities. If S_{cc} is positive, and S_{ha} is negative, their contribution is calculated as

$$Contribution\ of\ climate\ change\ (\%) = \frac{|S_{cc}|}{|S_{cc}| + |S_{ha}|} \tag{5.7}$$

$$Contribution\ of\ human\ activities\ (\%) = \frac{|S_{ha}|}{|S_{cc}| + |S_{ha}|} \tag{5.8}$$

The above two equations allow the quantification of the contribution of each type of force (Figure 5.8), and the same analysis can be repeated over different periods to assess whether the effect of each cause has lessened or worsened over the years. Such changes can reveal whether desertification control measures put in place have borne tangible fruits. However, the mechanism of how these factors trigger desertification exactly remains mostly unknown, and the relationship between them is likely to be non-linear. When multiple variables simultaneously affect desertification, the caustic relationship of desertification with individual triggers is impossible to determine.

Another method of quantifying desertification causes is to make use of the convergent cross-mapping (CCM) model after the spatiotemporal change of desertified areas has been detected from time-series desertification maps (Feng et al., 2016). This method allows the influence of individual factors, such as precipitation, temperature, wind speed, and population, to be assessed. The CCM model is a statistical test for the cause–effect relationship between two time-series variables, such as desertified area and population. It can overcome the problem associated with correlation that does not imply causation. CCM is based on the theory of dynamic systems and is applicable to systems where causal variables have synergistic effects. CCM results indicate that there is weak human forcing of aeolian desertified areas in North China. Instead, desertification is affected by temperature. There is no obvious forcing by precipitation for the aeolian desertified areas that bear a significant bidirectional causality relationship with wind speed ($p < 0.01$) (Feng et al., 2016).

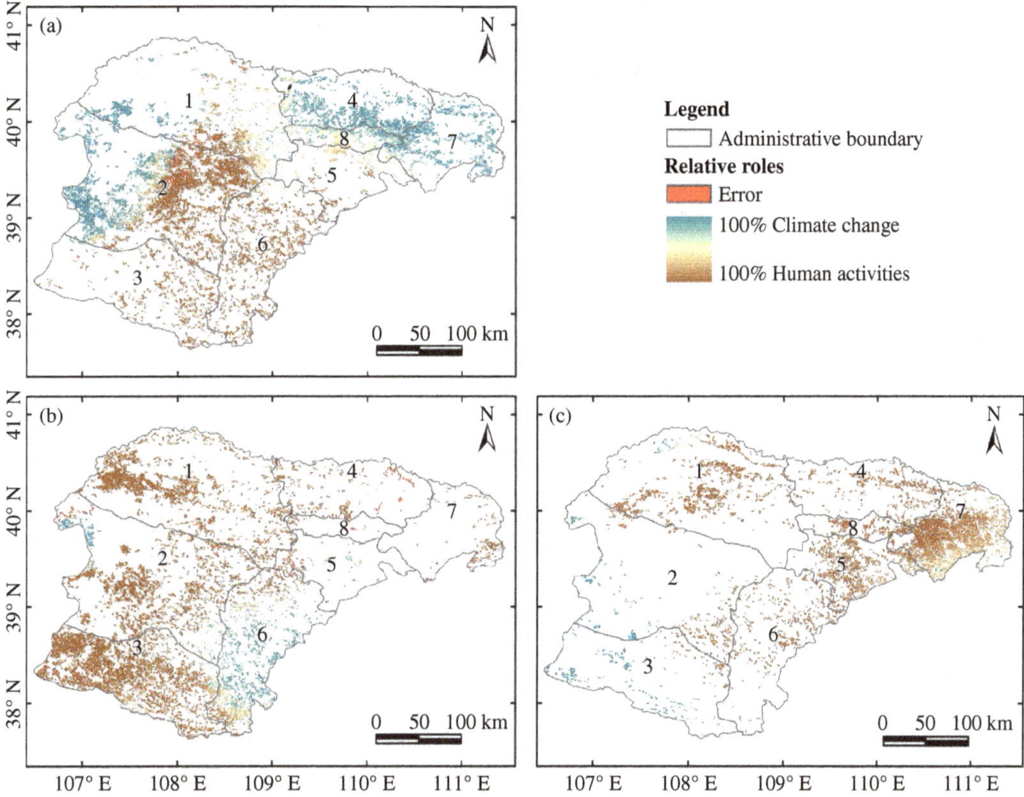

FIGURE 5.8 The relative roles of climate change and human activities in desertification reversion during three periods. (a) 2000–2006; (b) 2006–2010 and (c) 2010–2015. (Source: Guo et al., 2017.)

5.5 ASSESSMENT OF SEVERITY, RISK, AND VULNERABILITY

5.5.1 Desertification Severity

Once initiated, desertification takes place at various severity levels. Initially, it may occur in the form of land degradation. As the degradation process continues, the affected land loses its fertility to such a degree that agriculture and forestry are no longer economically or ecologically viable, causing the land to be taken over by desertification. Desertification severity is commonly expressed verbally as a few categories of slight, moderate, and severe (Table 5.8). Instead of focusing on quantitative severity, Costantini et al. (2004) simply graded the land into two types: sensitive and vulnerable. Such grouping is prompted by the need to illustrate the spatial pattern of desertification and its spatiotemporal change over time. The criteria used to grade desertification severity vary with the nature of desertification (e.g., sandification vs rockification) and the local environmental settings. In sandy grassland, the proportion of bare sand is used to grade desertification severity in arid and semi-arid regions (Li et al., 2013). Normally, the grading is confined to four to five classes (Table 5.9). The evidence for the grading is based on the joint consideration of indicator species (e.g., vegetation cover and biomass) and the portion of sandy land.

Apart from vegetation, the proportion of mobile dunes is also used in grading the severity of aeolian desertified land (Wang et al., 2012). No matter what and how many criteria are used in the grading, they must be quantifiable objectively from remotely sensed data. Naturally, the thresholds for a given level of severity vary with the number of severity classes. There is no agreement regarding the exact criteria that should be adopted to define a given severity level. For instance, a vegetative cover

TABLE 5.8

Criteria Used to Grade Grassland Desertification Severity in Ningxia, West China

	Vegetation Community Characteristics			
Grassland Desertification Intensity	**Composition**	**Coverage (%)**	**Bare Sand Ratio (%)[a]**	**Geomorphological Features**
Slight	Psammophytes become the main accompanying species.	40–55	30–50	Relatively moderate sand, fixed dunes
Moderate	Psammophytes become the dominant species.	30–40	50–65	Moderate sand, small blowout pits or semi-fixed dunes
Severe	Vegetation is sparse, only a few psammophytes remaining.	<30	>65	Medium and large dunes, large blowout pits, semi-mobile or mobile sand dunes

Source: Li et al. (2013). Used with permission.

[a] Percentage cover of sandy soil in each sampling plot.

of 30–50% is considered as slight desertification in grassland areas by Li et al. (2013) (Table 5.8), but a cover of 60% qualifies as slight degradation in aeolian desertified areas by Fang et al. (2008) (Table 5.9). The disparity between the thresholds stems likely from the differential fragility of the local environment of the area under study.

Apparently, this grouping of desertification severity is imprecise and subjective as each of the criteria is assessed only qualitatively from the remote sensing image used. More rigorous and objective

TABLE 5.9

Criteria Used to Grade the Severity of Land Desertification to Four Levels Based on Vegetation Cover and the Portion of Sandy Land

Severity of Desertification	Vegetation Cover (%)	Proportion of Sandy Land (%)	Description
Slight	>60	<5	Quicksand points sited on sandy surface; Only small, sparse, scattered patches of dunes or blowouts (erosion pits) are present; most of the area still resembles the original landscape. Mostly covered by vegetation, but some areas suffer from degradation resulting from a lack of water or from human activity
Moderate	30–60	5–25	Patches of quicksand dune with grass and shrub on sandy surface; Mobile sand sheets, dunes with coppice, and eroded land over the whole area. Blowouts (erosion pits) or sand sheets are sparsely distributed. Vegetation interspersed with sand sheets, dunes, or blowouts
Severe	10–30	25–50	Patches of quicksand or inter-distributing dunes, grass and shrub sand piles on sandy surface; Semi-fixed dunes over most area and some scattered mobile dunes or sand sheets around semi-fixed dunes, widely distributed blowouts (erosion pits) or sand sheets; sparse vegetation
Extreme	≤10	≥50	Concentrated, connected, and intensively active dunes on sandy surface, dominant mobile dunes or wind erosion widely distributed, little or no vegetation cover

Source: Fang et al., 2008. Used with permission.

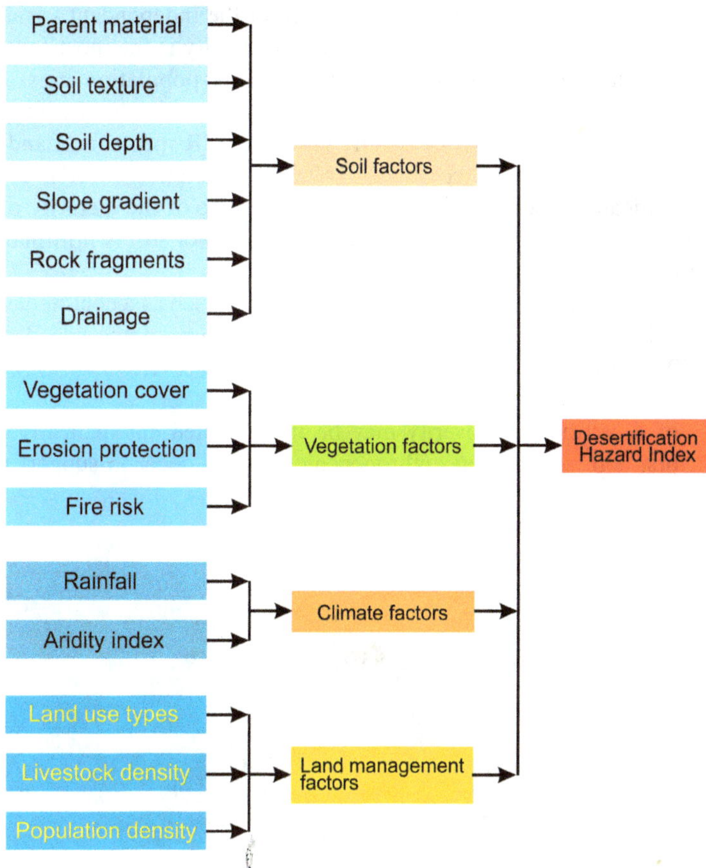

FIGURE 5.9 Parameters considered in defining and mapping environmentally sensitive areas to desertification measured by the desertification hazard index. (Source: Kosmas et al., 1999.)

grading of desertification severity is possible by quantitatively grading all the factors that play a role in the desertification process. These variables may be physical, concerned with soil quality, climate, and vegetation condition, as well as managerial aspects (Figure 5.9). Thus, not all of them are derivable from remotely sensed data. Only such parameters as NDVI, albedo, land use/land cover, and DEM can be derived directly from satellite data. Slope and aspect are then calculated from a DEM and used to estimate soil erosion. Other ancillary data such as soil maps, climate data, surface and groundwater data, and socioeconomic data are needed to derive the climate and water stress indices, as well as the population pressure index (PPI). It measures population pressure on land resources and is calculated as follows:

$$PPI = \frac{P_i - P}{P} \tag{5.9}$$

where P_i = population density of the final year. Through multiple regression analysis, desertification severity is expressed quantitatively as follows:

$$\begin{aligned}\text{Severity} &= 17.5\ \text{NDVI} + 16.2\ \text{LULC} + 13.7\ \text{albedo} + 13.5\ \text{USLE} + 10.5\ \text{MD} \\ &+ 9.2\ \text{WE} + 9.1\ \text{DMAI} + 6.2\ \text{GWD} + 4.1\ \text{PPI}\ \left(R^2 = 0.61\right)\end{aligned} \tag{5.10}$$

where MD = meteorological drought; USLE = universal soil loss equation; WE = wind erosion; DMAI = De Martonne's aridity index; GWD = groundwater depletion.

The coefficients in Eq. 5.10 indicate the importance of the independent variables to desertification risk. According to this equation, desertification severity is the most sensitive to albedo, followed by land use/land cover change and USLE, while population pressure is the least important. This is because albedo is directly related to surface cover. If a vegetative cover is desertified, its reflectance decreases dramatically in the near IR spectrum. Land use/land cover represents the manifestation of surface change. When a vegetative cover is replaced by bare sand, this change indicates desertification. USLE is indicative of the quantity of soil lost from the land. If the surface is well covered by vegetation, the loss of soil is minimal, implying that the risk of desertification is also low.

The severity score calculated from Eq. 5.10 is initially expressed as a continuous value. In order to show the spatial distribution of desertification risk, it is discretized into five classes of degradation severity: no (100–150), mild (150–250), moderate (250–350), severe (350–450), and very severe (450–500) (Figure 5.10).

The diagnosis of desertification severity may be implemented automatically from a number of considered factors, both natural and anthropogenic, using a decision support tool. To a large

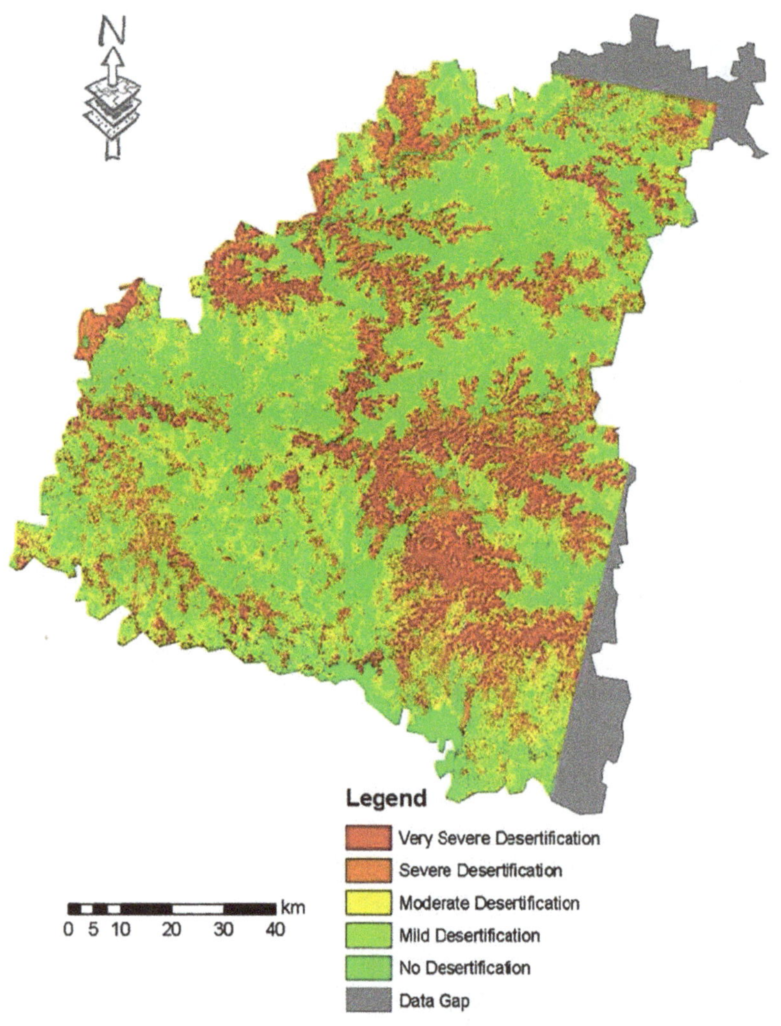

FIGURE 5.10 Desertification severity in the semi-arid Aurangabad district of central India derived from ten parameters and grouped into five levels. (Source: Khire and Agarwadkar, 2014.)

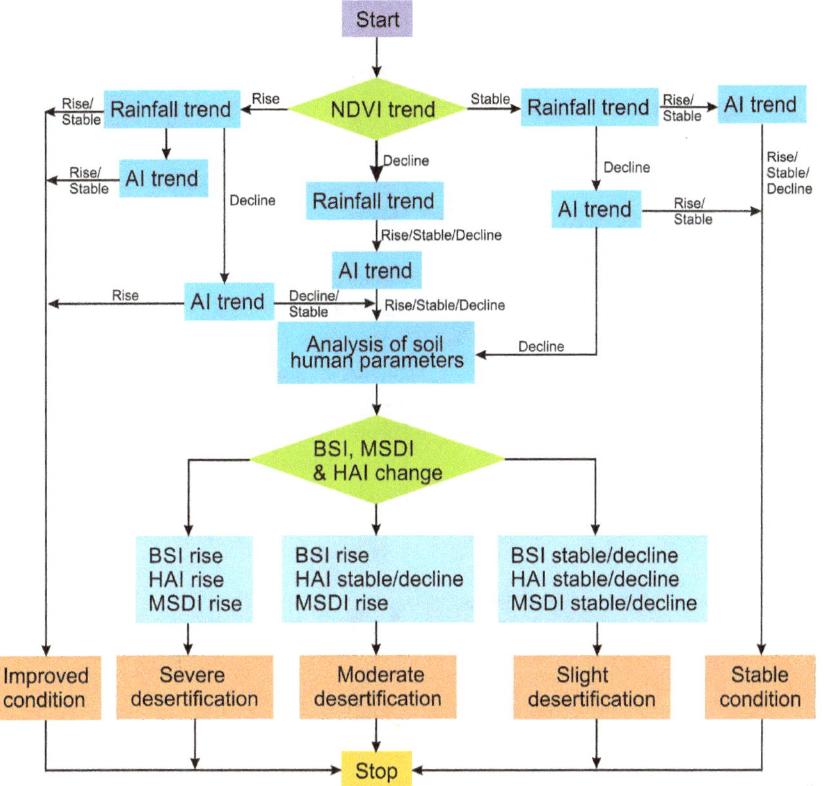

FIGURE 5.11 The decision support tool for diagnosing the severity of desertification from the consideration of NDVI trend, rainfall trend, AI trend, soil parameters (i.e., BSI and heterogeneity of the landscape as measured by moving standard deviation index – MSDI), and human activities impact – HAI. (Source: Modified from Elhag, 2006.)

degree, the diagnosis resembles the decision-tree classification of desertified land in structure but involves more sophisticated decision-making by identifying the trend of the variables instead of using a simple threshold, such as rainfall trend, NDVI trend, and aridity index trend, as well as the change in BSI, human activities impact (HAI), and the moving standard deviation index or MSDI (Figure 5.11). Therefore, it is able to not only grade desertification severity but also indicate whether the desertification condition is stable, has improved, or worsened. It must be pointed out that the yielding of such results requires time-series rainfall and vegetation condition data.

5.5.2 Desertification Risk

Desertification risk refers to the propensity of land to desertify under declined external conditions. Risk is synonymous with potential that may not eventuate in the end. It differs from desertification susceptibility that refers to the likelihood of an area to be desertified. Desertification risk is defined by vegetation type, plant cover, and land management strategies, such as tillage practice, tillage depth, controlled grazing, period of existing land-use type, and erosion control measures that have been put in place (Kosmas et al., 2003). It cannot be directly observed from remote sensing images as it stems from the interactions of multiple variables, among which hydrological conditions have to be studied using other measures than remote sensing (e.g., well water table). To most researchers, desertification risk, susceptibility, and severity are interchangeable without obvious distinctions among them. In other words, those areas facing a high risk of desertification are prone or susceptible

TABLE 5.10
Parameters and Factors Commonly Considered in Deriving Desertification Risk

Parameters	Factors
Biophysical	Land use/land cover changes
	Various vegetation indices
	Erosion due to rainfall and wind
	Albedo changes
Climate	Meteorological drought
	Aridity index
Hydrological	Surface and groundwater depletion
	Surface and groundwater quality changes
Socioeconomic	Population pressure index
	Livestock pressure index
Current desertification state	Portion of desertified areas
	Rate of desertification expansion
	Severity of desertification

Source: Modified from Khire and Agarwadkar (2014).

to desertification. Actually, desertification risk differs slightly from desertification vulnerability that is defined as the degree to which an area is likely to suffer from desertification-induced damage and loss. Those areas that have been severely desertified face a high risk of desertification in the future. For this reason, all those factors identified in Figure 5.9 are perfectly suited to assess desertification risk.

Desertification risk has been assessed from four categories of parameters, including biophysical, hydrological, climate, and socioeconomic (Table 5.10). Each category of parameters contains a varying number of factors, the highest being four in biophysical parameters. These comprehensive factors reveal the likelihood of desertification in light of climate change for areas that have not suffered desertification yet. For areas that have suffered desertification already, one more category of parameters must be taken into consideration, namely, the current state of desertification. The realistic assessment of desertification risk requires the considered indicators to be relevant and derivable from various data sources, preferably remotely sensed data, at different scales, in a straightforward manner.

What factors should be selected to evaluate the potential risk of desertification is subject to the scale of study. At the regional level, desertification risk is quantified using the degradation risk index (DRI) proposed by Dragan et al. (2005) (Eq. 5.11). It is derived from temperature, precipitation, and vegetative cover as measured by NDVI. This derivation is underpinned by an implicit assumption that areas with a high temperature, low precipitation, low NDVI, and where rainfall does not spread evenly over the year face the highest risk of desertification.

$$DRI = \frac{t/t_{max}}{\left[P \times J'(P) \times NDVI\right]} \tag{5.11}$$

where t = mean annual temperature; $t_{\max}$ = mean maximum temperature; $P = p/p_{\max}$ (p = mean precipitation; $p_{\max}$ = mean maximum precipitation); NDVI = the NDVI of the driest period; $J'(P)$ = Pielou's evenness index, calculated as

$$J' = \frac{-\Sigma_{i=1}^{S} P_i \times Ln(P_i)}{Ln(S)} \tag{5.12}$$

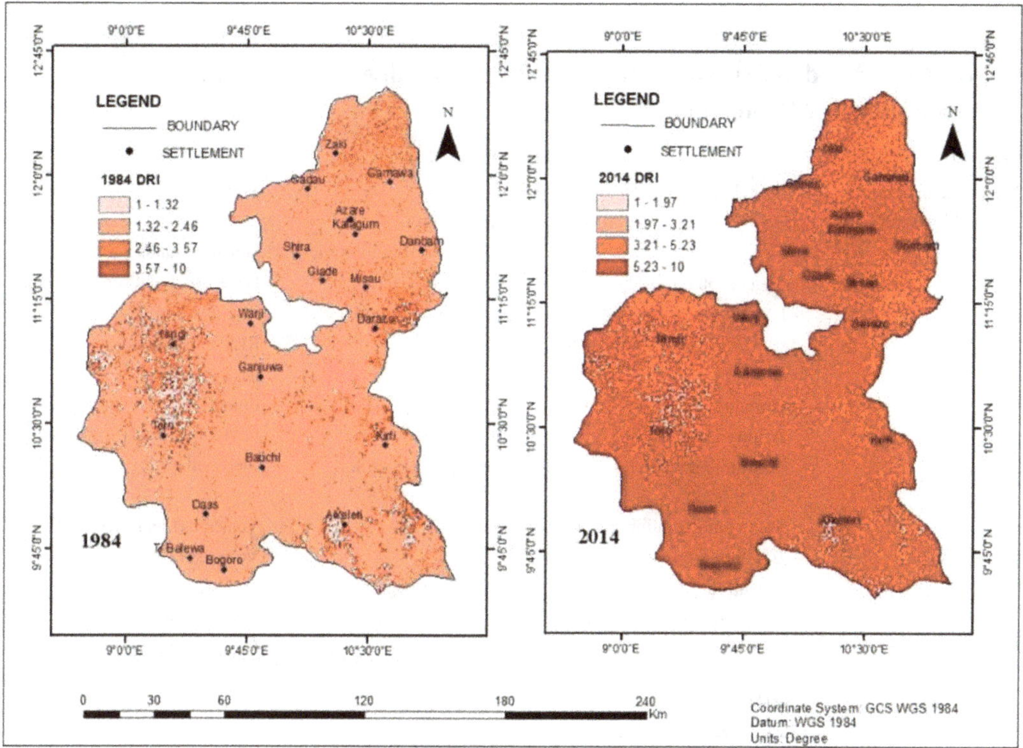

FIGURE 5.12 The worsening trend of degradation risk in Bauchi State, North Nigeria from 1984 to 2014. (Source: Joseph et al., 2018. Used with permission.)

where P_i = the proportion of rainfall in the ith month; S = mean monthly precipitation averaged over 12 months. The data needed to generate DRI include satellite images and meteorological data over a period that must be decades-long. The images are classified to detect land use/land cover changes using supervised image classification and are used to derive NDVI. The thermal band of the image is used to map LST. Another index similar to DRI called the DTRI is computed from Landsat images, population data, and climatological data (Becerril-Piña et al., 2015). DTRI is a low-cost and easy-to-use tool for assessing and monitoring desertification. Although both DRI and DTRI are initially generated quantitatively, they are commonly categorized into a few groups (e.g., 4) to show the spatial distribution of degradation risk (Figure 5.12).

At the national level, a whole different set of factors is used to derive desertification risk. They fall into three groups: pressures, state, and response/consequence (Table 5.11). Consequence factors are related to desertified land, and the land is sensitive and vulnerable. Pressure factors are associated with the desertification process, such as soil erosion, land salinization, drought, and exploitative land-use practices. Four groups of factors in this category and three environmental parameters (climate, soil, and vegetation) have been used to assess the desertification risk along the Mediterranean coast of Italy (Costantini et al., 2004). Apparently, some factors in Table 5.11 such as urbanization are not applicable at the local and even regional levels. Besides, land salinization is a factor not universally applicable. It is only surface denudation and climate that can be used at all scales around the globe.

The integral consideration of so many biophysical and socioeconomic aspects of desertification in assessing desertification vulnerability requires a suitable, robust, and scientific framework, which is easily achievable using a GIS, but not remote sensing. The role of remote sensing in the assessment is confined to mapping those variables needed in assessing and modeling

TABLE 5.11
Factors Considered in Assessing Areas Sensitive and Vulnerable to the Risk of Desertification and Drought in Italy

Pressure	State		Response
Surface denudation caused by water erosion	Soil characteristics	Slope	Vulnerable lands
		Rooting length	
		Presence of rills and gullies	Sensitive lands
	Vegetation cover	NDVI	Desertified and sensitive lands
	Human activities	Livestock population	Risk increase
		Protected areas	Risk mitigation
		Burnt areas	Risk increase
Drought	Climatic region		Potentially vulnerable lands (national scale)
	Aridity index		
	Soil moisture and temperature regimes (following Soil taxonomy)		
	Mean annual number of days when the soil is dry		Vulnerable lands (regional scale)
Salinization	Distance from the sea		Vulnerable lands
	Elevation		
	Irrigated vulnerable lands		Sensitive lands
Urbanization	Urban areas		Desertified lands

Source: Costantini et al. (2004).

desertification risk and susceptibility. The same desertification risk assessment method may be repeated using remote sensing data from different years. Just like desertification severity, desertification risk can be analyzed using time-series remote sensing data. A comparison of multi-temporal degradation risk maps reveals whether it has improved or worsened. To ease the comparison, it is better to convert DTRI to a few levels, such as low (0–3), medium (4–8), high (9–13), and extreme (14–27) risk, prior to the comparison (Becerril-Piña et al., 2015). Comparison of multi-temporal desertification risk maps reveals how it has changed over the years spatially.

5.5.3 Vulnerability/Susceptibility

At the regional or local level, such as in a district in Nigeria, much more detailed topographic features (e.g., slope, aspect, and curvature) are considered in assessing the areas sensitive and vulnerable to the risk of desertification (Table 5.12). The derivation of desertification vulnerability is commonly accomplished using two methods: modeling and regression analysis. The former can be implemented in the EPIC (Environmental Policy Integrated Climate) calculator. It enables drought risk to be evaluated at the national level, via the estimation of soil moisture and temperature, and at the regional level, by estimating the long-term mean number of days when the soil is dry per annum. This indicator is particularly useful for assessing land vulnerability in agricultural areas, where NDVI is not so useful.

One specific type of regression analysis is called logistic regression (LR). It is a multivariate statistical approach for establishing the regression relationship between desertification risk and a given indicator based on the selected training samples. This flexible method allows as many variables as possible to be considered, but only those significant to desertification vulnerability are selected based on the user-specified confidence level. Furthermore, LR can yield information on the strength and direction of the association between the dependent variable and all the considered independent variables. The dependent variable is the calculated DRI or a simple binary variable of desertification presence/absence in the form of

$$P = \frac{1}{1+e^{-z}} \tag{5.13}$$

where P = the probability of desertification occurrence expressed on a scale of 0–1; z = the linear logistic model, expressed as

$$z = \alpha + \Sigma \beta_i X_i \tag{5.14}$$

where α = intercept of the model; β_i (i = 1, 2, …, n, n = number of variables considered) = the estimated coefficients of the LR model, and X_i = the ith independent predictor variable. In Eq. 5.13, P can be the observed value, expressed as a ratio of the number of desertified pixels to the total number of pixels of the same attribute value. The number of specific attribute values varies among the considered factors, such as from 2 for NDVI and drainage density to 11 for land cover and topography (Table 5.12).

In the analysis, the probability of desertification (%) is regressed against all potential indicators in a process known as incremental growth. LR is strong at revealing the importance of all variables considered and can minimize their joint and interactive effects. Whether the considered variables are important to desertification is judged by the p -value. As shown in Table 5.13, 13 of the 16 independent variables are deemed influential to desertification susceptibility because their p-value is smaller than 0.05. LR analysis reveals that biological factors, such as NDVI and rangeland cover,

TABLE 5.12

The Proportion of Desertified Areas (%) at a Specific Property in 16 Factors Used to Assess Desertification Vulnerability

Factor	Class	Desertified (%)	Factor	Class	Desertified (%)
TPI	−9.56–0	46.33		Agriculture	19.32
	0–8.78	53.67		Grazed cropland	7.81
HI	<0.35	0.00		Cultivated rangelands	0.59
	0.35–0.6	93.89		Rangelands	64.27
	>0.6	6.11	**Land cover**	Forest	1.04
Slope (°)	<5	84.62		Afforestation	0.45
	5–10	14.28		Esparto grass	5.76
	>10	1.10		Dune	0.25
Curvature (1/m)	<0	41.31		Sebkha	0
	0	13.31		Uncultivated area	0.14
	>0	45.38		Urban area	0.38
	Flat	2.90		<100	17.51
	N - (337.5–22.5)	12.19		100–200	26.36
	S - (22.5–67.5)	14.07		200–300	28.75
	W - (67.5–112.5)	10.30	**Distance to drainage (m)**	300–400	17.15
Aspect	E - (112.5–157.5)	12.09		400–500	8.16
	NW - (157.5–202.5)	12.18		500–600	1.91
	NE - (202.5–247.5)	11.98		>600	0.15
	SW - (247.5–292.5)	10.17	**LST (°C)**	<38	36.43
	SE - (292.5–337.5)	14.73		>38	63.57

(Continued)

TABLE 5.12 (*Continued*)

The Proportion of Desertified Areas (%) at a Specific Property in 16 Factors Used to Assess Desertification Vulnerability

Factor	Class	Desertified (%)	Factor	Class	Desertified (%)
Lithology	Alluvium	27.46	**Rural population**	<0.49	65.00
	Conglomeratic clay-sand, lacustrine and limestone	2.13		>0.49	35.00
	Clay	31.40	**Livestock density**	0.48–2.88	20.56
	Schist-marls, few limestone and gypsum intercalations	12.66		2.89–9.94	60.34
	Massive limestone and clay	2.75		9.95–17	4.99
	Alternating sandstone and clay	20.49		>17	14.11
	Chotts	0.00	**Evapotranspiration**	1,300–1,400	7.19
	Sand	3.12		1,400–1,500	25.76
Elevation	700–800	19.33		1,500–1,600	65.58
	800–900	33.11		1,600–1,700	1.47
	900–1,000	7.51	**Precipitation**	<300	43.92
	1,000–1,100	11.22		300–400	54.54
	1,100–1,200	10.68		400–500	1.54
	1,200–1,300	17.06		>500	0.00
	>1,300	1.09	**Drainage density**	1–2	55.31
NDVI	<0.22	96.21		2–4	44.69
	>0.22	3.79			

Source: Djeddaoui et al. (2017).
Note: TPI – topographic position index; HI – hypsometric integral

TABLE 5.13

Relationship between Desertification Likelihood (Expressed As a Ratio) and Its Contributors (Predictive Factors) in the Final Logistic Model Derived from the Training Dataset

Category	Factor	β_i	Std Error of Estimate	Wald χ^2 Value	*p*-Value	Exp(β_i) (Odds Ratio)
Land cover	Rangelands	0.800	0.054	14.805	0.000	2.22
	Cultivated rangelands	0.473	0.238	1.986	0.047	1.61
	Grazed cropland	0.405	0.088	4.628	0.000	1.50
Slope aspect	South aspect	0.481	0.084	5.663	0.000	1.62
	Northwest aspect	–0.166	0.068	–2.462	0.014	0.85
Lithology	Sand	0.909	0.138	6.583	0.000	2.48
	Massive limestone in banks or platelets	0.412	0.168	2.448	0.014	1.51
Slope		–0.209	0.009	–22.617	0.000	0.81
NDVI		–52.391	1.270	–41.242	0.000	~0
HI		–2.982	0.373	–8.006	0.000	0.05
Precipitation		–0.420	0.054	–7.763	0.000	0.66
Drainage density		0.176	0.073	2.416	0.016	1.19
TPI		0.198	0.016	12.263	0.000	1.22

Source: Djeddaoui et al. (2017).

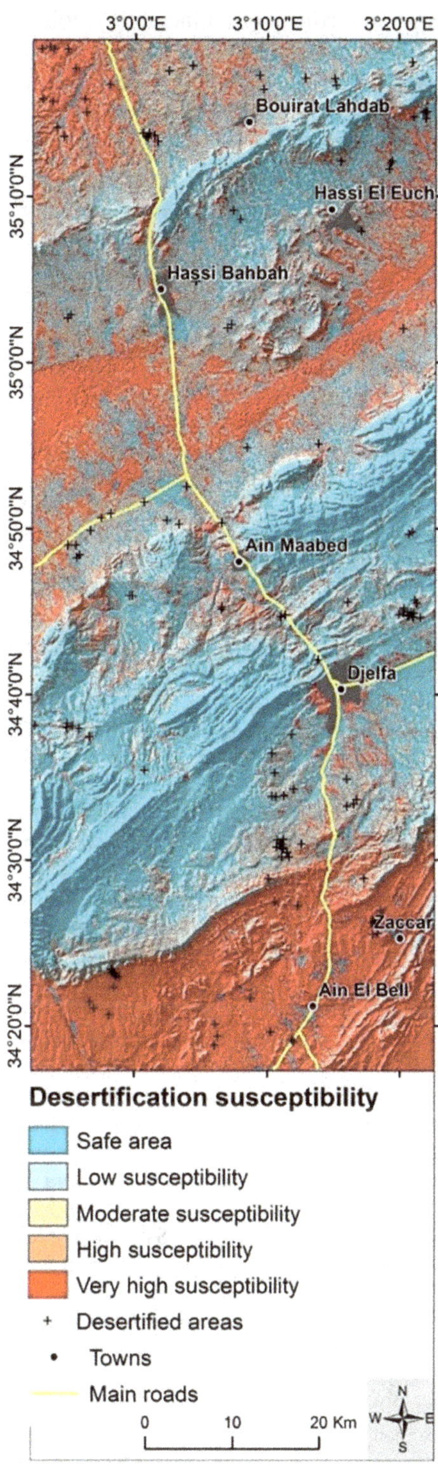

FIGURE 5.13 Desertification susceptibility in the Djelfa area, Algeria derived from 16 variables listed in Table 5.12 using logistic regression analysis. (Source: Djeddaoui et al., 2017.)

bear the closest relationship with desertification vulnerability, followed by climatic factors, while geomorphological factors are not so important to desertification (Djeddaoui et al., 2017).

Identical to desertification risk, the regressed desertification vulnerability is also grouped into four or five classes, such as safe (0–0.14 probability), low (0.15–0.36), moderate (0.37–0.60), high (0.61–0.83), and very high (0.84–1.00) (Figure 5.13). Such desertification vulnerability analysis is carried out usually at the regional or local level, but not at the national or global level due to data unavailability at broader scales.

REFERENCES

Abahussain, A.A., A.S. Abdu, W.K. Al-Zubari, N.A. El-Deen, and M. Abdul-Raheem. 2002. Desertification in the Arab region: Analysis of current status and trends. *J. Arid. Environ.* 51: 521–545.

Albalawi, E.K., and L. Kumar. 2013. Using remote sensing technology to detect, model and map desertification: A review. *J. Food Agric. Environ.* 11: 791–797.

Becerril-Piña, R., C.A. Mastachi-Loza, E. Gonzalez-Sosa, C. Diaz-Delgado, and K.M. Ba. 2015. Assessing desertification risk in the semi-arid highlands of Central Mexico. *J. Arid Environ.* 120: 4–13.

Collado, A.D., E. Chuvieco, and A. Camarasa. 2002. Satellite remote sensing analysis to monitor desertification processes in the crop-rangeland boundary of Argentina. *J. Arid Environ.* 52: 121–133.

Costantini, E.A.C., M. Bocci, G. L'Abate, A. Fais, G. Leone, S. Loj, S. Magini, R. Napoli, P. Nino, and F. Urbano. 2004. Mapping the state and risk of desertification in Italy by means of remote sensing, soil GIS and the EPIC model: Methodology validation in the Sardinia island, Italy. In *Proc. International Symposium: Evaluation and Monitoring of Desertification, Synthetic Activities for the Contribution to UNCCD.* Tsukuba, Japan: NIES Publication.

Djeddaoui, F., M. Chadli, and R. Gloaguen. 2017. Desertification susceptibility mapping using logistic regression analysis in the Djelfa area, Algeria. *Remote Sens.* 9(10): 1031. DOI: 10.3390/rs9101031

Dragan, M., T. Sahsuvaroglu, I. Gitas, and E. Feoli. 2005. Application and validation of a desertification risk index using data for Lebanon. *Manag. Environ. Qual.: Int. J.* 16(4): 309–326.

Duan, H.C., T. Wang, X. Xue, S.L. Liu, and J. Guo. 2014. Dynamics of aeolian desertification and its driving forces in the Horqin Sandy Land, Northern China. *Environ. Monit. Assess.* 186(10): 6083–6096. DOI: 10.1007/s10661-014-3841-3

Elhadi, E.M., N. Zomrawi, and G. Hu. 2009. Landscape change and sandy desertification monitoring and assessment. *Am. J. Environ. Sci.* 5(5): 633–638.

Elhag, M.M. 2006. Causes and impact of desertification in the Butana area of Sudan. Ph.D. dissertation, University of the Free State, Bloemfontein.

Fang, L., Z. Bai, S. Wei, H. Yanfen, W. Zongming, S. Kaishan, L. Dianwei, and L. Zhiming. 2008. Sandy desertification change and its driving forces in western Jilin Province, North China. *Environ. Monit. Assess.* 136: 379–390.

Feng, L., Z. Jia, and Q. Li. 2016. The dynamic monitoring of aeolian desertification land distribution and its response to climate change in northern China. *Sci. Rep.* 6: 39563. DOI: 10.1038/srep39563

Ghebrezgabher, M.G., T. Yang, X. Yang, and C. Wang. 2019. Assessment of desertification in Eritrea: Land degradation based on Landsat images. *J. Arid Land* 11(3): 319–331. DOI: 10.1007/s40333-019-0096-4

Guo, Q., B. Fu, P. Shi, T. Cudahy, J. Zhang, and H. Xu. 2017. Satellite monitoring the spatial-temporal dynamics of desertification in response to climate change and human activities across the Ordos Plateau, China. *Remote Sens.* 9(6): 525. DOI: 10.3390/rs9060525

Harahsheh, H., and R. Tateishi. 2001. Desertification mapping. *GIS Develop.*, July 2001, 15 p.

Joseph, O., A.E. Gbeng, and D.G. Langyit. 2018. Desertification risk analysis and assessment in Northern Nigeria. *Remote Sens. Appl.: Soc. Environ.* 11: 70–82.

Khire, M.V., and Y.Y. Agarwadkar. 2014. Qualitative analysis of extent and severity of desertification for semi-arid regions using remote sensing techniques. *Int. J. Environ. Sci. Dev.* 5: 238–243.

Kosmas, C., M. Tsara, N. Moustakas, and C. Karavitis. 2003. Identification of indicators for desertification. *Annals Arid Zone* 42(3&4): 393–416.

Kosmas, C., M.J. Kirkby, and N. Geeson (ed). 1999. *Medalus Project: Mediterranean Desertification and Land Use: Manual on Key Indicators of Desertification and Mapping Environmentally Sensitive Areas.* Brussels, Belgium: European Commission, Energy, Environment and Sustainable Development.

Krugmann, H. 1996. Grassroots indicators and scientific indicators: Their role in decentralized planning in the arid lands of Uganda. In *Grassroots Indicators for Desertification: Experience and Perspectives from Eastern and Southern Africa*, eds. B. Hambly and T. O. Angura. Ottawa, Canada: International Development Research Centre.

Kundu, A., N.R. Patel, S.K. Saha, and D. Dutta. 2015. Monitoring the extent of desertification processes in western Rajasthan (India) using geo-information science. *Arab. J. Geosci.* 8: 5727–5737. DOI: 10.1007/s12517-014-1645-y

Li, J., X. Yang, Y. Jin, Z. Yang, W. Huang, L. Zhao, T. Gao, H. Yu, H. Ma, Z. Qin, and B. Xu. 2013. Monitoring and analysis of grassland desertification dynamics using Landsat images in Ningxia, China. *Remote Sens. Environ.* 138: 19–26. DOI: 10.1016/j.rse.2013.07.010

Liang, S.L. 2001. Narrowband to broadband conversions of land surface albedo I algorithms. *Remote Sens. Environ.* 76: 213–238.

Othman, A.A., Y.I. Al-Saady, A.K. Al-Khafaji, and R. Gloaguen. 2014. Environmental change detection in the central part of Iraq using remote sensing data and GIS. *Arab. J. Geosci.* DOI: 10.1007/s12517-013-0870-0

Perez-Trejo, F. 1994. Desertification and land degradation in the European Mediterranean, EUR 14850 EN, Office for Official Publications of the European Communities, Luxembourg.

Sharma, K.D. 1998. The hydrological indicators of desertification. *J. Arid Environ.* 39: 121–132.

Symeonakis, E., and N. Drake. 2004. Monitoring desertification and land degradation over sub-Saharan Africa. *Int. J. Remote Sens.* 25(3): 573–592. DOI: 10.1080/0143116031000095998

Wang, T., C.Z. Yan, X. Song, and J.L. Xie. 2012. Monitoring recent trends in the area of aeolian desertified land using Landsat images in China's Xinjiang region. *ISPRS J. Photogramm.* 68: 184–190.

Wu, B., and L.J. Ci. 2002. Landscape change and desertification development in the Mu Us Sandland, Northern China. *J. Arid Environ.* 50: 429–444. DOI: 10.1006/jare.2001.0847

Xiao, J., Y. Shen, R. Tateishi, and W. Bayaer. 2006. Development of topsoil grain size index for monitoring desertification in arid land using remote sensing. *Int. J. Remote Sens.* 27: 2411–2422.

Zhang, Y., Z. Chen, B. Zhu, X. Luo, Y. Guan, S. Guo, and Y. Nie. 2008. Land desertification monitoring and assessment in Yulin of Northwest China using remote sensing and geographic information systems (GIS). *Environ. Monit. Assess.* 147: 327–337. DOI: 10.1007/s10661-007-0124-2

Zhou, W., C. Gang, F. Zhou, J. Li, X. Dong, and C. Zhao. 2015. Quantitative assessment of the individual contribution of climate and human factors to desertification in northwest China using net primary productivity as an indicator. *Ecol. Indic.* 48: 560–569.

6 Land Subsidence

6.1 INTRODUCTION

Land subsidence refers to the lowering of the ground surface below its normal height, stemming from sinking or settling. This process takes place gradually or suddenly, depending on how the underlying ground is altered. Subsidence is caused by a variety of factors, such as the removal of substances within the underlying earth in the forms of groundwater extraction and underground mining, compaction of loose soil, erosion of materials during rain, formation of sinkholes, and underground construction (e.g., tunneling). Large-scale subsidence can occur in coal mining areas after the coal pillars have burned or collapsed, or as a result of coal fires in conjunction with coal extraction. The compaction of loose soil is commonly associated with engineering works during which the ground is not allowed to sink sufficiently or the newly added materials are not compacted adequately before construction starts. Subsidence in newly developed urban areas is triggered by the added weight of human-made features, such as buildings and infrastructure atop newly laid foundations.

The magnitude of land subsidence is strongly associated with certain types of land cover (Ahmad et al., 2019). Urban areas and orchard vegetation are prone to excessive cumulative land subsidence due to the extraction of a large quantity of groundwater, but seasonally cultivated and barren land is barely affected by land subsidence. Subsidence triggered by different forces occurs at varying paces. Sinkhole-related subsidence takes place rather slowly and gradually and is always confined to the local scale. However, mining-triggered subsidence takes a long time to complete at variable paces, affecting extensive areas. For instance, longwall mining-induced subsidence is rather slow at only a few mm per day in the first 6–8 months, but much faster at around 1 cm·day^{-1} in the next 6–12 months, then almost non-existent afterward (Pawluszek-Filipiak and Borkowski, 2020), affecting 100s km^2 of land.

Surface subsidence is a worldwide problem. It is especially prevalent in China (Hu et al., 2004) and takes place over different scales, depending on how the underground materials are excavated or affected. Subsidence is a serious hazard common in underground mining areas or in areas where groundwater is extracted at an unsustainable level. Once initiated, subsidence spreads to adjacent areas, potentially affecting agricultural, industrial, and urban land (Figure 6.1). Because subsidence can expose the underground coal to air by fracturing (e.g., the pillar left after coal extraction), it raises the risk of initiating new coal fires. As a natural, and to some degree, anthropogenic in origin, hazard, land subsidence does not cause as much damage as earthquakes or floods. The damage caused by surface subsidence does vary spatially and results in different consequences. If taking place in agricultural areas, it affects only crop yield and hampers tillage and harvesting. Subsidence of river deltas causes aquifer salinization, infrastructure damage, and increases the vulnerability to floods and storm surges and permanent inundation of low-lying land. In urban areas, land subsidence impacts infrastructure, the environment, economy, and society. Surface subsidence has disastrous consequences for human lives if it underlies important roads and railways. Apart from the infrastructure itself, abrupt subsidence does additional damage to those who make use of it. For instance, the sudden subsidence of a paved road can suck in vehicles traveling in the collapsed part.

In order to minimize the risk of subsidence, it is important to monitor it using the same remote sensing methods as those for monitoring landslide-triggered deformation, in addition to field methods. Conventional field-based monitoring methods include leveling, layerwise marks, bedrock benchmarks, total stations, and GNSS, all of which can yield highly accurate measurements but are labor-intensive and time-consuming except for fixed GPS stations. Despite this drawback, field

DOI: 10.1201/9781003354321-6

FIGURE 6.1 Agricultural land subsidence on the Loess Plateau of West China caused by underground coal mining. (Source: Wang et al., 2017. Used with permission.)

surveys are still carried out to verify the accuracy of remote sensing-derived measurements of subsidence. For instance, GPS is able to yield spot elevations that can be converted to a field of subsidence distribution in the regular grid form (i.e., a DEM) essential in correcting the topographic phase in some remote sensing methods. The non-remote sensing methods are also included in the comparison of all elevation-yielding methods in this chapter.

In comparison, remote sensing-based monitoring is much more efficient, especially over large areas. Remote sensing can extend the monitoring of subsidence over a long time. The yielded time-series information is crucial for calibrating other data in modeling and predicting subsidence. So far, several imagining and non-imagery methods have been developed for this purpose, with satisfactory results. These remote sensing methods and/or data include optical images, InSAR, and non-imagery ALS.

This chapter first explains the principles of detecting subsided ground using radar sensing. Then the strengths and limitations of a wide range of radar data in sensing ground subsidence are compared and contrasted. A substantial component of this chapter is dedicated to discuss how to quantify ground subsidence using several imagery and non-imagining methods, particularly interferometric SAR (InSAR) methods. The discussion emphasizes the retrieval of surface height from various manners of implementing InSAR-based detection. How to detect subsidence using non-imagery GPS and LiDAR data forms the third component of this chapter. Although GPS and leveling are not remote sensing, they are still covered as the results are usually used to validate remotely sensed subsidence. This chapter ends with a comparison and assessment of all the imagery and non-imagery methods in terms of their pros, cons, and accuracy achievable in sensing subsidence, with the causes of inaccuracies identified wherever possible.

6.2 OPTICAL IMAGERY SENSING

Most optical images are obtained via nadir scanning, by positioning the sensor right above the sensed area to minimize image geometric distortion. As such, optical images are not sensitive to surface relief in general, and surface subsidence in particular as subsidence is manifested virtually as vertical displacement. Besides, most optical Earth resources satellite images have a spatial resolution of around 10 m or coarser (Table 1.6), inadequate to discern local-scale subsidence, let alone quantify it. In some cases, the vertical displacement in elevation in subsided land is too small to be perceived spectrally from even VHR images. Subsidence is mostly visible from morphological features on aerial photographs. For instance, subsidence caused by coal mining and coal fires is mapped from 1:10,000 color infrared aerial photographs based on its two pronounced characteristics of scarps and cracks (Chen, 1997). The presence of scarps facilities manual delineation of subsided areas. In the absence of scarps, subsidence has to be inferred from cracks. If subsidence triggered by underground coal mining is excessive, it can be easily detected from morphological features, but not if it is too subtle.

While not directly useful in studying subsidence, a 3D view of the surface can be constructed from two types of space-borne optical images: SPOT and ASTER. A stereoscopic view of the ground surface is obtained from consecutively acquired overlapping images from either two adjacent paths such as with SPOT or the same path as with ASTER. ASTER has a spatial resolution ranging from 15 to 30 m, too coarse to study subsidence. These medium-resolution data can be used to study the relationship between land subsidence and urban expansion (Cao et al., 2021), and to identify and categorize the spatial distribution of areas affected by land subsidence. Both SPOT and ASTER images allow the construction of DEMs over subsided areas. The subtraction of DEMs before and after subsidence is able to reveal the vertical change in surface elevation. This method is viable only for severe subsidence exceeding meters due to the coarse spatial resolution and high uncertainty in the data quality. The quantification of the pace of land subsidence requires much finer spatial resolution data, such as IKONOS and WorldView-3 (Table 1.5). VHR images have a vertical accuracy of up to a few decimeters. They are incapable of detecting subsidence over the range of <10 mm·year^{-1} in most urban areas. Although WorldView-3 images have a sub-meter spatial resolution, they do not have the 3D viewing capability. Thus, only microwave images are a viable data source capable of the task.

6.3 SYNTHETIC APERTURE RADAR

6.3.1 Principles of Sensing

The generation of fine-resolution radar imagery requires the use of a long antenna, which is technically challenging to accommodate. This requirement is circumvented by using SAR. It makes use of a small aperture antenna to receive backscattered radar signals, the phase and amplitude of which are stored and later assembled to create an aperture length that is much larger than the physical antenna dimension, thus leading to a superior spatial resolution. The synthetic antenna length is directly proportional to the range, producing a synthetic beam with a constant width in the azimuth direction (Ra), usually made to equal the range resolution (Rr). Inherently, SAR images are acquired via side-looking scanning, sensing the ground to either side of the flight path (Figure 6.2).

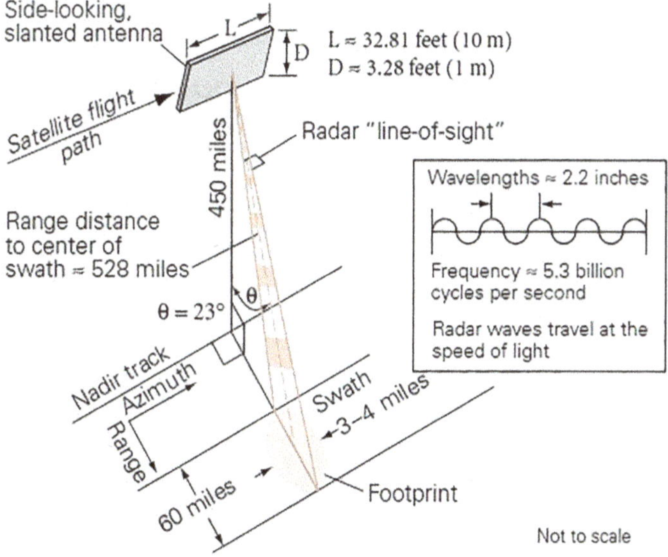

FIGURE 6.2 Simplified geometry for the European Remote Sensing satellite (cross-section of side-looking radar scanning) to separate two features on the ground. (Source: USGS, 2000.)

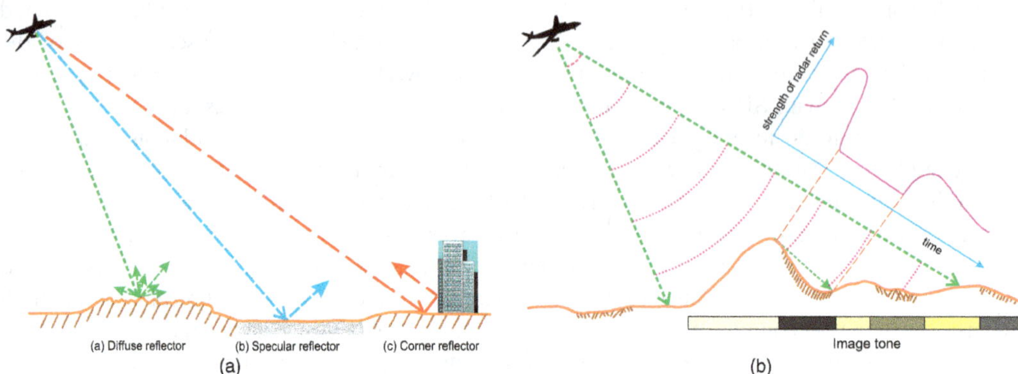

FIGURE 6.3 (a) Three representative types of surface scattering of incident radar radiation, and (b) the intensity of the bounced signal in relation to topographic relief.

This peculiar way of scanning makes the images not only sensitive to surface relief but also prone to topographic shadow and radar layover. A stereoscopic view of the ground being sensed is established from two slide-looking SAR images recorded in adjoining tracks, or from the same track but at two slightly different viewing angles. Two adjacent objects on the ground are resolved on a radar image based on their different ranges to the radar antenna. This difference is reflected by a phase difference between the two returned radar pulses.

Dissimilar to optical imagery that captures reflected energy, SAR imagery records the energy scattered back from the target. The intensity of the backscattered radar signal is a function of surface roughness or micro-relief, relative to the wavelength of the radar radiation. A smooth surface like water usually specularly reflects the incident radar pulses, while a rough surface diffuses the incoming radiation (Figure 6.3a). The intensity of the received signal is also subject to the orientation of the surface slope in relation to the incoming radiation if the surface has a relief. Those slopes facing the antenna tend to have a strong return, while slopes facing away from the antenna hardly scatter any radiation as they lie in the radar shadow (Figure 6.3b).

6.3.2 Data Utility

All the SAR images that can be used to detect surface subsidence are listed in Table 1.7. Due to the space limit, it is not possible to list all the properties of an image in detail. With the new generation of radar missions, the radar sensor is so versatile that it can operate in multiple modes, albeit not simultaneously. The mode of sensing is changeable by reprograming the sensor. Table 6.1 details the operating modes of the latest generation of radar sensors, each of which can sense the target in at least three modes, with RadarSat-2 having the most modes totaling 11. The mode of sensing affects both image swath width and spatial resolution. There exists an inverse relationship between these two critical parameters, which means there must be a compromise between them. Which mode should be selected for sensing depends on the target of sensing and its physical dimension. The fine-resolution mode is vital to study slow-paced subsidence, but coarser resolution modes are essential to study land subsidence over an extended area. However, the interferometric wide swath mode of Sentinel-1 should be the default choice for detecting ground subsidence in most cases.

VHR SAR imagery is excellent for studying subsidence because of its fine spatial resolution, repetitive ground coverage, and the ability to penetrate clouds. Different satellite SAR missions have a variable capability of detecting and mapping subsidence. Sentinel-1 imagery is a valuable source of data for mapping the spatiotemporal trend of land subsidence. The high phase gradient issue can be successfully addressed by tackling image spatial resolution, incidence angles, revisit cycle, and wavelength (Ng et al., 2009), all available from the new generation of satellite missions.

TABLE 6.1
The Main Properties of SAR Images That Have the Potential to Quantify Ground Subsidence

Satellite Mission	Mode of Sensing	Swath Width (km)	Spatial Resolution (m) Range	Azimuth	Incidence Angle (°)
Sentinel-1A/1B	StripMap	80	5	5	18.3–46.8
	Interferometric wide swath	250	5	20	29.1–46.0
	Extra-wide swath model	400	20	40	18.9–47.0
	Wave mode	20	5	5	21.6–25.1
					34.8–38.0
	Interferometric wide swath	250	5	20	29.1–46.0
TerraSAR-X	Spotlight	10	2	1	15–60
	StripMap	30	3	3	15–60
	ScanSAR	100	16	16	20–60
COSMO-SkyMed	Spotlight	10	0.6–0.8	0.3–0.8	20–60
	StripMap	40	3	3	20–60
	ScanSAR	100–200	4–6	20–40	20–50
RadarSat-2	**Wide ultra-fine**	**20**	**1.6–3.3**	**2.8**	**30–50**
	Wide (multi-look) fine	**50**	**3.1–10.4**	**4.6–7.6**	**29–50**
	Wide fine	**50**	**5.2–15.2**	**7.7**	**20–45**
	Wide fine Quad-Polar	**25**	**5.2–17.3**	**7.6**	**18–42**
	Wide standard Quad-Polar	**25**	**9.9–30.0**	**7.6**	**18–42**

Note: **Bold**: in the beam mode

However, these problems are minimal with ALOS PALSAR interferograms due to their superior spatial resolution up to 1 m and longer wavelength of 22.9 cm (L-band). ALOS data have been used to derive height information (Ge et al., 2007), with accuracies reported in the range of 10s m. ALOS-2 is more suited for mapping mine subsidence than Sentinel-1 and RadarSat-2. However, ALOS PALSAR is found to overestimate deformation, probably as a result of large baselines and the limited number of images used (Grzovic and Ghulam, 2015). Hence, they are suitable for monitoring ground subsidence in areas experiencing a high rate of deformation and covered by vegetation that radar can penetrate. In vegetation-covered areas, JERS-1 imagery of a long wavelength (23.5 cm, L-band) is more robust for measuring mine subsidence than ERS-1 and ERS-2 of a short wavelength (5.6 cm, C-band). It is also more suited to detect large ground subsidence deformation (Ge et al., (2007) because images of a short wavelength are more sensitive to mine subsidence.

In practice, it may be beneficial to make use of multi-sensor data to extend the duration of monitoring to several decades. If the SAR images have a reasonably short temporal interval between them, they allow the processes of land subsidence to be studied.

6.3.3 InSAR

InSAR is a method of producing interference by superimposing electromagnetic waves from a pair of SAR images. It has proved to be the best method for detecting vertical changes in general and subsidence in particular.

6.3.3.1 Fundamental Principles

Radar sensing differs from optical sensing in that the sensor captures the two components, amplitude and phase, of the received microwave signals. The former records the radar signal intensity received at the radar antenna. The phase component is proportional to the range (the distance

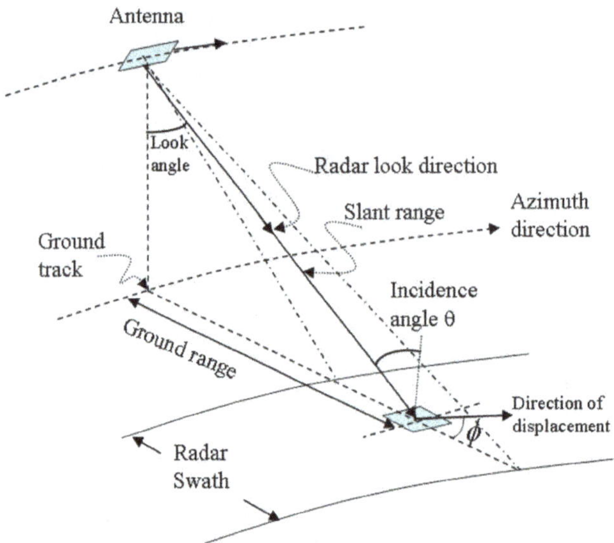

FIGURE 6.4 The geometry of InSAR sensing. (Source: Zhou et al., 2009.)

between the target and the sensor) difference. The "interferes" (differences) of these images or phase difference is known as an interferogram, the phase change between two SAR images. Any motion on the ground (e.g., subsidence) causes a slightly different portion of the wavelength to be scattered back to the radar antenna. Any change in a target's phase is related to the change in its position. Interferograms are established from a pair of side-looking SAR images of the same area obtained consecutively in the same path (Figure 6.4). If obtained from different passes, they must "view" the same ground from a slightly different perspective at the same spot but at different times (as with the use of one satellite) or from two different spots at the same time (as with the use of two satellites). Single-pass interferometry is formed from SAR images obtained using two antennas, one for transmitting the signal toward the target and another for receiving the signal backscattered from the target. Repeat-pass interferometry is formed from SAR images that are obtained from a similar position in the second pass to the previous one in the first pass. Thus, the same ground is sensed from the same antenna twice, at a time interval from each other.

InSAR tracks the difference in the phase information contained in SAR images. If the scattering phase is unchanged in the image pair, the interferometric phase $\Delta\phi_{disp}$ is a function of the range difference, $r2 - r1$, or

$$\Delta\phi = \phi 1 - \phi 2 = \frac{4\pi(r2 - r1)}{\lambda} \tag{6.1}$$

where $\phi 1$ and $\phi 2$ = the phase of the first and second SAR images, $r1$ and $r2$ = the ranges to the sensor, and λ = wavelength of the radar radiation used (e.g., 5.6 cm for C-band or 22.9 cm for L-band). The similarity of ground scattering on the two images is measured by coherence. According to the simplified geometry illustrated in Figure 6.4, the look direction of the radar antenna is perpendicular to its motion. In this case, the differential phase between the two repeat-pass SAR images is expressed as

$$\Delta\phi = \frac{4\pi \cdot B \cdot sin(\theta - \alpha)}{\lambda} + \frac{4\pi \cdot \Delta r}{\lambda} \tag{6.2}$$

where θ = look angle at the time of sensing; α = angle of the baseline vector measured from the orbital horizontal; B = distance between the positions of the two antennas known as the baseline; Δr = displacement of the pixel from one pass to the next in the line-of-sight (LOS) direction.

If the natural variation in the reference surface is ignored (e.g., $\Delta\phi = 0$), Eq. 6.2 can be rewritten as follows:

$$\Delta\phi_{flat} = \frac{4\pi}{\lambda} B cos(\theta_0 - \alpha) \frac{\Delta h}{r_0 sin\theta_0} + \frac{4\pi}{\lambda} \Delta r \tag{6.3}$$

where θ_0 = the look angle to the reference surface; r_0 = the range to the reference surface; Δh = height of the point above the reference surface. $\Delta\phi$ is inversely related to λ. In the same image, it is proportional to Δh.

Subtraction of the two phases effectively eliminates random scatters in the phase. The interferometric phase is then converted to displacement along the LOS direction by multiplying by the correction factor of $\lambda/4\pi$, if the incidence angle of the radar is quite small, and most of the surface deformation associated with land subsidence is assumed to be only vertical. If the two SAR images are acquired simultaneously at different positions, the phase difference is attributed mostly to the ground surface. If they are captured from the same position but at different times (e.g., after revisit), the phase difference is caused mostly by surface change. If the ground is imaged from different positions at different times, the phase difference is reflective of both topography and its change. These maps reveal the LOS ground surface displacement (range change) between the two times. The computed interferograms allow subsidence-triggered surface deformation to be measured (i.e., the negative interferogram values). Depending on the SAR images used, InSAR can detect cm-level surface subsidence, over large areas at a spatial resolution of 90 m or finer.

The quality of an interferogram is subject mostly to the influence of temporal and spatial baseline, in addition to ground cover, atmospheric conditions, and surface relief. Any changes to the scatterers caused by relative motion or in the look angle of sensing alter the scatterers' contribution to the sum, an effect known as decorrelation. It can be caused by an increase in the baseline between satellite passes. Decorrelation can be spatial and temporal. Temporal decorrelation rises rapidly with temporal baseline. If it exceeds four months, very low coherence exists, especially in vegetated areas. Dense vegetation weakens coherence and prevents the determination of vertical displacements to a sufficiently high accuracy level. Hence, interferograms should be generated from image pairs of a small temporal and spatial baseline to minimize decorrelation. In order to detect subsidence to the highest accuracy level, it is advisable to adopt a short baseline by using several interferograms generated from multi-temporal SAR images (Ge et al., 2007). In this way, the variation in the atmospheric conditions is minimized. Other factors that may affect InSAR signal quality include sensing position (angle and range), the timeline of the interferogram, and the precision of controlling the look angle and position of the radar satellite.

InSAR has been widely used to detect ground subsidence because of widely available space-borne SAR data and data processing software, mostly free of charge. The success of using InSAR to detect land subsidence is contingent on several factors, such as sensing the same ground from the same position in space to minimize the horizontal distance between each satellite pass or the perpendicular baseline. Robust InSAR results are possible by addressing several InSAR issues, such as temporal decorrelation, baseline decorrelation, external DEM inaccuracy, and atmospheric effects (Zhao et al., 2009). Atmospheric artifacts and noise generated from radar data processing (e.g., phase unwrapping) cause positive values in the interferogram. Atmospheric artifacts add an atmospheric phase to SAR images that can severely degrade the accuracy of the detected subsidence. This is because the atmospheric phase screen's spectral behavior is subject to the effect of water vapor in the troposphere and cannot be estimated from the coherence map produced from an interferogram (Ferretti et al., 2001). Such noises can be detected easily as they are temporally and spatially random, differing from genuine subsidence that is always geographically immobile (Dehghani et al., 2009). Although mostly unpredictable, the impact of atmospheric variations such as clouds and water vapor can be removed via filtering. Image smoothing enhances the quality of the detected subsidence. With the assistance of a DEM, the phase difference may be further processed to remove the component caused by topography.

InSAR is an ideal method for assessing the spatial extent, magnitude, and temporal evolution of land subsidence rapidly, efficiently, and cost-effectively in urban areas underlain by loose loess soil. InSAR is good at detecting the temporal and spatial pattern of surface subsidence over extensive coverage (10,000 km^2) at a fine spatial resolution of around 10 m and high accuracy around 1 cm, which is impossible with leveling and GPS. Its accuracy is the highest over a horizontal distance ranging from meters to tens of kilometers. InSAR is capable of mapping mm-scale vertical ground deformation and can potentially detect cm-scale ground motion cost-effectively, subject to the influence of slant range, baseline, phase, and height (Liu et al., 2007). But it is limited by its inability to measure deformation with a footprint wider than a swath width of 50–300 km for most SAR satellites. This limitation is due to slight inaccuracies in satellite orbital knowledge, which produce a linear or quadratic signal across most interferograms. Another limitation of InSAR is that it can yield only 1D deformation between the sensor and the target in the LOS direction, which is sufficient for detecting subsidence, but not 3D deformation. It must be noted that high accuracy is not possible with the classic InSAR technology in monitoring land subsidence because of the highly variable atmospheric conditions under which multi-temporal SAR images are recorded. Thus, it is imperative to consider the atmospheric effects.

6.3.3.2 Atmospheric Effects

If a ground motion is detected from a set of interferograms constructed from a common SAR image, the detected apparent motion stems likely from atmospheric phase delay. Atmospheric variations in vapor, pressure, temperature, and humidity affect the propagation of radar signals. Atmospheric conditions such as fog and clouds slow down the radar signal propagation, while the non-uniform distribution of water vapor interferes with the propagation (Sneed and Brandt, 2013), even though microwave radiation is able to penetrate clouds to some degree. The delay variation in the repeat interval results in a "phase screen" that makes SAR images highly prone to the influence of spatiotemporal decorrelation. This effect is independent of baseline and has been tackled differently using a few methods. The simplest way of addressing the impact of tropospheric delay on phase is to spatiotemporally filter time-series interferograms by assuming a temporal deformation model, but it may not be accurate in some complex situations such as mining-induced subsidence. High-pass and low-pass filtering of the interferograms is commonly known as atmospheric phase screen. It is stochastic and works under the assumption that the atmospheric component is highly correlated in space but poorly in time due to fast changes in weather conditions (Pawluszek-Filipiak and Borkowski, 2020). At a local scale, an empirical relationship between stratified tropospheric delay and topography may be used for this purpose. At the global scale, accurate atmospheric models are available for correcting the interferograms. A linear model with respect to height can be used to estimate the vertically stratified atmospheric path delay and remove it from the interferograms to unwrap the phase accurately. Phase unwrapping is the process of estimating integer phase-cycle ambiguities due to the phase being measured modulo 2π radians (Hooper, 2008). Phase unwrapping is vital to derive spatially continuous displacements.

Alternatively, this delay can be calculated based on the propagation of radar radiation in the troposphere. It comprises hydrostatic delay and wet delay. Caused by dry air, the former depends mainly on surface air pressure. Caused by water vapor, the latter depends mainly on vapor pressure. Thus, the total atmospheric delay (ΔL_{tot}) in the scanning look direction (θ) can be expressed as the sum of hydrostatic (ΔL_{hyd}) and wet (ΔL_{wet}) components if the negligible liquid term from water droplets and precipitating clouds is ignored (Ahmad et al., 2019), or

$$\Delta L_{tot} = \frac{\Delta L_{hyd} + \Delta L_{wet}}{cos\theta} \tag{6.4}$$

$$\Delta L_{hyd} = 77.6 \times 10^{-6} \frac{R_d}{g_m} P_h \tag{6.5}$$

$$\Delta L_{wet} = 10^{-6} \left\{ \frac{(23.3T_m + 3.75 \times 10^5) R_d e_a}{T_0 \left[g_m (p+1) - 6.5 \times 10^{-3} R_d \right]} \right\} \left(1 - \frac{65}{T_0} \right)^{\frac{1}{T_0 - T_m}} \tag{6.6}$$

where P_h = surface air pressure (mbar), R_d = the ideal gas constant with a value of 287.058 J·K^{-1}·kg^{-1}, e_a = vapor pressure, derived from relative humidity, T_0 = surface temperature (°K), and p = mean vapor pressure decrease, the value of which varies seasonally between 2.92 and 3.04. T_m = vapor temperature, and g_m = the acceleration (cm·s^{-2}) at the mass center of the air column due to gravity. T_m and g_m are calculated as

$$T_m = T_0 \left[1 - \frac{R_d}{g_m (p+1)} \right] \tag{6.7}$$

$$g_m = 97.84 \left[1 - 0.0026 \cos(2\varphi) - 0.00028 \Delta h \right] \tag{6.8}$$

where ϕ = the latitude of the area under study; Δh = surface elevation (km). With the use of climate data, ΔL_{tot} can be determined and removed from the radar data. This method takes into account atmospheric humidity, temperature, and pressure, so the correction should be more accurate than those sub-optimal ones proposed by Ferretti et al. (2000) because they are based on simple statistical assumptions without requiring *a priori* information on the spatiotemporal displacement. Atmospheric correction reduces the mean absolute error of the measured subsidence from 9.8 to 5.7 mm and increases the coefficient of estimation between the two variables from 0.90 to 0.94 (Ahmad et al., 2019).

6.3.3.3 InSAR Generation

InSAR-based estimation of subsidence is a highly complex and lengthy process involving a large number of processing steps. The exact steps vary with the source data used (e.g., their level of processing and format), the method of data processing (e.g., how the atmospheric effects are addressed), and the computing environment adopted. The general procedure of estimation comprises data preparation, creation of single look complex (SLC) images, image co-registration, terrain geocoding, interferogram generation, phase unwrapping, and derivation of subsidence rate (Figure 6.5). SLC images are created from raw SAR images if facilitated with auxiliary instrument characterization data and external calibration data, such as look-up tables for initial processing. Each SLC image contains three layers: the real and imagery parts of the complex radar echo and radar intensity. Data preparation may include conversion of the raw SAR data to the SLC images, maintaining full resolution for each acquisition; de-bursting each sub-swath to merge the individual bursts to form a continuous and seamless tile if the image swath width is shorter than the dimension of the study site (for Sentinel-1 SAR images only) and delineation of the study area from the image dataset using clipping. If the area of study encompasses water, it has to be masked out as water diffuses the incoming radar radiation (Figure 6.5).

Some of the important steps are connection graph creation, interferogram generation and filtering, orbital refinement and re-flattening, removal of the atmospheric and topographic error, phase unwrapping, and phase-to-displacement conversion. Additional steps may be required for different types of implementation. Three typical and most crucial steps in processing time-series InSAR data to derive subsidence rate are elaborated below (Figure 6.6):

i. Interferogram generation. After the baseline has been computed for every pair of SLC images, the image pair having the lowest temporal baseline is selected to generate an interferogram. If there are N SAR images, $N - 1$ full-resolution interferograms are created, all based on the same master image;

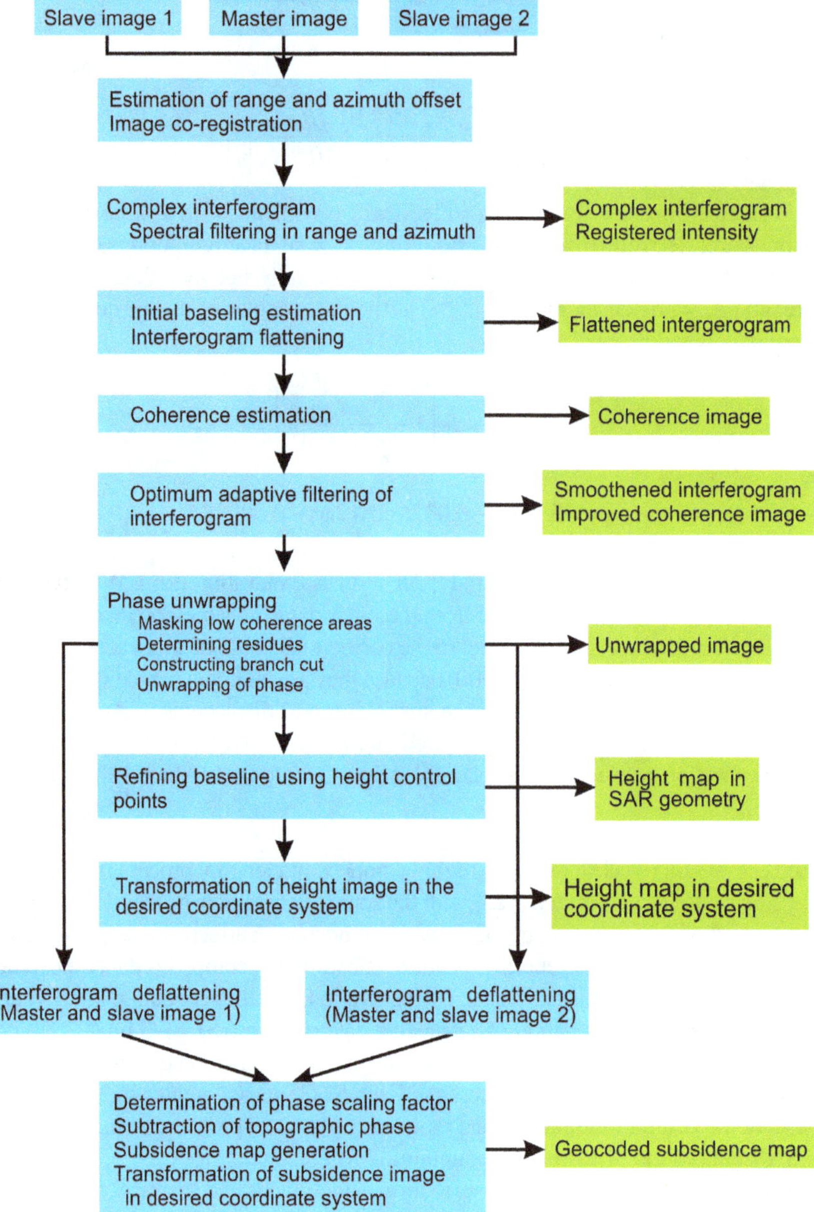

FIGURE 6.5 The procedure of processing SLC radar images to produce interferograms and eventually the subsidence map using the branch cut method. Green boxes – derived products. (Source: Modified from Bhattacharya et al., 2012.)

ii. DInSAR formation based on an externally sourced DEM. Of all the generated interferograms, the one with the largest baseline is selected as the master image, with which all other slave images are co-registered. The master image should have a relatively short spatial baseline to other slave images and temporally adjoin the first and the last images. If the dataset is small, a number of master images should be selected. Multi-master-based InSAR is necessitated by a shortage of multiple images (e.g., >15–20) and is an appropriate supplement to the single-master-based InSAR; and

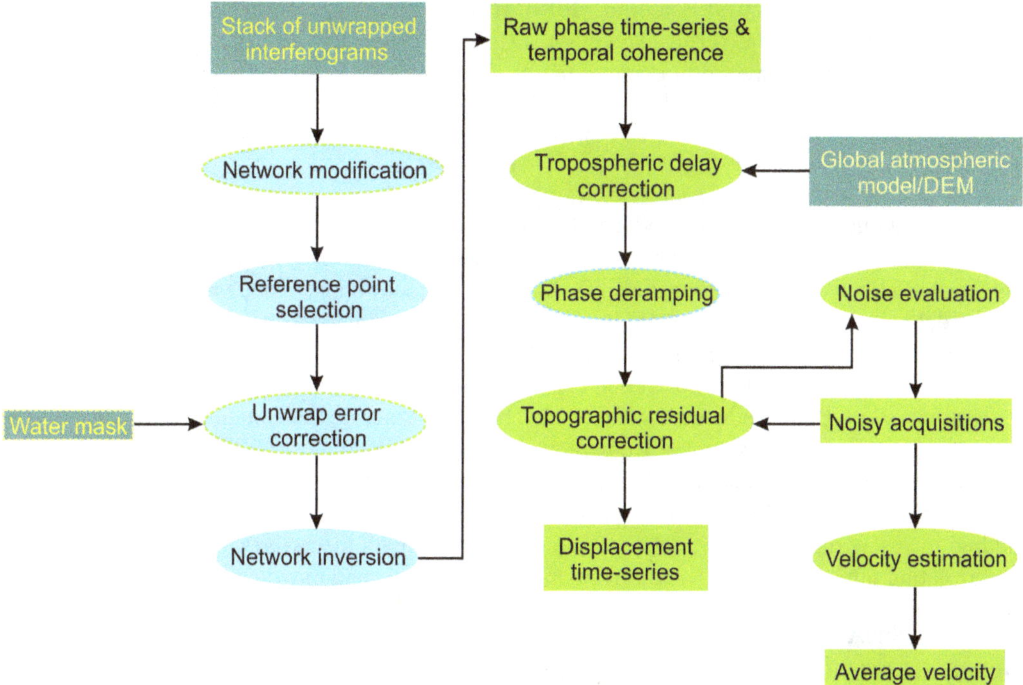

FIGURE 6.6 Routine workflow of InSAR time-series analysis. Blue ovals: interferogram domain; green ovals: time-series domain; dark green rectangles: input data; green rectangles: output results; dashed border: optional steps/data. Note: The order of empirical tropospheric delay correction, topographic residual correction, and phase deramping is inconsequential and has a negligible influence on the noise-reduced displacement time series. (Source: Modified from Zhang et al., 2019. Used with permission.)

iii. Preliminary estimation of LOS motion, topographic error, and atmospheric contribution. This may consist of five sub-steps:
   a. Correction of the interferograms for the synthetic topographic phase using external DEMs generated from various remote sensing data (see Table 6.1).
   b. Extraction of the displacement phase component for each of the raw interferograms and corresponding coherence maps for all paired images.
   c. Adaptive filtering of the differential interferograms to reduce phase noise, improve subsequent phase unwrapping, and increase the overall accuracy of phase unwrapping. Alternatively, the interferometric phase may be screened to increase the SNR and decrease residuals in phase unwrapping.
   d. Differencing the deformation and topographic interferograms for all paired images. After a differential interferogram has been produced, it is necessary to unwrap the phase. Common methods of phase unwrapping are minimum-cost flow and its extended version, weighted least mean squares, and branch and cut. Compared with these methods, the Statistical-Cost, Network-Flow Algorithm for Phase Unwrapping (SNAPHU) is able to resolve ambiguously wrapped phase data. The unwrapped phase is converted to the LOS displacement as

$$D_{LOS} = \frac{\lambda \cdot \Delta\phi}{4\pi} \tag{6.9}$$

   e. Conversion of the slant range deformation rate (D_{LOS}) to vertical subsidence by multiplying the LOS deformation by the cosine of the incidence angle (θ_{inc}) using Eq. 6.10

(Ahmad et al., 2019; Hsu et al., 2015). If the horizontal displacement is ignored, the subsidence in the vertical direction (D_{ver}) is calculated from the LOS displacement using a trigonometric function of θ_{inc} as

$$D_{ver} = \frac{D_{LOS}}{cos\theta_{inc}} \tag{6.10}$$

More accurate results can be produced by further refining step (c) for those pixels that have a high SNR in the azimuthal direction using spatial smoothing, but this method works for small areas only (e.g., <5 km × 5 km) (Ferretti et al., 2000). If the two radar images used in the detection have a known date of acquisition, the detected magnitude of subsidence can be converted to the rate of subsidence by dividing it by the temporal delay and the results are expressed in cm·year^{-1} (Figure 6.7).

It is worth noting that phase correction can be carried out in the time-series domain or the interferogram domain. Both produce identical results, but the former implementation enjoys two advantages: faster computation using $N - 1$ unwrapped phases instead of $N * (N - 1)/2$, and the ability to evaluate the impact of correction both spatially and temporally. InSAR data products have been produced from EnviSat ASAR and Sentinel-1 images. Sentinel-1 data products are openly accessible, ridding the users of the necessity of processing the InSAR data themselves. However, the tasks of decorrelating time-series InSAR, removing atmospheric noise, and unwrapping errors still remain.

6.3.3.4 DInSAR

Differential InSAR or DInSAR aims to separate the topography-related and displacement-related phases so that the latter can be mapped by differencing two interferograms of the same area but of different times. It takes advantage of phase information of SAR images. The subtraction of the second interferogram from the first reveals the displacement-related phase that can be converted to surface displacement. The phase difference of individual pixels of DInSAR indicates the changed height of the ground surface or subsidence. Accurate detection of phase changes induced by surface movements must take into consideration several factors, including topography, atmospheric conditions, and noises. The time-series differential interference phase ϕ is expressed as

$$\phi = \phi_{disp} + \phi_{air} + \phi_{topo} + \phi_{orb} + \phi_{noise} \tag{6.11}$$

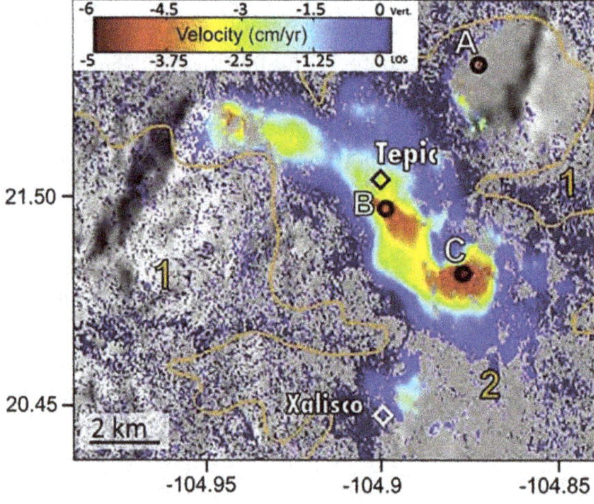

FIGURE 6.7 Distribution of mean subsidence velocity in both the vertical and LOS (line-of-sight) directions in Tepic, central Mexico detected from ALOS InSAR time-series images. (Source: Chaussard et al., 2014. Used with permission.)

where ϕ_{disp} = displacement phase; ϕ_{air} = phase contributed by the atmospheric conditions, in particular, the delay of radar pulse propagation caused by atmospheric water vapor; ϕ_{topo} = residual topographic phase that can be removed using a DEM. ϕ_{noise} = the noise phase introduced by decorrelation. The deterministic phase components of the tropospheric delay can be corrected using global atmospheric models or the delay-elevation ratio. The topographic residual and/or phase ramp are used to obtain the noise-reduced displacement time series. The topographic contribution to the phase difference (ϕ_{topo}) detected from a pair of SAR images is expressed as

$$\phi_{topo} = \frac{4\pi}{\lambda} \cdot \frac{\Delta h}{R sin\theta_{inc}} B_{\perp} \tag{6.12}$$

where $B_{\perp}$ = the perpendicular baseline, θ_{inc} = the incidence angle, and Δh = surface height. For DInSAR, Δh denotes the error caused by the DEM height inaccuracy (Ng et al., 2017). The value of the phase difference ranges from 0 to 2π and is used to generate the geometric displacement (ϕ_{defo}). The maximum subsidence (mm) between two adjacent pixels is estimated using the following simple model:

$$S_{max,\ vert} = \frac{365}{\Delta T_{max}} \cdot \frac{\lambda}{4 cos\theta_{inc}} \tag{6.13}$$

where ΔT_{max} = the maximum temporal baseline of all interferograms, $S_{max,\ vert}$ = the maximum detectable subsidence rate between two coherent points in the vertical direction. The limit of π on the differential phases is equivalent to the maximum differential deformation of $\lambda/4$ over the revisit interval. Thus, the maximum deformation rate measurable from EnviSat, TerraSAR-X, and ALOS is 14.7, 25.7, and 46.8 cm·year^{-1}, respectively. For Sentinel-1A/1B images, the maximal rate is 1.4 cm every six days, or 85 cm per annum. This rate is halved to 42.6 cm·year^{-1} if derived from either of the satellite images. In reality, the actual rate will be lower than these theoretical ones due to the effects of noises by other factors and phase-unwrapping ambiguities. A linear deformation model is required to successfully unwrap the interferometric phases, which forms a source of subsidence estimates if it behaves non-linearly. A centimeter precision of DInSAR results is achievable by selecting the rational interferometric pairs to mitigate baseline decorrelation and the atmospheric effects.

DInSAR is commonly produced from space-borne SAR images because of the satellite's repetitive orbits. Since DInSAR from SAR images of a short wavelength is much more sensitive to subsidence (see Eq. 6.1), C-band SAR images are preferable to longwave L-band SAR images. They are four times more sensitive than L-band, while X-band is twice as sensitive as C-band (Ge et al., 2007). Nevertheless, L-band SAR data are more advantageous than C-band data for quantifying landslide-induced subsidence in rural mountainous areas as the number of measurement points detected from PALSAR data is nearly five times those derived from ASAR data (Dong et al., 2018). The exact manner of implementing DInSAR varies with the availability and quality of DEM, the SAR images and their spatial baseline, the image acquisition time difference and coherence, the rate and shape of displacement, the surface land cover, and the topography of the area under study (Strozzi et al., 2001).

DInSAR can be two-pass or multi-pass. In two-pass DInSAR, the topography-related phase term is calculated from an external DEM. In multi-pass (e.g., three-pass) DInSAR, the topography-related phase in Eq. 6.11 is internally estimated from an independent interferometric pair of the shortest time interval (e.g., a small temporal baseline). If one image is common to both interferograms, it is known as a three-pass interferogram. The difficulty with this multi-pass approach is how to resolve the phase unwrapping operationally, which is rather challenging for low-coherence areas. Inconsistent, discontinuous, or noisy coherence increases the difficulty of phase unwrapping, even though such noises can be reduced via spatial filtering. In comparison, the use of an externally generated DEM is more operational and robust.

Repeat-pass InSAR exploits SAR images recorded at different times and with different baselines. It has the potential to monitor large-scale surface subsidence but is limited by temporal and spatial (geometrical) decorrelation and atmospheric artifacts. Temporal decorrelation disables the applicability of InSAR to vegetated areas or areas where the scattering behavior of the target changes with time. Geometrical (spatial) decorrelation restricts the number of image pairs useable for interferogram generation, which may indirectly impact the temporal resolution and temporal span in detecting long-term trends of subsidence. DInSAR extracts the deformation signal from individual interferograms that allow the detection of fast deformation or displacements concentrated in time. DInSAR may be implemented as consecutive and cumulative. In the former implementation, differential interferograms are produced from adjacent SAR images and accumulated to yield full time-series interferometric results. It uses low temporal baselines to minimize temporal decorrelation. If an error occurs in any of the estimated subsidence (e.g., residual terrain, atmospheric delay, and other phase inaccuracy) in the interferograms, it will propagate to the subsequent time-series subsidence results. The latter makes use of a fixed master image, and all others are treated as slave images. This implementation is sensitive to temporal decorrelation that tends to rise in the slave images.

DInSAR allows quasi-continuous monitoring of surface deformations over broad terrestrial areas. This method is best used to study large-scale deformations induced, for instance, by earthquakes or volcano activities. For small-scale subsidence, it is highly limited because of the phase contribution by the atmosphere. If combined with a DEM, it is capable of extracting the phase difference between multi-temporal SAR images, from which ground subsidence is detected to the sub-centimeter level in the LOS direction. DInSAR has proven to be a viable alternative in monitoring mine subsidence with high spatial resolution and accuracy. In monitoring slow subsidence over a long period, the performance of DInSAR is compromised by decorrelation if the temporal/spatial baseline is large, and by phase distortion caused by atmospheric heterogeneity. For instance, water vapor in the atmosphere can introduce an error of up to 10–20 cm to the InSAR-derived deformation measurements.

The quality of DInSAR-estimated subsidence is degraded mainly by noises stemming from the spatial and temporal decorrelation between the interferometric pair of images and the phase discontinuities in the interferogram (Ng et al., 2009). The results are reliable if the displacement rate is much higher than the atmospheric artifacts. Higher accuracies are still achievable via synergetic and innovative use of multiple data and methods (e.g., InSAR and GPS). With the use of an external DEM derived from GPS, the deviation of the DInSAR-derived subsidence profile is improved to an RMSE of only 1.4 cm in comparison with ground survey results, the largest discrepancy being up to 4 cm in areas of subsidence (Ge et al., 2007). In detecting ground subsidence or uplift, DInSAR measurements are consistent with a mean displacement rate of about 0.4 cm per month (Sarti et al., 2003). DInSAR can still be made more accurate in a number of ways, such as via the use of multi-temporal or time-series (TS)-InSAR, PS, and Stanford Method for Persistent Scatterers (StaMPS).

6.3.3.5 TS-InSAR

The limitations of the traditional interferometric approach and the optical systems, in general, can be overcome with TS-InSAR because the strong spatial correlation of atmospheric artifacts within individual SAR images is not correlated temporally. On the other hand, subsidence, the target of motion study, is usually strongly correlated in time, and the degree of spatial correlation depends on the pace of movement. TS-InSAR is an advanced form of DInSAR that takes advantage of long TS-InSAR images to enhance detection accuracy. The simultaneous processing of multiple images increases the density of measurements of deformation rates with reduced errors. The analysis of long TS-InSAR images (e.g., >20) enables the estimation and removal of the atmospheric effects (Zerbini et al., 2007). TS-InSAR achieves higher temporal and spatial sampling than InSAR but does not improve the level of detection precision over its traditional counterpart (Grzovic and Ghulam, 2015).

The accuracy of classic TS consecutive DInSAR is affected by such factors as spatial and temporal decorrelation, atmosphere-induced signal delay, and orbital or topographic uncertainties.

It must be pointed out that TS-InSAR is unable to detect "fast" subsidence due to the ambiguously wrapped interferometric phases. If the differential phase contributed by subsidence between two consecutive images exceeds the interval of $[-\pi, \pi]$, the true rate of subsidence cannot be entirely captured. This ambiguity in phase unwrapping restricts the maximum differential rate between two neighboring pixels to $\lambda/4$ (λ– wavelength) in the interim of the two images. In addition, TS-InSAR also faces the issue of temporal decorrelation, which can be redressed using two groups of methods. The first group exploits full-resolution pixels of single-point scatterers, the backscattering properties of which remain stable over a long period. This group of methods includes permanent scatterer (PS), StaMPS, SqueeSAR, interferometric point target analysis, and spatiotemporal unwrapping network. The second group of methods aims to optimize pixels containing distributed scatterers (DS), which will be presented separately in the following section.

6.3.3.6 PS-InSAR

The pixel value of a SAR image represents the total backscattering contributed by all scatterers within the sensed ground area and the atmosphere. Pixels in a SAR image change their phase due to the changed satellite-scatterer relative position, and temporal changes of the target over time. However, the phase of a persistently dominant scatterer due to decorrelation hardly varies with time, even if the subordinate scatterers do move relative to it. This variation is small if the target is sensed from different look angles. Thus, this scatterer is termed a persistent (permanent) scatter or PS. Pixels with reliable phase measurements can serve as PS, such as buildings or natural objects (e.g., rock outcrops) that have a brighter tone than the background scatterers. PS have a high temporal coherence as they hardly change with time and are stable over a long spatial perpendicular baseline separation. Such stable scatterers can be identified on the coherence maps associated with interferograms and selected based on the amplitude dispersion index. They are discrete and temporally coherent at long temporal intervals commonly found in urban areas and are identifiable from long TS-InSAR images.

If the PS's dimension is smaller than the pixel size of the SAR images used, strong coherence exists in the interferograms constructed from them if they have a baseline longer than the decorrelation one. This means that all the available SAR images can be used, and SAR satellite-based monitoring of subsidence can be made operational. In the absence of a dominant scatterer, phase variation due to decorrelation is frequently so large that it overwhelms the underlying signal. However, this signal is still detectable from interferograms formed by two SAR images acquired at a short-time interval from a similar look angle so long as decorrelation is sufficiently minimized (Hooper, 2008). In addition, decorrelation can be further minimized via spectral filtering in range and by excluding non-overlapping Doppler frequencies in azimuth.

PS-InSAR is an advanced form of DInSAR, taking advantage of multiple SAR images of the same area, all being stacked and co-registered with one another. Analysis of such images aims to identify point-like pixels or PS that dominate the returning phase. PS retain the phase information with time but have variable acquisition geometries. The stability of phase-stable pointwise targets enables the spatially uniform atmospheric phase screen to be eliminated, so their relative motion can be estimated at a precision on the order of 1 mm·year^{-1} for each scatterer (Zerbini et al., 2007). PS act as a kind of "natural geodetic network" offering a very high spatial density of measurements. PS-InSAR substantially increases the density of measurements to several hundreds per km^2, a density enabling detailed analysis of the local pattern of surface subsidence. If the density is not high enough, additional processing such as spatial interpolation may be applied to generate such measurements. Apart from the high density, PS-InSAR is rather robust and accurate. Land subsidence detected from Sentinel-1 PS-InSAR closely matches that measured at GPS stations with a mean absolute error of 5.7 mm and a coefficient of determination of 0.94 (Ahmad et al., 2019).

In order for PS-InSAR to function well, a large number of images (e.g., 15–20) must be used. It allows large areas of subsidence to be monitored for a long period. This approach applies to urban, peri-urban, and built-up areas, or even landslides, but not rural or wooded areas where coherent PS points are almost non-existent because variations in vegetation geometry or wetness cause phase decorrelation. Thus, it is very difficult to find coherent scatters. One major drawback of PS-InSAR is the low density of PS points (Figure 6.8a). Some pixels have no data due to the failure of InSAR caused by a low coherence, large subsidence gradient, unsuccessful unwrapping, or the use of improper deformation models. These gaps can be filled using spatial interpolation from the successfully estimated pixels (Figure 6.8b). This difficulty can also be overcome with a phase-decomposition-based PS-InSAR processing method (Cao et al., 2016).

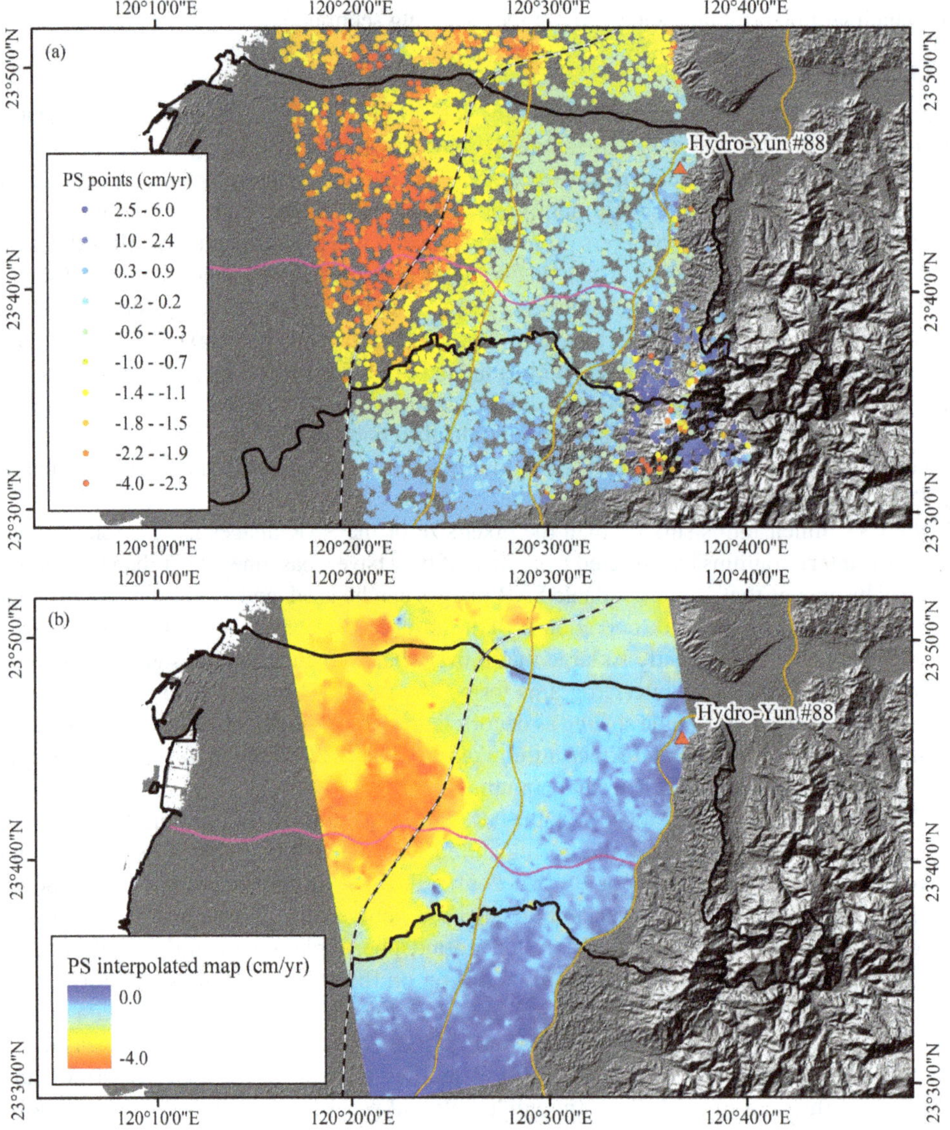

FIGURE 6.8 Distribution of PS-InSAR-derived mean LOS velocity of ground deformation from 2007 to 2010. (a) vertical deformation of all PS points; (b) raster map interpolated from (a) using kriging. (Source: Hsu et al., 2015.)

PS-InSAR differs from other versions of InSAR (e.g., DS-InSAR) in that a single-master image is used to generate a stack of differential interferograms without any restrictions imposed on the temporal or spatial baseline. The interferograms are analyzed at the single-look resolution to maximize the SNR of pixels containing a single dominant scatterer. In comparison, DS pixels are formed by the coherent summation of multiple scatterers within the pixel. These pixels adjoin those representative areas of moderate coherence in some interferometric pairs, and they all share a similar pixel value and likely belong to the same target. PS differ from DS in that they focus on point scatterers with a stable phase and are suitable for urban areas. DS make use of small baseline interferograms and a relaxed limit on the phase stability by including areas affected by decorrelation through the redundant network of interferograms (Zhang et al., 2019).

Both PS and DS are optimizers of different models of backscattering, even though the scattering behavior of real ground surfaces lies between these two idealized models. They represent the efforts to overcome the limitations of conventional DInSAR by minimizing the phase errors by tackling the atmospheric effects and temporal decorrelation. With proper data processing and analysis, they allow the displacement phase to be separated from other phase components. PS-InSAR and DInSAR are compared and contrasted in Table 6.2 in terms of baseline, atmospheric artifacts, coherence, DEM quality, and the number of SAR images required. PS-InSAR is more flexible without any restrictions on the baseline with the European Remote Sensing (ERS) image database, but DInSAR has a limit of 200 m. It can strongly reduce atmospheric artifacts, but DInSAR cannot. Besides, coherence is much higher on a single pixel with PS-InSAR, more than twice higher than 0.3 on several nearby pixels associated with DInSAR. Nevertheless, PS-InSAR does require more than 30 images, which is reduced to only more than two with DInSAR. The plus side is that they allow non-linear subsidence to be detected, while DInSAR can detect only linear subsidence if PS density and coherence are sufficiently high.

PS-InSAR analysis is rather complex. Typical processing steps may consist of master image selection, co-registration, differential interferogram generation, selection of PS candidates, atmospheric phase screen estimation and removal, PS selection, mean displacement rate estimation, topographic error removal, PS point displacement time-series analysis, and average deformation estimation. The procedure of analyzing classic InSAR data outlined in Section 6.3.3.4 previously can also be used to implement PS-InSAR. The two additional and important steps are elaborated on below:

i. Selection of PS candidate points based on the coherence maps. Various measures have been proposed to select PS pixels to increase the density and quality of PS-InSAR measurements. The easiest way of selecting PS pixels is to make use of a threshold. If a target

TABLE 6.2

Comparison of PS-InSAR Properties with Those of Classic DInSAR in Detecting Subsidence

Properties	PS-InSAR	DInSAR
Baseline	No limits with the ERS image database	Very small (<200 m)
Atmospheric artifacts	Strongly reduced	No reduction
Coherence	>0.7 on a single pixel	>0.3 on several nearby pixels
DEM quality	100 m	Baseline dependent
No. of SAR images required	>30	≥3
Nature of inverted subsidence	Non-linear	Linear if PS density and coherence are high

Source: Modified from Ferretti et al. (2000).

shows a coherence always higher than this threshold, it qualifies as a PS candidate. A candidate pixel is selected if the coherence of all identical pairs exceeds the pre-defined amplitude dispersion threshold (e.g., >0.4). However, the coherence map is not reliable if not compensated for other non-motion introduced phases. So the coherence threshold needs to be optimized. Whatever threshold is selected, it represents a trade-off between the false alarm rate and detection probability. Other better methods of selection are the identification of candidate PS pixels based primarily on their temporal phase variation, and the use of primary correlation of their phase in space. After PS pixels have been selected, PS-InSAR can be implemented in various forms, one of which is called Cousin PS. Characterized by a moderate spatial phase variation, it ensures correct phase unwrapping; and

ii. Computation of PS-InSAR using the phase information of each candidate PS, and extraction of PS points and their range deformation at each pixel after atmospheric correction to convert phase to linear units, during which low accuracy pixels are excluded from the interferograms. Derivation of the PS deformation velocity and time series takes into account the atmospheric phase screen (Mateos et al., 2017). After the 2π phase ambiguity in the temporal domain is resolved, height and deformation rate are determined via 2D linear regression iteratively from the candidate point targets. This process may be iterated several times so as to maximize the number of selected points. The interferograms having the original SLC radar geometry are transformed to multi-look images during which non-square pixels are made square. The main products of PS-InSAR analysis comprise estimated linear deformation rates in the scanning LOS direction and deformation time series or history at the measured points.

So far many methods and algorithms have been developed to process PS-InSAR data. The four most significant approaches are identified as PSInSAR™, (extended) small baseline subset (SBAS), StaMPS, and SqueeSAR™ (Crosetto et al., 2016). They are compared for their pros and cons in Table 6.3. SqueeSAR™ extends the PS-InSAR™ algorithm with the ability to jointly process PS and DS pixels, taking into account their statistical behavior. This algorithm concentrates on pixels in areas of moderate coherence. The neighboring pixels belonging to the same object share the same reflectivity values and are also selected as the candidates, which results in an improved density and quality of the PS-InSAR results. This implementation is particularly suitable for rural areas (Crosetto et al., 2016). SBAS will be introduced in detail in the next section. StaMPS will be covered in Section 6.3.3.8.

PS-InSAR is a more accurate and improved analytical algorithm than other InSAR methods. It allows the use of more flexible and accurate deformation models to estimate non-linear deformation and improved algorithms to estimate the atmospheric phase component and generate deformation results over vast areas (Crosetto et al., 2016). Consequently, it is able to yield an estimate of relative motion at a precision on the order of 1 $mm \cdot yr^{-1}$ for each scatterer because the stability of phase-stable pointwise targets enables the spatially uniform atmospheric phase screen to be eliminated (Zerbini et al., 2007). Nevertheless, this performance is subject to the influence of a plethora of factors, including the type of SAR images used, their quantities, temporal (B_t) and geometrical ($B_\perp$) baselines, PS density, nature of subsidence being detected, and the surface cover of the area under study. Probably the most important limiting factor is the temporal change in the complex reflectivity of the ground surface due to changed moisture content or vegetation (Ye et al., 2004). The ground cover type and density also significantly affect interferogram quality. Densely forested areas are difficult to penetrate by microwave radiation, especially short-wavelength bands. They strongly absorb the incident radar radiation and weaken the scattered signals. Non-uniform penetration of the canopy leads to incoherent signals. A rugged terrain, together with the slopes squarely facing the radar antenna, lowers the PS density.

TABLE 6.3
Comparison of Four Major Systems for Running PS-InSAR

Algorithm	Authors	Pros	Cons
PSInSAR™	Ferretti et al. (2001)	Consideration of both temporal and geometric decorrelation; flexible and accurate models possible to estimate non-linear subsidence	Scatterers must exhibit sufficiently high coherence and low PS density in non-urban areas
(Extended) SBAS	Berardino et al. (2002)	Small baseline to limit spatial decorrelation, multi-look data to reduce phase noise, increase spatial and temporal samples, coherence-based selection of PS	*a priori* deformation model essential, not suitable for detecting local subsidence
StaMPS	Hooper et al. (2004)	Amplitude-based selection of PS, suitable for low-amplitude natural targets; *prior* deformation models not needed	local subsidence smoothed out if the filtering window size is too large, external DEM essential for removing residual topographic phase
SqueeSAR™	Ferretti et al. (2001)	Combination of PS and DS reduces phase noise and residual atmospheric effects in areas of moderate coherence, improved density, and quality of PS-InSAR results, excellent for rural areas	A large number of images are needed, complex, DEM is essential

6.3.3.7 SBAS

The SBAS is one of the DS methods that use the network of interferograms with a small temporal and spatial baseline. The principles of SBAS are almost identical to PS-InSAR except that it targets slowly decorrelating filtered phase (SDFP) pixels or pixels, the phase of which scarcely decorrelates over short intervals following spectral filtering (Hooper, 2008). So the difference lies in whether spectral filtering is applied. All SBAS methods attempt to minimize the perpendicular baseline (image acquisition position relative to the LOS), temporal separation, and Doppler baselines using the most highly correlated areas to derive the deformation signal from multi-look interferograms. A network of redundant interferograms, the perpendicular baseline values of which fall below the prescribed threshold, is used to restrict the effects of geometric decorrelation (Shanker et al., 2011). The SBAS method has been widely and successfully used to monitor ground deformation and subsidence. It is implemented in various manners, each with its unique features, pros, and cons.

SBAS-InSAR is one of the two main techniques of time-series interferometric analysis aiming to overcome the limitations of classic DInSAR, such as temporal and geometrical decorrelation, and compensate for error contributions from atmospheric delays, inaccurate terrain modeling, and uncertain satellite orbits. It aims to minimize the temporal and spatial separation of the interferograms, as well as the frequency range of acquisition pairs to increase the correlation of the interferograms. Single-look slowly decorrelating filters phase for images of a short temporal separation has been developed to isolate SDFP pixels based on pre-defined thresholds, such as the maximum perpendicular absolute baseline difference and the maximum absolute temporal difference. Such obtained subsidence results agree with leveling results at six benchmarks with a correlation coefficient of 0.997 at the 95% confidence level (Zhou et al., 2017). SBAS is good at generating reliable and dense results in regions experiencing a relatively high rate of subsidence. As illustrated in Figure 6.9, it is created in three main steps (Chen et al., 2018):

i. Generation of a set of multi-look differential interferograms by selecting SAR image pairs with the specified constraints of spatial and temporal separation (e.g., <150 m and <150 days). Their SNR may be improved via filtering;
ii. Phase unwrapping of the wrapped phase signal derived from the selected interferograms, followed by removal of those interferogram pairs of low coherence and unwrapped phase error. The unwrapped phase may be corrected using the calculated residual phase content and phase ramps; and
iii. Confirmation of close-to-zero subsidence by interpolating the unwrapped phase and comparison with field survey results.

Standard SBAS methods work with interferograms that must be multi-looked first, followed by individual phase unwrapping. This strategy misses two potential benefits offered by time-series SAR images:

i. The opportunity to process the data at the highest possible resolution so as to identify isolated SDFP pixels surrounded by completely decorrelated ones; and
ii. The ability to unwrap the phase more robustly in 3D. A number of algorithms have been developed to fulfill this objective, such as the statistical cost flow algorithm applicable to multiple-master time-series InSAR images.

SBAS can reveal targets with relatively stable scattering properties and good coherence and accurately quantify slow deformations, but not rapid displacements such as subsidence at active longwall mines, the surface displacement of which exceeds the sensor's gradient limit (Ng et al., 2017). The surface of mining areas can subside by up to 600 mm during the first one to two months after the roof of the longwall panel collapses, and over 800 mm one year after mining has ceased operation. Such rapid subsidence cannot be quantified from SAR images of a short wavelength owing to phase discontinuity. This issue can be overcome by combining InSAR image pairs with the shortest temporal baseline. SBAS examines a DS within the resolution cell, different from PS-InSAR that searches for a point-like scatterer. Besides, the selected PS pixels are spectrally filtered. SBAS can yield a continuous temporal pattern of subsidence rate spatially without the linearity assumption, while the PS approach is able to reveal small-scale subsidence disparities between various man-made structures.

In order to properly estimate ϕ_{disp}, SBAS requires a pre-defined *a priori* deformation model. If the subsidence is characterized by high non-linearity and a varying rate as with mining-induced subsidence, it cannot be depicted by a single model. An inaccurate model increases the difficulty of selecting sufficient PS and/or DS. SBAS is advantageous over other TS-InSAR processing in that it generates interferograms from appropriately selected pairs of InSAR to minimize the spatial and temporal baseline between two acquisition orbits to avoid the decorrelation problem associated with a long baseline. However, SBAS interferograms cannot be easily unwrapped spatially in the presence of a large subsidence gradient and temporal decorrelation. In comparison, the PS temporal unwrapping strategy is well suited to study linear subsidence, but not over extensive areas. Over a large area, multiple-master coherent target SBAS-InSAR is proposed to process time-series SAR images. This improvement has the provision of adjustment for wide-area monitoring and offers an integrated solution to calibrating InSAR-derived deformation and harmonizing the deformation estimates from overlapping SAR frames (Zhang et al., 2016). An adjustment is made to detect subsidence from overlapping frames after it has been detected from multiple frames already.

Both PS-InSAR and SBAS-measured subsidence are in close agreement with each other, and the differences between the two sets of measurement fall statistically close to the nominal rate of

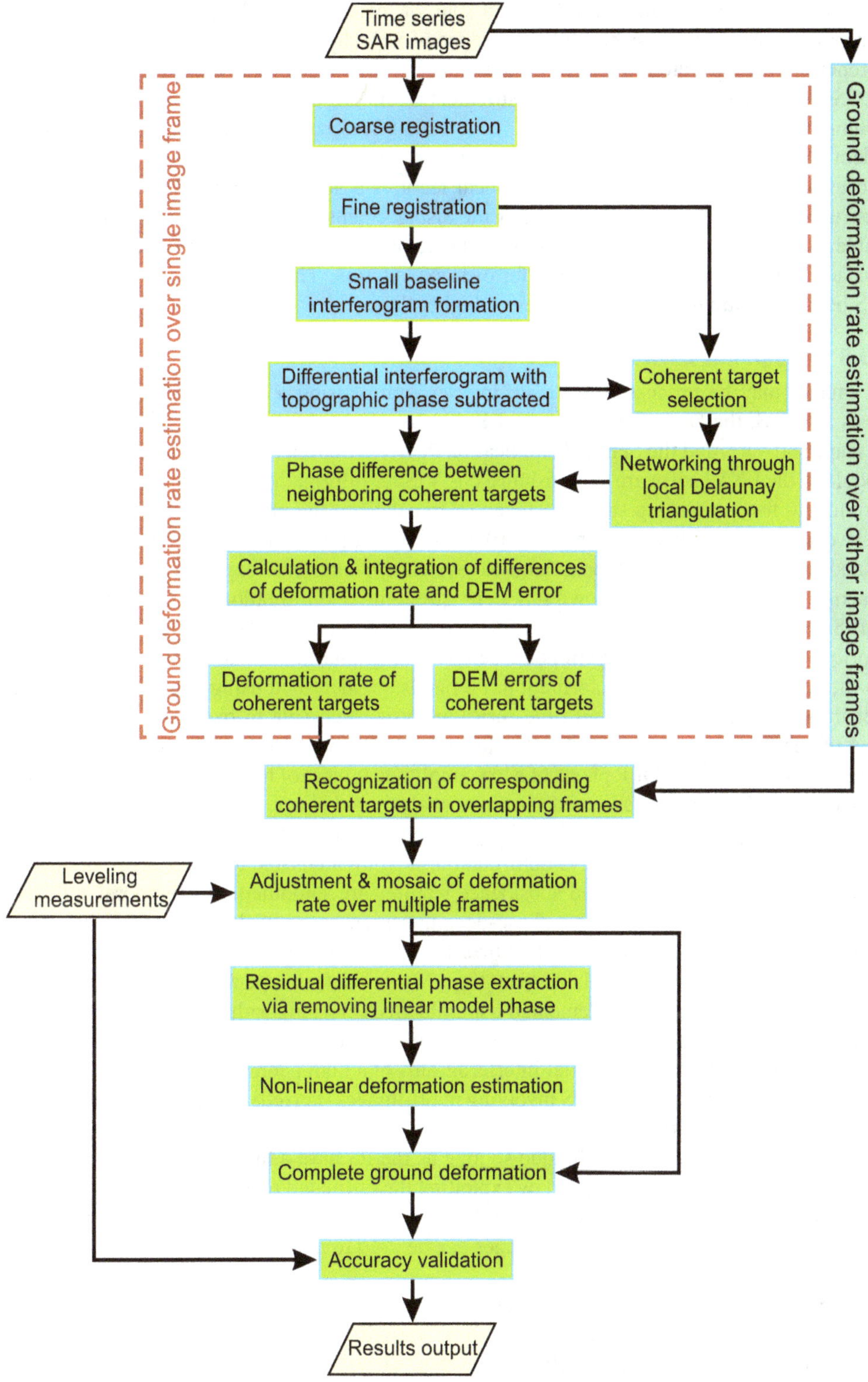

FIGURE 6.9 The workflow of time-series InSAR data processing to produce the complete ground deformation field using a small baseline subset. Blue boxes – data pre-processing. (Source: Modified from Zhang et al., 2016.)

subsidence uncertainty (Yan et al., 2012). In spite of this agreement, the former produces results of a coarser density due probably to the presence of vegetation (Grzovic and Ghulam, 2015). The difference can be attributed to the different number of reference points used. PS-InSAR measures the LOS displacement of a pixel. The measurement errors originate chiefly from phase unwrapping. If combined with atmospheric correction, coherence SBAS helps to minimize the low coherence problem and reduces atmospheric- and orbital-related errors. The differential subsidence map produced using the SBAS approach displays local features associated with urban constructions and infrastructures. The PS approach quantifies the displacement of individual targets, which makes it possible to further quantify the relative importance of various variables on land subsidence, such as surface loads and drying due to aquifer over-exploitation. The combination of PS-InSAR with SBAS helps to obtain more PS and achieve a higher SNR.

6.3.3.8 Computing Environments

SAR images may be processed to create interferograms using a number of software packages, such as StaMPS, the Delft Object-oriented Radar Interferometric Software, the PSI Toolbox, the SARscape software, and the Sentinel Application Platform Sentinel-1 toolbox by ESA. SARscape is a module of the ENVI software suite from Exelis VISual Information Solutions. In the SARscape© environment, five processing steps are taken (i) to calculate the phase difference between master and slave images, (ii) to flatten the phase, (iii) to unwrap the phase, (iv) to convert the phase to elevation, and (v) to geocode in the World Geodetic System 1984 latitude/longitude cartographic system.

Each of these packages has its own functions, strengths, and limitations, as well as best uses. The StaMPS software is designed to process SAR images and perform PS-InSAR. It does not involve any prior assumptions about the temporal behavior of ground deformation. Instead, it is assessed from the spatial correlation of deformation. PS pixels are identified using phase spatial correlation rather than amplitude analysis and time-series analysis, each able to overcome one of the weaknesses of the first generation of InSAR. The selected PS pixels have 2D phases of modulo 2π. StaMPS can detect PS or SDFP pixels of a low amplitude typical of natural terrains. The probability of a pixel being PS/SDFP is estimated via phase analysis and is iteratively and successively refined. Filtering may be performed in StaMPS using a combination of spatial and temporal low-pass filters on the unwrapped data to derive deformation. It is not advisable to use a large window in the filtering. Otherwise, some localized subsidence features are smoothed out. Temporal and baseline decorrelation can be overcome with a phase-decomposition-based PS-InSAR processing method to improve deformation analysis (Cao et al., 2016). The PS, SB, and combined StaMPS/multi-temporal InSAR (MTI) algorithms are available from http://www.hi.is/~ahooper/stamps.

Phase unwrapping can be implemented using the SNAPHU algorithm. It aims to come up with the most likely unwrapped solution with the maximum posterior probability for the given set of input data. Phase unwrapping of differential interferograms of a high geometrical and temporal baseline is possible only on a sparse grid of PS involving multi-images (Ferretti et al., 2000). Mathematically, the unwrapping is easier to realize if a constant velocity model is adopted. Non-linear target motion can be handled by a stochastic technique so long as the PS have a high enough density and coherence. Phase differences are unwrapped by exploiting the estimated relative velocity (Δv) and relative elevation (Δh) of each pair of scatterers. The SNAP (Sentinel Application Platform) and SNAPHU software dedicated to processing SAR images can be used to map the location of hazard zones and their ongoing subsidence rate.

PS-InSAR may be created using the Gamma Software or the PSIG Cousin package. Cousin PS are PS characterized by a moderate spatial phase variation that guarantees the correctness of the unwrapped phase. Derivation of the PS deformation velocity and time series takes into account the atmospheric phase screen. SBAS-InSAR can be carried out using a number of existing packages, such as StaMPS/MTI and the Generic InSAR Analysis Toolbox (GIAnT). This toolbox encompasses three small baseline interferometric strategies, including the generic SBAS, new small baseline subset (NSBAS), and a temporal analysis method (generic TimeFun) adapted from the multi-scale

TABLE 6.4

Comparison of the Main Features of Four Packages for Implementing SBAS-InSAR Based on SB Network of Unwrapped Interferograms and Linearly Optimized Time-Series Estimator

Package	G-SBAS	G-NSBAS	G-TimeFun	MintPy
Initial pixel selection	Coherent in all	A total min no. of coherent interferograms	Coherent in all interferograms	A min no. of coherent interferograms for every acquisition
Weighted inversion	No	No	No	Yes
Unwrapping error correction	No	No	No	Bridging/phase closure
Posterior quality check	No	No	No	Yes
Prior deformation model	No	Yes	Yes	No
Phase correction	Interferogram domain	Interferogram domain	Interferogram domain	Time-series domain

Source: Zhang et al. (2019). Used with permission.

InSAR time-series algorithm. Their main characteristics and differences are summarized in Table 6.4. The table also compares their performance with MintPy, the Miami InSAR Time-series software in Python for analyzing SB InSAR time-series data. It is available at https://github.com/insarlab/MintPy. They differ from each other in the selection of pixels, inversion, unwrapping error correction, and posterior quality check. Only G-NSBAS and G-TimeFun require *a prior* deformation model. Phase correction is carried out in the interferogram domain with all packages except MintPy that carries it out in the time-series domain.

Table 6.5 compares the performances of the four versions of SBAS in detecting surface subsidence, benchmarked against GPS results. In both cases, G-TimeFun achieves the lowest residual among the four implementations, while G-NSBAS is the worst performer in the first case and the second-worst in case 2. STaMPS-SB has the largest residual and offsets in case 2, but the smallest offset in the first case. It is less accurate than only G-TimeFun in the first case. The first two methods produced slightly more accurate detection in case 2 than in case 1, which is just the opposite of

TABLE 6.5

Accuracy Comparison of Four Versions of SBAS against GPS in Detecting Surface Subsidence in Two Cases

Arc			Mean Distance (m)		Residual (mm)	
Point 1	**Point 2**	**Method**	**Point 1**	**Point 2**	**Offset**	**σ**
OKCE	OKFG	G-NSBAS	196.75	145.68	5.92	14.47
		G-SBAS	196.72	224.00	1.91	12.92
		G-TimeFun	196.72	224.00	4.77	8.57
		StaMPS-SB	168.35	453.61	1.37	9.00
OKCD	OKSO	G-NSBAS	58.41	72.84	0.77	13.20
		G-SBAS	88.88	72.84	0.96	12.32
		G-TimeFun	88.88	72.84	0.04	11.42
		StaMPS-SB	112.32	123.57	7.18	15.68

Source: Gong et al. (2016).

the last two methods. Overall, all four methods detected deformation at an accuracy level ranging from mm to cm if the radar dataset has good temporal sampling and the area of study is distributed with stable backscatters (e.g., urban areas) (Gong et al., 2016).

Since *a prior* deformation model or temporal filtering is not relevant, SBAS-InSAR time-series analysis is well suited to detect non-linear subsidence. Although this method has a similar performance to other methods in high coherent areas, it is more accurate in low coherent areas and in high coherent areas with phase-unwrapping errors or complex displacements because of unwrapping error correction, weighted network inversion, temporal coherence-based selection of the initial and reliable PS pixels (Zhang et al., 2019). Temporal coherence is a more robust and reliable measure than mean spatial coherence as it takes into account phase unwrapping errors. Since it cannot capture temporal variations of the reliability of the phase time series, it is not useful for partially coherent scatters. This method is still able to detect non-linear subsidence but requires a stochastic model. The phase contributions by different factors (e.g., subsidence, atmospheric phase components, and decorrelation noise) are separated by means of the least mean square estimator, taking into account the spatiotemporal data distribution and the correlation between samples (Ferretti et al., 2000).

6.4 NON-IMAGERY SENSING

Although InSAR-based methods are able to retrieve the field of surface subsidence, they are rather complex and require special software. In comparison, non-imagery methods such as GPS and LiDAR are much simpler in detecting ground subsidence via the subtraction of time-lapsed height at the same location. The difference between the two non-imagery methods is the density of heights used. GPS offers a limited number of point heights that fail to yield a holistic field of subsidence accurately, while LiDAR point heights are dense that can be converted to a field view of subsidence at high confidence. The resolution of detection depends on the grid size of DEMs produced from the non-imagery data.

6.4.1 GPS for Subsidence Monitoring

GPS has been used to monitor subsidence for over two decades, both as a complementary tool and an alternative to the conventional leveling method. GPS is able to record continuously geodetic observations and provide absolute coordinates with respect to a well-defined global geocentric reference system. This monitoring is accomplished via a network of GPS stations designed primarily for studying seismic and tectonic activities, or for national geodetic survey. It is exemplified by the well-known GEONET and the nationwide network of Japan comprising 947 stations (Sato et al., 2003). The GPS receivers used in these networks are highly sophisticated, receiving the L1 and L2 carrier phase signals (see Section 1.3.3.2). These dual-frequency, geodetic-quality receivers log coordinates at a typical vertical accuracy of 1.5 cm. These stationary, continuously operating reference GPS stations or geodetic monuments are suitable for long-term monitoring of subsidence. These stations supply GPS observations and site velocities relative to a stable reference system instead of actual positions. As such, they are subject further to the availability of the geodetic infrastructure of the area under study. Therefore, GPS data from a single station are incapable of monitoring site stability.

In some cities around the world, a sound GPS-based geodetic infrastructure has been constructed specifically for monitoring subsidence (Figure 6.10). For instance, Houston, Texas has a network of 230 permanent GPS stations for this purpose, collecting data continuously for a long time (Agudelo et al., 2020). The maintenance cost of a dense network, however, can be prohibitively high for developing countries. The cost may be brought down using semi-permanent GPS stations. They are identical to permanent GPS stations except the same receivers being deployable in multiple permanent GPS monuments (sites) to save cost because GPS receivers are the most expensive in establishing a permanent station. So, continuous data are not available at all stations. The use of semi-permanent

FIGURE 6.10 A representative GPS observation station in the GEONET system. The GPS is fixed at a height of 5 m in the pillar. (Source: Sato et al., 2003. Used with permission.)

stations densifies a GPS network at a lower cost. Additional processing such as spline-based smoothing may be performed to best estimate the pace of subsidence from such GPS data.

The use of network-based observations requires additional adjustments to the measured horizontal coordinates and ellipsoid heights to improve accuracy (Sneed and Brandt, 2013). Apart from a dense GPS network, successful and precise local geodetic infrastructure should also include a stable local reference frame, and the GPS-logged data have to be processed using sophisticated software packages (Wang et al., 2015). It is a judicious practice to transfer the GPS coordinates to a local, stable, and consistent reference frame so that GPS data logged by different agencies using different receivers at different times are all directly comparable to one another.

Since the primary objective of the GPS stations is to monitor long-term drift and rotations of tectonic plates, local or temporal ground subsidence may be overwhelmed by the monitored tectonic movements. These motions can be excluded by adopting a stable regional reference frame. The 3D coordinates of a monitoring station expressed as (x, y, z), of which z (the vertical coordinate) is known as the ellipsoid height, are measured from a geodetic reference ellipse. Initially, these positional coordinates are referenced within a global frame, such as the International GNSS Service reference frame of 2008 (IGS08). The World Geodetic System 1984 (WGS-84) ellipsoid is another geodetic system commonly used to express height differences in some parts of the world. Since a global reference frame aims to minimize the overall movements of numerous widely distributed stations, the majority of sites keep moving within a global reference frame (Wang et al., 2015). Such variations must be taken into account when interpreting the GPS-derivation subsidence trends.

GPS-derived height is referenced to a smooth ellipsoid different from orthometric height used in leveling surveys. Orthometric height is referenced to the surface of the geoid (Wang and Soler, 2014).

It cannot be determined using a GPS. GPS-derived ellipsoid heights can be used directly to survey long-term subsidence without the need of performing leveling. In detecting surface subsidence, it does not matter which height, ellipsoid or orthometric, is measured so long as the same height is adhered to throughout the monitoring period, as subsidence is detected from the differences in height over time. The temporal change in ellipsoid height is assumed to be caused by vertical displacement or subsidence. It must be emphasized that GPS-derived time-series positions exhibit periodic variations, such as annual and seasonal cyclical sinusoids. Cyclical variations stem from periodic ground motion, and seasonal variations may be attributed to the extraction of groundwater for irrigation, one of the major causes of ground subsidence. It can be determined by correlating the mapped land subsidence with different land cover types and different underlying geological formations to identify the causes of subsidence.

GPS-based monitoring of subsidence can be implemented for decades. Geodetic network-based GPS monitoring offers high accuracy (about a few centimeters) in detecting ground subsidence. The ellipsoid heights at 11 geodetic monuments have an inaccuracy ranging from ±2 to ±5 cm at the 95% confidence level (Sneed and Brandt, 2013). Higher accuracy is achievable via differential carrier phase measurements, and by correcting or eliminating systematic residuals and adopting precise ephemeris. In this way, ground subsidence can be detected at sub-centimeter accuracy. For instance, Sato et al. (2003) achieved an accuracy of 9.5 mm standard deviation in monitoring land subsidence using GPS, demonstrating that land subsidence can be accurately quantified if the GPS network is well-calibrated. The GPS results can be made more accurate if integrated with leveling and InSAR.

The GPS method is suitable for determining spot heights (e.g., where the receiver is placed or at the permanent GPS stations). This method yields only the location and magnitude of the vertical land surface change, but not its extent nor the spatial variability of subsidence owing to the coarse spacing of permanent GPS stations over 10 km. Such scarce and sparse observations cannot paint a detailed picture of the spatial distribution of subsidence. In spite of this deficiency, periodic GPS measurements supply supplementary information for correcting the subsidence maps produced from InSAR images. More importantly, continuous GPS-derived subsidence is regarded as the ground truth (see Table 6.5), against which image-derived results are compared to assess their accuracy (Ahmad et al., 2019). For instance, the InSAR-retrieved subsidence rate is checked against GPS measurements to evaluate its reliability, but not against leveling data for three reasons (Grzovic and Ghulam, 2015):

i. InSAR-derived displacement is referenced to the ground areas that are deemed to have no deformation;
ii. The subsidence rate estimated from InSAR applies to the entire pixel, the area of which varies with pixel size. On the other hand, leveling-derived subsidence refers to a specific point; and
iii. InSAR deformation results are valid in the LOS direction. When it is converted to vertical subsidence, a fixed incidence angle is adopted which is a source of inaccuracy. In comparison, leveling always measures vertical (orthometric) displacement.

6.4.2 LiDAR-based Monitoring

Identical to GPS, LiDAR is a means of quickly acquiring dense 3D coordinates of the ground surface remotely from a laser scanner mounted in a moving platform (Figure 1.1). Thus, LiDAR-based detection of surface subsidence shares the same principles with GPS except that LiDAR data are not temporally continuous but spatially extensive. This detection requires two-time LiDAR datasets of the same area that must be interpolated to construct local-scale DEMs at the same fine spatial resolution. The detection is performed on grid cells rather than at the GPS stations, so is able to yield a field view of subsidence (Figure 6.11). Compared with long-term GPS observations, ALS data

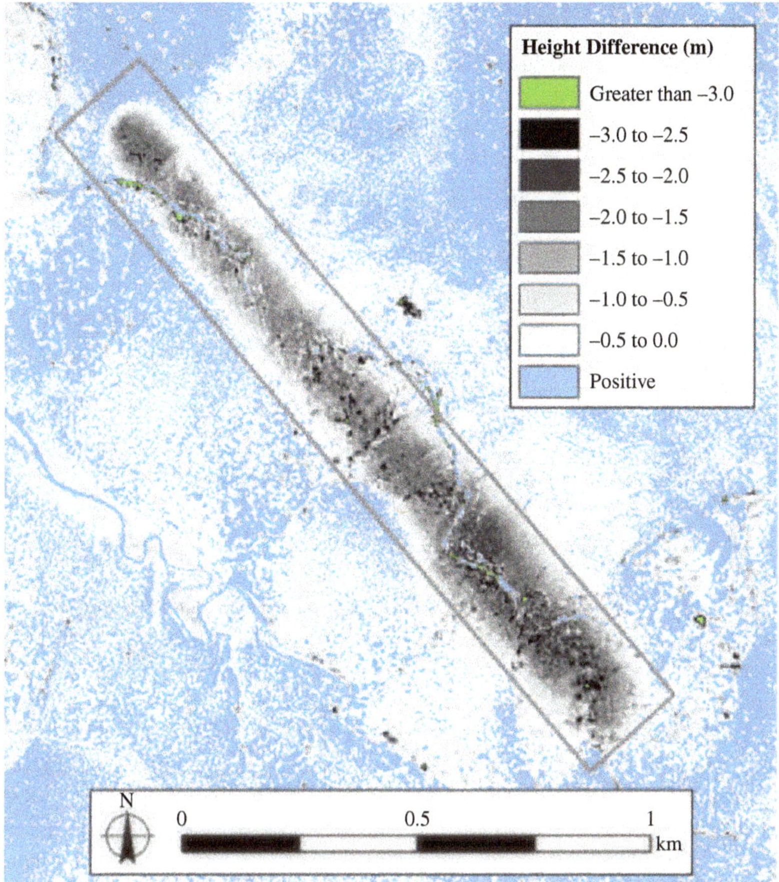

FIGURE 6.11 Spatial distribution of land subsidence in a coalmine area near Wollongong, Eastern Australia, detected from differencing of two DEMs interpolated from two airborne laser scan surveys. (Source: Palamara et al., 2007.)

are less accurate than GPS, achieving a vertical accuracy of about 0.26 m in deciduous forests, and 0.17–0.19 m in low grass and evergreen forests, even though the accuracy is higher (typically around 0.15 m, the nominal accuracy) in bare ground areas devoid of vegetation (Palamara et al., 2007). When the pre- and post-event data are compared with each other to detect subsidence, the vertical accuracy of detection is further degraded. This inaccuracy is easily ascertained from the height changes between the two DEMs in areas of no subsidence. The mean difference of these areas (e.g., offset) should be subtracted from the detected subsidence as a means of calibration. An average accuracy of around 0.23 m is achieved from ALS in mapping coalmine subsidence if the data are interpreted to a proper spatial resolution, even though this error may be exaggerated by the presence of a systematic 0.15 m offset between the two surveys (the error is much larger in areas of a steep terrain) (Palamara et al., 2007). Such an accuracy level is acceptable for studying mine subsidence but may not be accurate enough to study urban subsidence smaller than 10 cm. In addition, ALS is not generally suitable for accurately mapping steep or narrow features because of the inaccurate horizontal positioning of ALS data. They do not allow the horizontal displacement of surface deformation to be mapped competently. Although ALS is a viable option in monitoring land subsidence at a high accuracy level, the use of small-footprint LiDAR data is constrained by a narrow swath width, a disadvantage that can be overcome by integrating InSAR with Landsat ETM+ data, both offering far wider coverage. How to integrate multi-sensor data for detecting ground subsidence is the topic to be covered in the next section.

6.5 COMPARISON AND COMBINATION

Leveling, RTK GPS, ALS, and airborne DInSAR all have the potential for obtaining centimeter-level accuracies sufficient to monitor ground subsidence. Different methods of height determination have different accuracy levels and associated ease of use. Although the leveling method is the slowest, it is still needed along lines as its high accuracy cannot be easily matched by other methods. Leveling results are essential in validating the measurements from other methods. ALS has reached an operational status for monitoring decimeter-range subsidence. In comparison with leveling surveys, it is relatively inexpensive and fast, and hence preferable. Airborne InSAR and LiDAR are suitable for detecting local-scale subsidence. Unlike these imaging methods, GPS holds a considerable potential for providing accurate height measurements over extensive distances. It is more accurate than conventional leveling at long ranges, and much more efficient at the shortest ranges (Dodson, 1995).

The various types of height-finding methods introduced previously are best at detecting subsidence in different directions. All of them can detect vertical displacement except InSAR that only detects change in the range (line of sight) direction (Table 6.6). With the assistance of the incident angle, this range-direction displacement can be converted to vertical subsidence. Spirit leveling is very similar to GPS in that they work in the network scale, while InSAR, ALS, and borehole all work in the point or pixel scale. Both sample density and the accuracy (resolution) of detection vary widely, especially with InSAR-based methods caused by the large variation in the sensing range.

The vertical displacements detected via leveling surveys and InSAR results extracted from the stacked interferogram along the same transect have a high degree of agreement between them in both magnitude and the trend of displacements (Motagh et al., 2007). The land subsidence rates derived from InSAR closely match the leveling-surveyed and GPS-derived rates, but InSAR-generated maps show more details about both the amplitude and spatial extent of subsidence. A series of land subsidence maps generated from Sentinel-1 SAR images using PS-InSAR compares favorably with the in situ observed subsidence at one GPS station. Leveling survey, continuous GPS, monitoring wells, and PS-InSAR all achieved a vertical accuracy lower than 1 cm (Hsu et al., 2015), of which monitoring wells are the most accurate with a vertical error <0.5 cm, less than half of the residuals achieved by all other methods (Table 6.7). It must be noted that these results are obtained through lengthy observations over a period up to a few decades. In reality, it is impossible to monitor subsidence of the same area for such a long period, so the actual accuracy of monitoring subsidence will deviate slightly from the reported accuracies. The user is at liberty to decide which method is the best and should be adopted based on its main features and best use for a given area that may differ from the spelled best usage in the table.

TABLE 6.6
Comparison of Main Methods in Detecting Land Subsidence

Method	Component Displacement	Resolution (mm)	Spatial Density[a] (samples/survey)	Spatial Scale (elements)
Spirit level	Vertical	0.1–1	10–100	Line-network
Borehole extensometer	Vertical	0.01–0.1	1–3	Point
GPS	Vertical	20	10–100	Network
InSAR	Range	5–10	100,000–10,000,000	Map pixel[b]

Source: USGS (2000).

[a] Number of measurements generally attainable under the optimal conditions to define the spatial extent of land subsidence at the scale of the survey

[b] A pixel on an InSAR displacement map is typically 40–80 m^2 on the ground

TABLE 6.7
Accuracy of Main Methods of Monitoring Subsidence

Method	Dense Array	Survey Frequency/ Duration	Vertical Accuracy (cm)	Best Usage
Leveling survey	1.5–2 km	Yearly (1992–2002)	<1	Individual benchmark and isopleth of annual subsidence rate over 3 cm
Continuous GPS stations	5–20 km	Weekly (2001–2012)	<1	Comparison with other observations to understand subsiding mechanism
Monitoring wells	5–15 km	Monthly (1996–2011)	<0.5	Comparison with other observations to understand subsiding mechanism
PS-InSAR (ALOS)	25 m[a]	46 days (2007–2010)	<1	Feasibility study by comparing with historical results

Source: Hsu et al. (2015).
[a] Spatial resolution

The vertical accuracy of space-borne radar varies widely with the images used (Ge et al., 2007). ERS Tendem, JERS-1, and RadarSat-1 all achieved a vertical accuracy measured over 10 m, too inaccurate for most subsidence monitoring. Both SRTM and photogrammetry methods are noticeably more accurate than the three radar imagery-based methods. While the photogrammetry method is reasonably accurate with an RMSE of less than 3 m, slightly more accurate than SRTM data in three cases. The accuracy will definitely be better if the InSAR is created from fine-resolution radar images such as RadarSat-2. In comparison to the aforementioned methods, ALS offers the highest accuracy of 0.15 m in elevation. Even such an accuracy is inadequate in studying urban subsidence, even though it may be of some use in detecting mining-induced ground subsidence. Better methods are needed to detect cm-level subsidence, such as InSAR.

The performance of various forms of InSAR has been comparatively evaluated. Both PS and SBAS achieved an RMSE of about 1 mm per annum and 5 mm, respectively, in estimating the mean velocity of deformation (Shanker et al., 2011). These rates are within the expected noise levels and characteristic of the pixel selection parameters for both TS techniques. These estimates agree with creep measurements from alignment arrays within 1.5 mm for LOS displacement. The subsidence results produced from SBAS and DInSAR are in close agreement with each other. SBAS can reliably monitor residual subsidence of longwall mines surrounding the subsidence trough (Pawluszek-Filipiak and Borkowski, 2020), at a cm-level vertical accuracy. This accuracy level is more accurate than that of the most accurate method of ALS reported by Ge et al. (2007). SBAS is applicable to detecting subsidence measured in the decimeter range or higher.

The best way of monitoring subsidence is to make use of multi-sensor data involving GPS stations, leveling surveys, monitoring wells, and PS-InSAR (Hsu et al., 2015). The integration of multiple remote sensing techniques (e.g., GPS, InSAR, LiDAR, and photogrammetry) under a unified geodetic reference should enrich the data and extend the monitoring period of subsidence. The combination of GPS with InSAR is advantageous to each of them as they have complementary strengths that overcome the limitations inherent in every single technique alone. GPS supplies time-continuous data, while InSAR provides spatially continuous results. Their integration enables the conversion of the relative InSAR LOS displacement rates to a geodetic reference frame. If combined with velocities derived from the GPS and gravity data, the PS-InSAR technique can monitor spatio-temporal land subsidence continuously (Zerbini et al., 2007). This combination boosts the density of measurements that is crucial to understanding the process of subsidence. If integrated via kriging, the DInSAR and SBAS combination is effective at monitoring mining-related subsidence with a vertical RMSE of 11 mm between estimated and ground-measured results, even if the subsidence rate is rather high (Ahmad et al., 2019).

After subsidence has been quantified, it is possible to further map subsidence hazards using two indices (Hu et al., 2004). The first is known as accumulative land subsidence (Z) calculated as

$$Z = \frac{\text{accumulative subsidence} - \text{critical subsidence}}{\text{critical subsidence}} \tag{6.14}$$

where critical subsidence refers to the threshold of land subsidence that starts to be hazardous (e.g., 300 mm). From Z, another index, the hazardous degree index (D), is calculated as

$$\mathrm{D} = \mathrm{Y} \times \mathrm{Z} \tag{6.15}$$

where Y = the vulnerability index to land subsidence related to soil looseness and groundwater level. Contour maps of both $\underline{Z}$ and D can be produced to illustrate the spatial distribution of land subsidence hazards in a municipality.

REFERENCES

Agudelo, G., Wang, G., Liu, Y., Bao, Y., and Turco, M. J., 2020. GPS geodetic infrastructure for subsidence and fault monitoring in Houston, Texas, USA, Proc. IAHS, 382, 11–18, https://doi.org/10.5194/piahs-382-11-2020.

Ahmad, W., M. Choi, S. Kim, and D. Kim. 2019. Detection of land subsidence and its relationship with land cover types using ESA Sentinel satellite data: A case study of Quetta Valley, Pakistan. *Int. J. Remote Sens.* 40(24): 9572–9603.

Berardino, P., G. Fornaro, R. Lanari, and E. Sansosti. 2002. A new algorithm for surface deformation monitoring based on small baseline differential SAR interferograms. *IEEE Trans. Geosci. Remote Sens.* 40(11): 2375–2383. DOI: 10.1109/TGRS.2002.803792

Bhattacharya, A., M.K. Aroram, and M.L. Sharma. 2012. Usefulness of synthetic aperture radar (SAR) interferometry for digital elevation model (DEM) generation and estimation of land surface displacement in Jharia coal field area. *Geocarto Int.* 2(1): 57–77. DOI: 10.1080/10106049.2011.614358

Cao, J., H. Gong, B. Chen, M. Shi, C. Zhou, K. Lei, H. Yu, and Y. Sun. 2021. Land subsidence in Beijing's sub-administrative center and its relationship with urban expansion inferred from Sentinel-1/2 observations. *Can. J. Remote Sens.* 47(6): 802–817. DOI: 10.1080/07038992.2021.1964944

Cao, N., H. Lee, and H.C. Jung. 2016. A phase-decomposition-based PSInSAR processing method. *IEEE Trans. Geosci. Remote Sens.* 54(2): 1074–1090.

Chaussard, E., S. Wdowinski, E. Cabral-Cano, and F. Amelung. 2014. Land subsidence in central Mexico detected by ALOS InSAR time-series. *Remote Sens. Environ.* 140: 94–106. DOI: 10.1016/j.rse.2013.08.038

Chen, G., Y. Zhang, R. Zeng, Z. Yang, X. Chen, F. Zhao, and X. Meng. 2018. Detection of land subsidence associated with land creation and rapid urbanization in the Chinese Loess Plateau using time series InSAR: A case study of Lanzhou New District. *Remote Sens.* 10(2), 270. DOI: 10.3390/rs10020270

Chen, L. 1997. Subsidence assessment in the Ruqigou coalfield, Ningxia, China, using a geomorphological approach. MSc thesis, International Institute for Aerospace Survey and Earth Sciences, Enschede, The Netherlands.

Crosetto, M., O. Monserrat, M. Cuevas-González, N. Devanthéry, and B. Crippa. 2016. Persistent scatterer interferometry: A review. *ISPRS J. Photogramm. Remote Sens.* 115: 78–89.

Dehghani, M., V. Zoej, M. Javad, A.E. Mansourian, and S. Saatchi. 2009. InSAR monitoring of progressive land subsidence in Neyshabour, northeast Iran. *Geophys. J. Int.* 178(1): 47–56. DOI: 10.1111/j.1365-246X.2009.04135.x

Dodson, A.H. 1995: GPS for height determination. *Survey Review* 33, 66–76.

Dong, J., Zhang, L., Tang, M., Liao, M., Xu, Q. Gong, J., and Ao, M., 2018. Mapping landslide surface displacements with time series SAR interferometry by combining persistent and distributed scatterers: A case study of Jiaju landslide in Danba, China. *Remote Sensing of Environment*, 205: 180–198.

Ferretti, A., C. Prati, and F. Rocca. 2000. Nonlinear subsidence rate estimation using permanent scatterers in differential SAR interferometry. *IEEE Trans. Geosci. Remote Sens.* 38: 2202–2212.

Ferretti, A., C. Prati, and F. Rocca. 2001. Permanent scatterers in SAR interferometry. *IEEE Trans. Geosci. Remote Sens.* 39: 8–20.

Ge, L., H.-C. Chang, and C. Rizos. 2007. Mine subsidence monitoring using multi-source satellite SAR images. *Photogramm. Eng. Remote Sens.* 73(3): 259–266.

Gong, W., A. Thiele, S. Hinz, F.J. Meyer, A. Hooper, and P.S. Agram. 2016. Comparison of small baseline interferometric SAR processors for estimating ground deformation. *Remote Sens.* 8(4): 330. DOI: 10.3390/rs8040330

Grzovic, M., and A. Ghulam. 2015. Evaluation of land subsidence from underground coal mining using TimeSAR (SBAS and PSI) in Springfield, Illinois, USA. *Nat. Hazards* 79: 1739–1751. DOI: 10.1007/s11069-015-1927-z

Hooper, A. 2008. A multi-temporal InSAR method incorporating both persistent scatterer and small baseline approaches. *Geophys. Res. Lett.* 35(L16302): 96–106. DOI: 10.1029/2008GL034654

Hooper, A., H. Zebker, P. Segall, and B. Kampes. 2004. A new method for measuring deformation on volcanoes and other natural terrains using InSAR persistent scatterers. *Geophs. Res. Lett.* 31(23): L23611. DOI: 10.1029/2004GL021737

Hsu, W.C., H.C. Chang, K.T. Chang, E.K. Lin, J.K. Liu, and Y.A. Liou. 2015. Observing land subsidence and revealing the factors that influence it using a multi-sensor approach in Yunlin County, Taiwan. *Remote Sens.* 7: 8202–8223. DOI: 10.3390/rs70608202

Hu, R.L., Z.Q. Yue, L.C. Wang, and S.J. Wang. 2004. Review on current status and challenging issues of land subsidence in China. *Eng. Geol.* 76: 65–77. DOI: 10.1016/j.enggeo.2004.06.006

Liu, G., Hao, X., Xue, H., Du, Z. 2007. Related analysis of effecting factors of height measurement accuracy of InSAR. *Geomatics and Information Science of Wuhan University*, 32(1): 55–58.

Mateos, R.M., P. Ezquerro, J.A. Luque-Espinar, M. Béjar-Pizarro, D. Notti, J.M. Azañón, O. Montserrat, G. Herrera, F. Fernández-Chacón, and T. Peinado. 2017. Multiband PSInSAR and long-period monitoring of land subsidence in a strategic detrital aquifer (Vega de Granada, SE Spain): An approach to support management decisions. *J. Hydrol.* 553: 71–87.

Motagh, M., Y. Djamour, T.R. Walter, H.J. Wetzel, J. Zschau, and S. Arabi. 2007. Land subsidence in Mashhad Valley, northeast Iran: Results from InSAR, levelling and GPS. *Geophys. J. Int.* 168(2): 518–526. DOI: 10.1111/j.1365-246X.2006.03246.x

Ng, A., H.-C. Chang, L. Ge, C. Rizos, and M. Omura. 2009. Assessment of radar interferometry performance for ground subsidence monitoring due to underground mining. *Earth Planets Space* 61: 733–745. DOI: 10.1186/BF03353180

Ng, A.H.M., L. Ge, Z. Du, S. Wang, and C. Ma. 2017. Satellite radar interferometry for monitoring subsidence induced by longwall mining activity using Radarsat-2, Sentinel-1 and ALOS-2 data. *Int. J. Appl. Earth Obs. Geoinf.* 61: 92–103.

Palamara, D.R., M. Nicholson, P. Flentje, E. Baafi, and G.M. Brassington. 2007. An evaluation of airborne laser scan data for coalmine subsidence mapping. *Int. J. Remote Sens.* 28(15): 3181–3203. DOI: 10.1080/01431160600993439

Pawluszek-Filipiak, K., and A. Borkowski. 2020. Integration of DInSAR and SBAS techniques to determine mining-related deformations using Sentinel-1 data: The case study of Rydułtowy mine in Poland. *Remote Sens.* 12: 242. DOI: 10.3390/rs12020242

Rateb, A., and A.Z. Abotalib. 2020. Inferencing the land subsidence in the Nile Delta using Sentinel-1 satellites and GPS between 2015 and 2019. *Sci. Total Environ.* 729. DOI: 10.1016/j.scitotenv.2020.138868, 1–11

Sarti, F., Arkin, Y., Chorowicz, J., Karnieli, A., Cunha, T. 2003. Assessing pre- and post-deformation in the southern Arava Valley segment of the Dead Sea Transform, Israel by differential interferometry. *Remote Sensing of Environment*, 86(2): 141–149.

Sato, H., K. Abe, and O. Ootaki. 2003. GPS-measured land subsidence in Ojiya City, Niigata Prefecture, Japan. *Eng. Geol.* 67: 379–390. DOI: 10.1016/S0013-7952(02)00221-1

Shanker, P., F. Casu, H.A. Zebker, and R. Lanari. 2011. Comparison of persistent scatterers and small baseline time-series InSAR results: A case study of the San Francisco bay area. *IEEE Trans. Geosci. Remote Sens.* 8(4): 592–596. DOI: 10.1109/LGRS.2010.2095829

Sneed, M., and J.T. Brandt. 2013. Detection and measurement of land subsidence using global positioning system surveying and interferometric synthetic aperture radar, Coachella Valley, California, 1996–2005. USGS Scientific Investigations Report 2007–5251, 30 p.

Strozzi, T., U. Wegmüller, L. Tosi, G. Bitelli, and V. Spreckels. 2001. Land subsidence monitoring with differential SAR interferometry. *Photogram. Eng. Remote Sens.* 67(11): 1261–1270.

USGS, 2000. Measuring Land Subsidence from Space, Fact Sheet-051-00, retrieved from https://pubs.usgs.gov/fs/fs-051-00/pdf/fs-051-00.pdf.

Wang, G., and Soler, T., 2014. Measuring land subsidence using GPS: Ellipsoid height versus orthometric height. J. of *Surveying Engineering*, 141(2)

Wang, J., P. Wang, Q. Qin, and H. Wang. 2017. The effects of land subsidence and rehabilitation on soil hydraulic properties in a mining area in the Loess Plateau of China. *Catena* 159: 51–59. DOI: 10.1016/j.catena.2017.08.001

Yan, Y., M.P. Doin, P. Lopez-Quiroz, F. Tupin, B. Fruneau, V. Pinel, and E. Trouve. 2012. Mexico City subsidence measured by InSAR time series: Joint analysis using PS and SBAS approaches. *IEEE Trans. Geosci. Remote Sens.* 5: 1312–1326.

Ye, X., Kaufmann, H., and Guo, X.F., 2004. Landslide monitoring in the three Gorges area using D-InSAR and corner reflectors. *Photogrammetric Engineering and Remote Sensing*, 70(10): 1167–1172.

Zerbini, S., B. Richter, F. Rocca, T. van Dam, and F. Matonti. 2007. A combination of space and terrestrial geodetic techniques to monitor land subsidence: Case study, the Southeastern Po Plain, Italy. *J. Geophys. Res.* 112: B05401. DOI: 10.1029/2006JB004338

Zhang, Y., F. Fattahi, and A. Amelung. 2019. Small baseline InSAR time series analysis: Unwrapping error correction and noise reduction. *Comput. Geosci.* 133: 104331. DOI: 10.1016/j.cageo.2019.104331

Zhang, Y., H. Wu, Y. Kang, and C. Zhu. 2016. Ground subsidence in the Beijing-Tianjin-Hebei region from 1992 to 2014 revealed by multiple SAR stacks. *Remote Sens.* 8: 675. DOI: 10.3390/rs8080675

Zhao, C.Y., Q. Zhang, X.L. Ding, Z. Lu, C.S. Yang, and X.M. Qi. 2009. Monitoring of land subsidence and ground fissures in Xian, China 2005–2006: Mapped by SAR interferometry. *Environ. Geol.* 58: 1533. DOI: 10.1007/s00254-008-1654-9

Zhou, C., H. Gong, B. Chen, J. Li, M. Gao, F. Zhu, W. Chen, and Y. Liang. 2017. InSAR time-series analysis of land subsidence under different land use types in the eastern Beijing Plain, China. *Remote Sens.* 9: 380. DOI: 10.3390/rs9040380

Zhou, X., N.-B. Chang, and S. Li. 2009. Applications of SAR interferometry in earth and environmental science research. *Sensors* 9(3): 1876–1912. DOI: 10.3390/s90301876

7 Droughts

7.1 INTRODUCTION

Drought is a meteorological phenomenon characterized by a lack of rainfall over an extended period that has resulted in soil moisture deficiency to such a level to have negatively impacted the bio-ecosystem. Of all the natural hazards covered in this book, drought is the most complex involving a number of processes and numerous factors, such as evaporation, transpiration, and surface cover, all of which directly affect the water balance of a region. Droughts impact agriculture, ecology, and socioeconomics. They may take place naturally, but their occurrence has been accelerated by human activities and climate warming, even though it is impossible to quantify the precise relationship between the two. As the global climate continues to warm, droughts are becoming more frequent and severer. Although not as destructive as other types of natural hazards such as hurricanes and earthquakes, droughts are the costliest hazard among all natural hazards covered in this book. Each year they cause a gigantic economic loss owing to the extensive areas affected over a prolonged period. In the worst case, droughts can threaten the livelihood of farmers and even national food security.

Droughts may be perceived from four perspectives of meteorology, hydrology, agriculture, and economy. *Meteorological drought* or climatological drought refers mainly to precipitation anomaly below the normal trend over an extended period and is related solely to atmospheric processes. It is characterized by a rainfall deficit that may be exacerbated by temperature anomaly-enhanced evapotranspiration (Figure 7.1). Meteorological drought can trigger other types of drought if persistent, such as *agricultural drought*. It refers to the deficit of soil moisture that affects normal farming, stunts plant growth (e.g., lowered biomass), reduces crop yield, and even results in crop failure, as a consequence of below-normal precipitation, and/or above-average evaporation and transpiration. *Hydrological drought* is the manifestation of meteorological drought in surface water features, in the forms of inadequate surface and subsurface water resources, a low channel flow, or a low groundwater level. It is related to land surface processes. *Socioeconomic drought* alludes to the failure of water resources systems to meet water demands and is thus associated with the supply of and demand for economic goods (water). It also pertains to the social, economic, and environmental impacts of drought. Since it is not related to remote sensing, it will not be covered in this chapter.

It is important to differentiate the first three types of drought as they are best studied using different means and data. Although meteorological drought is commonly studied from rain gauge data, such data are point-based and highly limited in their spatial distribution. For places where no such data are available, remote sensing becomes the only means to study it. In turn, meteorological drought serves as a reference to assess the validity of remote sensing-derived drought results. Of the three types of drought, it is the agricultural drought that benefits the most from remote sensing. In contrast, hydrological drought is difficult to study from remotely sensed data, especially channel flow due to the coarse spatial resolution of imagery.

In studying agricultural drought, remote sensing has significantly enhanced the observations and estimation of major drought-related variables over broad spatial and long-term temporal scales (West et al., 2019). A vast variety of remotely sensed data products have been produced for studying nearly all aspects of drought propagation. They play a vital role in facilitating our understanding of drought and assessing its severity for all types of drought. This chapter first presents an overview of the vast variety of remotely sensed data products that are excellent sources for drought monitoring and assessment and demonstrates how remote sensing can be used to fulfill different needs in drought monitoring and assessment. Then it introduces various indicators and diverse spectral

DOI: 10.1201/9781003354321-7

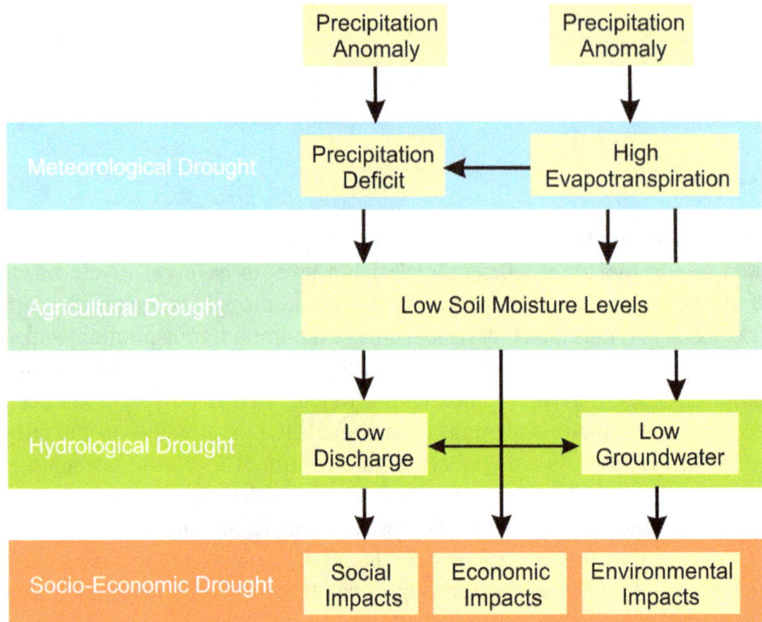

FIGURE 7.1 Four types of drought, their interactions, and impacts. (Source: Modified from West et al., 2019. Used with permission.)

indices that have been proposed to study different types of drought. In particular, how to sense agricultural drought based on indicators and indices forms the bulk of the discussion. Also explained in depth in this chapter is how to assess the ecological response to droughts and monitor drought effectively using various indices, followed by an exploration of how to assess the impacts and risks of drought. Finally, this chapter expounds on how to predict drought from diverse types of remotely sensed data using several analytical methods.

7.2 USEFUL REMOTE SENSING DATA (PRODUCTS)

The growing number of continuously improving Earth observation satellite missions, along with the increasingly powerful and diverse remote sensing data processing methods and techniques, presents ever more opportunities and raises new capabilities in monitoring and assessing droughts. Drought monitoring requires imagery data of broad spatial coverage and a coarse spatial resolution spanning over a long period, preferably decades. Of all the satellite data available, three have proved the most useful. They are Tropical Rainfall Measuring Mission (TRMM), Advanced Microwave Scanning Radiometer – Earth Observing System (AMSR-E), and MODIS.

7.2.1 TRMM and GPM

TRMM is a joint program between the National Aeronautics and Space Administration (NASA) of the US and the National Space Development Agency of Japan. Launched in 1997 but decommissioned in 2015, TRMM senses tropical and subtropical rainfall (35°S–35°N). It is the first satellite to carry a specific microwave precipitation radar. Its other payload comprises several sensors for sensing precipitation, such as the TRMM microwave imager, and the visible and infrared radiometer system. TRMM data products are released at various processing levels. The level 3 product 3B43 contains monthly precipitation rates ($mm \cdot h^{-1}$) at a spatial resolution of 0.25° × 0.25° generated from various sensors, such as high-resolution radar, visible-infrared radiometer, and passive microwave radiometer.

TRMM has been succeeded by the Global Precipitation Measurement (GPM) mission launched in February 2014. This mission operates in a non-polar, low inclination orbit completing 16 revolutions per day. It extends the geographic coverage of TRMM to 65°S–65°N. GPM data products have a temporal resolution of 1–2 hour, with an improved spatial resolution of 0.1–0.25°. The Global Satellite Mapping of Precipitation is another source of precipitation data. Daily average gauge-calibrated 0.1° grid data (daily0.1_G, version 6) are available from the Global Rainfall Watch of the Japan Aerospace Exploration Agency (JAXA) Earth Observation Research Center website (https://sharaku.eorc.jaxa.jp/GSMaP/index.htm). This global dataset covers 60°N–60°S.

7.2.2 SMAP

The Soil Moisture Active Passive (SMAP) mission was launched in 2015 to sense soil moisture remotely. It aims to generate a product of a relatively high spatial resolution (3, 9, and 36 km) and a temporal resolution of two to three days by merging high-resolution active radar data with coarse-resolution, highly sensitive, passive radiometer data. However, only operational for nine months, the radar sensor malfunctioned. Despite the loss of the onboard radar, SMAP products have the potential for large-scale agricultural drought monitoring. SMAP data over 13–26 week intervals are able to accurately capture changing drought intensity levels.

7.2.3 AMSR-E

Designed by JAXA, the AMSR-E sensor is a payload of NASA's Earth Observing System (EOS) Aqua spacecraft launched on 4 May 2002. The instrument stopped working properly on 4 October 2011 and has been superseded by AMSR-2, launched on 18 May 2012. AMSR-E is equipped with a passive microwave radiometer that acquires dual-polarized data in 12 channels at six frequencies. This sensor is designed to measure precipitation rate, cloud water, water vapor, sea surface winds, sea surface temperature (SST), ice, snow, and soil moisture under all weather conditions. The measured geophysical parameters support several global change studies and monitoring efforts. AMSR-E data have a demonstrated potential for effective drought monitoring. AMSR-E historic products have been reanalyzed to derive various agricultural drought indices. The data record is found to have a good potential for representing long-term drought events over large spatial scales. Microwave remotely sensed data from AMSR-E, in the form of the soil moisture condition index (SMCI), is well correlated with the short-term standardized precipitation index (SPI) in regions of low vegetation cover (Zhang and Jia, 2013).

7.2.4 Data Products

A variety of data products have been derived from the aforementioned space missions, and they have considerably eased drought monitoring. These diverse products serve different purposes in studying droughts. Summarized in Table 7.1 are the most commonly used data products. They are by no means exhaustive. Many more products are available for niche applications. So far, MODIS data have yielded the most comprehensive range of data products covering all aspects of the Earth's biological and surface zones, such as rainfall, LST (MYD11C3), albedo (MCD43A3), vegetation indices, LAI (MCD15A2H), and evapotranspiration, some of which are time-integrated 8-day or 16-day products. The LAI and albedo datasets are delivered at 500 m spatial resolution and eight-day temporal resolution. Released in 2017, MODIS-ET version 6 has an improved spatial resolution from 1,000 to 500 m, delivered as eight-day composites. MYD16A2 is the main product from the Aqua satellite data. Level-4 global LAI and the Fraction of Absorbed Photosynthetically Active Radiation (FPAR) (MOD15A2) datasets are available every eight days at a 1-km resolution on a sinusoidal grid and span from February 2000 to the present. They are useful as proxies for vegetation activity and can be used to calculate meteorological drought indices directly.

TABLE 7.1
Common Satellite Programs and Their Data Products Useful for Studying Droughts

Satellite	Data Product	Parameter	Spatial Resolution	Temporal Resolution
MODIS	MOD11A1 (A2)	LST	1 km	Daily (eight-daily)
	MOD13A3	NDVI	1 km	Monthly
	MOD16A2	Actual ET	0.5 km	Eight days
	MCD43B2 (A3)	Albedo	1 km (500 m)	Eight days
	MOD15A2	LAI, FPAR	1 km	Eight days
	MYD11C3	Surface reflectance	0.05°	Eight days
	MYD13C3	LST (day, night)	0.05°	Monthly
PERSIANN-CDR	PERSIANN-CDR	Precipitation	0.25°	Daily
TRMM	TRMM3B42	Precipitation	0.25°	Three hourly
	TRMM3B43	Precipitation	0.25°	Monthly (mm/hour)
GPM	GPM IMERG	Precipitation	0.1°	Daily and monthly[a]
SMAP	ESA CCI	Soil moisture	0.25°	Daily
	EASE-Grid Surface	Soil moisture	9 km	Three-hourly
	SPL4SMGP	Root zone soil moisture		
	ERA-Interim/Land	Soil moisture	0.25°	Daily
	GLEAM	ET	0.25°	Daily
	ERA-Interim	Temperature and P	0.25°	Daily
	ESA CCI	Land cover	300 m	Static
	NDVI3g	NDVI	1/12°	Biweekly
GLDAS	NOAH025_M	Soil moisture	0.25°	Monthly from 3-hourly data

[a] Hourly data are available, but their spatial resolution is much coarser (e.g., 10°)

Other more recent and long-term precipitation data products include the CPC (Climate Prediction Center) Morphing Technique (CMORPH) Climate Data Record (CDR), Precipitation Estimation from Remotely Sensed Information using Artificial Neural Networks–Climate Data Record (PERSIANN-CDR), and the Integrated Multi-satellitE Retrievals for GPM (IMERG). CMORPH is a technique for combining existing rainfall retrieval algorithms to generate precipitation estimates that have been retrieved from low orbiter passive microwave and infrared sensors aboard such satellites as NOAA (AMSU-B), NASA's Aqua, and TRMM spacecraft (Joyce et al., 2004). The satellite precipitation estimates have been bias-corrected and reprocessed to form a global (60°N–60°S), high-resolution (0.07277° or 8 km at the equator) precipitation data product. It has a temporal resolution of 30 minutes and is available from January 1998 (the post-TRMM period) to the present. CMORPH performs well, especially when compared to other high-resolution products (Joseph et al., 2009).

PERSIANN-CDR is a daily, quasi-global precipitation product spanning from 1983 to 2020, ideal for long-term hydroclimate studies at the regional and global scales. This data product is derived from the infrared brightness temperature (T_B) data of multi-satellite observations (i.e., SSM/I, AMSR-E, and TRMM) using the PERSIANN algorithm, adjusted using the Global Precipitation Climatology Project monthly product to maintain consistency. It covers the area between 60°S and 60°N at a spatial resolution of 0.25°. This long-term, high-resolution data product is a very important data source for supporting climatologists, hydrologists, and hydrometeorologists to study global changes and trends in daily precipitation and extreme events (e.g., floods and droughts). This dataset has been found to provide reasonably accurate rainfall estimates (Ashouri et al., 2015).

IMERG is an algorithm for inter-calibrating, merging, and interpolating precipitation estimates from all microwave satellites, together with microwave-calibrated infrared satellite estimates, precipitation gauge analyses, and potentially other precipitation estimators. The system has three "Runs" – Early (4 hours after observation time), Late (14 hours afterward), and Final (3.5 months afterward) (Huffman et al., 2020). The early run produces a quick estimate and the late runs successively provide better estimates as more data become available. The final run uses monthly gauge data to create a research-level product every 30 minutes. At a grid size of 0.1° × 0.1°, this gridded dataset has a fine temporal and spatial scale for the TRMM and GPM eras over the entire globe between 60°N and 60°S (and partially outside of that latitude band).

Compared with precipitation data products, evaporation data products are much less common. A major one is the Global Land Evaporation Amsterdam Model (GLEAM) product. This set of algorithms separately estimates the various components of land evaporation, including transpiration, bare-soil evaporation, interception loss, open water evaporation, and sublimation, at a 0.25° spatial resolution and a daily temporal resolution. It comprises a series of C- and L-band measurements of vegetation, soil moisture, precipitation, and LST, from sensors such as MODIS and the Soil Moisture Ocean Salinity mission. This is the only global-scale evaporation product derived purely from remotely sensed data. GLEAM version 3.5 products total ten in number, supplying information about surface and root-zone soil moisture, potential evaporation, and evaporative stress conditions. GLEAM data products (version 3.3) have a long history of existence, spanning from 1980 to 2018.

7.3 DROUGHT INDICATORS AND INDICES

The most direct and perceivable manifestation of drought is vegetation conditions, followed by soil conditions, such as moisture. Naturally, the most effective way of monitoring drought is via vegetation and soil conditions that are best studied using remote sensing. The assessment of drought relies heavily on its indicators. In agricultural and vegetated regions, the most useful and reliable indicator derivable from remote sensing data is NDVI. Although its value varies considerably across different geographic regions and seasonally even for the same geographic area, the temporal and yearly variation of NDVI of the same geographic area at the same time can reveal the degree of drought. Apart from vegetation indices, a large number of drought-specific indices have also been developed and used to characterize drought and its severity.

7.3.1 Meteorological Drought

Meteorological drought is best studied from meteorological records logged at ground stations that maintain a comprehensive range of data consistently collected since their inception. The commonly used meteorological parameters are precipitation, temperature, and evaporation, both potential and actual, as well as evapotranspiration (EP), from which a number of indices have been developed to characterize meteorological drought. They include aridity index (AI), precipitation condition index (PCI), and SPI. AI is a climate factor that controls the spatial distribution of vegetation in an area. Calculated by ratioing the annual mean precipitation to the annual air temperature, this index is useful for assessing whether the spatial distribution of AI has affected the spatial differentiation of vegetation activity. PCI is calculated as follows:

$$PCI = \frac{CHIRPS_{current} - CHIRPS_{min}}{CHIRPS_{max} - CHIRPS_{min}} \times 100 \tag{7.1}$$

where CHIRPS = Climate Hazards Group Infrared Precipitation with Station monthly precipitation data. Min and max precipitation are determined from long-term (e.g., ten years) records. SPI is based solely on precipitation data (McKee et al., 1993), and precipitation anomaly is flexibly

TABLE 7.2
SPI and SPEI Value Ranges Representing Different Drought Levels

Index Value	Drought Severity
≥2.00	Extremely wet
≥1.50 to 2.00	Very wet
≥1.00 to 1.50	Moderately wet
≥0.99 to –0.99	Near normal
≥–1.00 to –1.50	Moderate drought
≥–1.50 to –2.00	Severe drought
≤–2.00	Extreme drought

Source: McKee et al. (1993).

quantified over a period ranging from one month to one year in accordance with the long-term precipitation record for a specific drought type. SPI is calculated as follows:

$$g(p) = \frac{p^{\alpha-1} e^{\frac{-p}{\beta}}}{\beta^{\alpha} \cdot \tau(\alpha)} \ (p > 0) \tag{7.2}$$

where $\alpha > 0$ = a shape parameter, $\beta > 0$ = a scale factor, p = precipitation, and $\tau(\alpha)$ = a gamma function. SPI is normally averaged monthly or seasonally, such as 3 months, 6 months, and 12 months in studying drought. An SPI value between –0.99 and 0.99 means near normality, a value below this range indicates dryness, and above it wetness. A value above 2 signifies extremely wet, below –2.0 extremely drought, of which a value between –1.5 and –1.0 indicates moderate drought, and between –2.0 and –1.5 severe drought (Table 7.2). SPI has been used extensively to study the complex interactions among drought, climate conditions, and vegetation response (Figure 7.2). SPI has an excellent performance in characterizing the spatial and temporal properties of drought (Zhu et al., 2020). Cumulative negative SPI shows more prominent effects than minimum SPI in characterizing different droughts. For this reason, SPI is commonly used to substantiate and validate the results about other types of droughts detected from remotely sensed data.

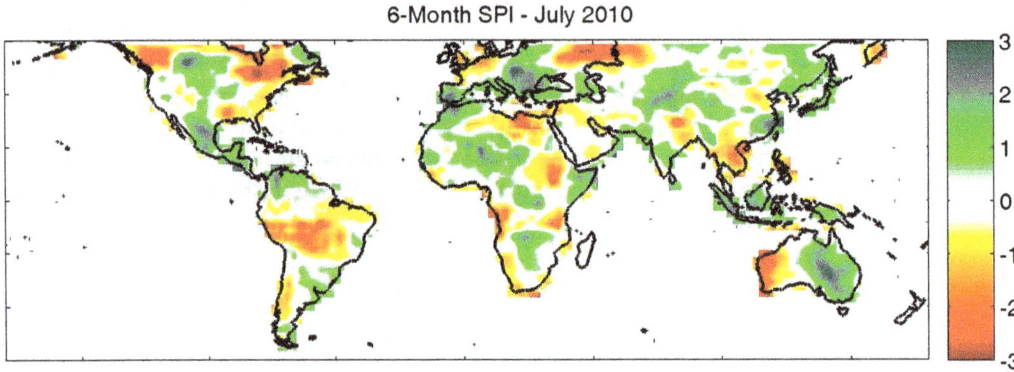

FIGURE 7.2 Global near-real-time drought monitoring based on SPI combining two remote sensing datasets of GPCP – Global Precipitation Climatology Project, and PERSIANN. (Source: AghaKouchak et al., 2015.)

Another meteorological index similar to SPI is the standardized precipitation evapotranspiration index (SPEI). It measures the difference between precipitation and potential evapotranspiration (PET), taking into account the role of temperature indirectly. Drought is determined by standardizing the disparity between precipitation and potential evapotranspiration (P-PET). Both SPI and SPEI are treated as the dependent variables in predicting drought from remotely sensed drought factors derived from the TRMM and MODIS satellite images.

Percentage of precipitation anomalies (P_a) is another meteorological indicator able to characterize the drought associated with precipitation anomalies over a certain period, such as monthly, seasonally, or yearly, by comparing the amount of rainfall in relation to the normal value. It manifests the occurrence of drought directly. P_a is normalized and standardized as follows:

$$P_{ai} = \frac{P_i - \bar{P}}{\bar{P}} \times 100\% \tag{7.3}$$

where P_{ai} = precipitation anomaly in month i. P_i = monthly precipitation in month i. $\bar{P}$ = average annual precipitation in a month.

More meteorological drought indicators have been developed based on ET. The first two are crop water stress index (CWSI) and water deficit index (WDI), calculated as follows:

$$CWSI = 1 - \frac{AET}{PET} \tag{7.4}$$

$$WDI = 1 - \frac{\Psi_{AET}}{\Psi_{PET}} \tag{7.5}$$

where AET = actual ET, a combination of transpiration from vegetation and evaporation from soil; ψ = rate of PET. As an important component of the hydrologic budget, CWSI indicates the exchange of mass and energy between the soil–water–vegetation system and the atmosphere. WDI is the same as CWSI except that it makes use of the rate of ET. The third index is evapotranspiration condition index (ETCI) calculated as follows:

$$ETCI = \frac{ET_{max} - ET_{current}}{ET_{max} - ET_{min}} \times 100 \tag{7.6}$$

Finally, the water budget drought index (WBDI) has also been proposed to study drought. It is calculated as follows:

$$\mathrm{WBDI} = \mathrm{z(P - E) = z(\Delta s + R)} \tag{7.7}$$

where z = the standardization coefficient; P = precipitation (mm), E = actual evaporation (mm), ΔS = soil moisture change (mm), and R = potential runoff (Kim et al., 2021).

Of all the meteorological drought indices, those based on ground weather station data are more accurate than those from remotely sensed ones but have limited spatial coverage, which is irrelevant to remote sensing drought indices. Most of these indices cannot be derived directly from remotely sensed data, so they are not explored further. Instead, the discussion will concentrate on agricultural drought indicators.

7.3.2 Agricultural Drought

Drought causes changes to the external appearance of vegetation, which can be detected from its changed spectral response and evaluated using satellite-derived vegetation indices. It is relatively easy to measure and quantify meteorological drought thanks to the availability of long-term data

TABLE 7.3
Remote Sensing-Derived Indices and/or Parameters That Have Been Used to Predict Agricultural Drought Measured by Ground-Based Drought Indices from MODIS Data

Index	Definition	Formula	Resolution
Pre	Precipitation	–	0.25°
PET	Potential evapotranspiration	–	500 m
ET	Evapotranspiration	–	500 m
LSTd/n/m	Land surface temperature (day, night, mean)	(LSTd + LSTn)/2	1 km
NDVI	Normalized difference vegetation index	(B2 – B1)/(B2 + B1)	1 km
EVI	Enhanced vegetation index	2.5(B2 – B1)/(B2 + 6B1 – 7.5B3 + 1)	1 km
NMVI	Normalized multi-band vegetation index	(B2 – B6 + B7)/ (B2 + B6 – B7)	500 m
NDWI15	Normalized difference water index[a]	(B2 – B5)/(B2 + B5)	500 m
NDD15	Normalized difference drought index[a]	(NDVI – NDWI5) /(NDVI + NDWI5)	500 m

Source: Modified from Feng et al. (2019). Used with permission.

[a] The computation is repeated twice, using b6 and b7, resulting in NDWI6 and NDWI7, as well as NDDI6 and NDDI7, respectively

records obtained from ground weather stations, but this is not the case with agricultural drought which has to be studied using remote sensing data in conjunction with climate data. Agricultural drought is commonly studied via vegetation conditions. A large number of drought indices have been developed based on VIs to characterize agricultural drought, relying on either climate data or remote sensing-derived drought proxies (Table 7.3). More variables can be created by averaging these parameters over different durations, such as one month and three months. They are correlated with the ground-observed data. The highest correlation between them indicates the time when drought is the most pronounced.

Of all the existing VIs, NDVI is the most popular in studying drought, followed by EVI. Apart from NDVI variation, LST has also been used to detect drought (Karnieli and Dall'Olmo, 2003). If a surface is drier, its temperature must be higher, hence LST is also an indirect indicator of drought, and its values represent seasonal climatic fluctuation. If a crop does not have sufficient water to evaporate on the surface of its leaves, it responds to the stress by closing some of the stomata, causing its canopy temperature to rise. So, LST is also related to agricultural drought. Remote sensing-derived time-series NDVI and LST (or T_B) data can reveal the response of vegetation to rainfall. The joint use of NDVI and LST allows the determination of three geometric expressions of the two extreme values of each indicator, namely, the difference between the minimum and maximum values. Therefore, drought indicators are derived from the three geometrical expressions based on the two extreme points in the NDVI-LST scatterplot (Figure 7.3). In the 2D space of NDVI vs LST, the area, distance, and angle of the $NDVI_{max}$–$NDVI_{min}$ and LST_{max}–LST_{min} are calculated and treated as indicators. Of the three parameters, only length can successfully separate drought years from wet years, but not angle (slope) and area.

From NDVI, more indices have been created, including vegetation condition index (VCI) (Kogan, 1995a), temperature condition index (TCI), and vegetation health index (VHI) (Kogan, 1995b) that is calculated from VCI and TCI. In turn, VCI and TCI are calculated from NDVI and LST, respectively, as

$$VCI = \frac{NDVI_{current} - NDVI_{min}}{NDVI_{max} - NDVI_{min}} \times 100 \tag{7.8}$$

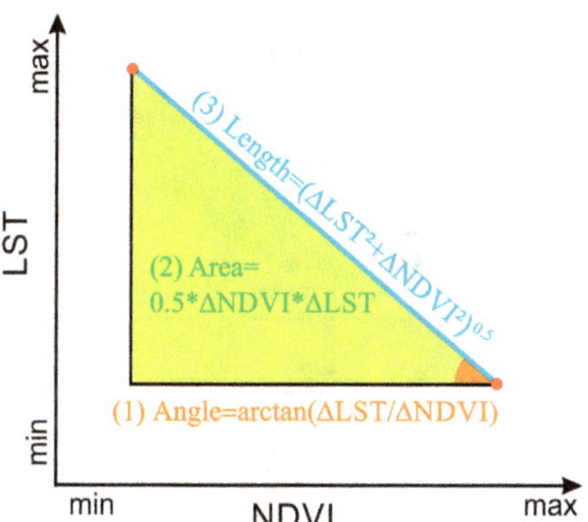

FIGURE 7.3 Derivation of three geometric parameters for indicating drought severity: (1) Angle = arctan(ΔLST/ΔNDVI), (2) Area = 0.5ΔNDVI × ΔLST, and (3) Length of hypotenuse= $[(\Delta LST)^2 + (\Delta NDVI)^2]^{0.5}$. (Source: Modified Karnieli and Dall'Olmo, 2003.)

$$TCI = \frac{LST_{max} - LST_{current}}{LST_{max} - LST_{min}} \times 100 \tag{7.9}$$

$$\text{VHI} = \rho \cdot \text{VCI} + (1 - \rho) \cdot \text{TCI} \tag{7.10}$$

where max and min = the highest and lowest NDVI over a period (e.g., a year). Current = the present month; ρ = land surface albedo. In addition to NDVI, VCI has also been calculated from EVI (Graw et al., 2017). VCI reflects the growth status of vegetation. A VCI value between 30 and 20 is considered to indicate moderate drought, between 20 and 10 severe, and <10 extreme (Figure 7.4). The existence of a strong positive correlation between VCI and rainy season crop yield testifies to the feasibility of using VCI to monitor the occurrence and development of agricultural drought. TCI is almost identical to VCI in its calculation except that it is derived from LST instead of NDVI. For this reason, TCI is commonly used to study agricultural drought, together with VCI.

LST has also been combined with EVI or NDVI to derive the vegetation supply water index (VSWI) calculated as follows:

$$\text{VSWI} = \frac{EVI\ (or\ NDVI)}{LST} \tag{7.11}$$

where EVI and LST are monthly data. According to this index, under normal water supply conditions, remotely sensed VI and crop canopy temperature lie within a certain range over a certain growing period. If a crop is stressed by drought or faces water shortage, its VI will drop.

It is possible to linearly combine both climate parameters and vegetation indices to create new indices such as the scaled drought condition index (SDCI) (Rhee et al., 2010). It is a linear combination of multi-variables (indices) related to drought conditions, such as LST, precipitation, and NDVI, all of which are derivable from remote sensing data. It is expressed as follows:

$$\text{SDCI} = \alpha \times \text{LST} + \beta \times \text{P} + \gamma \times \text{NDVI} \tag{7.12}$$

where P = precipitation obtainable from the TRMM data product. α, β, and γ = coefficients of the parameters (indices) reflecting their importance ($\alpha + \beta + \gamma = 1$). Prior to the summation, all three variables must be scaled from 0 to 1. SDCI outperforms VHI and NDVI in monitoring agricultural drought in both arid and humid/sub-humid regions (Rhee et al., 2010). It is considered the optimal

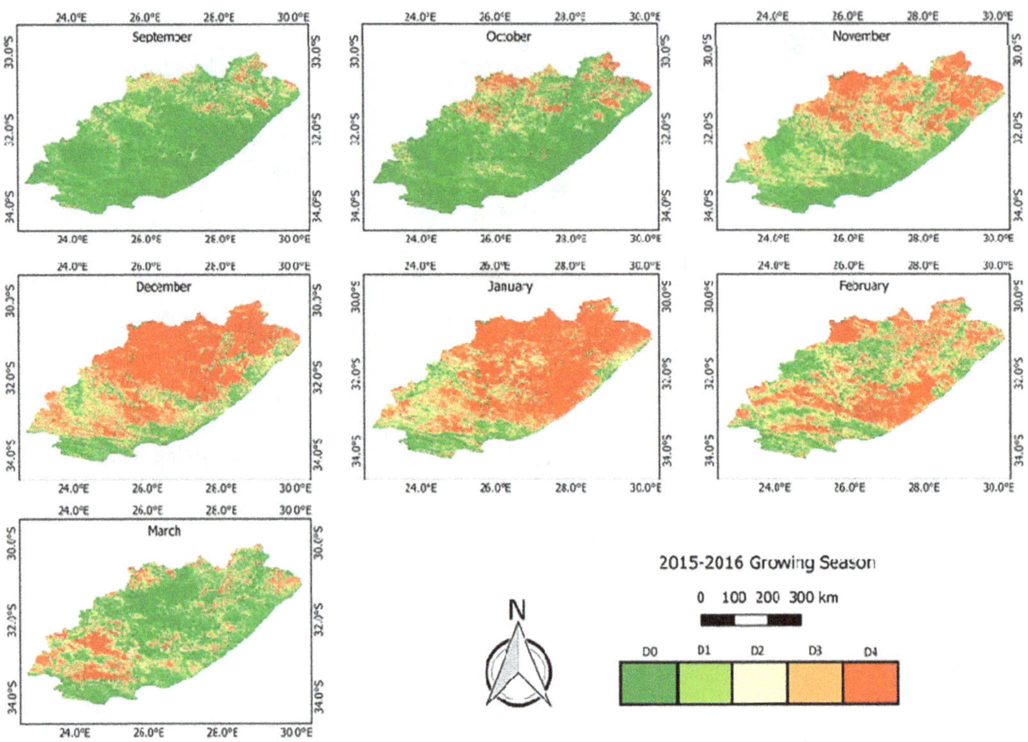

FIGURE 7.4 VCI-based drought classification of Eastern Cape Province in South Africa over the growing season in the drought year of 2015/2016. (Source: Graw et al., 2017.)

remote sensing-based drought index that can be used to monitor agricultural drought in (sub-) humid and (semi-) arid regions.

The most comprehensive agricultural drought index is the agricultural drought condition index (ADCI) proposed by Badamassi et al. (2020). It has been used to monitor drought in the agricultural area of Niger. ADCI is calculated using principal component analysis (PCA), namely, the first principal component of PCI, VCI, TCI, and ETCI. It can also be calculated from the weighted averaging of the soil moisture saturation index (SMSI), VCI, and TCI (Sur et al., 2019) as follows:

$$\text{ADCI} = 0.6\ \text{SMSI} + 0.2\ \text{VCI} + 0.2\ \text{TCI} \tag{7.13}$$

$$SMSI = \frac{ATI - ATI_{min}}{ATI_{max} - ATI_{min}} \tag{7.14}$$

where ATI = apparent thermal inertia calculated from day and nighttime LST as follows:

$$ATI = \frac{1-\rho}{LST_{day} - LST_{night}} \tag{7.15}$$

where ρ = land surface albedo (reflectance). Since these indices are derived from different sensors, ADCI virtually integrates multi-source remote sensing data in its derivation. It takes into account the effect of precipitation deficits, vegetation growth status, soil thermal stress, and crop water stress in the drought process. Predictably, it outperforms VHI (Badamassi et al., 2020). ADCI is strongly correlated with one-month SPI, the variation of millet yield, and VHI, suggesting that it is a competent drought monitoring indicator. This competency is attributed to the fact that it not only contains meteorological drought information but also reflects drought influence on crops owing to the integration of a few types of drought information from multiple sources, including precipitation, vegetation health, soil moisture, and crop water requirements (Badamassi et al., 2020). For these reasons, it can yield valuable information on droughts, such as the onset, intensity, affected areas, and end date.

7.3.3 Hydrological Drought

Hydrological drought has not been studied as extensively as meteorological and agricultural droughts in that only a few indices or methods have been developed (Table 7.4). One of them is the surface water supply index (SWSI) that replaces the Palmer drought severity index (PDSI) in areas where local precipitation is not the sole or primary source of streamflow, such as in mountainous areas where snowmelting (and even permafrost thawing) also contributes to surface runoff.

The remote sensing-based reconnaissance drought index (RDI) is computed from hydro-meteorological parameters, such as precipitation and PET (Dalezios et al., 2012). It enables the assessment of hydro-meteorological drought. The standardized RDI (RDI_{st}) is calculated as follows:

$$RDI_{st}(k) = \frac{y_k - \overline{y_k}}{\widehat{\sigma_k}} \tag{7.16}$$

where $y_k = \ln(a_k)$; $\overline{y_k}$ and $\widehat{\sigma_k}$ = the arithmetic mean and standard deviation of y_k, respectively. a_k is calculated from precipitation (P) and PET as follows:

$$a_k = \frac{\sum_{j=1}^{k} P_j}{\sum_{j=1}^{k} PET_j} \tag{7.17}$$

TABLE 7.4
Overview of Generic Drought Indices vs Type-Specific Drought Indices that Are Commonly Derived from Satellite Images

Type of Drought	Name of Indices
Generic drought	Drought severity index (DSI)
	Reconnaissance drought index (RDI)
	Reclamation drought index (RDI)
	Microwave integrated drought index (MIDI)
	Normalized difference drought index (NDDI)
	Scaled drought condition index (SDCI)
Agricultural drought	Enhanced vegetation index (EVI)
	(monthly) Vegetation condition index (VCI)
	Vegetation health index (VHI)
	Temperature-vegetation dryness index (TVDI)
	Normalized difference vegetation index (NDVI)
	Deviation NDVI index
Hydrological drought	Crop water stress index (CWSI)
	Soil water deficit index (SWDI)
	Normalized difference water index (NDWI)
Meteorological drought	Temperature condition index (TCI)
	Normalized difference temperature index (NDTI)

where P_j and PET_j = precipitation and potential evapotranspiration, respectively, of the jth month of the hydrological year. PET is calculated from temperature as follows:

$$PET_j = k_c\left[0.46T_{air} + 8.16\right] \times \text{hr} \tag{7.18}$$

where k_c = the crop coefficient, the value of which varies with crop types, hr = monthly daytime sunshine hours that are a function of the latitude of the study area, and T_{air} = mean monthly air temperature of month j. It is calculated from LST as follows:

$$T_{air} = 0.6143 \text{ LST} + 7.3674 \tag{7.19}$$

RDI offers renewed capabilities in assessing and monitoring the spatiotemporal variability of drought episodes in agricultural areas based exclusively on remotely sensed data. RDI enables hydro-meteorological droughts to be assessed, as it is calculated from hydro-meteorological parameters. Through the analysis of remotely sensed monthly RDI images, it is possible to identify the severity, duration, areal extent, onset, and end time of a hydro-meteorological drought.

7.3.4 Generic Drought Indices

All of the drought indices introduced so far are derived from meteorological or remote sensing data. They focus mainly on monitoring single drought response factors such as soil or vegetation, so have a limited capability of depicting drought comprehensively. This drawback can be overcome by integrating multiple indices/climate variables (Table 7.5), such as NDVI, LST, ET, and PET, in addition to individual spectral bands. Since global data products of these components are widely available, these indices are easy to derive and have been commonly used to monitor and predict droughts. So far numerous generic indices have been developed from the aforementioned drought indices and climate variables for studying general droughts, involving diverse rationales, assumptions, and physical quantities. These indices include the synthesized drought index (SDI) and the microwave integrated drought index (MIDI) (Zhang and Jia, 2013). Similar to ADCI, SDI is also a principal component product combining VCI, TCI, and PCI. MIDI combines three components/indices of PCI, soil moisture condition index [SMCI = $(SM_i - SM_{min})/(SM_{max} - SM_{min})$], and TCI or

$$\text{MIDI} = \alpha \times \text{PCI} + \beta \times \text{SMCI} + (1 - \alpha - \beta) \times \text{TCI} \tag{7.20}$$

where α and β = the scaling factors that reflect the weight assigned to the respective component indices. Their value is commensurate with their importance in measuring drought. MIDI integrates precipitation, soil moisture, and LST derived from radar imagery data such as TRMM and AMSR-E. It is the best at timely monitoring short-term drought, especially meteorological drought in semi-arid regions of the world.

Sandholt et al. (2002) proposed two other indices of drought called temperature vegetation drought index (TVDI) and soil water deficit index (SWDI), calculated as follows:

$$TVDI = \frac{LST_{NDVI} - LST_{NDVI-min}}{LST_{NDVI-max} - lST_{NDVI-min}} \tag{7.21}$$

$$SWDI = \frac{\theta - \theta_{FC}}{\theta_{FC} - \theta_{WP}} \times 10 \tag{7.22}$$

where θ_{FC} and θ_{WC} = field capacity and available water capacity, respectively. A negative SWDI represents drought, with a value below –2 signifying moderate and worsened drought. TVDI can indicate the drought condition and is calculated from existing soil moisture products such as the

TABLE 7.5

Common Remote Sensing Drought Indices and the Parameters from Which They Are Derived

Index	Components	Formula	Authors
VCI – vegetation condition index	NDVI	$\frac{NDVI_{current} - NDVI_{min}}{NDVI_{max} - NDVI_{min}} \times 100$	Kogan (1995a)
TCI – temperature condition index	LST	$\frac{LST_{max} - LST_{current}}{LST_{max} - LST_{min}} \times 100$	Kogan (1995b)
VHI – vegetation health index	NDVI, LST	$\rho \times \text{VCI} + (1 - \rho) \times \text{TCI}$	Kogan (1995b)
NDWI – normalized difference water index	Green, NIR	$= (\rho_{0.86} - \rho_{1.24})/(\rho_{0.86} + \rho_{1.24})$	McFeeters (1996)
TVDI – temperature vegetation dryness index	NDVI, LST	$\frac{LST_{NDVI} - LST_{NDVI-min}}{LST_{NDVI-max} - lST_{NDVI-min}}$	Sandholt et al. (2002)
PDI – perpendicular drought index	NIR, red	$\frac{1}{\sqrt{M^2+1}}\left(R_{red} + M \cdot R_{NIR}\right)$	Ghulam et al. (2007)
DSI – drought severity index	NDVI, ET, PET	$\frac{z - \overline{z}}{\sigma_z}$	Mu et al. (2013)
NDDI – normalized difference drought index	NDVI, NDWI	(NDVI – NDWI)/(NDVI + NDWI)	Gu et al. (2007)
MIDI – microwave integrated drought index	PCI, SMCI, TCI	$\alpha \times \text{PCI} + \beta \times \text{SMCI} + (1 - \alpha - \beta) \times \text{TCI}$	Zhang and Jia (2013)
NMDI – normalized multiband drought index	Reflectance of three bands	$[\rho_{0.86} - (\rho_{1.64} - \rho_{2.13})]/[\rho_{0.86} + (\rho_{1.64} - \rho_{2.13})]$	Wang and Qu (2007)
CDIR – comprehensive drought index of remote sensing	PCI, VCI, TCI	$a + b \times \text{PCI} - c \times \text{PCI}^2 - d \times \text{VCI} - e \times \text{TCI} + f \times \text{TCI}^2$	Yu et al. (2019)

time-series CLSMDAS (0–5 cm surface). Both drought indices are commonly used as the dependent variable in predicting drought (see Section 7.4.4 below).

In addition, the normalized difference drought index (NDDI) (Gu et al., 2007) and the normalized multiband drought index (NMDI) (Wang and Qu, 2007) are also available for studying general drought. They are calculated as follows:

$$NDDI = \frac{\text{NDVI} - \text{NDWI}}{\text{NDVI} + \text{NDWI}} \tag{7.23}$$

$$NMDI = \frac{\rho_{0.86} - \left(\rho_{1.64} - \rho_{2.13}\right)}{\rho_{0.86} + \left(\rho_{1.64} - \rho_{2.13}\right)} \tag{7.24}$$

where ρ = surface albedo (reflectance); subscription = wavelength in micron. 0.86 µm is NIR, 1.64 and 2.13 µm are SWIR wavelengths. NDDI was initially conceived for assessing grassland drought of the central Great Plains in the US using MODIS data. The MODIS-based NDDI is a better indicator of drought than either NDVI or NDWI, or their difference. It is easy to calculate from only a few spectral bands.

Finally, drought can be monitored using a simple, effective index called the perpendicular drought index (PDI) proposed by Ghulam et al. (2007). It is calculated as follows:

$$PDI = \frac{1}{\sqrt{M^2+1}}\left(\rho_{red} + S \cdot \rho_{NIR}\right) \tag{7.25}$$

where S = slope of the soil line, ρ_{red} and ρ_{NIR} = the atmospherically corrected reflectance of the red and NIR bands, respectively. This index is based on the spatial characteristics of moisture distribution in NIR–red bands. Compared with other recognized drought monitoring methods such as LST/NDVI and vegetation TCI, the PDI graph demonstrates very similar trends with ground truth drought data. PDI is correlated with in situ drought values calculated from 0 to 20 cm mean soil moisture at a coefficient of 0.75 ($R^2 = 0.49$).

Nevertheless, none of the aforementioned drought indices are able to indicate the severity of drought directly except PDSI. Another drought severity index (DSI) is calculated from ET, PET, and NDVI as follows:

$$DSI = \frac{z - \bar{z}}{\sigma_z} \tag{7.26}$$

where z = standardized Z ($Z_{ratio} + Z_{NDVI}$). Z_{NDVI} is the standardized NDVI calculated using the same equation as Eq. 7.26, so is Z_{ratio}. Here ratio means ET to PET (Mu et al., 2013). DSI can be implemented fully from operational satellite data at the pixel level. It is suitable to assess and monitor drought severity and mitigation efforts in real time on a global scale.

7.3.5 Comprehensive Indices and Comparison

Naturally, it is desirable and rather common to merge multiple types of indices (e.g., meteorological and agricultural indices) to take advantage of their respective strengths. The integration of such diverse indices is commonly achieved statistically via linear combination, PCA, and advanced data fusion approaches. Linear combination is easily implemented using the same methods as those used to derive the original index from multiple climate variables by weighting each of the considered indices. Two indices are produced in this way, the comprehensive index (CI) and the comprehensive drought index of remote sensing (CDIR). The former is calculated as follows:

$$CI = \alpha SPI_{30} + \beta SPI_{90} + \gamma RMI_{30} \tag{7.27}$$

where SPI_{30} and SPI_{90} = short-term (the most recent 30 days) precipitation and long-term (nearly 90 days) climatic characteristics (SPI). RMI = relative moisture index [(P – PE)/PE] of the last 30 days, and it reflects the water balance between precipitation and evaporation (Huang et al., 2020). α, β, and γ = the weights assigned to the three indices. They have a value of 0.4, 0.4, and 0.8, respectively. CI is reflective of the effects of precipitation anomalies on the short-term (monthly) and long-term (seasonal) scales and the short-term (monthly) water deficit that affects crop growth. It is suited to monitor meteorological drought and assess historical meteorological drought in real time and has been adopted to produce the national meteorological drought monitoring maps in China.

CDIR is produced by integrating VCI, TCI, and PCI non-linearly as some of them can be squared (Eq. 7.28), all of which are acquired from satellite data products such as the TRMM satellite, and VCI and TCI from MODIS (Yu et al., 2019). The combination of the three indices is via a linear regression equation in the form of

$$\mathrm{CDIR} = \omega + \alpha \times \mathrm{PCI} - \beta \times \mathrm{PCI}^2 - \gamma \times \mathrm{VCI} - \eta \times \mathrm{TCI} + \nu \times \mathrm{TCI}^2 \tag{7.28}$$

where α, β, γ, η, and ν = coefficients determined via linear regression analysis. Their exact value varies with the month. A CRID value <1 means mild drought, or moderate if CRID <0 (severe if CRID < –1). It has been used to assess spring drought in the Beijing–Tianjin–Hebei region of northeast China (Figure 7.5). CDIR is closely correlated to three-month SPI at a coefficient between 0.45 and 0.86. The R^2 of the regression equations ranges from 0.74 in April to 0.70 in September (Yu et al., 2019). Its correlation with the area of drought-affected cropland is similarly high at

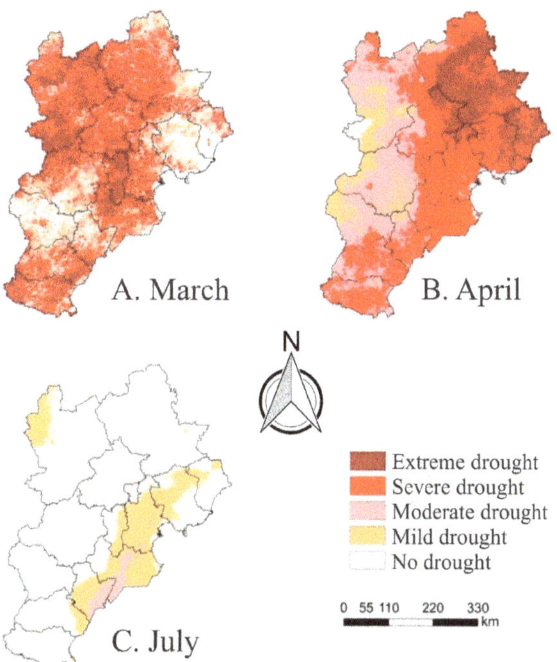

FIGURE 7.5 Distribution of spring drought in the Beijing–Tianjin–Hebei Province of China as measured by the CDIR index in three months (A) March; (B) April and (C) July. (Source: Modified from Yu et al., 2019.)

−0.86 ($p < 0.01$), higher than that of VCI and TCI. The correlation coefficient between accumulative CDIR and the standardized unit crop yield in the growing season is lower between integrated drought monitoring at 0.60 and 0.76, respectively, but is still significant at $p < 0.05$. Thus, CDIR is a reliable indicator useful for integrated drought monitoring because it comprehensively reflects meteorological and agricultural drought information.

Single drought indices may be combined linearly with remote sensing-derived climate variables to formulate the integrated drought condition index (IDCI) through proper weighting of each factor at different times of the year (Meng et al., 2016). This novel remote sensing index is designed to monitor short-term drought. The index sets different weights for each month of the growing season to three components of precipitation, vegetation growth condition, and LST, calculated as follows:

$$\mathrm{IDCI} = \alpha \times \mathrm{TRMM} + \beta \times \mathrm{TCI} + \gamma \times \mathrm{VCI} \tag{7.29}$$

where α, β, and γ = the coefficients or weights of the three considered factors determined through the PCC with SPEI using monthly or quarterly data (Table 7.6). It is very similar to SDCI (Eq. 7.12) except that the three coefficients do not sum to 1. Their value varies with the month, such as (4, 1, 0), (3, 2, 0), (5, 2, 1), (4, 1, 1), and (4, 0, 1) (Table 7.6). The best weight combination is judged by the highest r value with the SPEI averaged over three time periods. Statistical analyses of IDCI and in situ observed SPEI of various time scales reveal a close correlation between them, demonstrating the effectiveness of IDCI in characterizing drought conditions and patterns (Meng et al., 2016). Its deficiency is the lack of consideration for soil moisture data. Besides, the linear combination of the three variables faces the issue of how to weigh them objectively instead of intuitively, even though the issue is non-existent if determined via regression analysis.

The second method of creating new comprehensive indices is via PCA. It outputs components from a number of inputs that may include drought indices and climate variables. Their information

TABLE 7.6

Correlation Coefficient *r* between the Integrated Drought Condition Index with Different Weights and the In Situ Three Time-Scale SPEIs in August (*n* = 1,230)

Weight			r		
α-TRMM	β-TCI	γ-VCI	SPEI-1	SPEI-3	SPEI-6
1	1	1	0.66	0.51	0.49
2	1	1	**0.75**	0.61	0.62
3	0	2	0.74	0.58	0.58
3	2	2	0.74	0.59	0.58
4	0	3	0.73	0.58	0.55
4	1	1	**0.75**	**0.62**	**0.65**
4	1	3	0.73	0.58	0.55
5	3	2	**0.75**	**0.62**	0.64

Source: Modified from Meng et al. (2016). Used with permission.

Note: In all cases, the *p* value <0.01. boldfaced figure – maximum *r* value of the column. SPEI-1, 1-month SPEI; SPEI-3, 3-month SPEI; SPEI-6, 6-month SPEI

is projected maximally to the first component. If this proves impossible, the remaining information is projected to the second component orthogonal to the first component, and so on. The first output component is considered the new comprehensive index. For instance, Kim et al. (2021) created a remote sensing-based integrated drought index from three indices of SPI, ADCI, and WBDI using Bayesian PCA. It can provide robust and comprehensive drought information by maintaining the inherent characteristics of the three drought indices. Analysis of this index shows that it can reflect the timing, severity, and evolutionary pattern of meteorological, agricultural, and hydrological droughts. The main problem with PCA component-based indices is their ambiguous physical meaning.

The third method of creating new indices is image fusion that is virtually multivariate data analysis, in which precipitation and soil moisture may be combined to derive the multivariate standardized drought index (MSDI) for multi-index drought assessment (Hao and AghaKouchak, 2013). This index is developed using the multi-index drought-modeling approach based on the concept of copulas.

$$P(X \le x,\ Y \le y) = C\left[F(X), G(Y)\right] = p \tag{7.30}$$

where C = the copula, and $F(X)$ and $G(Y)$ = the marginal cumulative distribution functions of random variables X (e.g., precipitation) and Y (e.g., soil moisture), respectively. The copula C offers flexibility in constructing the joint distribution of random variables in terms of their marginal distributions. It can be used to produce the cumulative joint probability density function of any two of three indices. Recently, this method has been applied to deriving MSDI from SPEI and standardized soil (storage) index (SSI) following a statistical approach (Forootan et al., 2019). MSDI probabilistically combines SPI and SSI and uniquely produces a composite index by integrating meteorological and agricultural drought indices (Figure 7.6). Different datasets, including satellite observations and model simulations, can be used as inputs to produce a composite drought map. Apart from SPEI and SSI, MSDI is also derivable by combining GRACE terrestrial water storage and net precipitation, as well as ERA-Interim soil moisture and net precipitation. It is better correlated to global SST data than those drought indices derived only from SSI or SPEI. MSDI is able to indicate the drought onset being dominated by SPI and termination. Nevertheless, this copula method can fuse only two indices and is complex to implement.

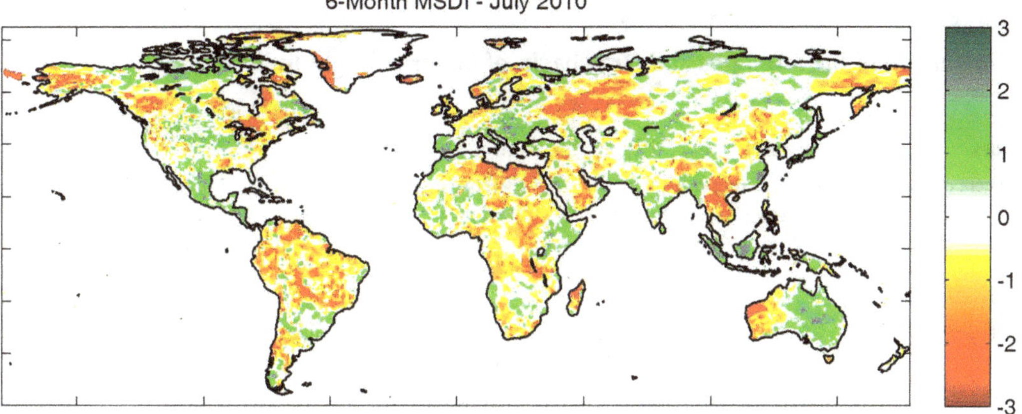

FIGURE 7.6 Global distribution of multivariate standardized drought index (MSDI) for July 2010, derived from NASA's Modern-Era Retrospective Analysis for Research and Applications (MERRA-Land [Reichle et al., 2011]) precipitation and soil moisture data. (Source: AghaKouchak et al., 2015.)

All of the proposed indices have their strengths and limitations, as well as the best uses (Table 7.7). So far, nobody has comprehensively evaluated all of them, even though researchers have compared the performance of some of them in monitoring agricultural droughts, such as VHI, TVDI, and DSI (Huang et al., 2020). DSI outperforms TVDI and VHI in monitoring agricultural drought on the regional scale, bearing a closer correlation with SPI and soil moisture than the other two indicators. DSI also corresponds more closely with soil moisture and the national meteorological drought monitoring maps than the other two indices (Figure 7.7). A comparison of two study areas reveals that these remote-sensing drought indices provide more accurate predictions of the impacts of drought in predominantly rain-fed agricultural areas. A comparison of monthly DSI values also sheds light on how crops respond to drought. For instance, the jointing and grain-filling stages of winter wheat are more sensitive to water stress, indicating that winter wheat requires more water during these stages.

7.4 REMOTE SENSING OF DROUGHTS

7.4.1 Detection and Monitoring

Different types of drought are detected and monitored most effectively from different data using different methods. Hydrological drought on various temporal scales in ungauged areas can be detected based on the percentiles of key variables of water and energy balance. They include precipitation, actual evapotranspiration, NDVI, LST, and soil moisture. Of these factors, the most important variable is identified as precipitation, followed by soil moisture (three-month time scale) or NDVI (longer time scale). The detection may be implemented as rule-based models such as decision trees, adaptive boosting of decision trees (Adaboost), RF, and extremely randomized trees (Rhee et al., 2020). Models with an ensemble of trees successfully detected hydrological drought despite the limited number of input variables used. The best performing method is Adaboost (correlation coefficients ≥0.85, mean absolute error ≤0.12, RMSE–observations standard deviation ratio ≤0.53, and larger Nash–Sutcliffe efficiency of drought severity ≥0.72). Its predicted severity of hydrological drought is in closer agreement with the observations than other models in most cases. With the assistance of the best model, drought severity maps can be produced from the input data to display the spatial distribution of hydrological drought conditions.

TABLE 7.7
Comparison of the Strengths and Drawbacks of Major Drought Indices

		Properties	
Index	**Purpose**	**Pro**	**Cons**
Percent of normal	For detecting drought by dividing actual P by normal P (%) over 30 years	Effective in a single region or season	Minimal information, e.g., frequency of drought, unable to compare different locations or drought factors
SPI	Simple based on long-term P record aggregated to multiple time scales	Able to provide early warning and severity, suitable for risk management	Reliable within 24 months only
PDSI	T and P considered, soil moisture calibrated to homogeneous climate zone, fixed at nine months	Able to identify abnormality of drought, showing historical information on the current drought	May lag a few months due to soil moisture data, not accurate during winter and spring due to frozen ground, underestimate runoff conditions
Palmer hydrological drought index	Derived from PDSI based on P, outflow, and storage	Able to quantify the long-term impact of hydrological drought	May be less sensitive than PDSI, sluggish response to drought
Crop moisture index	A derivative of PDSI, a form of monthly moisture (top 5 ft soil) anomaly or Z index as a product of PDSI	Effective at detecting short-term agricultural drought (on a monthly scale if using the Z index). More responsive to drought than PDSI and PHDI	Limited to the growing season only, unable to determine the long-term period of drought
Surface water supply index	Frequency analysis to normalize long-term data of P, snowpack, streamflow, and reservoir level	Useful for indicating snowpack conditions in mountainous areas	Results incomparable across basins as there is a component of seasonality
Reclamation drought index (RDI)	Similar to SWSI by combining the functions of supply, demand, and duration, T	Able to be used as the trigger to evaluate drought reclamation plans	Same as those of SWSI

Source: Modified from Hayes (2006). Used with permission.

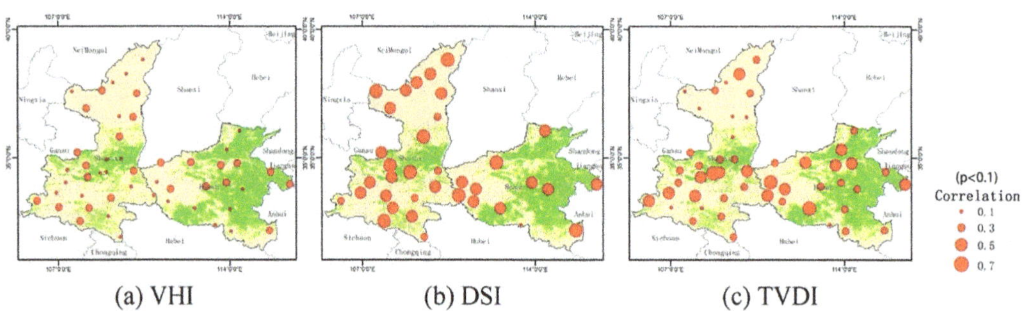

FIGURE 7.7 Comparison of three drought indices (a - VHI, b - DSI, and c - TVDI) based on their correlation with the three-month standardized precipitation index. (Source: Huang et al., 2020). *Note:* The direction of correlation (e.g., negative correlation) is ignored. No similar results can be produced for winter wheat yield because it has a temporal component (e.g., The correlation varies at different growing stages).

It is significant to accurately monitor drought for the sustainable development of regional agriculture in light of increasingly complex and erratic global climate change. In theory, drought can be easily monitored from climate data recorded at ground stations. In practice, these stations may not be sufficiently dense to provide a detailed spatial perspective on the extent and severity of drought. In comparison, remote sensing-based monitoring has the advantage of being macroscopic, rapid, and able to provide continuous spatiotemporal information. Drought years can be detected from the change in total annual vegetation productivity, while seasonal dynamics of drought are better monitored using VCI (Sur and Lunagaria, 2020). So far drought has been monitored mostly from coarse-resolution MODIS data based on a variety of indices derived from them. The efficiency of time-series TRMM and MODIS data products for spatiotemporally monitoring the status of meteorological drought has been assessed in conjunction with its impact on vegetation conditions and crop yield. The yield anomaly index, statistically calculated from the data records, bears the highest correlation of only 0.46 with VCI for paddy, while the correlation lies between −0.3 and 0.3 for most other crops (Sur and Lunagaria, 2020). Thus, drought indices derived from coarse-resolution remote sensing images provide a synoptic view of crop yield but fail to reveal the precise relationship between drought and small-scale crop yield owing to the coarse resolution of the satellite images used. In spite of this failure, remote sensing can still be an effective way of monitoring and understanding the dynamics of drought and agriculture patterns over any regions.

The effective characterization of the spatial extent and intensity of drought requires consideration of a comprehensive range of drought-causing and manifesting factors, such as precipitation, vegetation growth condition, and LST. The most effective parameter for monitoring drought is normalized VCI that represents vegetation conditions for the assessment of relative changes, detects drought conditions in a dynamic manner, and therefore is suitable for monitoring the seasonal dynamics of drought conditions (Graw et al., 2017). Regional drought can also be effectively monitored using a comprehensive model constructed from various hazard factors derived from satellite data, or the comprehensive meteorological drought index (CI) measured at the site scale (monthly). It is derived from eight variables in the form of

$$\mathrm{CI} = \mathrm{f}\left(\mathrm{VCI},\ \mathrm{TCI},\ \mathrm{VSWI},\ \mathrm{Pa},\ \mathrm{TRMM-Z},\ \mathrm{AWC},\ \mathrm{LC},\ \mathrm{DEM}\right) \tag{7.31}$$

where AWC = available water-holding capacity. LC = land cover. This composite drought monitoring model is theoretically grounded on the fact that drought is jointly determined by a comprehensive range of factors related to precipitation, soil moisture stress, vegetation growth status, available water capacity, land cover, and landform (Shen et al., 2019). The complex relationship between drought and these factors may be established using deep learning algorithms or one of the neural network-based machine learning methods, such as the deep learning feed-forward neural network (DFNN) model. Dissimilar to the traditional ANNs that usually have two or three hidden layers, a deep learning model has at least five layers or even more neural networks, all of which can be more effectively trained than the classic networks. Their contribution to CI is shown in Table 7.8. A CI value between −1.8 and −1.2 (inclusive) represents moderate drought. A value between −2.4 and −1.8 (inclusive) indicates severe drought (Figure 7.8).

The drought index of the model output is significantly correlated with the in situ measured CI, achieving an R^2 of 0.89 for the training dataset (consistency rate = 86.6%), and 0.83 for the test dataset (consistency rate of the drought grade of the two models = 79.8%) (Shen et al., 2019). The high correlation coefficient (0.772–0.910) of CI with SPEI suggests that it can reflect both the degree of meteorological drought and monitor agricultural drought. This is not surprising given that CI is a comprehensive index that takes into account all major factors related to drought, such as vegetation temperature factors (VCI, TCI, and VSWI), soil factors (AWC, LC, and DEM), and precipitation factors (Pa and TRMM-Z).

Despite this impressive performance, no operational systems have been established for regional or even local-scale monitoring except the national US Drought Monitoring (USDM) and the global

TABLE 7.8
The Contribution Rate of Various Factors to the Simulated Drought Index Determined via Deep Learning

Drought Factor	Contribution Rate
TRMM-Z	0.957
Pa	0.935
VCI	0.899
TCI	0.894
VSWI	0.889
AWC	0.823
LC	0.646
DEM	0.569

Source: Shen et al. (2019). Used with permission.

integrated drought monitoring and prediction system (GIDMaPS). USDM (https://droughtmonitor.unl.edu/) is a state-of-the-art drought monitoring tool operational since 1999, based on a number of factors, such as short- and long-term meteorological drought indicators, hydrologic indices, and remote sensing-derived data. It produces a drought map that displays the areas currently experiencing abnormal dryness or drought across the US at five levels: abnormally dry, moderate drought, severe drought, extreme drought, and exceptional drought. The online map is updated weekly. GIDMaPS has both monitoring and prediction functions, plus a data dissemination interface. The monitoring component is near real time, and a seasonal probabilistic prediction module is able

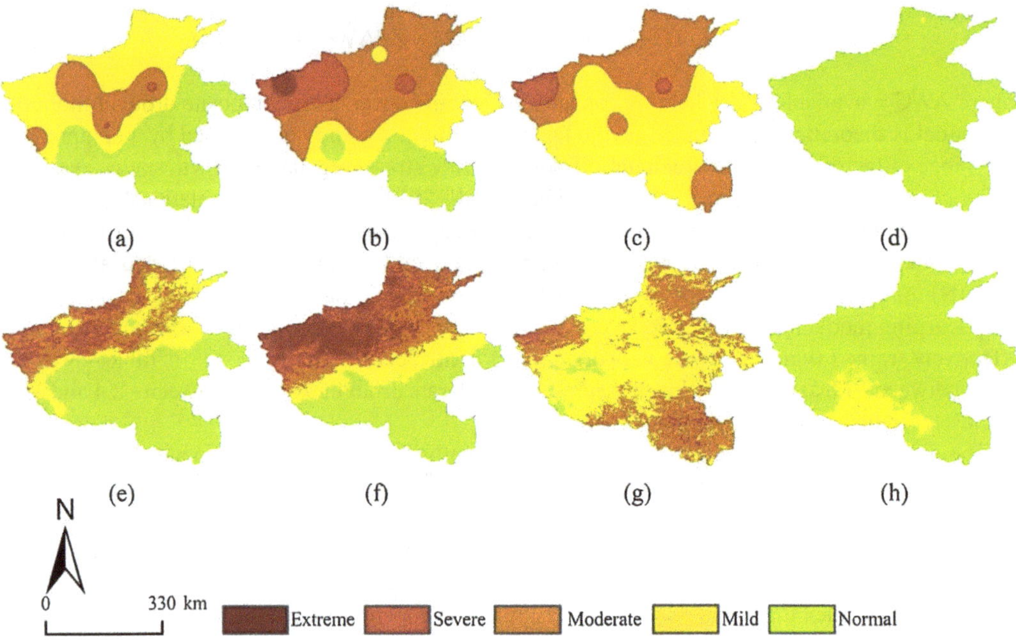

FIGURE 7.8 Drought severity (a–d) and its spatial distribution in Henan Province of China from August (e) to September (f), and in October (g) and November (h) 2013 generated using a deep learning model. (Source: Shen et al., 2019. Used with permission)

to provide composite drought information based on MSDI (Hao and AghaKouchak, 2013; Hao et al., 2014). The probabilistic forecasts offer essential information for early warning, preventive measures, and mitigation strategies. The GIDMaPS datasets considerably extend the current capabilities and datasets for global drought assessment and early warning. Its inputs include precipitation and soil moisture, from which three indicators are produced, SPI, SSI, and MSDI, all of which can be used to monitor meteorological and agricultural drought and provide integrated drought information. It can also predict seasonal SPI, MSDI, and SSI based on six-month accumulated precipitation and soil moisture of one month.

7.4.2 Drought Response Assessment

7.4.2.1 Via Indices

Vegetation conditions and soil moisture are the most sensitive and immediately responsive to drought. The latter's response is much harder to detect than the former's response remotely, so is not commonly studied. The response of vegetation to drought has been assessed using different methods based on different concepts, including the use of comprehensive indices (see Section 7.3.5), regression analysis, and modeling. Index-based response assessment relies on a comprehensive drought index and its relationship with vegetation condition indicators or vegetation indices. The correlation between the two is used to reveal the level of response and assess the impact of different droughts, such as short and severe vs prolonged and moderate droughts. Changes in total annual vegetation productivity can detect drought years, while seasonal drought dynamics can be monitored via VCI. However, the vegetation drought response index (VegDRI) is the most useful and frequently used for this purpose. It is developed by the US Geological Survey and the National Drought Mitigation Center, based on the National Weather Service multi-sensor precipitation (radar and USGS rain gauge) data (Brown et al., 2008). This index integrates eight climate-based drought indicators and satellite-derived vegetation indices with plant biophysical information, including SPI, PDSI, average seasonal greenness (%), the start of season anomaly, land cover, soil water available capacity, irrigated agriculture, and ecological regions.

It is a lengthy process to generate VegDRI maps involving many steps. They fall into three major groups: (i) data preparation, standardization, and summation by the ground station, and separation into three seasons of spring, summer, and fall; (ii) generation of an empirical model for each phase using supervised classification and regression-tree analysis if certain conditions are met; and (iii) application of the empirical model to the observed geospatial data to produce the VegDRI map at 1 km resolution (Figure 7.9). VegDRI is implemented as a combination of rules. This indexing method is objective and can produce replicable results so long as the same rules are conformed to. It enables droughts to be monitored in near-real-time.

7.4.2.2 Via Regression Analysis

The easiest way of assessing how vegetation (e.g., chaparral and coastal sage scrub) and crops have responded to a prolonged drought is to make use of time-series NDVI. NDVI anomalies are then regressed against a drought index. The regression may be implemented as bootstrapped multiple regression in which only a certain portion of the input data is selected randomly to create a subset that is used to construct the model. The same regression is repeated numerous times, each time using the newly selected subset inputs, and the final model is the average of all the repetitions. In the regression analysis, environmental factors, AWC, soil texture measured by the percentage of sand and clay, and terrain parameters (elevation, slope, and aspect) are all treated as the explanatory variables (Okin et al., 2018). The regression slope between NDVI anomalies and PDSI at the pixel level is regarded as the response variable. A positive and statistically significant (e.g., $p < 0.05$) slope indicates vulnerability to the effects of a profound, multi-year drought. The steeper the slope, the more profound the influence.

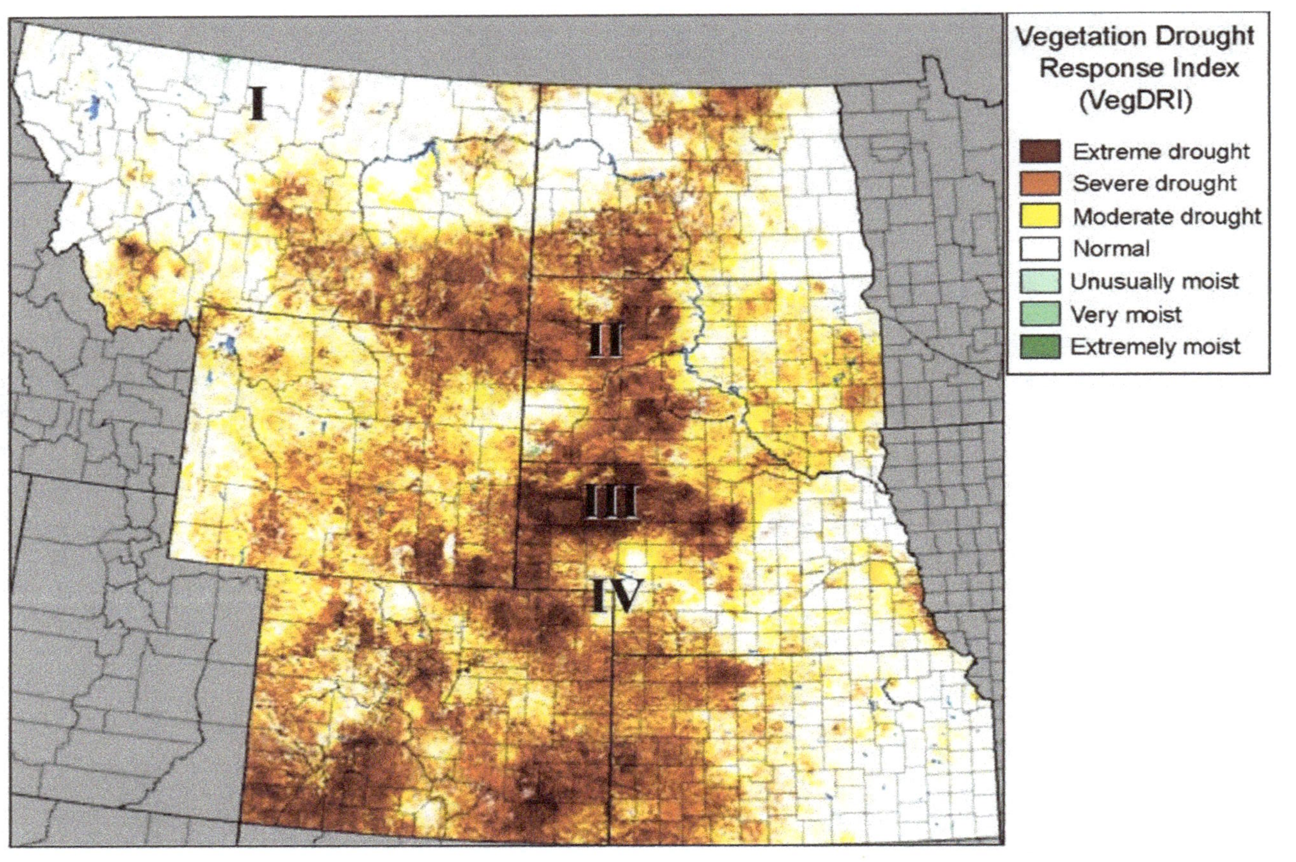

FIGURE 7.9 Response of vegetation to drought in six mid-western states of the US, calculated using the VegDRI index based on eight variables related to climate, satellite-observed vegetation conditions, and biophysical characteristics of the environment. (Source: Brown et al., 2008). I, II, III, IV = four typical levels of response.

On the basis of the amount of inter-annual variability in NDVI and its correlation with the DSI, it is possible to identify how different vegetation communities have responded to the drought on longer time scales (e.g., large inter-annual variability). The established relationship also allows the exploration of factors that affect drought indirectly, such as topography and soil texture because the addition to or the removal of an additional variable from the established regression model triggers changes to the slope. Its correlation with a given explanatory variable is used to determine the variables and types of vegetation that are sensitive to drought and the spatial distribution of drought response of different vegetation communities. If the model's slope and R^2 hardly change after a variable has been excluded from it, this variable does not exert a noticeable influence on the vegetation's response to drought.

The aforementioned assessment of vegetation response to drought can be extended or adapted to different land management and land tenure systems after vegetation productivity trends have been identified and compared with the monitored intensity, frequency, and distribution of drought (Graw et al., 2017). Depending on the focus of the assessment, the independent variables may change, but the dependent variable is always a drought index or its proxy such as VCI derived from EVI and CHIRPS precipitation data. Time-series EVI data can also be analyzed to identify changing trends in relation to surface cover. Whether a trend does exist is tested using the Mann–Kendall trend test frequently used to test time-series data of temperature, runoff, and precipitation. Essential in the assessment is a land cover map. It should show at least the land covers, the impact of which on drought is to be studied, such as the impact of drought events on vegetation productivity in grasslands and commercial and communal croplands via cumulative EVI over time.

The response of vegetation to drought can also be gauged from the correlation of total vegetation productivity measured by cumulative VI with the growing season precipitation. If an area experiences a significant vegetation productivity trend but has a low and even negative correlation coefficient with rainfall, it can be inferred that other factors are driving the productivity change and also exert an impact, such as land management strategies (e.g., use of irrigation and fertilizer) or rainfall variability (Graw et al., 2017). This method is less effective than multiple regression models as it cannot pinpoint the factors that affect the vegetation response to the drought most directly and intuitively. Still, it is possible to repeat the same analysis for different land covers. In this way, how different land management systems affect vegetation response to drought can be enlightened.

7.4.2.3 Via Modeling

How vegetation has responded to drought can also be studied by building a vegetation-drought multi-dimensional response model using the copula method (Zhu et al., 2020). This copula joint distribution model allows the exploration of multiple non-linear responses of vegetation to drought conditions. Both drought conditions and vegetation growth are two-dimensional time-series data that can be represented by indices, such as SPI for drought and NDVI (EVI) for vegetation growth. In practice, both indices can be defined by different statistical measures, such as the cumulative negative NDVI or minimum EVI. The association between these two random spatial variables is measured by the Kendall correlation coefficient with a range of –1 to 1, and the p-value of the correlation (goodness-of-fit) calculated using the bootstrap method. In the assessment, the drought index (e.g., SPI) must be discretized to a few levels, such as mild, moderate, severe, and extreme. Then the conditional probability expectation of vegetation growth at each drought level is calculated and treated as the likely vegetation change value (Zhu et al., 2020). This method has the potential to identify the internal association between VI and SPI and has produced results to indicate that the growth of the grassland vegetation in a plateau environment is subject more to the impact of persistent drought episodes than extreme droughts of short duration because the change in vegetation conditions bears a closer correlation with the cumulative negative SPI than the quarterly minimum SPI. This method is limited in that only one external variable is allowed at a time, and it cannot accommodate the influence of other climate variables such as rainfall or terrain.

7.4.3 Impact Assessment

Remote sensing plays an indispensable role in assessing the impact of drought on vegetation, the conditions of which can be effectively characterized using NDVI data products derived from satellite data. These data have a coarse spatial resolution of around 1 km and repeatedly cover the same geographic area at varying temporal resolutions over several years, if not decades. Such time-series NDVI data of 20 or even 30-year duration are valuable in assessing not only drought severity but also its spatial extent, duration, and onset time. Apart from NDVI, the assessment may be based on SPI and NDVI-derived VCI. VCI enables the assessment of climate impact on different types of vegetation and ecosystems comparatively. Although SPI is not derivable remotely, it can be produced from climate records at ground meteorological stations and then interpolated to the same spatial extent as NDVI and VCI data layers of the same resolution. Also vital in the assessment are land cover maps that are routinely produced from remotely sensed data. These maps are essential to assess how different types of land covers have responded to the same drought. The drought impact is revealed through the correlation between SPI and VCI. A positive, close relationship indicates the existence and impact of drought on vegetation, while the absence of a correlation shows no drought (e.g., for irrigated farmland and deciduous forest). Varying seasonally, the correlation is stronger in summer but rather weak in winter. Those regions having high aridity, a low vegetation cover, and a high coefficient of variability are the most impacted by drought (Vicente-Serrano, 2007). Humid regions that are characterized by a higher vegetation cover and low inter-annual variability in their NDVI values are less impacted, as well. The correlation also reveals how the effect of drought on vegetation varies temporally from month to month, and from season to season. For instance, during each drought episode, the most severe impact occurs usually in the summer and lasts until the end of the hydrological year (Dalezios et al., 2012). The start of severe and extreme droughts usually coincides with the beginning of the hydrological year (October), whereas moderate droughts start usually in spring (April), both last until the end of the hydrological year (September).

Drought impact assessment may target drought severity, spatial extent, duration, periodicity, onset, and end time. Severity refers to the intensity of drought that can be qualitatively described as mild (e.g., $-0.99 < \text{RDI} < 0.99$), moderate ($-1.00 < \text{RDI} < -1.49$), severe ($-1.50 < \text{RDI} < -1.99$), and extreme ($\text{RDI} < -2.00$). Duration means the temporal span from the onset to the end of a drought episode, usually measured in months. It can be determined by counting the number of negative indices in each of the drought index value ranges. Onset refers to the month when a drought event is initially detected, such as the month when the negative drought index value is first encountered. The end time is the month when the negative value is last detected. Areal extent refers to the spatial coverage of a drought episode, determined from the number of pixels, the values of which fall into the pre-determined negative drought index value range. Not all droughts take place at the same time. The onset of severe droughts coincides with the beginning of the hydrological year, whereas the onset of mild droughts is spring (Ribeiro et al., 2019). Periodicity refers to the recurrence interval of drought.

One of the most critical impacts of drought is on crop yield. Drought-caused yield loss cannot be estimated directly from remotely sensed data, so it has to be modeled based on multi-scalar SPEI, VCI, TCI, and VHI. The caustic relationship between these drought indicators and regional rain-fed winter cereal yield is established via MLR and ANN models (Ribeiro et al., 2019). This modeling approach reveals that yield decreases with moisture depletion (e.g., lower VCI values) during early spring and with too high temperatures (lower TCI values) close to the harvest time. All drought indicators exert the most influence during the plant stages in which the crop is photosynthetically more active (spring and summer), rather than during the earlier stages of the plant life cycle in autumn/winter.

The modeling accuracy varies with the explanatory variables used. In general, those models involving more explanatory variables are more accurate than those involving fewer variables. ANN produces slightly better predictions than MLR models, but its performance varies with network

TABLE 7.9

Comparison of MLR Model Performance for Estimating Wheat (W) and Barely (B) Yield Using Leave-One-Out Cross-Validation

Model	Model Equation	$R^2_{adj\_no\_CV}$	R^2_{adj}	RMSE	SS_{RMSE} (%)
W1a	$0.67VCI_{21} + 0.76TCI_{23}$	0.79	0.73	0.49	69.31
W1b	$0.24VHI_{1} + 0.88VHI_{22}$	0.78	0.73	0.49	68.87
B1a	$0.47VCI_{18} + 0.82TCI_{20}$	0.76	0.69	0.52	66.86
B1b	$0.23VHI_{50} + 0.91VHI_{22}$	0.81	0.76	0.47	70.41
W2a	$-0.39VCI_{35} + 0.63VCI_{25}$	0.50	0.39	0.74	53.42
W2b	$0.70VHI_{20}$	0.47	0.39	0.75	52.40
W2c	$0.94SPEI2_{2-5} + 1.05SPEI_{4-1} - 0.53SPEI_{5-12} - 0.32SPEI_{6-9}$	0.80	0.71	0.49	69.23
B2a	$-0.63VCI_{40} + 0.52VCI_{52} + 0.65VCI_{25}$	0.67	0.56	0.61	60.99
B2b	$-0.34VHI_{43} + 0.49VHI_{49} + 0.76VHI_{19}$	0.68	0.58	0.60	61.69
B2c	$-0.39VCI_{40} + 0.34VCI5_{51} + 1.07SPEI_{2-5} + 0.91SPEI_{4-1} - 0.84SPEI_{6-6}$	0.91	0.85	0.34	78.20
B2d	$1.14SPEI_{2-5} + 0.86SPEI_{4-1} - 0.78SPEI_{6-6}$	0.83	0.75	0.46	70.74

Source: Ribeiro et al. (2019). Used with permission.

Note: Boldface: the model with the highest accuracy for a given type of crop.

configuration, as well as the crop type. An improperly configured ANN is less accurate than MLR (Table 7.9). Of the 11 established models, 5 are more accurate than their counterparts established using MLR. The best prediction is established from two predictor variables using a network configuration of 2-3-1 (three layers) for wheat yield (Table 7.10). Similarly high accuracy is achieved for predicting barley yield, but from five predictor variables. If a sufficient number of predictor variables are included in the model, MLR can achieve a prediction accuracy comparable to that of ANN.

TABLE 7.10

Comparison of the Performance of ANN Models for Estimating Wheat (W) and Barley (B) Yields and Model Architecture Based on Three Layers, Using Leave-One-Out Cross-Validation

Model	N	Architecture	$R^2_{adj\_no\_CV}$	R^2_{adj}	RMSE	SS_{RMSE} (%)
W1a[a]	2	2-3-1	0.93	0.85	0.36	77.30
W1b	2	2-1-1	0.80	0.71	0.51	67.85
B1a[a]	2	2-4-1	0.90	0.80	0.42	73.30
B1b[a]	2	2-4-1	0.89	0.83	0.39	75.20
W2a	2	2-3-1	0.65	0.36	0.76	52.04
W2b	1	1-2-1	0.57	0.12	0.90	43.20
W2c[a]	4	4-2-1	0.89	0.73	0.47	70.34
B2a	3	3-1-1	0.67	0.49	0.66	57.76
B2b[a]	3	3-3-1	0.89	0.75	0.46	70.56
B2c	5	5-1-1	0.91	0.84	0.36	76.96
B2d	3	3-1-1	0.83	0.74	0.47	69.92

Source: Ribeiro et al. (2019). Used with permission.

Note: Boldface: the model with the highest accuracy for each type of crop. N – The number of predictor variables

[a] Models, the performance of which is better than their MLP counterparts in Table 7.9

Nevertheless, it is still uncertain how the yield has been affected by droughts (e.g., what is the relationship between climate variables and yield reduction).

7.4.4 Drought Prediction

It is important to predict when a drought will occur so as to mitigate its economic loss (e.g., reduced planting of water-hungry crops). Drought forecasting attempts to predict what is going to happen to rainfall in the near future based on what has happened in the past. This prediction is made from a variety of inputs, and the predicted (dependent) variable is one of the drought indices presented in Section 7.3.4. In particular, soil moisture and its corresponding drought index (e.g., SWDI) are commonly predicted (Table 7.11). SPI and SPEI of 3-, 6-, 9-, and 12-month durations are other dependent variables used for forecasting with 1–6 month lead times. Common predictor variables may include field observed precipitation, temperature, PET, solar radiation, relative humidity, and wind speed. Apparently, the exact variables used to predict drought vary with its nature. Meteorological drought in ungauged areas is predicted mostly from climate and vegetation factors, such as LST_{day}, LST_{night}, NDVI, NDWI, and large-scale climate indices, such as multivariate El Niño-Southern Oscillation index and arctic oscillation index (Rhee and Im, 2017). In the prediction, remote sensing data products of precipitation and soil moisture (e.g., CMORPH-CRT, IMERG V05, and TRMM 3B42V7) serve as the inputs or independent variables (Zhu et al., 2021). These long-range climate data may be combined with remote sensing-derived precipitation, PET, temperature, and topographical information (month, elevation) to achieve better predictions.

The prediction is accomplished via MLR or using machine learning methods, such as bias-corrected RF, decision trees, extremely randomized trees, SVM, and multi-layer perceptron neural networks (Feng et al., 2019; Zhu et al., 2021). MLR is strong at tackling the complex, nonlinear relationship between drought causers and its indicators but is unable to handle the collinearity problem among the considered input variables. This inability disappears with machine learning algorithms. They are superior because they can also indicate the importance of each considered factor in isolation, not just the joint importance of multiple predictor variables. The best prediction is made using in situ precipitation as input for soil moisture prediction, followed by TRMM 3B42V7, IMERG V05, and CMORPH-CRT (Table 7.11). In situ precipitation and IMERG V05 are more suitable for drought prediction based on indirect SWDI, while precipitation data products (CMORPH-CRT and TRMM 3B42V7) are more suitable for direct SWDI prediction. The combination of in situ soil moisture with in situ precipitation or CMORPH-CRT improves direct SWDI prediction. Typically, the combination of remotely sensed soil moisture with in situ precipitation produces the best predictions, much better than the remote sensing precipitation data products.

TABLE 7.11
Influence of Input Variables on Model Accuracy in Which In Situ Measured and Remotely Sensed Moisture Products Are the Dependent variable to Be Predicted

		Training Dataset		Validation Dataset	
Dependent Variable	**Independent Variables**	*r*	MSE	*r*	MSE
Soil moisture	P, Rn, Rh, Ws, PET	0.70	0.04	0.04	0.26
	P, T, PET, Rh, Rn, Ws	0.48	0.07	0.56	0.16
SWDI	P, Rn, Rh, Ws, PET	0.68	0.07	0.63	0.19
	P, T, PET, Rh, Rn, Ws	0.70	0.07	0.65	0.17

Source: Zhu et al. (2021). Used with permission.

Of decision trees, RF, and extremely randomized trees, the last method is the best performer in most cases in terms of producer's drought accuracy, defined as the number of correctly classified samples in moderate, severe, and extreme drought classes over the total number of samples in the corresponding classes. Its accuracy reaches up to 64% in forecasting SPEI, and up to 56% in forecasting SPI from long-term climate data (Rhee and Im, 2017). This method is competent for forecasting high-resolution droughts in ungauged areas from climatological data. However, spatial interpolation outperforms them in terms of the user's accuracy (up to 44%), defined as the number of correctly classified samples in drought classes over the total number of samples classified in those classes (i.e., false-positive prediction). In addition to SVM, other machine learning algorithms such as ANN, least squares support vector regression (LSSVR), and adaptive neuro-fuzzy inference system (ANFIS) have also been applied to predicting drought. Their prediction accuracy has been comparatively evaluated based on three meteorological drought indices of VCI, LST/NDVI, and SPI of three durations (3, 6, and 12 months) (Table 7.12). They are derived from three types of data (product): NDVI, time-series MODIS LST data products, and time-series TRMM rainfall data products (Khosravi et al., 2017). SVR has the lowest RMSE, and NN has the lowest efficiency among all the methods (Table 7.12). In addition, LSSVR is the fastest, followed by ANFIS.

In comparison with drought itself, it is much more challenging to forecast its effects, such as reduced crop yield which is much more important than drought in terms of food production and security. This forecast can be made accurate by combining multiple datasets and/or assimilating satellite observations into a model to generate long-term climate data records via simulation (AghaKouchak et al., 2015), such as the ecohydrological Land Data Assimilation System (LDAS). It has the potential to predict agricultural drought by assimilating remote sensing data sensitive to surface soil moisture and vegetation water content (both being important variables for agricultural drought monitoring), such as microwave T_B data using an existing land surface model called EcoHydro-SiB (Sawada et al., 2020). It can simultaneously simulate surface soil moisture, root-zone soil moisture, and vegetation state. The estimates of vegetation growth and senescence are modeled from the carbon pools of plant roots and leaves. Their NPP is calculated from the SiB2 photosynthesis-conductance model. Water-related stress is derived from the vertical distribution of soil moisture from soil moisture data products. LDAS output provides the appropriate initial conditions for Land Surface Model (LSM) prediction. If applied to an LSM, LDAS can simultaneously simulate surface soil moisture, root-zoon soil moisture, and vegetation growth.

LDAS can be combined with a radiative transfer model to convert land surface conditions (e.g., reflectivity) to microwave T_B to form the Coupled Land and Vegetation Data Assimilation System (CLVDAS). Its coupling with the Geophysical Fluid Dynamics Laboratory (GFDL) hindcast product

TABLE 7.12

Comparison of Four Machine Learning Methods in Predicting Drought Measured by Three Indices

		ANN		SVR		LSSVR		ANFIS	
Index	**Data**	**MAE**	**RMSE**	**MAE**	**RMSE**	**MAE**	**RMSE**	**MAE**	**RMSE**
VCI	All	13.06	20.49	3.59	6.63	11.21	14.31	9.50	12.66
	Test	17.48	31.14	8.89	12.06	26.72	28.80	24.36	26.06
LST/NDVI	All	17.30	21.43	10.34	12.81	15.06	19.63	8.94	13.32
	Test	16.44	20.00	11.47	13.26	13.61	17.01	20.77	23.73
SPI	All	0.95	1.16	0.12	0.20	0.12	0.21	0.37	0.57
	Test	0.98	1.24	0.19	0.21	0.36	0.38	1.10	1.36

Source: Adapted from Khosravi et al. (2017). Used with permission.

TABLE 7.13

Comparison of Three Drought Monitoring Frameworks

Item	Sawada	Princeton	FLDAS
Forecast	Yes (three-month lead)	Yes (seven-day lead)	No
Sequential DA	Yes (microwave)	No	No
Estimated land surface variables used to monitor and predict drought	Soil moisture, evapotranspiration, runoff, LAI	Soil moisture, evapotranspiration, runoff, streamflow	Soil moisture, evapotranspiration, runoff, streamflow
LSM used	EcoHydro-SiB	VIC	VIC, Noah
Meteorological forcing dataset	GLDAS (monitoring), GFDL (prediction)	TMPA + NCEP GFS + empirical regressions	RFE2 + GDAS, CHIRPS + MERRA – 2

Source: Sawada et al., 2020

can simulate microwave T_B. The coupled ocean-atmospheric model allows the prediction of atmospheric forcing that is vital to predicting agricultural drought. CLVDAS is superior to LDAS in that it optimizes the unknown parameters of the radiative transfer model and EcoHydro-SiB via a parameter optimization module. It searches for the optimal parameters that minimize the cost function defined as the squared difference between simulated and observed T_B. CLVDAS also sequentially adjusts the model states (e.g., soil moistures of all soil columns and LAI). The coupled framework can be used to both monitor and predict agricultural drought. In monitoring, CLVDAS is run with data analysis (e.g., reanalysis of AMSR-2 T_B data from $t_{0-1\ \text{month}}$ to t_0), with the initial conditions determined via reanalysis of data at $t = t_{0-1\ \text{month}}$. In prediction, the state at $t_0 +_{3.5\ \text{months}}$ is predicted from the t_0 state based on the past record of meteorological forcing and GFDL-predicted seasonal precipitation. This framework can simulate soil moisture and LAI that are closely correlated with the declined national wheat yield induced by drought ($r = 0.65 - 0.72$). A general circulation model (GCM)-based seasonal meteorological prediction significantly contributes to the accurate prediction of LAI and agricultural droughts in a two- to three-month lead time (Sawada et al., 2020).

At present, there are no early warning systems for drought in existence. A remotely resembling early warning system of drought is the Famine Early Warning Systems Network Land Data Assimilation System (FLDAS) that has been used to predict agricultural drought (Table 7.13). For an early warning system to function well, it must meet three essential requirements (Sawada et al., 2020). (i) It needs to explicitly capture crop states and accurately predict regional crop yield with the assistance of an LSM. Remote sensing-derived vegetation indices such as LAI can serve as the input to the LSM to develop the empirical relationship between regional and national crop production and vegetation states. (ii) The agricultural drought prediction system needs to dynamically forecast the conditions of soil moisture and vegetation in a few months of lead time using GCM-based seasonal prediction. (iii) The initial conditions of vegetation and soil moisture should be set realistically to accurately predict agricultural drought. As indicated by Table 7.13, FLDAS does not have the forecast capacity. More research is needed in this area. FLDAS is very similar to the other two systems but without the ability to forecast drought with lead time. To some degree, the Sawada system can be used to issue early warnings of drought.

REFERENCES

AghaKouchak, A., A. Farahmand, F.S. Melton, J. Teixeira, M.C. Anderson, B.D. Wardlow, and C.R. Hain. 2015. Remote sensing of drought: Progress, challenges and opportunities. *Rev. Geophys.* 53(2): 452–480. DOI: 10.1002/2014RG000456

Ashouri, H., K. Hsu, S. Sorooshian, D.K. Braithwaite, K.R. Knapp, L.D. Cecil, B.R. Nelson, and O.P. Prat. 2015. PERSIANN-CDR: Daily precipitation climate data record from multisatellite observations for hydrological and climate studies. *Bull. Am. Meteorol. Soc.* 96(1): 69–83.

Badamassi, M., M. Barkawi, A. El-Aboudi, and P.G. Gbetkom. 2020. A new index to better detect and monitor agricultural drought in Niger using multisensor remote sensing data. *Prof. Geogr.* 72(3): 421–432. DOI: 10.1080/00330124.2020.1730197

Brown, J.F., B.D. Wardlow, T. Tadesse, M.J. Hayes, and B.C. Reed. 2008. The vegetation drought response index (VegDRI): A new integrated approach for monitoring drought stress in vegetation. *GISci. Remote Sens.* 45(1): 16–46. DOI: 10.2747/1548-1603.45.1.16

Dalezios, N.R., A. Blanta, and N.V. Spyropoulos. 2012. Assessment of remotely sensed drought features in vulnerable agriculture. *Nat. Hazards Earth Syst. Sci.* 12(10): 3139–3150. DOI: 10.5194/nhess-12-3139-2012

Feng, P., B. Wang, D.L. Liu, and Q. Yu. 2019. Machine learning-based integration of remotely-sensed drought factors can improve the estimation of agricultural drought in South-Eastern Australia. *Agric. Syst.* 173: 303–316. DOI: 10.1016/j.agsy.2019.03.015

Forootan, E., M. Khaki, M. Schumacher, V. Wulfmeyer, N. Mehrnegar, A.I.J.M. van Dijk, L. Brocca, S. Farzaneh, F. Akinluyi, G. Ramillien, C.K. Shum, J. Awange, and A. Mostafaie. 2019. Understanding the global hydrological droughts of 2003–2016 and their relationships with teleconnections. *Sci. Total Environ.* 650(2): 2587–2604.

Ghulam, A., Q. Qin, and Z. Zhan. 2007. Designing of the perpendicular drought index. *Environ. Geol.* 52: 1045–1052. DOI: 10.1007/s00254-006-0544-2

Graw, V., G. Ghazaryan, K. Dall, A.D. Gómez, A. Abdel-Hamid, A. Jordaan, R. Piroska, J. Post, J. Szarzynski, Y. Walz, and O. Dubovyk. 2017. Drought dynamics and vegetation productivity in different land management systems of Eastern Cape, South Africa – A remote sensing perspective. *Sustainability* 9(10). DOI: 10.3390/su9101728

Gu, Y., J.F. Brown, J.P. Verdin, and B. Wardlow. 2007. A five-year analysis of MODIS NDVI and NDWI for grassland drought assessment over the central Great Plains of the United States. *Geophy. Res. Lett.* 34: L06407. DOI: 10.1029/2006GL029127

Hao, Z., and A. AghaKouchak. 2013. Multivariate standardized drought index: A parametric multi-index model. *Adv. Water Resour.* 57(2013): 12–18. DOI: 10.1016/j.advwatres.2013.03.009

Hao, Z., A. AghaKouchak, N. Nakhjiri, and A. Farahmand. 2014. Global integrated drought monitoring and prediction system. *Sci Data* 1: 14000. DOI: 10.1038/sdata.2014.1

Hayes, M.J. 2006. Drought Indices. Van Nostrand's Scientific Encyclopedia, John Wiley & Sons. doi:10.1002/0471743984.vse8593 (accessed May 20, 2022).

Huang, J., W. Zhuo, Y. Li, R. Huang, F. Sedano, W. Su, J. Dong, L. Tian, Y. Huang, D. Zhu, and X. Zhang. 2020. Comparison of three remotely sensed drought indices for assessing the impact of drought on winter wheat yield. *Int. J. Digital Earth* 13(4): 504–526. DOI: 10.1080/17538947.2018.1542040

Huffman, G.J. et al. (2020). Integrated multi-satellite retrievals for the global precipitation measurement (GPM) mission (IMERG). In *Satellite Precipitation Measurement. Advances in Global Change Research*, eds. V. Levizzani, C. Kidd, D.B. Kirschbaum, C.D. Kummerow, K. Nakamura, and F.J. Turk, vol 67. Cham: Springer. https://doi.org/10.1007/978-3-030-24568-9_19

Joseph, R., T.M. Smith, M.R.P. Sapiano, and R.R. Ferraro. 2009. A new high-resolution satellite-derived precipitation dataset for climate studies. *J. Hydromet.* 10: 935–952. DOI: 10.1175/2009JHM1096.1

Joyce, R.J., J.E. Janowiak, P.A. Arkin, and P. Xie. 2004. CMORPH: A method that produces global precipitation estimates from passive microwave and infrared data at high spatial and temporal resolution. *J. Hydromet.* 5: 487–503.

Karnieli, A., and G. Dall'Olmo. 2003. Remote-sensing monitoring of desertification, phenology, and droughts. *Manag. Environ. Qual.: An Int. J.* 14(1): 22–38.

Khosravi, I., Y. Jouybari-Moghaddam, and M.R. Sarajian. 2017. The comparison of NN, SVR, LSSVR and ANFIS at modeling meteorological and remotely sensed drought indices over the eastern district of Isfahan, Iran. *Nat. Hazards* 87(3): 1507–1522. DOI: 10.1007/s11069-017-2827-1

Kim, J.S., S.Y. Park, J.H. Lee, J. Chen, S. Chen, and T.W. Kim. 2021. Integrated drought monitoring and evaluation through multi-sensor satellite-based statistical simulation. *Remote Sens.* 13: 272. DOI: 10.3390/rs13020272

Kogan, F.N. 1995a. Droughts of the late 1980s in the United States derived from NOAA polar orbiting satellite data. *Bull. Am. Meteorol. Soc.* 76: 655–668.

Kogan, F.N., 1995b. AVHRR data for detection and analysis of vegetation stress. In *Meteorological Satellite Data Users Conference*, Winchester, UK, pp. 155–162.

McFeeters, S.K. 1996. The use of the normalized difference water index (NDWI) in the delineation of open water features. *Int. J. Remote Sens.* 17: 1425–1432. DOI: 10.1080/01431169608948714

McKee, T.B., N.J. Doesken, and J. Kleist. 1993. The relationship of drought frequency and duration of time scales. In *Proc. of the 8th Conf. of Applied Climatology*, American Meteorological Society, Anaheim, CA, pp. 179–184.

Meng, L., T. Dong, and W. Zhang. 2016. drought monitoring using an integrated drought condition index (IDCI) derived from multi-sensor remote sensing data. *Nat. Hazards* 80(2): 1135–1152. DOI: 10.1007/s11069-015-2014-1

Mu, Q., M. Zhao, J. Kimball, N. McDowell, and S. Running. 2013. A remotely sensed global terrestrial drought severity index. *Bull. Am. Meteorol. Soc.* 94: 83–98. DOI: 10.1175/BAMS-D-11-00213.1

Okin, G., C. Dong, K.S. Willis, T.W. Gillespie, and G.M. MacDonald. 2018. The impact of drought on native southern California vegetation: Remote sensing analysis using MODIS-derived time series. *J. Geophy. Res. Biogeosci.* 123(6): 1927–1939. DOI: 10.1029/2018JG004485

Reichle, R.H., R.D. Koster, G.L.M. De Lannoy, B.A. Forman, Q. Liu, S.P.P. Mahanama, and A. Toure. 2011. Assessment and enhancement of MERRA land surface hydrology estimates. *J. Clim.* 24: 6322–6338. DOI: 10.1175/JCLI-D-10-05033.1

Rhee, J., and J. Im. 2017. Meteorological drought forecasting for ungauged areas based on machine learning: Using long-range climate forecast and remote sensing data. *Agri. Forest Meteorol.* 237–238: 105–122. DOI: 10.1016/j.agrformet.2017.02.011

Rhee, J., J. Im, and G.J. Carbone. 2010. Monitoring agricultural drought for arid and humid regions using multi-sensor remote sensing data. *Remote Sens. Environ.* 114(12): 2875–2887.

Rhee, J., K. Park, S. Lee, S. Jang, and S. Yoon. 2020. Detecting hydrological droughts in ungauged areas from remotely sensed hydro-meteorological variables using rule-based models. *Nat. Hazards* 103(3): 2961–2988. DOI: 10.1007/s11069-020-04114-5

Ribeiro, A.F.S., A. Russo, C.M. Gouveia, and P. Páscoa. 2019. Modelling drought-related yield losses in Iberia using remote sensing and multiscalar indices. *Theor. Appl. Climatol.* 136(1–2): 203–220. DOI: 10.1007/s00704-018-2478-5

Sandholt, I., K. Rasmussen, and J. Andersen. 2002. A simple interpretation of the surface temperature/vegetation index space for assessment of surface moisture stress. *Remote Sens. Environ.* 79(2): 213–224.

Sawada, Y., T. Koike, E. Ikoma, and M. Kitsuregawa. 2020. Monitoring and predicting agricultural droughts for a water-limited subcontinental region by integrating a land surface model and microwave remote sensing. *IEEE Trans. Geosci. Remote Sens.* 58(1): 14–33. DOI: 10.1109/TGRS.2019.2927342

Shen, R., A. Huang, B. Li, and J. Guo. 2019. Construction of a drought monitoring model using deep learning based on multi-source remote sensing data. *Int. J. Applied Earth Obs. Geoinfo.* 79: 48–57. DOI: 10.1016/j.jag.2019.03.006

Sur, K., and M.M. Lunagaria. 2020. Association between drought and agricultural productivity using remote sensing data: A case study of Gujarat state of India. *J. Water Clim. Change* 11(1S): 189–202. DOI: 10.2166/wcc.2020.157

Sur, C., S.Y. Park, T.W. Kim, and J.H. Lee. 2019. Remote sensing-based agricultural drought monitoring using hydrometeorological variables. *KSCE J. Civ. Eng.* 23: 5244–5256.

Vicente-Serrano, S.M. 2007. Evaluating the impact of drought using remote sensing in a Mediterranean, semi-arid region. *Nat. Hazards* 40(1): 173–208. DOI: 10.1007/s11069-006-0009-7

West, H., N. Quinn, and M. Horswell. 2019. Remote sensing for drought monitoring and impact assessment: Progress, past challenges and future opportunities. *Remote Sens. Environ.* 232. DOI: 10.1016/j.rse.2019.111291

Yu, H., L. Li, Y. Liu, and J. Li. 2019. Construction of comprehensive drought monitoring model in Jing-Jin-Ji region based on multisource remote sensing data. *Water* 11(5). DOI: 10.3390/w11051077

Zhang, A., and G. Jia. 2013. Monitoring meteorological drought in semiarid regions using multi-sensor microwave remote sensing data. *Remote Sens. Environ.* 134: 12–23. DOI: 10.1016/j.rse.2013.02.023

Zhu, Q., Y. Luo, D. Zhou, Y.P. Xu, G. Wang, and Y. Tian. 2021. Drought prediction using in situ and remote sensing products with SVM over the Xiang River Basin, China. *Nat. Hazards* 105(2): 2161–2185. DOI: 10.1007/s11069-020-04394-x

Zhu, S., Z. Xiao, X. Luo, H. Zhang, X. Liu, R. Wang, M. Zhang, and Z. Huo. 2020. Multidimensional response evaluation of remote-sensing vegetation change to drought stress in the Three-River Headwaters, China. *IEEE J. Sel. Top. Appl. Earth Obs. Remote Sens.* 13: 6249–6259. DOI: 10.1109/JSTARS.2020.3027347

8 Dust Storms

8.1 INTRODUCTION

A dust storm, also known as a sandstorm, refers to strong winds that blow loose sand and dust particles from scarcely vegetated surfaces into the air. The large quantity of dust and fine sandy materials are suspended in the atmosphere and transported tens or even hundreds of kilometers from their source, reducing air visibility. This natural meteorological phenomenon is ephemeral, lasting a few minutes to hours. Normally, this kind of disastrous weather phenomenon takes place on a regional or even continental scale. Dust and sandstorms are considered a natural hazard because they degrade air quality, accelerate land desertification, reduce horizontal visibility to less than 1 km, disrupt transportation and communication, hamper normal aviation, and pose a grave threat to human health, especially aggravating respiratory ailments. Besides, they also exert feedback effects on regional and even global climate change by interfering with solar radiation. Dust storms are frequently occurring natural hazards in arid and semi-arid regions of the world, such as Northern Africa, the Middle East, and central Asia where the climate is rather dry with little vegetation cover. Northern and northwestern China is also vulnerable to highly disruptive sandstorms that occur frequently in late winter and early spring (Figure 8.1). These regions are prone to the attacks of sandstorms because of an abundant supply of sand on the dry ground.

Remote sensing offers an excellent source of data for determining the source regions of dust storms and tracking their transport pathways, assessing their intensity, and identifying their spatial extent of influence so as to better understand sandstorm distributions and development processes and help to devise effective means of mitigating sandstorm damage. In addition, remote sensing is good at monitoring, tracking, and forecasting dust storms. Remote sensing of dust storms is based on the spectral behavior of dust and the usual ground cover. The difference in the spectral reflectance of suspended sand and the underlying ground cover is the only clue useable in delineating the extent of a dust storm.

This chapter first describes the spectral behavior of dust and other features to demonstrate the feasibility of sensing dust storms remotely from space, as well as the best wavelength(s) for the detection of dust storms. Also demonstrated is the feasibility of sensing different properties of dust storms. Next, this chapter assesses the utility of four major types of satellite images that are the best for sensing dust storms and dust thickness. Also included in the discussion is a comprehensive assessment of the data products derived from them that are useful for monitoring dust storms. This chapter then introduces, compares, and evaluates the large number of spectral indices that have been developed to detect dust storms effectively. Afterward, this chapter details the procedure of implementing the detection based on these indices. Finally, this chapter expounds on how to monitor and assess the risk of dust storms, including ascertaining dust sources and transport pathways.

8.2 FEASIBILITY OF SENSING

Whether sand and dust storms can be successfully detected from remotely sensed data and the accuracy of detection depend on the unique spectral behavior of dust and its disparity to other ground features present in the same image. These features may include variations in dust properties, such as thickness and spatial variability, clouds, the underlying surface, as well as ocean water (if applicable). Dust storms are reliably identifiable from true-color images of aerosol optical thickness (AOT) data owing to the significant spectral distinctions between dust storms and typical land surfaces (Figure 8.1). The spectral disparity between dust and the ground cover is maximal in the visible and

DOI: 10.1201/9781003354321-8

FIGURE 8.1 A dust storm on 24 April 2009 in northern China (left), and the same area on a cloud-free color composite of MODIS bands 1, 4, and 3 (right). (Source: NASA and Google Earth.)

NIR regions of the spectrum under clear skies. A close and detailed scrutiny of the spectral behavior of dust over different spectral zones reveals that apparent radiation disparities exist between dust storms and clouds in visible and NIR bands due to different scattering caused by particle size and the differential absorption of radiation by water vapor (Figure 8.2). There are also apparent differences between the spectral behavior of dust and other four features in the scatterplot of visible light and SWIR bands (Figure 8.2a) and in the T_B domain of the MIR and TIR bands (Figure 8.2b). Both diagrams illustrate that dust pixels have a spectrally compact distribution that seldom overlaps with that of other features. It is thus theoretically feasible to detect and map dust storms reliably from these spectral bands. However, whether it is possible to further differentiate dust into multiple thickness classes remains unresolved.

Apart from the commonly used visible (VIS) light and NIR bands, SWIR and even TIR bands can also be used for dust/sandstorm detection. Sand reflectance peaks in MODIS band 6 (SWIR wavelength of 1.628–1.652 μm) and is much higher than in band 3 (visible light of 0.459–0.479 μm). As illustrated in Figure 8.3, a dust storm shows up as yellow over land that appears to be pink while

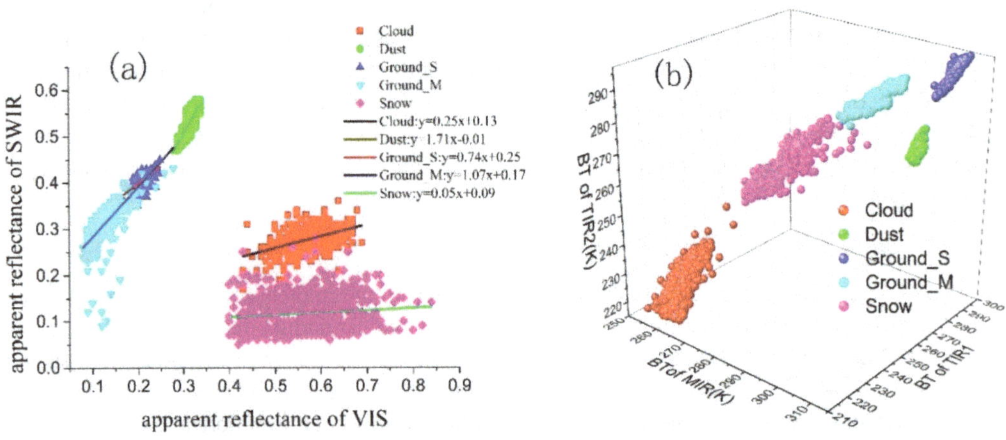

FIGURE 8.2 (a) Disparity between dust and other features in the spectral reflectance domain of VIS (0.65 μm) vs SWIR (1.6 μm), (b) the T_B domain of MIR (3.9 μm) vs TIR (10.8 μm). (Source: Di et al., 2016.)

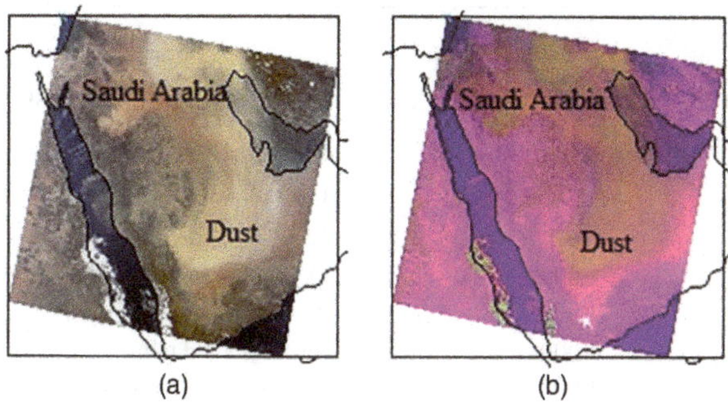

FIGURE 8.3 Enhancement of dust on satellite images via color compositing. (a) MODIS true-color RGB image, and (b) pseudo-color RGB image. (Source: Kunte and Aswini, 2015. Used with permission.)

water has a color of dark blue in pseudo-color MODIS images (Kunte and Aswini, 2015). In the MIR wavelength of 3.9 μm, both dust and water clouds scatter the incoming radiation more strongly than ice clouds. Dust and water clouds often have a similar single scatter albedo, but the ratio of particle reflectance at 3.7 μm to that at 0.65 μm is frequently higher for dust than that for clouds (Mishra et al., 2015). The T_B of dust and clouds is best detected from TIR bands. Since TIR algorithms enhance dust accurately, the detection based on the enhanced TIR bands or their combination with reflective bands produces better indicators of dust storms than a reflectance-based algorithm (Jafari and Malekian, 2015). Moreover, dust signatures on satellite imagery can be enhanced using various techniques such as color compositing, band ratioing, and indexing of multiple spectral bands. With normalization, dust is best discriminated from MODIS bands 6 and 3 in the form of (band 6 − band 3)/(band 6 + band 3) > 0 (Su et al., 2017).

8.3 USEFUL SATELLITE DATA

Since dust storms are spatially extensive but compositionally monotonous comprising the same sand/dust substance with little spectral heterogeneity apart from variation in thickness, they are best studied from space-borne images of a coarse spatial resolution. Satellite remote sensing is especially suited to detect dust storms over extensive areas at a high temporal frequency. These satellite images have a broad ground cover with a very short revisit period that enables the same dust storm to be tracked repeatedly from time-series images. Dust storms are commonly detected from various meteorological satellite images, including those ultraviolet sensors such as Total Ozone Mapping Spectrometer and ozone monitoring instrument (OMI) based on the ultraviolet absorption aerosol algorithms. Other commonly used visible and TIR images include MODIS and GEOS. Microwave sensors such as TRMM, AMSR-E, and AMSR-2 have also been used, even though ultraviolet and microwave sensors are significantly compromised by their coarse spatial resolution in detecting dust storms. In contrast, visible light and infrared images obtained from MODIS, Indian National Satellite System (INSAT), Visible Infrared Imaging Radiometer Suite (VIIRS), and Advanced Himawari Imager (AHI) do not face this problem and have been widely used.

8.3.1 MODIS

MODIS is one of the five sensors aboard the Terra/Aqua satellites. This multispectral radiometer measures biological and physical processes in the terrestrial and maritime spheres using 36 spectral bands, the wavelengths of which span from visible light to TIR (Table 8.1). These bands are designed to sense different targets optimally. For instance, bands 1 and 2 are best for separating land/cloud/aerosols boundaries, bands 3–7 for studying land/cloud/aerosols properties, bands

TABLE 8.1
MODIS Spectral Bands, Spatial Resolution, and Primary Uses

Bands	Resolution (m)	Primary Uses
1–2	250	Land/cloud/aerosols boundaries
3–7	500	Land/cloud/aerosols properties
8–16	1,000	Ocean color/phytoplankton/biogeochemistry
17–19	1,000	Atmospheric water vapor
20–23	1,000	Surface/cloud temperature
24–25	1,000	Atmospheric temperature
26–28	1,000	Cirrus clouds/water vapor
29	1,000	Cloud properties
30	1,000	Ozone
31–32	1,000	Surface/cloud temperature
33–36	1,000	Cloud top altitude

8–16 for sensing ocean color phytoplankton biogeochemistry, bands 17–19 for sensing atmospheric water vapor, and so on, on the regional or global scale. All MODIS bands have a fine temporal resolution of up to 6 hours to timely capture short-lived sand and dust storms and track their movement and transport route. Fine temporal resolution MODIS data are excellent for large-scale, continuous, and dynamic monitoring of the sandstorm migration process, from the onset, development, to final dissipation. Time-series MODIS images enable the detection of the route of travel and delineation of the sandstorm-affected areas.

MODIS data have been widely used to study sandstorms because of their extensive spatial coverage and long history of data availability. More importantly, a large range of data products have been derived from them (Table 8.2), some of which are vital in improving the reliability and verifying the accuracy of remote sensing-retrieved dust storms. For instance, the MODIS aerosol optical depth (AOD) and OMI aerosol index products are effective for the initial dust detection, and the AOD and aerosol index images are highly correlated with the dust storms at the provincial scale (p-value <0.001), but not at the local scale due to their coarse spatial resolution (Jafari and Malekian, 2015).

TABLE 8.2
Typical Data Products Produced from Combined MODIS Images and Their Properties that Are Useful for Detecting Dust Storms

Product Name	Description	Spatial Resolution (m)	Temporal Resolution
MCD12C1.006	Land cover	500	Yearly
MCD12Q1.006	Land cover	500	Yearly
MCD12Q2.006	Land cover	500	Yearly
MCD12Q2.005	Land cover	500	Yearly
MCD15A2H.006	FPAR, LAI	500	Multi-day
MCD15A3H.006	FPAR, LAI	500	Multi-day
MCD18A1.006	Surface radiance	500	Daily
MCD19A2.006	Aerosol optical depth	1,000	Daily
MOD09GQ	Band 1–2 reflectance	250	Daily
MOD09A1	Band 1–7 reflectance	500	Eight-day

TABLE 8.3
Properties of INSAT-3D Imager Bands

Bands	Wavelength Range (μm)	Central Wavelength (μm)	Resolution (km)
VIS	0.52–0.72	0.65	1
SWIR	1.55–1.70	1.625	1
MWIR	3.80–4.00	3.9	4
Water Vapor	6.50–7.00	6.8	8
TIR1	10.2–11.2	10.8	4
TIR2	11.5–12.5	12.0	4

8.3.2 INSAT

INSAT is a long series of multi-purpose satellites. The most useful satellite for studying dust storms is INSAT-3D launched on 26 July 2013. This geostationary meteorological satellite images the Earth's surface every 30 minutes. One of the payloads aboard this satellite is the Imager that senses the Earth in six multispectral channels of visible (0.55–0.75 μm), TIR (10.5–12.5 μm), and Water Vapour (5.7–7.1 μm) at three resolutions of 1, 4, and 8 km, respectively (Table 8.3). The SWIR and mid-wave IR (MWIR) bands with a resolution of 1 and 4 km, respectively, are good at discriminating land and cloud and detecting surface features like snow. INSAT-3D images cover at least 50°N to 40°S, acquired in 23 minutes per scene. A large number of INSAT-3D data products are publically available at mosdac.gov.in/satellite-catalog, such as LST (3DIMG_L2B_LST), daytime cloud microphysical parameters (3DIMG_L2C_CMP), SST, and vapor, all being an excellent source of data for detecting sandstorms.

8.3.3 VIIRS

One of the key payloads onboard the Suomi National Polar-Orbiting Partnership (S-NPP) spacecraft, the VIIRS sensor was successfully launched on 28 October 2011. This new generation of satellite with operational moderate resolution-imaging capabilities expands the legacies of AVHRR and MODIS sensors by sensing the Earth's surface and atmosphere in 22 imaging and radiometric bands over the wavelength range of 0.41–12.5 μm, including 5 high-resolution I-channels (spatial resolution = 375 m), 16 medium-resolution M-channels (resolution = 750 m), and 1 panchromatic Day/Night channel (resolution = 750 m) (Table 8.4). In total, 43 data products are produced from VIIRS data at level-0 and level-1. They can be used to study cloud and aerosol properties, ocean color, ocean and land surface temperature, and surface albedo. Some of the products useful for dust storm detection are NDVI, thermal anomalies, and bidirectional reflectance distribution function (BRDF)/albedo. This satellite imagery is disadvantaged by its long revisit period of 16 days in tracking dust storms.

8.3.4 AHI

Himawari-8 is a satellite launched on 7 October 2014 by the Japan Meteorological Agency, and its payload comprises the AHI sensor. AHI imagery has 16 channels, including 3 visible (0.47, 0.51, and 0.67 μm), 3 NIR, and 10 mid- and thermal IR channels (Table 8.5). All of them have a spatial resolution of 2 km except bands 1, 2, and 4, the spatial resolution of which is 1 km, and band 3 has a resolution of 500 m. AHI can scan the full disk of 120° × 120° centered at (0°, 140°E) every 10 minutes, making it the ideal data source for tracking dust storm movement, even though the temporal resolution degrades for areas outside Southeast Asia, such as the Mid-East and Africa.

TABLE 8.4
Properties of 16 M-Channels of NPP VIIRS Imagery (Resolution = 750 m)

Channels	Spectral Region	Wavelength Range (μm)	Primary Use
M1	Visible	0.402–0.422	Ocean color aerosol
M2	Visible	0.436–0.454	Ocean color aerosol
M3	Visible	0.478–0.498	Ocean color aerosol
M4	Visible	0.545–0.565	Ocean color aerosol
M5	NIR	0.662–0.682	Ocean color aerosol
M6	NIR	0.739–0.754	Atmospheric correction
M7	SWIR	0.846–0.885	Ocean color aerosol
M8	SWIR	1.23–1.25	Cloud particle size
M9	SWIR	1.371–1.386	Cirrus cloud cover
M10	SWIR	1.58–1.64	Snow fraction
M11	MWIR	2.23–2.28	Clouds
M12	MWIS	3.66–3.84	Sea surface temperature
M13	LWIR	3.97–4.13	Sea surface temperature/fires
M14	LWIR	8.4–8.7	Cloud top properties
M15	LWIR	10.26–11.26	Sea surface temperature
M16	Day/night	11.54–12.49	Sea surface temperature

AHI data products useful for dust storm detection include clear-sky radiance, high-resolution cloud analysis information, and AOT.

For studying those short-lived dust storms outside Southeast Asia, the temporal resolution of the aforementioned satellite images measured by hours may not be sufficiently fine. They are better studied using geostationary satellite images such as GOES that has a temporal resolution of 30 minutes (hemisphere view, and 5 minutes in mode 4), although they are designed primarily for meteorological applications.

8.4 DETECTION METHODS AND ALGORITHMS

Dust storm detection is a prerequisite for analyzing dust storm trajectories and determining their sources. The detection and recognition of the extent of dust storms are very critical to designing an early warning system and managing and mitigating the risk of this natural hazard. Under normal

TABLE 8.5
Channel Designations of the Advanced Himawari Imager (AHI)

Channel No.	Central Wavelength (μm)	Channel No.	Central Wavelength (μm)
1	0.47	9	6.94
2	0.51	10	7.35
3	0.64	11	8.60
4	0.86	12	9.64
5	1.61	13	10.41
6	2.25	14	11.24
7	3.89	15	12.38
8	6.24	16	13.28

Source: www.jma-net.go.jp/msc/en/.

TABLE 8.6
Thresholds for Separating Dust Storms from Other Features Based on Three MODIS TIR Bands at 8, 11, and 12 μm

Threshold	Mask Flag	Feature Detected
$BTD_{11-12} < -0.5$ and $BTD_{8-11} > 0$	1	Relative strong dust region
$BTD_{11-12} < -0.5$ and $BTD_{8-11} < 0$	2	Relative weak dust region
$BTD_{11-12} > 0$ and $BTD_{8-11} > 0$	3	Ice cloud
$BTD_{11-12} > 0$ and $BTD_{8-11} < 0$	4	Low cloud or surface
$0 > BTD_{11-12} > -0.5$	5	Uncertain

Note: BTD = brightness temperature difference

circumstances, dust storms are identifiable using AOD rather reliably owing to the significant spectral disparity between dust storms and typical terrestrial surfaces. However, the identification fails if a sandstorm shares a similar spectral behavior to that of bright surfaces. They cause the reliability of the retrieved AOD to drop noticeably, which can be avoided with the use of multispectral images. Their spectral sensing region spans from visible light to thermal wavelengths. These images have been processed using various algorithms. They fall into four categories of spectral indexing, temporal anomaly detection, spatial coherence testing (physical-based algorithms), and machine learning (Li et al., 2021). Of the four types of methods, only indexing and machine learning will be covered below as they are the most commonly used and useful methods.

8.4.1 Indexing Methods

Spectral indexing is the simplest and easiest image processing to undertake as it involves only two or more spectral bands. This empirical approach is based on the physical properties and spectral signature of dust particles in multispectral bands and has found wide applications in detecting dust storms. A few indices may be combined to increase the reliability of detection. An index may be produced from individual bands or their transformation, such as T_B. The number and nature of spectral bands used to produce an index are subject to the target of detection, whether it is a dust storm or its properties, as well as the imagery used. While it is relatively easy to derive indices from multispectral bands, it is a thorny issue to set an appropriate threshold of dust and non-dust pixels on the same image (Table 8.6). Setting an appropriate threshold is the key to the success of spectral indexing. The thresholds are empirically determined from the histogram of a single spectral band or the index layer of multiple spectral bands. They can also be determined from the scatterplots of dust storm samples selected from the image (Yang et al., 2017). These thresholds are then used to extract dust pixels.

8.4.1.1 Brightness Temperature Difference*

Brightness temperature difference (BTD) is grounded on the distinct spectral response of dust in three TIR bands centered around 8.5, 11, and 12 μm. Initially developed to detect volcanic aerosols based on the BTD of 11 and 12 μm bands, it has also been used to detect mineral dust and identify dust storms due to their similarities (Ackerman, 1997). Dust particles extinct more solar radiation at 11 μm than at 12 μm, resulting in a negative BTD_{11-12} for dust, but this relativity is reversed for cloud. BTD_{11-12} has a value close to 0 for most underlying surfaces except bright surfaces like deserts. More importantly, the value of negative BTD_{11-12} rises as the dust layer thickens. The thicker

* In this section, BTD_{i-j}, BT_i, and ρ*i* stand for BTD between bands *i* and *j*, brightness temperature, and reflectance of band *i*, respectively.

the dust layer, the higher the negative BTD value. This relationship is true only for dry dust because atmospheric water vapor decreases its BTD value. Nevertheless, BTD_{11-12} is highly sensitive to dense and high-altitude dust storms, but less so to low-density, low-altitude ones. This relationship makes it possible to further differentiate dust by its thickness and altitude. Negative BTD_{11-12} can also be caused by strong surface-based temperature inversions, barren surfaces such as deserts, volcanic ashes, and clouds over the tropopause. In tropical areas, water absorbs more radiation at 12 μm than at 11 μm, causing BTD_{11-12} to be positive over areas with dense water vapor. Such characteristics can definitely mask the negative BTD_{11-12} value of dust when viewing an actual dust cloud.

Apart from the aforementioned bands, BTD has also been derived from 3.7 and 11 μm bands. Dust storms have a higher BTD between 3.7 and 11 μm than clouds and the ground surface under clear-sky conditions. BTD_{3-11} is sensitive to dust loading, making it potentially viable for AOT retrieval, from which dust is identified indirectly. Other versions of BTD make use of three bands of 8, 11, and 12 μm, and they are used to produce two differences, BTD_{11-12} and BTD_{8-11}. BTD_{11-12} alone may confuse bright surfaces such as deserts in arid or semi-arid regions with dust storms, but it is robust in identifying dust if combined with BTD_{8-11} that is sensitive to dust loading over sandy surfaces, but not to dust height (Ackerman, 1997).

BTD enables not only the detection of dust but also the separation of dust into various thickness groups (Figure 8.4). The two scatterplots of the three BTDs clearly demonstrate that ground features and thin and thick dust over them are mostly clustered, even though there is a slight overlap at the interfaces of thin and thick dust clusters. These scatterplots can guide the selection of the appropriate thresholds for detecting dust over arid and semi-arid regions using AHI data (She et al., 2018), such as $BTD_{11-8.6}$ <8, BT_{11-12} <1.2, and $BT_{3.9-11}$ >18 for detecting dust over a desert. For detecting dust over relatively dark surfaces, the thresholds change to $BT_{11-8.6}$ <5, BT_{11-12} <1.4, and $BT_{3.9-11}$ >10. The rules for detecting high-altitude dust are $BT_{11-8.6}$ <5, BT_{11-12} <0, and $BT_{3.9-11}$ >18. In spite of the stated simplicity of BTD, however, its sign and magnitude are subject to the influence of various factors, such as dust mineral composition, dust optical thickness, dust layer altitude, surface emissivity, and even the viewing zenith angle. Thus, it is rather challenging to set a universal BTD_{11-12} threshold to differentiate dust storms except for a threshold value <0 representing dust.

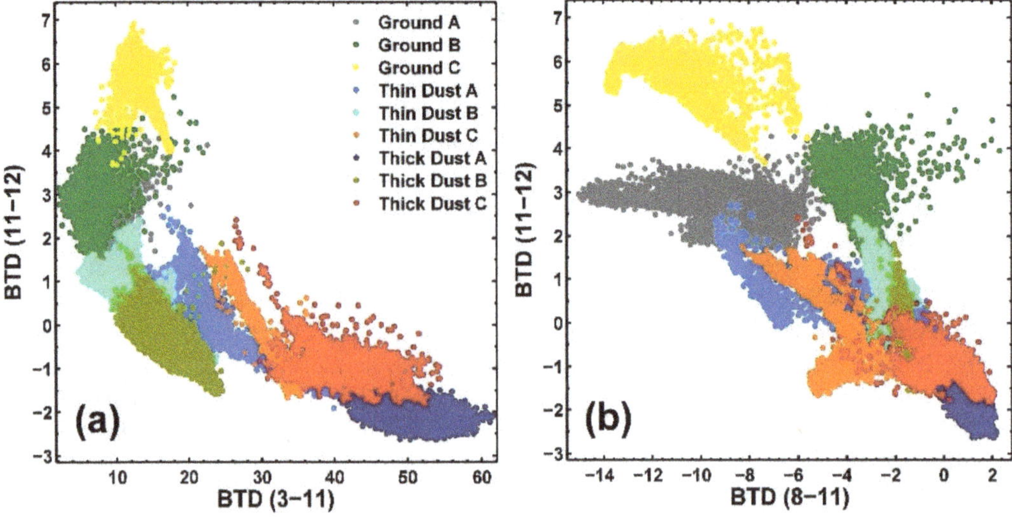

FIGURE 8.4 Scatterplots of the brightness temperature difference (BTD) for diverse ground covers: (a) BTD_{11-12} vs BTD_{3-11}, (b) BTD_{11-12} vs BTD_{8-11}. Ground A – Gobi desert, Ground B – relatively dark land surface, and Ground C – Taklimakan. (Source: She et al., 2018.)

Apart from the atmospheric conditions (e.g., water vapor, aerosols, clouds, and temperature profiles), BTD is also subject to the influence of the underlying surface. The extra layer of dust particles suspended in the atmosphere boosts the spectral reflectance of the ground cover. The use of all the aforementioned indices faces the challenge of how to account for this influence and the effect of dust aerosol properties and thickness on the BTD thresholds. If no universal relationship between the two can be established, the determined thresholds are most likely scene-specific, which considerably undermines the utility of this method. Besides, BTD is sensitive to temperature, which reduces its effectiveness in detecting dust over deserts. Over water and densely vegetated surfaces, temperature has a negligible effect on BTD. However, in desert and semi-desert regions, the temperature of the clear-sky surface is much higher than that of suspended dust, and this temperature disparity degrades the reliability of the BTD method but is useful for separating dust storms from clear land covers. Spatially, dust storms are more homogeneous than the underlying land surface. Another method of eliminating the influence is to subtract the observed BTD from a reference BTD under clear-sky conditions. Alternatively, the temperature's influence may be removed by using a dynamic reference BTD or DRBTD (Liu et al., 2013). The observed BT at 12 μm (BT_{12}) and 11 μm (BT_{11}) is expressed as a scaled (*a*) BT_{11} and BT_{12} under clear-sky conditions with an offset (*b*), and this relationship is independent of temperature, namely,

$$\mathrm{BT8.6} = a_{8.6} \cdot BT11_{obs} + b_{8.6} \tag{8.1}$$

$$\mathrm{BT812} = a_{12} \cdot BT11_{obs} + b_{12} \tag{8.2}$$

where $a_{8.6}$, $b_{8.6}$, a_{12}, and b_{12} are related to the emissivity of the underlying surface at wavelengths of 8.6 and 12 μm. They can be determined via simulating the atmospheric transfer model if the emissivity is known. A linear regression relationship eliminates the contribution of the underlying land surface, leaving only the effect of dust and aerosols if they are differentiated. The difference between the observed BTD and the reference BTD at 8.6 and 11 μm (DRBTDI8.6) and at 12 and 11 μm (DRBTDI12) is calculated as follows:

$$\begin{aligned} DRBTD8.6 &= \left(BT8.6_{obs} - BT11_{obs}\right) - \left(a_{8.6} \times BT11_{obs} + b_{8.6} - BT11_{obs}\right) \\ &= BT8.6_{obs} - \left(a_{8.6} \times BT11_{obs} + b_{8.6}\right) \end{aligned} \tag{8.3}$$

$$\begin{aligned} DRBTD12 &= \left(BT12_{obs} - BT11_{obs}\right) - \left(a_{82} \times BT11_{obs} + b_{12} - BT11_{obs}\right) \\ &= BT12_{obs} - \left(a_{12} \times BT11_{obs} + b_{12}\right) \end{aligned} \tag{8.4}$$

where $BT8.6_{obs}$, $BT11_{obs}$, and $BT12_{obs}$ = the BT observed in the 8.6, 11, and 12 μm bands, respectively.

Given that DRBTD8.6 is able to differentiate dust and clouds from the underlying surface, and DRBTD12 can separate dust from clouds, the combination of these two indices can effectively distinguish dust from both clouds and land surfaces. So the DRBTD index is defined as the sum of both indices, or

$$\mathrm{DRBTDI} = \mathrm{DRBTD8.6} + \mathrm{DRBTD12} \tag{8.5}$$

Thin cloud has a small negative DRBTDI12 value and a small positive DRBTDI value close to zero. Dust pixels have a positive DRBTDI value, and both of their DRBTD12 and DRBTD8.6 are larger than −0.5. Pixels, the DRBTD12 or DRBTD8.6 of which is lower than −0.5, are deemed to represent clouds. The DRBTD algorithm is effective at distinguishing mineral dust from clouds and land surfaces. The detected results have an agreement of 78% with ground observations (Liu et al., 2013). Higher accuracy is not possible because DRBTD8.6 and DRBTD12 have difficulty differentiating airborne dust from thin cirrus over barren surfaces. The algorithm can be used to track the transport of airborne dust during both day and nighttime, a capability extremely valuable for monitoring global dust transport and assessing diurnal cycles of dust emission. In addition, the presence of a large amount of vapor in the atmosphere also introduces a large uncertainty to the detected dust, especially in those regions where dust events are strongly associated with deep convection and high levels of atmospheric moisture (e.g., Western Sahara). Since the DRBTD method requires long-term observations as the input, it is suitable for regions where surface conditions scarcely change with time, such as deserts, evergreen forests, and oceans. If the dust storms take place in the dry season, changes in the surface cover have a subdued impact on the detected dust storms, and the surface cover does not need to be stable.

BTD has been extended to include four thermal bands of 3.7, 8.6, 11, and 12 μm to derive the thermal infrared integrated dust index (TIIDI) (Liu and Liu, 2011). It is calculated from $BTD_{3.7-11}$, $BTD_{8.6-11}$, and BTD_{12-11} (Eq. 8.6). BTD_{12-11} is mainly used to identify clouds, $BTD_{8.6-11}$ is meant to detect airborne dust and surface sand, and $BTD_{3.7-11}$ is aimed to discriminate dark surfaces and quantify the intensity of dust storms. TIIDI has a reliable performance over bright surfaces and oceans. It is competent at monitoring atmospheric dust storms over various surfaces, including oceans, vegetation (dark objects), and bright surfaces (e.g., deserts) from MODIS data. This index can distinguish mineral dust from cloud and land surface and is functional both daytime and nighttime globally.

$$TIIDI = BTD_{12-11} \times \exp\left(\frac{BTD_{8.6-11}}{a}\right) \times BTD_{3.7-11} \tag{8.6}$$

where a has a binary value of either 10 if $BTD_{8.6-11}$ is positive, or 5 otherwise.

No matter how many bands are used to derive BTD, this approach is utterly useless in detecting dust under cirrus clouds because it is impossible to discriminate the two, given that dust-free regions and cloud over dust regions have a similar BTD histogram to each other with a peak at 0.7 and 0.1°K, respectively (Huang et al., 2007). This difficulty is surmountable using microwave sensing as radar radiation can penetrate clouds. The usual BTD is hence adapted as microwave polarized index (MPI) calculated from the difference of two channels of 89 and 23.8 GHz or $\Delta T_{b89}-\Delta T_{b23.8}$. ΔT_{b89} and $\Delta T_{b23.8}$ are the differences ($\Delta T_b = T_{bv} - T_{bh}$) between the vertical and horizontal components of the BT at 89 and 23.8 GHz. They are distinct for dust-free and dust groups. Dust-free cloud regions have a small ΔT_b at both 89 and 23.8 GHz, ranging from 0 to 15°K. Both ΔT_{b89} and $\Delta T_{b23.8}$ are much higher for cloud over dust regions and pure dust regions. An MPI threshold of –7.0°K identifies about 85% of the cloud over dust and pure dust pixels.

In dust storm detection, the combined use of multi-sensor data is superior to the use of sole optical data based on infrared BTD, especially in those dust systems covered by ice clouds. For instance, MPI may be combined with BTD. This combination takes advantage of the strength of each index. Namely, MPI is good at detecting clouds over dust while infrared BTD is powerful for monitoring cloud-free dust storms. Since BT measures the thermal properties of dust, it can also be combined with those indices that measure the reflectance properties of dust such as MODIS bands to improve the reliability of dust storm detection. In this combination, MODIS reflective bands 3 and 6 differentiate sand from clouds and snow based on the threshold of $(B_6 - B_3)/(B_6 + B_3) > 0$ while $(BT_{20} - BT_{31}) > 20K$ is able to distinguish sandy dust from the underlying surface under the conditions of $BT_{31} < 290K$ and $BT_{20} > 300K$ (Su et al., 2017). These conditions may have to be modified in case the sandstorms are so strong that visibility is low and the air stagnant. The thresholds are relaxed to $(BT_{20} - BT_{24}) > 80K$ and $(BT_{20} - BT_{31}) > 30K$ when $BT_{31} < 290$ and $BT_{20} > 300K$. The joint applications of these thresholds to the respective BT images effectively extract dust pixels that are 96.3% identical to ground measurements. The causes of incorrect extraction are identified as variable dust thickness and the underlying ground cover. Specifically, the edges of the sandstorm are too thin to identify, while some of the thin sandstorms over deserts are also omitted.

In sensing dust storms, it may be desirable to detect dust by its thickness. In the domain of spectral reflectance vs BT, thin dust is spectrally clustered right next to thick dust (Figure 8.5). However, thick dust partially overlaps with thin dust, but both of them are rather distinct from thin clouds in

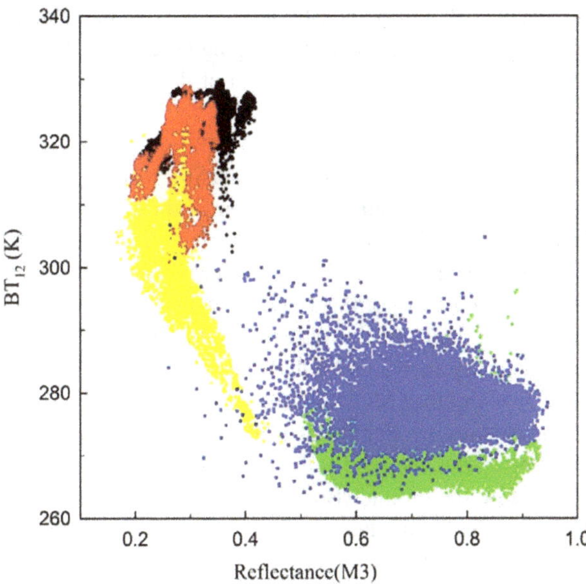

FIGURE 8.5 Scatterplot of brightness temperature (BT_{12}) vs reflectance (M3) for thick dust (black), thin dust (red), thick cloud (green), thin cloud (yellow), and ice/snow (blue). (Source: Yang et al., 2017. Used with permission.)

terms of BT. According to this scatterplot, it is feasible to differentiate dust into various thickness layers based on the BT value in multiple bands. Thick clouds and ice/snow are easily separable from the remaining three features (Yang et al., 2017). However, the slight overlap between these groups means that the accuracy of separation can never be perfect.

8.4.1.2 NDDI

Proposed by Qu et al. (2006), the normalized dust difference index (NDDI) is derived from the spectral reflectance of the visible (near 0.469 μm) and NIR (near 2.13 μm) bands, or MODIS bands 3 and 7, respectively. It is calculated as follows:

$$NDDI = \frac{B7 - B3}{B7 + B3} = \frac{\rho_{2.13} - \rho_{0.469}}{\rho_{2.13} + \rho_{0.469}} \tag{8.7}$$

This index is based on the fact that dust reflectance peaks near 2.13 μm, in sharp contrast to cloud, the peak reflectance of which occurs at 0.469 μm but is the lowest at 2.13 μm. Thus, dust pixels have a positive NDDI value, but the NDDI of cloud is negative while other covers have an NDDI value close to 0. Positive NDDI values can be further grouped to differentiate surface features from sand and dust features. For instance, an NDDI value ≤0.28 represents surface features ($0.0 < NDDI \leq 0.28$), and a value higher than 0.28 represents dust and sand particles in the Gobi region of China (Qu et al., 2006). It must be emphasized that these thresholds for separating dust storms and sand from surface features vary geographically. For instance, in Saudi Arabia, an NDDI value >0.23 represents sand and dust features while a value between 0 and 0.23 represents surface features (Butt and Mashat, 2018). The main cause for such a slight variation is surface temperature. Deserts in general are warmer than the Gobi.

Although NDDI can successfully identify and differentiate sand and dust storms from clouds in central Asia (Qu et al., 2006), this index is insensitive to dust density and altitude. Its effectiveness is drastically reduced with bright surfaces or when the dust is rather dense because both have a similar spectral reflectance pattern. In the Middle East where there are bright ground surfaces, its performance is much inferior to four other dust detection algorithms due to the difficulty of setting a clear-cut threshold to extract dust (Jafari and Malekian, 2015). Reliable detection of sand and dust storms over arid and semi-arid regions is more challenging because of the bright source areas. This challenge can be overcome through stepwise separation of dust clouds from bright underlying surfaces and water/ice clouds in three ways:

i. Separation of dust/dust cloud from water/ice cloud using NDDI to refine results from cloud masking. This improved method soundly separates sand/dust storms from bright surfaces;
ii. Separation of cloud, including dust cloud and water/ice cloud, from surface pixels using a cloud mask based on existing cloud mask data products (Xu et al., 2011). It is implemented as threshold-based, namely, $\rho 1 + \rho 2 > 0.70$ ($\rho 1$ and $\rho 2$ = the reflectance of MODIS bands 1 and 2, respectively) and $BT_{32} < 285K$ or $\rho 1 + \rho 2 > 0.63$ and $BT_{32} < 275K$. These thresholds may be determined via statistically analyzing MODIS data over bright surfaces such as deserts; and
iii. Combination of NDDI with other MODIS bands to improve the detection and detect more properties of sand and dust storms. For instance, after sand and dust storms have been detected from MODIS data using NDDI, the sand and dust suspended in the air are discriminated from the sand and dust lying on the ground (Butt and Mashat, 2018). In this combination, NDDI successfully differentiates sand and dust storms from clouds, and MODIS band 31 discriminates atmospheric and surface sand and dust based on an additional condition of $BT_{31} = 290$. All pixels meeting this condition represent airborne sand and dust.

As illustrated in Figure 8.6, NDDI itself is unable to differentiate the dust storm from other features, but the BT image can. Naturally, NDDI is used jointly with the aforementioned BTD_{11-12} index, and this combination effectively enhances the detection of dust over land and the ocean (Kunte and Aswini, 2015). This combination creates more indices to take advantage of their respective strengths, such as the brightness temperature adjusted dust index (BADI) (Yue et al., 2017), the normalized dust layer index (NDLI) (Kazi et al., 2019), the thermal infrared dust index (TDI) (Hao and Qu, 2007), and the Middle East dust index (MEDI) (Karimi et al., 2012), all involving the use of the dust-sensitive bands with a wavelength around 0.86, 2.3, 3.7, 8.6, 9.6, 11, and 12 μm (Table 8.7). Most of these indices are derived from thermal bands except the *D*-parameter, which is derived from both thermal and visible bands. It will be discussed in the next section.

TABLE 8.7

Comparison of the Major Properties of Popular Spectral Indices for Detecting Dust Storms from Multispectral Bands

Index Name	Best Use	Reference
BTD_{11-12}	Most effective under most conditions but limited over deserts	Ackerman (1997)
BTD_{3-11}	Effective daytime, potential to infer dust optical depth	Ackerman (1989)
BTD_{8-11}	Sensitive to dust loading over the desert, but insensitive to dust height	Ackerman (1997)
NDDI	Effective at separating dust pixels from cloud pixels	Qu et al. (2006)
BADI	Sensitive to both dust presence and intensity	Yue et al. (2017)
NDLI	Able to indicate dust storm phase (e.g., originating, blowing)	Kazi et al. (2019)
TDI	Sensitive to dust intensity	Hao and Qu (2007)
MEDI	Effective over deserts, esp. Middle East	Karimi et al. (2012)
D parameter	Accurate differentiation of dust from cirrus cloud	Roskovensky and Liou (2005)
IDDI	Effective over desert	Legrand et al. (2001)

Source: Simplified from Li et al. (2021). Used with permission.

Note: NDLI – normalized dust layer index; BADI – brightness temperature adjusted dust index; MEDI – middle east dust index; IDDI – infrared difference dust index

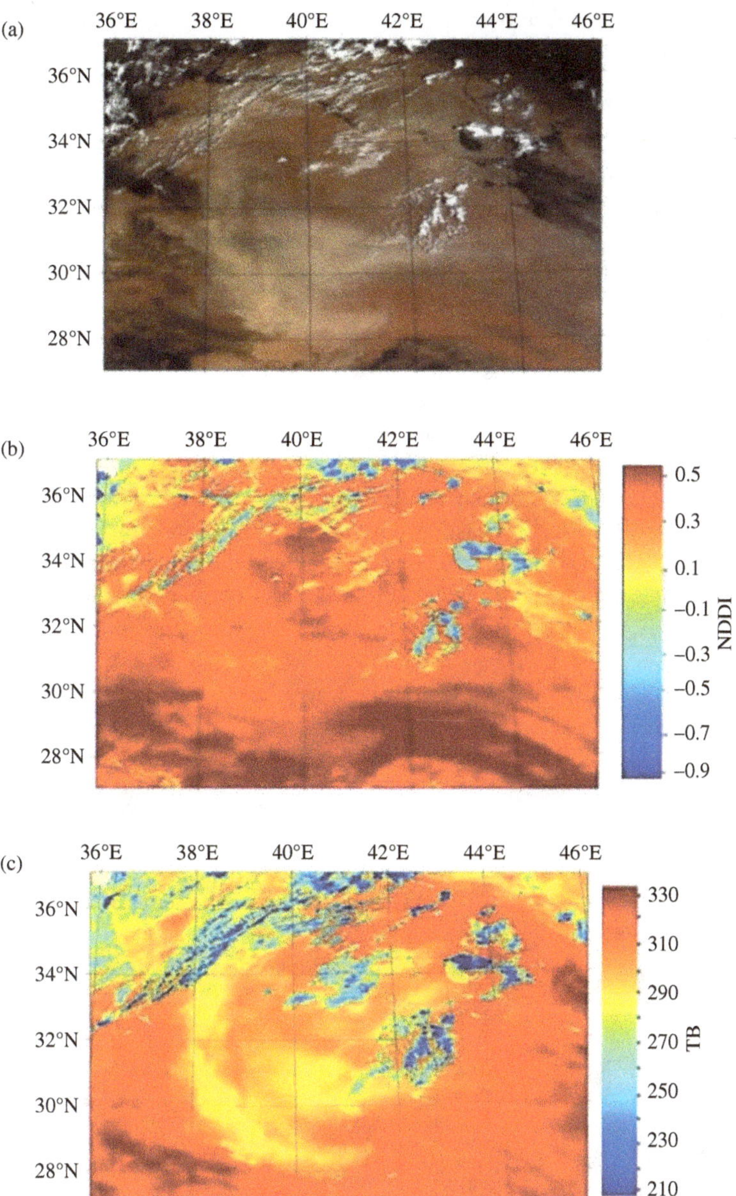

FIGURE 8.6 (a) A sand and dust storm event on 29 May 2011 over South Saudi Arabia, East Iraq, North Syria, and West Jordan captured on a true-color MODIS image, (b) the NDDI image showing sand–dust storms and clouds, and (c) the brightness temperature image from MODIS band 13. (Source: Butt and Mashat, 2018.)

8.4.1.3 GDDI and *D*-Parameter

The global dust detection index (GDDI) is actually a complex method of detecting dust storms based on the use of two indices, BTD and NDDI. It comprises a number of conditional statements to separate all possible covers present in the input image (Figure 8.7). The entire process is implemented in two parts: detection of dust over land and detection of dust over water. The terrestrial area is further divided into dark and bright surfaces after water and clouds have been stripped. The first step is to separate bright surfaces such as deserts and plains from dark surfaces such as topographic shadows in mountainous areas and vegetation based on the following conditions (Samadi et al., 2014):

$$0.01 \leq \rho_{2.13} \leq 2.5 \tag{8.8}$$

where $\rho_{2.13}$ = reflectance of the 2.13 μm MODIS band. This rule is devised to extract dark surfaces. Then bright and dark land covers and water bodies are extracted from MODIS images, and their mean and standard deviation in all MODIS bands are computed statistically, together with five other features that include clouds, bright surfaces, dust over bright surfaces, dust over dark surfaces, and dust over water. Their comparison reveals that (B7 – B4)/(B7 + B4) is good at differentiating dust from non-dust pixels over bright surfaces. In conjunction with BTD of MODIS bands 31 (11 μm) and 32 (12 μm) or BT_{31}–BT_{32}, dust is differentiated from clouds. Based on the fact that dust over bright surfaces has a higher BT than bright surfaces in band 20, but the relativity is reversed in band 31, BT_{20}–BT_{31} is used to separate dust from bright surfaces. BTD_{20-31} > 20K and >15K represent dust over bright and dark surfaces, respectively (Figure 8.7).

GDDI is able to detect dust storms over land surfaces and water bodies simultaneously from the same image. It is robust for forecasting dust storms if incorporated into a dust warning system (Samadi et al., 2014). Applied to detecting 14 sand and dust storm events over Saudi Arabia, it achieved an estimated accuracy of 76%, with the rate of correct detection being 93%, and the

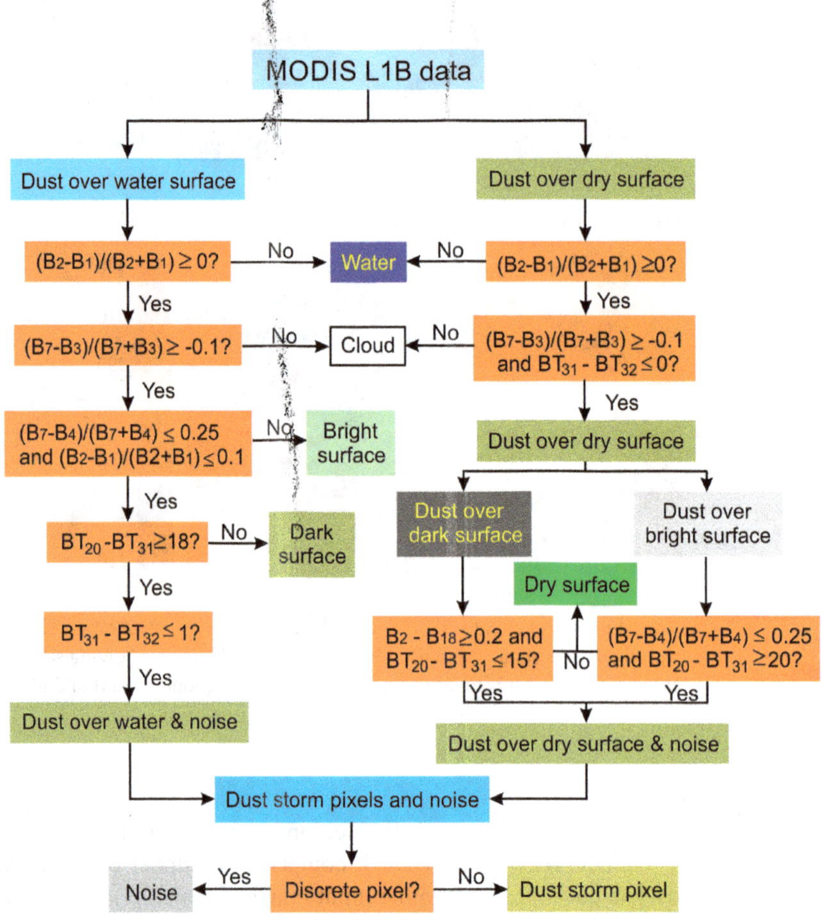

FIGURE 8.7 The procedure of implementing the GDDI automatic method of detecting dust storms from MODIS data. (Source: modified from Zandkarimi et al., 2020.)

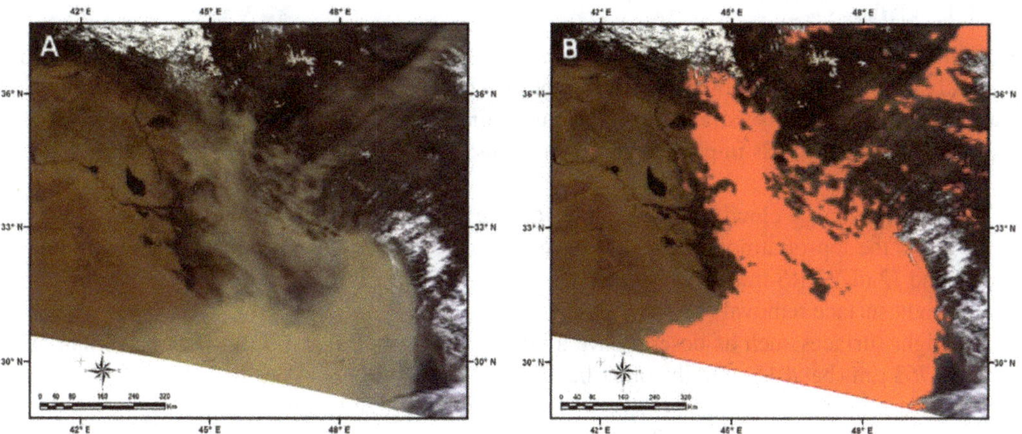

FIGURE 8.8 (A) A dust storm on 13 April 2011 shown on a MODIS RGB image, (B) the dust storm detected using the GDDI method. (Source: Samadi et al., 2014.)

portion of false detection being 28% (Alghamdi et al., 2021). If proper thresholds are set, GDDI enables the identification of various features, such as cloud cover, sand and dust mixed with clouds, low sand and dust in the atmosphere, moderate sand and dust storms, and heavy sand and dust storms (Figure 8.8).

The separation of a dust storm from other ground covers is hampered by the presence of clouds in the sky. The presence of clouds, particularly cirrus, in the same image is a major impeder to the accurate extraction of dust storms. They are situated at an altitude of around 5 km, almost the same height as dust. Besides, atmospheric dust is commonly sourced from deserts where the atmospheric column vapor is low. The emissivity difference of fine dust particles between 0.86 and 1.1 μm wavelengths causes the BTD of dust storms to resemble that of thin cirrus (Roskovensky and Liou, 2005), but the two are still separable at 0.54, 0.65, 0.86, and 1.38 μm. The ratio of visible (0.54 μm) to shortwave (0.86 μm) bands is effective at the separation. This ratio decreases as cirrus and dust increase in the atmosphere. The ratio for cirrus and dust diverges at an optical depth over 1, but BTD_{11-12} of dust and cirrus changes in the opposite direction. Based on these changes, two parameters are calculated to detect dust, the *D*-parameter, and the *P*-parameter. They are produced by combining the shortwave reflectance ratio with longwave BTD, taking advantage of both reflective and emissive properties of both features, namely,

$$\mathrm{P} = \exp\left[-\left(a_1 \cdot \rho_{1.38} / \rho_{0.65} + BTD_{8.6-11} - b_1\right)\right] \tag{8.9}$$

$$\mathrm{D} = \exp\left[-\left(a_2 \cdot \rho_{0.54} / \rho_{0.86} + BTD_{11-12} - b_2\right)\right] \tag{8.10}$$

where a_i and b_i = the scaling factor and the 8.6–11 μm BTD offset, respectively. *P* is devised to extract both dust and cirrus; *D* is intended to detect dust specifically. With the use of the *D*-parameter, cirrus clouds are separated from high-altitude dust. The *P*-parameter differentiates cirrus clouds from other low clouds, clear sky, and aerosol-filled pixels. Cirrus is detected if the *p*-value is ≥1 and the *D* value is <1. If both the *P* and *D* values are <1, neither cirrus nor dust is present. The two parameters, if used jointly, take advantage of both the infrared signal and the visible reflectance ratio. Their combination is genuinely robust enough to accurately decipher dust and cirrus from the satellite perspective.

8.4.1.4 SDDA

The simplified dust detection algorithm (SDDA) proposed by Yang et al. (2017) is developed using VIIRS bands. Figure 8.9 shows the scatterplot of VIIRS BTD_{13-15} vs BTD_{15-16}. According to this diagram, dust is readily separable into thick and thin layers quite reliably from thin clouds (Yang et al., 2017). The SDDA is implemented in four steps:

i. Identification of cloud/ice/snow based on the thresholds of $\rho_3^* \geq 0.44$ or $BT_{12} \leq 300$ (ρ_3^* = the top-of-atmosphere reflectance of band 3. BT_{12} = the brightness temperature of band 12 of VIIRS imagery).

ii. Bright surface removal
Bright surfaces such as deserts have a noticeably lower BT at 8.5 μm (band 14) than at 10.763 μm (band 15). On the other hand, the BTD between 8.5 and 10.763 μm is positive for dust due to the extinction effect of dust particles. These two types of features are separable using the criterion of $BTD_{14-15} \leq -5.0$.

iii. Dark surface removal
Dark surfaces are commonly associated with vegetation in winter and spring images. Its dark tone is similar to that of dust on a true-color image. The BT of dust decreases noticeably from 3.7 μm (band 12) to 12.013 μm (band 16), in stark contrast to vegetation, the BT of which hardly changes between these two wavelengths. The rule of $BTD_{12-16} \leq 17.0$ effectively separates them.

iv. Dust detection
Precise detection of dust requires the separation of thin clouds from dust, which is achieved based on the emissivity of dust at 11 μm that is lower than at 12 μm. Analysis of the split window BTDs of dust and thin clouds in bands 13, 15, and 16 demonstrates that they can be easily separated (Figure 8.10). The slight overlap between thin dust and clouds is resolved using two rules: $3.0 < BTD_{13-15} < 17.0$ and $BTD_{15-16} \leq -1.5$. Thick dust is detected based on three criteria of $17 \leq BTD_{13-15} < 25.0$, $BTD_{15-16} \leq -0.5$, and $BTD_{13-16} \geq 25.0$.

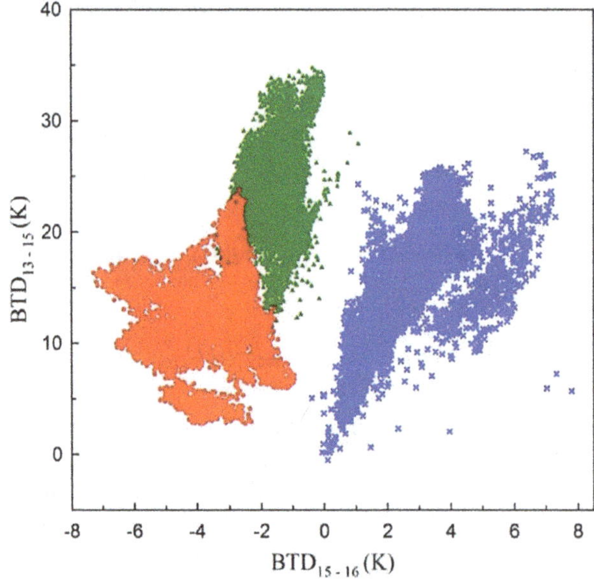

FIGURE 8.9 The BTD scatterplot of NPP VIIRS bands 13 and 15 vs the VIIRS bands 15 and 16 for thick dust (green), thin dust (red), and thin cloud (blue). (Source: Yang et al., 2017. Used with permission.)

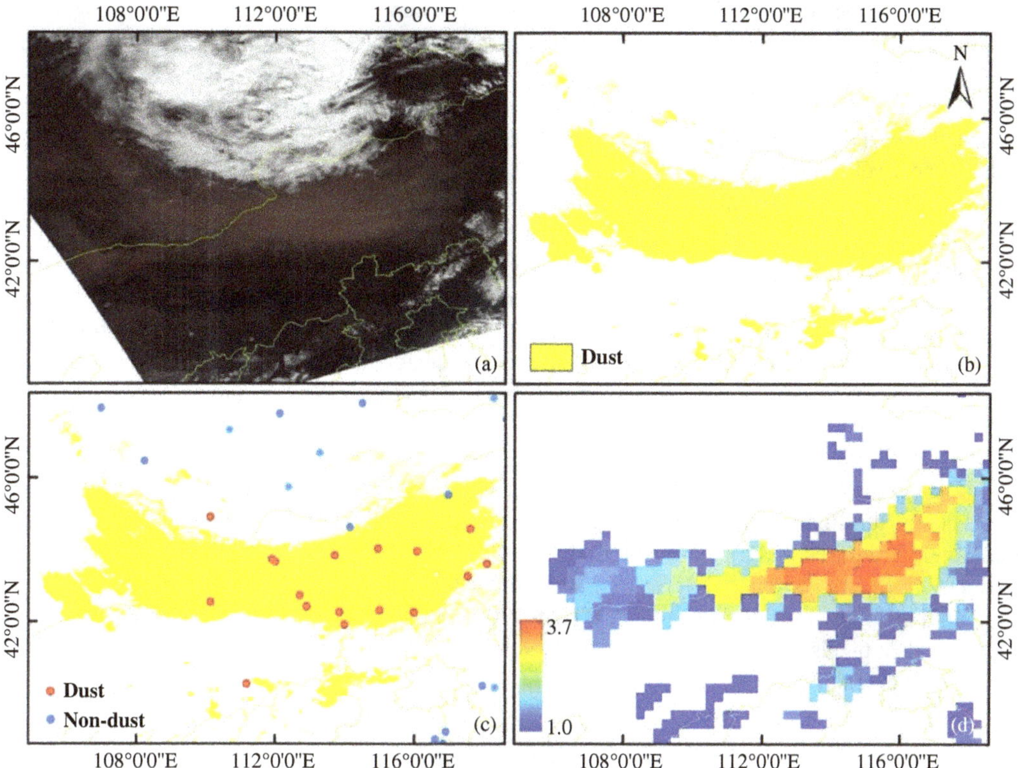

FIGURE 8.10 (a) A VIIRS RGB image of 6 May 2016 showing dust and cloud; (b) results of dust detected from (a) using a simplified dust detection algorithm; (c) in comparison with ground measurements; and (d) the ozone monitoring instrument aerosol index product. (Source: Yang et al., 2017. Used with permission.)

As shown in Figure 8.10, the SDDA algorithm effectively detects most thin and thick dust storm areas over both dark and bright surfaces. The detected results match mostly with the ground measurements except for a few mix-ups at the margin of dust storms (Figure 8.10c). The spatial pattern of the detected dust storm resembles that of the OMI aerosol index product (Figure 8.10d). The overall average dust-station correct rate is rather high at 83.10% and the error rate is rather low at only 4.88%. This relatively robust algorithm can be used to automatically monitor dust storms from VIIRS data. However, its utility with other satellite data remains to be explored because of different bandwidth designation.

8.4.1.5 TDI

The TDI is calculated from four MODIS spectral bands of 20, 30, 31, and 32. Instead of ratio, they are linearly combined after weighing to detect and monitor dust storms based on temperature (Hao and Qu, 2007). It is mathematically expressed as

$$\text{TDI} = \text{c}0 + \text{c}1BT_{3.7} + \text{c}2BT_{9.7} + \text{c}3\ BT_{11} + \text{c}4BT_{12} \tag{8.11}$$

where c_0, c_1, c_2, c_3, and c_4 = coefficients determined via regression analysis of AOD against BTs. Their typical values are −7.9370, 0.1227, 0.0260, −0.7068, and 0.5883, respectively. Thus, the highest weight is assigned to BT_{11}, followed by BT_{12} and $BT_{3.7}$. $BT_{9.7}$ receives the lowest weight because it is also sensitive to ozone. Since there is a huge difference in temperature between dust storms and water, this index is applicable to detecting dust storms over oceans. Given that surface temperature is genuinely detected from nighttime images, this index can detect dust storms during nighttime, and the detected result can serve as a proxy for dust intensity as it can also derive AOD at 0.55 μm. TDI is able to discriminate between Sahara dust storms and background aerosols. Compared with other detection algorithms, it has a better performance over water surfaces and dust sources, and TDI-detected dust accounts for approximately 93% and 90% of variations in the AOD and OMI data, respectively (Jafari and Malekian, 2015).

All of the aforementioned indices face three major drawbacks: (i) most of them are competent in detecting dust storms over either water bodies or lands, but not both simultaneously; (ii) the thresholds are not fixed and universal. Instead, they vary with seasonality, environmental complexity, and heterogeneity; and (iii) they are unable to discriminate dust storms from other features such as clouds, water, and land surface (Samadi et al., 2014). Thus, there is a need to develop a global index that is able to discriminate dust from other features and detect dust storms in all seasons without the need to specify a threshold. The answer is the improved dust identification index (IDII).

8.4.1.6 IDII

The IDII is based on MODIS bands 7, 8, 20, 31, and 32, of which BT_{31} and BT_{32} are used to mask cloud pixels. It is highly similar to GDDI in that dust pixels are detected separately over land surface and water. It requires the calculation of modified NDWI (MNDWI) that is used to differentiate land (MNDWI < 0) from water (MNDWI > 0). Then the detection relies on BT in MODIS bands 32 and 31 (Zandkarimi et al., 2020). Over land, $BT_{32} > 290$ suggests cloud pixels, otherwise $BTD_{20-31} \leq 17$ indicates dust pixels. Over water, $BT_{32} > BT_{31}$ suggests cloud pixels. Otherwise $\log(R7/R8) > -0.3$ means dust pixels (Figure 8.11).

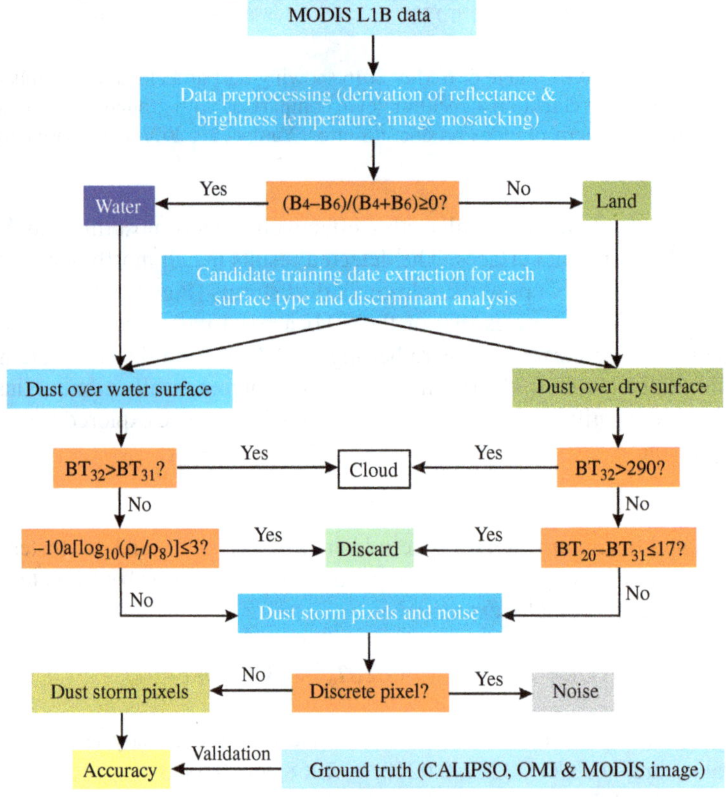

FIGURE 8.11 The procedure of implementing IDII-based dust detection. (Source: Modified from Zandkarimi et al., 2020.)

This versatile index overcomes the limitations of other common dust detection algorithms, particularly in detecting dust storms over different surfaces (e.g., both land and water), in different seasons, and in resolving the spectral similarity between dust and other features such as clouds, land, and water at a higher accuracy than other approaches, much better than GDDI in all seasons, achieving an accuracy of 82%. The main cause of inaccuracy is very thin clouds over water and at cloud edges.

8.4.1.7 EDI and EDII

The enhanced dust index (EDI) is based on the apparent reflectance in two spectral bands of 0.65 μm (visible light) and 1.63 μm (SWIR). It is derived by combining the apparent top-of-atmospheric reflectance (ρ) of these two bands with BT derived from the MIR and TIR bands, and the retrieved AOD, or

$$EDI = \ln\left[a \times \frac{\rho_{SWIR} + \rho_{vis}}{\rho_{SWIR} - \rho_{vis}} + b \times \frac{BT_{MIR} - BT_{TIR1}}{BT_{MIR} + BT_{TIR1}} + c \times AOD\right] \tag{8.12}$$

where a, b, and c = coefficients for scaling the three components to an equivalent range of 0–1. This is virtually a method of standardization or weighting. They are commonly set as $a = c = 0.1$, and $b = 10$. An EDI value >1 means the presence of dust, and a value <1 means non-dust areas.

EDI is implemented in three steps: (i) detection of cloud and snow pixels using the criterion of $\rho_{VIS} - \rho_{SWIR} < 0$; (ii) detection of dust using EDI; and (iii) homogeneity test within a window for EDI. There exists a positive correlation between EDI and dust intensity, suggesting that EDI can potentially detect dust events and extract dust extent. A good linear correlation ($R^2 = 0.78, p < 0.01$) exists between EDI and in situ measured visibility. Furthermore, EDI is a rather competent method in extracting dust intensity at rather high accuracy under the assumption that the dust is spatially homogeneous (Di et al., 2016). This index works with INSAT images only. Its applicability with other satellite images remains unknown except for the AHI data due to variability in the spectral band wavelengths.

Adapted to the AHI data, EDI has been turned into a new index called enhanced dust intensity index (EDII) by She et al. (2018). It is based on a dynamic threshold of BTD derived from three BTDs of four bands: BTD_{11-12}, BTD_{3-11}, and BTD_{11-8} of AHI data. This index can identify dust quickly by dynamically adapting the threshold scheme to different land surfaces to achieve accurate dust identification over different surface conditions. It can also serve as a semi-quantitative measure of dust intensity (a correlation coefficient of 0.81 with visibility data). It is computed as follows:

$$EDII = \ln\left[a \times \frac{\rho_{1.6} - \rho_{0.47}}{\rho_{1.6} + \rho_{0.47}} + b \times \frac{BT_{3.9} - BT_{11}}{BT_{11}} + c \times AOD\right] \tag{8.13}$$

where a, b, and c have a value of 1, 10, and 0.1, respectively. This equation is grounded on three facts: (i) the minimum and maximum reflectance of dust occurs at 0.47 and 1.61 μm, respectively. Thus, the reflectance difference between these two AHI bands is able to indicate the dust intensity which is reflected by the term $(\rho_{1.6} - \rho_{0.47})/(\rho_{1.6} + \rho_{0.47})$ similar to NDDI; (ii) dust strongly increases the AHI BT at 3.9 μm but reduces the AHI BT at 11 μm. The difference between these two bands is positively related to dust intensity, and (iii) AOD is also positively related to dust intensity.

This index requires AOD to compute, but the computation is simple. EDII is accurate, achieving a detection agreement of 84% with ground measurements. It can be implemented quickly and is ideal for monitoring dust in real-time using geostationary satellite data (She et al., 2018).

8.4.1.8 DSSI

The dust spectral similarity index (DSSI) is derived from the accumulative BTD between 16 AIRS (atmospheric infrared sounder) channels in the NIR spectrum of 0.80–1.25 μm. Its derivation requires hyperspectral infrared bands. Virtually, this index is an extension of the BTD algorithm to hyperspectral data, calculated as follows:

$$DSSI = \frac{\sum_{i=1}^{k} \sum_{j=i+1}^{k} N \cdot M_{i,j}}{0.5k(k-1)} \times \frac{\sum_{i=1}^{k} \sum_{j=i+1}^{k} P \cdot M_{i,j}}{0.5k(k-1)} \tag{8.14}$$

where $k = 8$, i and j = the row and column number, respectively; $0.5k(k - 1)$ = the total number of calculated BTD in each matrix; $M_{i,j}$ has a binary value of 0 (if $M_{i,j} \leq 0$) or 1 (if $M_{i,j} > 0$). $N(M_{i,j})$ and $P(M_{i,j})$ represent the matric elements, calculated as

$$N(M) = \begin{vmatrix} 0 & BTD_{526}^{572} & BTD_{526}^{663} & BTD_{526}^{752} & BTD_{526}^{830} & BTD_{526}^{879} & BTD_{526}^{925} & BTD_{526}^{973} \\ 0 & 0 & BTD_{572}^{663} & BTD_{572}^{752} & BTD_{572}^{830} & BTD_{572}^{879} & BTD_{572}^{925} & BTD_{572}^{973} \\ 0 & 0 & 0 & BTD_{663}^{752} & BTD_{669}^{830} & BTD_{663}^{879} & BTD_{663}^{925} & BTD_{663}^{973} \\ 0 & 0 & 0 & 0 & BTD_{752}^{830} & BTD_{752}^{879} & BTD_{752}^{925} & BTD_{752}^{973} \\ 0 & 0 & 0 & 0 & 0 & BTD_{830}^{879} & BTD_{830}^{925} & BTD_{820}^{973} \\ 0 & 0 & 0 & 0 & 0 & 0 & BTD_{879}^{925} & BTD_{879}^{973} \\ 0 & 0 & 0 & 0 & 0 & 0 & 0 & BTD_{925}^{973} \\ 0 & 0 & 0 & 0 & 0 & 0 & 0 & 0 \end{vmatrix} \tag{8.15}$$

$$P(M) = \begin{vmatrix} 0 & BTD_{1292}^{1254} & BTD_{1292}^{1239} & BTD_{1292}^{1222} & BTD_{1292}^{1201} & BTD_{1292}^{1186} & BTD_{1292}^{1171} & BTD_{1292}^{1152} \\ 0 & 0 & BTD_{1254}^{1239} & BTD_{1254}^{1222} & BTD_{1254}^{1201} & BTD_{1254}^{1186} & BTD_{1254}^{1171} & BTD_{1254}^{1152} \\ 0 & 0 & 0 & BTD_{1239}^{1222} & BTD_{1239}^{1201} & BTD_{1239}^{1186} & BTD_{1239}^{1171} & BTD_{1239}^{1152} \\ 0 & 0 & 0 & 0 & BTD_{1222}^{1201} & BTD_{1222}^{1186} & BTD_{1222}^{1171} & BTD_{1222}^{1152} \\ 0 & 0 & 0 & 0 & 0 & BTD_{1201}^{1186} & BTD_{1201}^{1171} & BTD_{1201}^{1152} \\ 0 & 0 & 0 & 0 & 0 & 0 & BTD_{1186}^{1171} & BTD_{1186}^{1152} \\ 0 & 0 & 0 & 0 & 0 & 0 & 0 & BTD_{1171}^{1152} \\ 0 & 0 & 0 & 0 & 0 & 0 & 0 & 0 \end{vmatrix} \tag{8.16}$$

The upper half of the matrix elements should all have a positive value in the presence of dust. The DSSI algorithm can effectively differentiate dust from clouds and clear-sky surfaces (Xu et al., 2015). Underlying surfaces covered with dust have a DSSI value close to 1.0. Clear-sky surfaces or clouds (ice and water) have a value lower than that of dust (Figure 8.12).

All of the aforementioned physical-based approaches directly link dust storms with spectral characteristics. As such, they are empirical, specific to the scenes selected for the area under study. They all suffer from certain drawbacks such as high variability of thresholds subject to the influence of the underlying surfaces, and even seasonality. Besides, weak or intense dust storms likely have a threshold different from the threshold values of regular dust storms. Not surprisingly, the published dust/no-dust thresholds in some algorithms fail to produce satisfactory results in a new study area (Jafari and Malekian, 2015). The threshold-based algorithms of NDDI and TDI by Ackerman (1997), Roskovensky and Liou (2005), and Miller (2003) work the best with dense dust, but less so in cloudy regions and over water and bright desert surfaces. They also confuse thick clouds with dust storms except for the Roskovensky and Liou (2005) algorithm, even though the scaling factor and offset values have to be re-defined. In addition, the dust thresholds have to be re-defined as shown in Table 8.8. The scene-specific thresholds are quite different from the recommended thresholds on

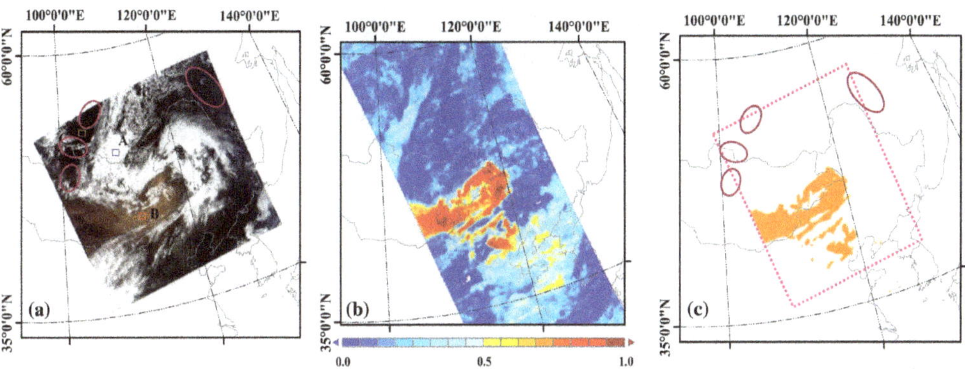

FIGURE 8.12 A dust storm on 11 May 2011 and its detection results. (a) MODIS RGB composite image, (b) DSSI, and (c) dust detection results based on DSSI. (Source: Xu et al., 2015.)

two occasions. Thus, new thresholds have to be established from scratch for every new image, and even an event-specific threshold is necessary to produce satisfactory detections.

No matter how the thresholds are obtained, the use of fixed thresholds is easily affected by mixed pixels and atmospheric conditions, resulting in poor detections, especially in the presence of clouds. Besides, the threshold value varies with the type of land and is problematic in discriminating clouds, water, and land. These limitations can be overcome using three methods. The first is dynamic thresholds (Sun et al., 2019). Their value is not pre-defined but estimated from the satellite images. This estimation requires the apparent reflectance on MODIS bands 1, 3, 6, and 7 that can be simulated using a model, such as the Second Simulation of the Satellite Signal in the Solar Spectrum (6S) atmospheric radiation transfer equation. It takes into account surface reflectance, the atmospheric model, the aerosol model, AOD, observation geometry, and solar zenith angle. The input to the model is the MODIS land surface reflectance data product (eight-day gridded MOD09A1). The simulated reflectance is compared with that recorded on the MODIS bands. If a pixel's simulated apparent reflectance ρ' is larger than its observed counterpart ρ in spectral band i, it represents a dust pixel, namely,

$$R_i = \rho_i - \rho_i' \ (\text{i} = \text{blue or B, red or R, SWIR1} - \text{band 6, SWIR2} - \text{band 7}) \tag{8.17}$$

$$\text{R} = R_B \cup R_R \cup R_{SWIR1} \cup R_{SWIR2} \tag{8.18}$$

TABLE 8.8

Comparison of Best Thresholds with the Author-Recommended Thresholds in Five Dust Detection Algorithms

Algorithm	Recommended Thresholds	2009 Event	2010 Event	Accuracy[a] (R^2)
Ackerman (1997)	Dust < 0	Dust < –0.5	Dust < 0	0.85
Miller (2003)	1.3 < Dust < 2.27	Dust > –0.2	Dust > 0.5	0.84
NDDI	Dust > 0.28	–0.5 < Dust < 0.3	Dust > –0.05	0.07
Roskovensky and Liou (2005)	Dust > 1	Dust > 0.12 (a = 0.024, b = 2.57)	Dust > 9 (a = 0.8, b = 0.5)	0.74
TDI	Dust > 1	Dust > 2.3	Dust > 3	0.93

Source: Jafari and Malekian (2015).

[a] In comparison with AOD

where R_i = the detection outcome based on band i; R = the detection results based on all four bands. The equations for estimating the simulated apparent reflectance in the four bands are given below:

$$\rho'_B = 0.664\rho_B + 0.056cos\alpha \cdot cos\beta + 0.099 \tag{8.19}$$

$$\rho'_R = 0.746\rho_R + 0.027cos\alpha \cdot cos\beta + 0.049 \tag{8.20}$$

$$\rho'_{SWIR1} = 0.899\rho_{SWIR1} + 0.008cos\alpha \cdot cos\beta + 0.015 \tag{8.21}$$

$$\rho'_{SWIR2} = 0.874\rho_{SWIR2} + 0.005cos\alpha \cdot cos\beta + 0.009 \tag{8.22}$$

where α and β = the solar and satellite zenith angle, respectively. Eqs. 8.19–8.22 are derived based on the pre-determined settings for the aerosol type, AOD, and chosen atmospheric model. The coefficients will vary with different settings adopted for another study area. In this sense, the threshold is dynamic. DSSI works well except for its inability to detect dust properties (e.g., thin or thick dust). Besides, it requires long-term land surface reflectance data that are assumed to be stable over the study period (Sun et al., 2019). Its performance is also affected by clouds, and the underlying surfaces, as is common with other previously introduced indices.

The second method is image classification using statistical methods such as the maximum likelihood classifier. After clouds have been masked out from a BTD composite of MODIS level 1B emissive bands, it is classified using the maximum likelihood method (Nisa et al., 2019). A dust spatial distribution map is then produced after the classified image is majority filtered. This method works the best only when the dust is thick and has a uniform thickness. Otherwise, the same dust storm will be mapped into multiple classes, not all of which will be genuine due to confusion with other similar covers. The last method of mapping dust storms is to use machine learning methods such as probabilistic neural networks. This topic is so complex that it will be presented in a separate section below.

8.4.2 Machine Learning Classification

Monitoring of dust storms over sandy surfaces from space-borne imagery has not met with complete success because of the two share similar spectral characteristics. The separation of the two can be made more reliable by taking into consideration more variables than mere spectral bands and their transformations, which prompts the use of machine learning algorithms. Rather than relying on the thresholds of dust and non-dust features, machine learning algorithms classify dust storms from the input image based on training samples of the dust storm, land, cloud, water, and vegetation manually extracted from true-color composites of multispectral bands, and various types of BTD. This new paradigm of image classification can be implemented as statistical-based algorithms such as weights of evidence (WOE), frequency ratio (FR), RF, SVM, and ANN. ANN-based detection of dust storms requires a large dataset generated from BTD (El-Ossta et al., 2013). The input image is classified into a binary map of dust and no-dust pixels in three steps: (i) selection of training samples for all the considered types of features; (ii) division of the selected samples into two parts: training (e.g., 70%) and test (e.g., 30%). The former is used to train the machine learning classifier, and the latter is used to assess the classification accuracy; and (iii) image classification, using one of the machine learning methods, such as feed-forward BP neural network. The ANN approach (accuracy = 94.6%) outperforms the TIR integrated dust index (accuracy = 73.5%) in detecting dust storms over the Sahara. The ANN trained using data from the Sahara desert achieved an accuracy of 88% when applied to detecting Gobi dust storms from MODIS imagery. Of the 96 dust storms classified as dusty images on NASA Earth Observatory, it successfully detected 90 of them, with the 6 missed ones all being rather weak (El-Ossta et al., 2013). Thus, this method is more universal than those index-based ones presented in the preceding section.

Known as ensemble classification, the RF classifier is a set of tree classifiers. A tree is established using training samples. They can be prior knowledge and experience of the study area, gained from existing maps or color composite images. Once an RF has been properly trained, it can be used to classify other images. Two parameters must be specified to run RF classification: N – number of trees to grow, and m – the number of attributes that can be binary dust and non-dust, or multiple categories, such as cloud, dust, soil, snow, vegetation, and water. Similar to all tree-based algorithms, RF is able to deal with large datasets using automatic variable selection. RF has demonstrated a competent capability of detecting dust over both water and land simultaneously (Souri and Vajedian, 2015). It outperforms physical-based approaches because they make use of empirically determined thresholds that are not always accurate. RF-generated dust is in qualitative agreement with the location and extent of dust observed in true-color satellite images. Quantitatively, the dust classes generated for eight dust outbreaks in the Middle East have a low probability of false detection (7%) and a low probability of missing detection (6%). This classification method seems to be more promising than the indexing methods.

8.5 IDENTIFICATION OF DUST SOURCES

It is significant to identify the sources of sand and dust storms as such information helps not only monitor and understand the processes of dust storms but also mitigate their adverse effects. This identification is ideally accomplished using space-borne images because of the vast extent of dust storms. They cannot be monitored via ground observations or from airborne data. Satellite imagery provides a synoptic view of dust distribution from its origin. Their bird's eye view of surface features is valuable in detecting the spatial pattern of a dust storm if they are captured at a sufficiently fine temporal resolution. Routes of dust transport are also identifiable from multi-temporal images. The use of satellite images such as MODIS in identifying local sand and dust storm sources is plagued by the fact that their acquisition rarely coincides with the initialization of a sandstorm, so it is almost impossible to study the origin of dust storms. Therefore, attention focuses only on those areas with a high potential for soil erosion. They are the likely dust sources because sand and dust storms are commonly associated with surfaces that are easily erodible or sandy.

This potential is commonly judged from vegetation status (structure, spatial pattern, height, and cover), soil moisture, surface roughness, and texture. In addition to three soil factors of moisture, roughness, and gravel content, one vegetation factor and one geological factor have also been considered in identifying the potential sources of sand and dust storms (Rayegani et al., 2020). Of these factors, soil moisture, gravel content, and soil roughness are critical to assessing soil aeolian erodibility. Wind erosion sensitivity is evaluated from soil moisture, land cover, and vegetation cover that exerts a considerable effect on wind erosion rate and emissions of dust. These variables are considered because they can be easily mapped using remote sensing. For instance, difference vegetation index (DVI) (bNIR–RED) is grouped into a binary map of erodible and non-erodible based on the non-erodible thresholds determined with the assistance of ground-simulated results. The potential dust source areas are mapped by integrating the erodibility and wind erosion sensitivity maps with geology and soil roughness layers through multi-criteria evaluation. Wind erosion sensitivity is assessed from rock types and constituent particles.

Apart from multi-criteria evaluation, dust sources have been identified from MODIS images using machine learning methods based on BTD_{31-32}, BTD_{29-31}, NDDI, and D-parameter (Boroughani et al., 2020). After 65 dust source points are selected, they are split into two parts for training and validating three statistical-based machine learning algorithms of WOE, FR, and RF. These algorithms are able to produce dust source susceptibility maps from the simultaneous consideration of seven conditioning variables, including land use, lithology, slope, soil, geomorphology, NDVI, and distance from the river. Their importance to the occurrence of dust storms is shown in Table 8.9. Of these variables, slope, geomorphology, and land use are more important than the remaining four in influencing dust sources area in Khorasan Razavi Province of Iran. The produced

TABLE 8.9

Importance of Effective Variables in Predicting Dust Storm Occurrence Estimated Using Frequency Ratio (FR) and Weight-of-Evidence (WOE)

Variable	FR	Variable	WOE
Geomorphology	0.993	Slope	0.421
Land use	0.914	Geomorphology	0.301
Slope	0.860	Land use	0.299
NDVI	0.758	Soil	0.296
Lithology	0.755	Lithology	0.252
Soil	0.744	NDVI	0.059
Distance from river	0.458	Distance from river	0.001

Source: Boroughani et al. (2020). Used with permission.

dust source susceptibility map may be discretized into a few categories to illustrate its spatial pattern. This map can also be overlaid with dust source data to validate its accuracy (Figure 8.13). The three algorithms have different performances. RF mistakenly predicted dust sources where they were absent (type I error) in 15 cases and predicted the absence of dust sources in places where they were actually present (type II error) on 16 occasions. On the other hand, in 31 and 30 cases the model correctly predicted the absence and the presence of dust sources, respectively. FR achieved an accuracy of 91% as judged by the AUC success rate (training). The accuracy drops slightly to 88% for the predictive rate (validation) (Boroughani et al., 2020). It is still higher than 82.5% of

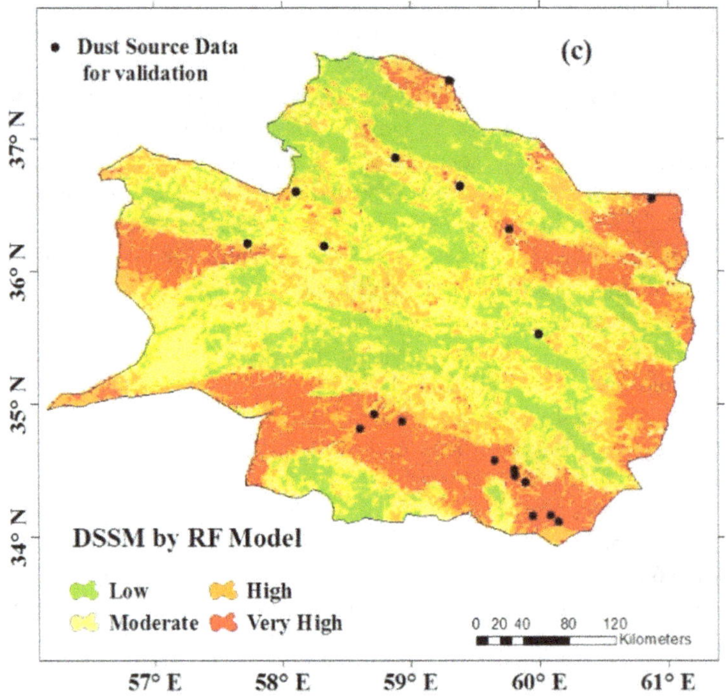

FIGURE 8.13 Dust source susceptibility at four levels produced from the seven variables shown in Table 8.9 using RF. (Source: Boroughani et al. (2020). Used with permission.)

WOE and 81.8% of FR. Higher accuracies are not possible because the analysis of dust emergence and transport does not take into account meteorological factors, such as airflow, surface temperature, and atmospheric pressure.

In addition to machine learning analysis, simulation also offers a promising solution to identify dust sources. Through modeling, it is possible to identify the transport direction and the locations of the dust sources and assess the synoptic situations during sandstorm events. The origin and pathways of dust storms are commonly predicted from physio-meteorological models, such as the Hybrid Single-Particle Lagrangian Integrated Trajectory (HYSPLIT) airflow simulation model. HYSPLIT backward or forward trajectory analysis allows the investigation of the source and patterns of sand and dust distribution. This model can trace satellite-observed dust particles back to their source area and ensure that airflow trajectory and erodible lands are in physical contact. In order to determine the transfer trajectories of sand and dust storms, the local events must be separated from regional ones, which can be achieved with the assistance of MODIS data. Wind erosion risk is mapped from the contact areas between airflow and the terrain, which can be identified via the HYSPLIT airflow simulation model. Back trajectory analysis ascertains that the dust storm converging over Kyrgyzstan and Tajikistan actually originated from Southwest Kazakhstan and Eastern Europe (Nisa et al., 2019).

The identification of dust sources requires a wind erosion risk map that can be produced based on the contact areas between airflow and the ground. This map may be combined with a wind erosion sensitivity map via fuzzy multi-criteria assessment after linear weighting. Probable sand and dust storm sources are identified from the wind erosion risk map and the wind erosion sensitivity map after they have been weighted and linearly combined. The identified sources are validated against the trend of vegetation cover, soil moisture, and LST over the specified period (e.g., use of time-series synoptic data), the identified source areas of sand and dust storms are highly accurate. In particular, LST as a climatic parameter is crucial to validating the identified sand and dust storm sources. Areas with frequent sand and dust storms experienced a significant decrease in LST. It must be emphasized that the multilayer analytical approach merely identifies areas that have the potential to be eroded. Whether or not the dust is blown away in a storm remains unknown.

8.6 MONITORING AND RISK ASSESSMENT

Monitoring of dust storms is virtually multi-temporal sensing of the same geographic area where they have emerged and tracking of their spatial change and movement qualitatively over their life span. Comparison of multi-temporal dust storm maps produced from the images with a separation of 12 hours or shorter can reveal the whole process of formation, development, transport, and dissipation of a dust storm, making it feasible to monitor large-scale sandstorms continuously and dynamically (Su et al., 2017). The transport route and the area influenced by a dust storm can all be revealed via analysis of time-series images, including the origin and source area of dust. In order to maintain consistency, usually, the monitoring should be based on multi-temporal images acquired from the same sensor, and these images are processed identically using one of the detection methods described previously. The use of such data also ensures that the sandstorm is monitored at the same time each day. Monitoring of dust storms relies on backward tracking, surface wind, surface pressure, wind speed and direction, and geo-potential height for different pressure levels that cannot be acquired from remote sensing data alone or using remote sensing methods. No quantitative information on any of these dust storm properties is available at present. Thus, the source, dust transport pathway, and dissipation of a storm have to be deduced intuitively (Kunte and Aswini, 2015). The visually deduced results are verified via examining the weather conditions at the time of sensing, such as air temperature and pressure difference. It is, however, feasible to predict the possible pathway of a sandstorm through simulation and investigate the impact of a super dust storm on the ocean by comparing SST and Chla before, during, and after the storm (Kunte and Aswini, 2015).

Similar to dust susceptibility, dust risk is also assessed from the main contributors and their influence mechanisms using a range of methods, including the AHP, weighted comprehensive method, and maximum likelihood classification. These factors may include severe dust storm frequency in the past, and meteorological factors such as the number of monthly strong wind days, monthly mean maximum temperature, monthly average relative humidity, monthly precipitation, and the underlying surface factors, such as monthly soil moisture and monthly vegetation coverage (Liu et al., 2012). Their PCC with the dust storm frequency is used to weigh them. The most important factors are identified as vegetation (0.337), the number of strong wind days (0.299), soil moisture (0.239), and mean maximum temperature (0.125). These important factors are then used to derive a composite severe dust storm index (I_{SDS}) after importance-based weighing. Then the transcendental probability curves of I_{SDS} or the hazard risk curves are drawn to assess the risk of severe dust storms at a ground weather station. Initially, a number of distribution curves are drawn (e.g., normal, Weibull, beta, gamma, and logistic), and the parameters of each distribution are calculated using the maximum likelihood estimation method. These parameters are used to determine the probability density distribution model of I_{SDS}, and it is compared with all the potential cumulative distributions to identify the closest match. The cumulative distribution function (CDF) of the best match is then used to determine the transcendental probability function (curve) (1–CDF) of the same station. The risk of the occurrence of severe dust storms is analyzed under scenarios of different return periods. There is no way of verifying whether or not the risk has been estimated at a reasonable accuracy due to the absence of independent ground truth data.

REFERENCES

Ackerman, S.A. 1989. Using the radiative temperature difference at 3.7 and 11 μm to track dust outbreaks. *Remote Sens. Environ.* 27(2): 129–133. DOI: 10.1016/0034-4257(89)90012-6

Ackerman, S.A. 1997. Remote sensing aerosols using satellite infrared observations. *J. Geophys. Res.* 102(D14): 17069–17079. DOI: 10.1029/96JD03066

Alghamdi, E.M., M. Assiri, and M.J. Butt. 2021. Application of global dust detection index (GDDI) for sand and dust storm monitoring over Kingdom of Saudi Arabia. DOI: 10.21203/RS.3.RS-161978/V1

Boroughani, M., S. Pourhashemi, H. Hashemi, M. Salehi, A. Amirahmadi, M.A.Z. Asadi, and R. Berndtsson. 2020. Application of remote sensing techniques and machine learning algorithms in dust source detection and dust source susceptibility mapping. *Ecol. Inform.* 56. DOI: 10.1016/j.ecoinf.2020.101059

Butt, M.J., and A.S. Mashat. 2018. MODIS satellite data evaluation for sand and dust storm monitoring in Saudi Arabia. *Int. J. Remote Sens.* 39(23): 8627–8645. DOI: 10.1080/01431161.2018.1488293

Di, A., Y. Xue, X. Yang, J. Leys, J. Guang, L. Mei, and Y. Che. 2016. Dust aerosol optical depth retrieval and dust storm detection for Xinjiang region using Indian National Satellite Observations. *Remote Sens.* 8: 702. DOI: 10.3390/rs8090702

El-Ossta, E., R. Qahwaji, and S.S. Ipson. 2013. Detection of dust storms using MODIS reflective and emissive bands. *IEEE J. Sel. Top. Appl. Earth Obs. Remote Sens.* 6(6): 2480–2485. DOI: 10.1109/JSTARS.2013.2248131

Hao, X., and J.J. Qu. 2007. Saharan dust storm detection using moderate resolution imaging spectroradiometer thermal infrared bands. *J. Appl. Remote Sens.* 1: 013510. DOI: 10.1117/1.2740039

Huang, J., J. Ge, and F. Weng. 2007. Detection of Asia dust storms using multisensor satellite measurements. *Remote Sens. Environ.* 110: 186–191.

Jafari, R., and M. Malekian. 2015. Comparison and evaluation of dust detection algorithms using MODIS Aqua/Terra Level 1B data and MODIS/OMI dust products in the Middle East. *Int. J. Remote Sens.* 36: 597–617.

Karimi, N., A. Moridnejad, S. Golian, J.M. Vali Samani, D. Karimi, and S. Javadi. 2012. Comparison of dust source identification techniques over land in the Middle East region using MODIS data. *Can. J. Remote Sens.* 38(5): 586–599. DOI: 10.5589/m12-048

Kazi, A., I. Nagatani, K. Kawano, and J.-I. Kudoh. 2019. Development of a new dust index NDLI for Asian dust extraction system based on Aqua MODIS data and monitoring of trans-boundary Asian dust events in Japan. *Int. J. Remote Sens.* 40(3): 1030–1047. DOI: 10.1080/01431161.2018.1524170

Kunte, P.D., and M.A. Aswini. 2015. Detection and monitoring of super sandstorm and its impacts on Arabian Sea – Remote sensing approach. *Atmos. Res.* 160: 109–125. DOI: 10.1016/j.atmosres.2015.03.003

Legrand, M., A. Plana-Fattori, and C. N'doumé. 2001. Satellite detection of dust using the IR imagery of Meteosat: 1. Infrared difference dust index. *J. Geophys. Res.* 106(D16): 18251–18274. DOI: 10.1029/2000JD900749

Li, J., M.S. Wong, K.H. Lee, J. Nichol, and P.W. Chan. 2021. Review of dust storm detection algorithms for multispectral satellite sensors. *Atmos. Res.* 250. DOI: 10.1016/j.atmosres.2020.105398

Liu, Y., and R. Liu. 2011. A thermal index from MODIS data for dust detection. *IEEE Int. Geosci. Remote Sens. Symp.* 3783–3786. DOI: 10.1109/IGARSS.2011.6050054

Liu, Y., R. Liu, and X. Cheng. 2013. Dust detection over desert surfaces with thermal infrared bands using dynamic reference brightness temperature differences. *J. Geophys. Res. Atmos.* 118: 8566–8584. DOI: 10.1002/jgrd.50647

Liu, X., N. Li, W. Xie, J. Wu, P. Zhang, and Z. Ji. 2012. The return periods and risk assessment of severe dust storms in Inner Mongolia with consideration of the main contributing factors. *Environ. Monit. Assess.* 184(9): 5471–5485. DOI: 10.1007/s10661-011-2354-6

Miller, S.D. 2003. A consolidated technique for enhancing desert dust storms with MODIS. *Geophys. Res. Lett.* 30: 2071. DOI: 10.1029/2003GL018279

Mishra, M.K., P. Chauhan, and A. Sahay. 2015. Detection of Asian dust storms from geostationary satellite observations of the INSAT-3D imager. *Int. J. Remote Sens.* 36(18): 4668–4682. DOI: 10.1080/01431161.2015.1084432

Nisa, Z.U., S. Atif, and M.F. Khokhar. 2019. Identification of dust transport patterns and sources by using MODIS: A technique developed to discriminate dust and clouds. *Int. J. Environ. Pollut.* 66(1–3): 80–97. DOI: 10.1504/IJEP.2019.104537

Qu, J.J., X. Hao, M. Kafatos, and L. Wang. 2006. Asian dust storm monitoring combining Terra and Aqua MODIS SRB measurements. *IEEE Trans. Geosci. Remote Sens.* 3: 484–486.

Rayegani, B., S. Barati, H. Goshtasb, S. Gachpaz, J. Ramezani, and H. Sarkheil. 2020. Sand and dust storm sources identification: A remote sensing approach. *Ecol. Indic.* 112. DOI: 10.1016/j.ecolind.2020.106099

Roskovensky, K.J., and K.N. Liou. 2005. Differentiating airborne dust from cirrus clouds using MODIS data. *Geophys. Res. Lett.* 32: L12809. DOI: 10.1029/2005GL022798

Samadi, M., A.D. Boloorani, S.K. Alavipanah, H. Mohamadi, and M.S. Najafi. 2014. Global dust detection index (GDDI): A new remotely sensed methodology for dust storm detection. *J. Environ. Health Sci. Eng.* 12(1): 1–14. DOI: 10.1186/2052-336X-12-20

She, L., Y. Xue, X. Yang, J. Guang, Y. Li, Y. Che, F. Fan, and Y. Xie. 2018. Dust detection and intensity estimation using himawari-8/AHI observation. *Remote Sens.* 10, 490. DOI: 10.3390/rs10040490

Souri, A.H., and S. Vajedian. 2015. Dust storm detection using random forests and physical-based approaches over the Middle East. *J. Earth System Sci.* 124(5): 1127–1141. DOI: 10.1007/s12040-015-0585-6

Su, Q., L. Sun, Y. Yang, X. Zhou, R. Li, and S. Jia. 2017. Dynamic monitoring of the strong sandstorm migration in northern and northwestern China via satellite data. *Aerosol Air Qual. Res.* 17(12): 3244–3252. DOI: 10.4209/aaqr.2016.12.0600

Sun, K., Q. Su, and Y. Ming. 2019. Dust storm remote sensing monitoring supported by MODIS land surface reflectance database. *Remote Sens.* 11(15): 1772. DOI: 10.3390/rs11151772

Xu, H., T. Cheng, X. Gu, T. Yu, Y. Wu, and H. Chen. 2015. New Asia dust storm detection method based on the thermal infrared spectral signature. *Remote Sens.* 7: 51–71. DOI: 10.3390/rs70100051

Xu, D., J.J. Qu, S. Niu, and X. Hao. 2011. Sand and dust storm detection over desert regions in China with MODIS measurements. *Int. J. Remote Sens.* 32(24): 9365–9373. DOI: 10.1080/01431161.2011.556679

Yang, Y.K., L. Sun, J.S. Zhu, J. Wei, Q.H. Su, W.X. Sun, F.W. Liu, and M.Y. Shu. 2017. A simplified Suomi NPP VIIRS dust detection algorithm. *J. Atmos. Solar-Terr. Phys.* 164: 314–323. DOI: 10.1016/j.jastp.2017.08.010

Yue, H., He, C., Zhao, Y., Ma, Q., Zhang, Q. 2017. The brightness temperature adjusted dust index: An improved approach to detect dust storms using MODIS imagery, *Int. J. Appl. Earth Obs. Geoinf.* 57: 166–176, DOI: 10.1016/j.jag.2016.12.016

Zandkarimi, A., P. Fatehi, and R. Shah-Hoseini. 2020. An improved dust identification index (IDII) based on MODIS observation. *Int. J. Remote Sens.* 41(20): 8048–8068. DOI: 10.1080/01431161.2020.1770366

Zhang, P., N.M. Lu, X.Q. Hu, and C.H. Dong. 2006. Identification and physical retrieval of dust storm using three MODIS thermal IR channels. *Glob. Planet. Change* 52: 197–206. DOI: 10.1016/j.gloplacha.2006.02.014

9 Hurricanes and Tornados

9.1 INTRODUCTION

Hurricanes, also known as typhoons, are one of the major natural hazards frequently devastating certain coastal areas of the world in their path. Originating from the Atlantic and Pacific Oceans, these tropical storms, once formed, produce ferocious gales, cause stormy surges of ocean waves, and are accompanied by downpours, all of which are hazardous and can cause enormous damage to properties and even loss of human lives. Remote sensing of hurricanes and their damage differs from studying other types of natural hazards in that they are not confined to a fixed location nor a small spatial extent. Once formed, they are in a state of constant motion over the ocean before striking the land. Thus, they are temporally variable, and their path of movement is influenced by the regional meteorological conditions, and are difficult to predict reliably, especially at a long lead time. The periodic tracking of the same hurricane over its life cycle is ideally accomplished using airborne sensing, supplemented with meteorological satellite data of a sufficiently fine temporal resolution. However, their coarse spatial resolution does not allow hurricane damage to be assessed at an adequate accuracy level. Besides, optical meteorological satellite sensors make use of visible and infrared radiations that are unable to penetrate dense clouds intrinsically associated with tropical storms, even though hurricanes are easily identifiable by their distinctive circular cloud patterns on satellite images. The sensing of cloud properties, which is important to predict the possible amount of rainfall, is commonly achieved using microwave radiation, particularly C-band radar that can penetrate clouds and rainbands to some degree. Microwave remote sensing enables the retrieval of cloud properties, such as their composition and moisture level. Such information is valuable for predicting the potential rainfall and intensity of a hurricane.

This chapter first scrutinizes the sensing platforms that have found applications specifically in monitoring hurricanes, and the purposely designed sensors for this application. This discussion is followed by an exploration of various hurricane parameters that have been successfully retrieved from various satellite data, including wind vectors, waves, and rainfall. Finally, this chapter expounds how hurricane effects and damage are monitored and assessed from a wide variety of remotely sensed data using diverse analytical methods. The covered effects include coastal erosion, damage to coastal ecosystems, and change in water quality. Explained in detail is how they are assessed from various satellite images, both very high-resolution and coarse-resolution data. Naturally, the downpour associated with hurricanes may also trigger landslides and coastal floods. These secondary natural hazards are separately covered in respective chapters elsewhere and are not repeated here. Instead, this chapter concentrates on how to sense tornado damage using UAV data.

9.2 SENSING PLATFORMS AND SENSORS

9.2.1 Sensing Platforms

The platforms ideal for sensing hurricanes fall into two broad categories of space-borne and airborne, each having its unique strengths and limitations. Airborne sensors can be deployed rapidly and flexibly over the affected areas days ahead of a fledgling cyclone. Such flexibility can never be matched by space-borne sensing platforms. The speedy deployment of airborne sensors allows widespread damage of hurricanes to be rapidly assessed digitally at the sub-meter spatial resolution (Corbley, 2006). Such data are valuable in detecting houses that have been flooded beyond repair, and thus their owners can be reached and assisted by the government emergency management agency.

DOI: 10.1201/9781003354321-9

Although meteorological satellite images have a fine temporal resolution that enables the same hurricane to be monitored at short intervals frequently over its lifespan, their spatial resolution is insufficiently fine to enable small-scale, fast-evolving, and rapidly intensifying hurricanes to be monitored. Thus, it is not feasible to comprehend the role of small-scale deep convective forcing in the intensification process of a hurricane from these images. So airborne sensors have to be relied. The best airborne sensing platform is exemplified by the NOAA's WP-3D Orion aircraft (Figure 9.1). The P-3 aircraft flies at a medium to high altitude near the top of the hurricane planetary boundary layer. It has a maximum navigation range of 4,630 km or an operation duration of 9.5 hours at low altitude. At high altitude, the distance rises to 7,037 km or a duration of 11.5 hours. This airborne sensing platform holds a great potential for safe and reliable measurements of hurricane parameters, judged by the fact that the surface wind speed, wind direction, and rain rate obtained by the aircraft are highly comparable to those inferred from radar data.

The WP-3D Orion aircraft is equipped with several sensors, including a lower fuselage C-band radar, a C-band nose radar, and a Tail Doppler radar (TDR). Mounted to the belly of an aircraft, the first sensor scans a hurricane horizontally, while the last sensor scans it vertically, yielding a 3D view of the storm in real time. In addition, the aircraft is also equipped with GPS dropwindsondes. They are dropped out of an aircraft as it flies over a hurricane. These sensors continuously transmit readings of air pressure, humidity, temperature, and wind direction and speed as they fall toward the sea. The collected data enable the reconstruction of a hurricane structure and intensity in detail. The aircraft also deploys probes named bathythermographs for measuring sea temperature.

A very important tailored space-borne platform is the Coriolis satellite carrying the WindSat Radiometer. This sensor operates at five frequencies, of which 10.7, 18.7, and 37.0 GHz are fully polarimetric, and 6.8 and 23.8 GHz are only dual-polarized. WindSat data have been used to derive a diverse range of data products, each having its own scale and valid range (Table 9.1). Since space-borne sensors cover a huge ground area, a feature complementary to the coverage of airborne data, naturally, it is sensible and advantageous to merge them with airborne sensors. This integration is especially crucial for studying those hurricanes beyond the reach of airborne platforms.

FIGURE 9.1 The newly upgraded and repainted NOAA Lockheed WP-3D Orion N42RF aircraft, equipped with GPS, lower fuselage and tail Doppler radar (TDR) systems specifically for sensing hurricanes. (Source: Mike Mascaro, NOAA.)

TABLE 9.1
Data Products from WindSat Ocean Measurements

Product Name	Description	Scale	Valid Range
Sea surface temperature	Top layer (skin) of water ~1 mm thick	0.15	−3 to 34.5°
10-m wind speed	From 10.7 GHz channel and above	0.2	0–50 m·s^{-1}
10-m wind speed	From 18.7 GHz channel and above	0.2	0–50 m·s^{-1}
Columnar atmospheric water vapor	Total gaseous water in a vertical column of the atmosphere	0.3	0–75 mm
Columnar cloud liquid water content	Total cloud liquid water contained in a vertical column of the atmosphere	0.01	−0.05 to 2.45 mm
Rain rate	Rate of liquid water precipitation	0.1	0–25 mm·h^{-1}
All-weather 10-m wind speed	Based on all channels and three separate algorithms	0.2	0–50 m·s^{-1}
10-m wind direction	Oceanographic-convention wind direction relative to north	1.5	0–360°

Source: https://www.remss.com/missions/windsat/. Used with permission.

9.2.2 Tailored Sensors

A number of airborne sensors have been designed specially to study hurricanes. One of them is the Doppler radar High-altitude Imaging Wind and Rain Airborne Profiler (HIWRAP) devised to study hurricanes and other precipitating systems. HIWRAP senses wind speed using dual-frequency, VH single-polarized Ka and Ku bands. These radar data allow the retrieval of a 3D wind vector over the entire radar-sampling volume, up to a horizontal grid and a vertical spacing of 1 km to an accuracy of 2.0 m·s^{-1} for horizontal wind components. Although some of the collected data are non-imagery, they can still document the intensification process of a tropical storm and facilitate a comprehensive understanding of the structure, evolution, and role of small-scale deep convective forcing in the storm (Guimond et al., 2016). If carried by the new Global Hawk unmanned aircraft at an altitude of 18–19 km, HIWRAP is operational airborne for one day, allowing long-lasting sampling of a hurricane for up to 24 hours. This feature is crucial for monitoring those hurricanes forming over remote regions of the ocean with important and fast physical processes occurring on a time scale that can be easily missed by conventional aircraft. This is because they can stay airborne for about a few hours in sensing the velocity and spatial patterns of hurricane wind (Figure 9.2).

HIWRAP is supported by data from the High-Altitude Monolithic Microwave Integrated Circuit Sounding Radiometer (HAMSR). This microwave sounder is designed to measure the atmospheric upwelling radiation at three frequencies of 50, 118, and 183 GHz. The first two frequencies are used to sense temperature and the last water vapor. In regions where upwelling radiation is scattered out of the beam by ice particles, the resultant anomalously low T_B at the sensor can detect the intensity of convective clouds. This sensor scans the ground ±60° across-track and covers a swath width of about 65 km if carried aboard the NOAA WP-3D aircraft. Along-track resolution is up to 250 m. HAMSR is synchronized with HIWRAP via coordinated flights with the NOAA WP-3D aircraft. The synergetically acquired data can be used to determine the location of deep convective bursts and their manner of initiation in a hurricane, as well as during the rapid intensification process. More importantly, such airborne data can reveal the main processes responsible for the formation and maintenance of the bursts, such as convergence of counter-rotating mesovortex circulations and the larger vortex-scale flow, and the turbulent transport of anomalously warm, buoyant air masses from the hurricane eye to the eyewall at low levels (Guimond et al., 2016).

The Stepped-Frequency Microwave Radiometer (SFMR) aboard the NOAA WP-3D or Air Force research aircraft is able to sense microwave emissions at six C-band frequencies of 4.55, 5.06, 5.64,

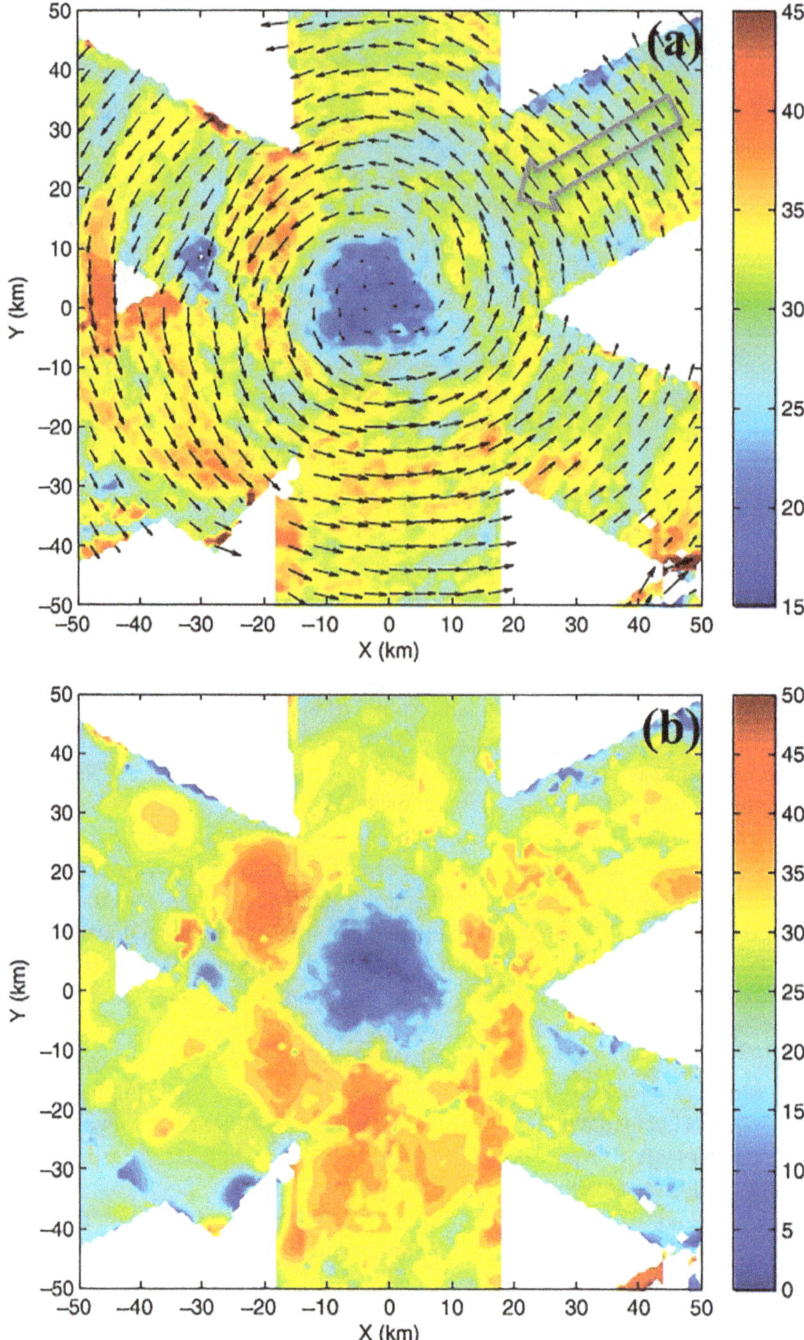

FIGURE 9.2 Wind speed and pattern of Hurricane Karl analyzed from HIWRAP data averaged over the period from ~1900 UTC 16 to 0800 UTC 17 September 2010 (height: 2 km). (a) Ku-band reflectivity (dBZ) overlaid with horizontal wind vectors. The large grey arrow: the large-scale vertical wind shear vector valid for this time interval with a value of about 5 $m \cdot s^{-1}$; (b) horizontal wind speeds in $m \cdot s^{-1}$. (Source: Guimond et al., 2016. © American Meteorological Society. Used with permission.)

6.34, 6.96, and 7.22 GHz, which are converted to 1-min sustained surface wind speed at the surface of the ocean via a Geophysical Model Function (GMF). It measures radiation emitted by sea foam produced by strong winds at the ocean surface. This sensor is designed to measure hurricane-force ocean surface winds. Validated against measurements from both GPS dropsonde and in situ instruments, its data products have an RMSE < 5 $m \cdot s^{-1}$ over the wind speed range of 10–70 $m \cdot s^{-1}$.

The sensing of hurricanes outside the US territory has to rely on commercial radar satellite images, such as RadarSat-2 SAR. Such images are suitable for monitoring hurricanes and typhoons over a broad spatial scale, allowing the retrieval of wind speeds and ocean waves. Waves are calculated from two C-band (5.4 GHz) dual-polarized images, one cross-polarized HV or VH, and another HH or VV polarized using an ocean surface wind retrieval model over the speed range of 3.7–39.7 $m \cdot s^{-1}$ (Zhang et al., 2018). From the near to far range, the incidence angle varies between 20° and 49°.

SAR-based wave products are also available from Sentinel-1A satellite images. If acquired in the wide-swath mode, cross-polarized Sentinel-1A images are good for detecting typhoons and hurricanes. The SAR wave mode of Sentinel-1A came into service in July 2015 for sensing global ocean waves. It produces images of 4 m resolution, covering a small footprint area of 20 km × 20 km. This mode yields almost continuous measurements of wind and waves simultaneously under stormy conditions. Wind and wave products routinely derived from Sentinel-1A data are released at several levels. The level-2 ocean products contain wind speed and significant wave height produced from wave-mode images.

The Soil Moisture and Ocean Salinity (SMOS) mission launched by the European Space Agency in November 2009 currently provides multi-angular L-band (1.4 GHz) T_B images of the Earth. In sensing ocean upwelling, SMOS is much less affected by rain and the atmospheric effects than images obtained at higher frequencies or shorter wavelengths. This is a marked advantage over existing ocean satellite high wind observations that are often contaminated by heavy rain and clouds (Reul et al., 2011). SMOS data have a large swath width of around 1,200 km, a spatial resolution of 30–80 km, and a revisit period shorter than 3 days. These traits make SMOS the ideal data to study the mesoscale evolution of surface winds and whitecap statistical properties commonly associated with hurricanes and severe storms. This sensor is capable of yielding quantitative and complementary surface wind information of interest for forecasting hurricane intensity operationally (Reul et al., 2011). Surface wind speeds estimated from SMOS T_B images agree well with the observed and modeled surface wind speed features.

Other radiometers aboard various satellites can serve a similar purpose as the aforementioned ones. The innovative combination of all of these sensors and platforms has yielded unprecedentedly abundant information on hurricanes that was unimaginable only years ago. These excellent sources of data allow operational retrieval of various hurricane parameters.

9.3 RETRIEVAL OF HURRICANE PARAMETERS

During the process of hurricane formation, it is important to track its evolution so as to accurately predict its path of movement, strength, ocean surface surge level, and landfall location. Such parameters can be retrieved from various remote sensing data to different accuracy levels.

9.3.1 Winds

A number of hurricane parameters can be retrieved remotely, including wind speed, path, and velocity of travel. Such information is critical to forecasting the landing location and potential damage, and studying the long-term effects of tropical storms on ocean circulation and heat transport. This forecasting is based on the air–sea exchange of energy, moisture, and momentum that dictates hurricane strength and intensity. A critical measure of hurricane intensity is its wind speed. The distribution of surface winds in a hurricane is estimatable using two types

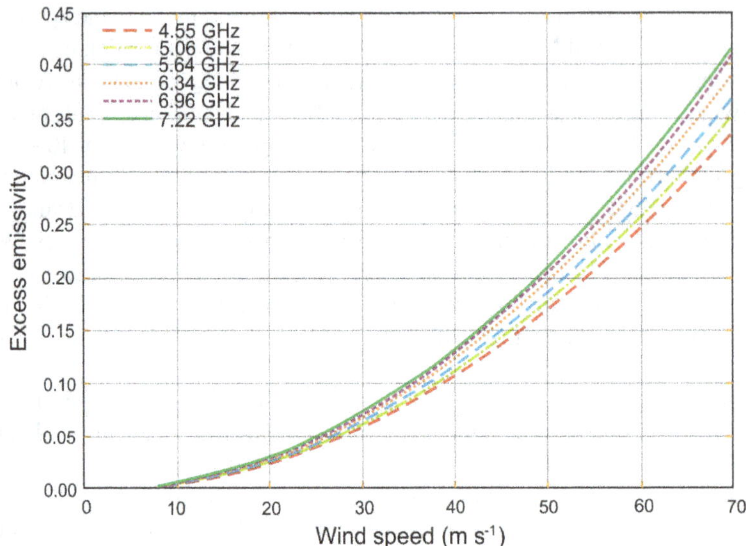

FIGURE 9.3 Empirical relationship between excess emissivity captured by a passive microwave radiometer and surface wind speed. (Source: modified and redrawn from Uhlhorn and Black, 2003. © American Meteorological Society. Used with permission.)

of sensors, microwave radiometer and GPS. The former is exemplified by the SFMR onboard NOAA hurricane hunters, or any radiometers aboard various meteorological satellites, such as WindSat, TRMM, and AMSR. The radiometer senses the nadir microwave emissions expressed in terms of surface T_B. It increases with foam coverage if the ocean surface is approximated as a microwave blackbody with an emissivity close to unity. Wind speeds are then calculated from the presumably linear relationship between T_B and wind speed along the flight track independent of frequency (Uhlhorn and Black, 2003). SFMR is a top performer in measuring hurricane-force winds at the sea surface.

With further refinement, winds over 60 m·s^{-1} can be estimated from the empirical emissivity-wind speed relationship (Figure 9.3) at 20 m above the water surface, but not winds below 10 m·s^{-1} due to low sensitivity. Over the range of 10–70 m·s^{-1}, average wind speeds from SFMR have an estimated RMSE of 4 m·s^{-1} near the surface (Uhlhorn and Black, 2003). Although this measurement is accurate, it is point-based and available only along the aircraft flight track. The SFMR equivalent 1-min mean 10-m level neutral stability winds are found to be biased by 2.3 m·s^{-1} relative to the 10-m GPS dropsonde computed from an estimate of the mean boundary layer wind. The 3D profile of hurricane winds may be reconstrured by combining measurements of the SFMR surface wind with flight-level GPS wind data.

The second approach is to use GPS dropsonde, a special device for sensing wind speed vertically. As its name implies, it is dropped out of a flying aircraft to record wind measurements during its descent, including wind speed, wind direction, and GPS position (Figure 9.4). A dropsonde also receives GPS navigation signals and measures the Doppler shift of each signal and converts this information to winds at different heights. Thus, it can profile atmospheric wind and thermodynamics. This method of sensing vastly improves the accuracy of the retrieved wind speed, including its magnitude and location, within the boundary layer. GPS dropsondes measure horizontal wind vector components, but not winds all the way to the sea surface, especially when the wind is unusually strong. A dropsonde has a near-surface fall speed of about 12–14 m·s^{-1}, yielding a vertical sampling interval of about 5–7 m, at the typical sampling rate of 2 Hz. It can potentially map along-track wind speeds at a relatively high spatial resolution of about 120 m and temporal resolution of 1 Hz.

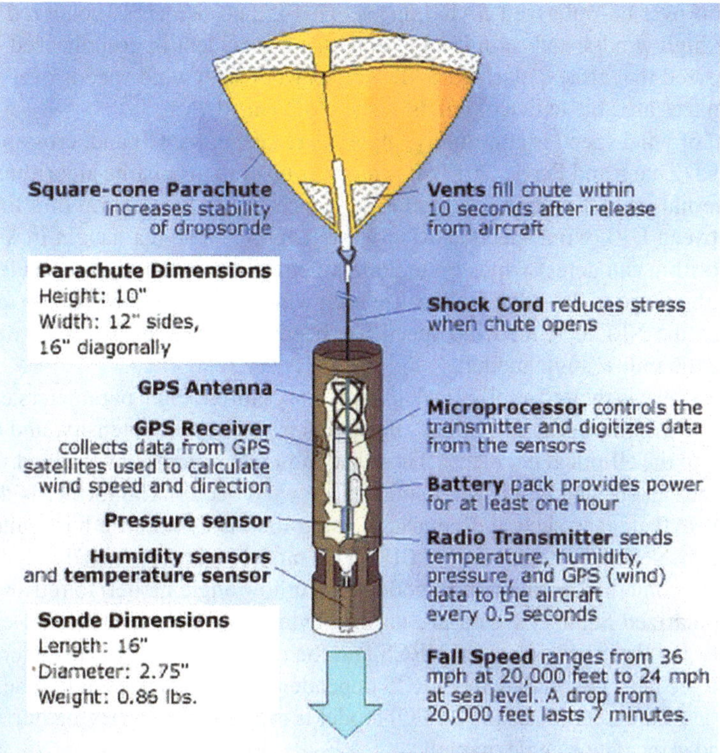

FIGURE 9.4 A sketch illustrating the various components and functions of a GPS dropsonde for profiling atmospheric wind and thermodynamics of a hurricane. (Source: L. J. Schmidt, NASA.)

Both sensors face the same limitation of being unable to yield the spatial distribution of wind parameters, which can be overcome with the use of space-borne SAR images. SAR-based estimation of wind speed relies on its relationship with small-scale surface roughness detected from radar images. On a scale comparable with the radar wavelength, the spectral density of surface roughness is proportional to the normalized radar cross-section (NRCS) of radar backscatter at a moderate incidence angle between 20° and 60°. However, the relationship between wind speed and the NRCS is complicated by wind direction without the saturation effect at high wind speeds, precipitating the wind speed ambiguity problem for SAR wind speed retrieval. This problem can be redressed with C-band cross-polarized ocean backscatter that is insensitive to wind direction and radar incidence angle. Wind and wave information is retrieved using empirical GMF algorithms depicting the relationship between ocean surface wind and the NRCS at C-band cross-polarization, or C-band model 4. This model relates wind vectors to the measured NRCS in VV polarization and the radar incidence angle to compute wind speed in certain directions. Over the wind speed of 10 $m \cdot s^{-1}$, the NRCS is significantly dependent on wind speed. The sensitivity of the backscatter decreases with wind speed (Horstmann et al., 2015).

Improved GMFs have been developed to convert cross-polarized SAR RCS to wind speed both with and without wind direction dependence, but the wind-direction dependent GMF produces better wind speed estimates than that without the dependence (Horstmann et al., 2015). Thus, it is necessary to simultaneously co-polarize SAR images to derive wind direction as it cannot be independently derived from only cross-polarized SAR images. Wind direction and strength are commonly retrieved from dual-polarized SAR images because the backscatter signal of single-polarized SAR images may experience saturation and become double valued. All of the GMFs based on cross-polarized SAR images with or without wind direction dependence significantly improve wind speed estimation over co-polarized SAR images. This is because cross-polarized RCS does not suffer from the high wind speed saturation effects that are evident in co-polarized RCS results. It must be emphasized that cross-polarized SAR images work with wind speeds over 10 m·s^{-1}, while co-polarized images are able to detect winds from 5 to 20 m·s^{-1}.

The retrieval of wind speed is commonly implemented using the C-band, cross-polarized ocean algorithm C-2PO (Zhang and Perrie, 2012). It is much simpler than existing algorithms as it does not require any external wind directions and radar incidence angles in the form of a linear regression relationship between U10 (wind speed at 10 m wave height) and sigma naught in VV polarization (σ^0_{VV}). This algorithm can detect winds up to 26 m·s^{-1} competently but tends to underestimate high wind speeds in the range of 30–38 m·s^{-1} because high winds and high rain rates up to 35 mm·hour^{-1} severely dampen the NRCS. After wind speed has been retrieved, surface wind direction can be simulated using the inflow angle model.

If the hurricane eye is treated as elliptical, a total of six morphologic parameters can be retrieved about it: major axis, minor axis, azimuthal angle, center, symmetric intensity, and decay, the first three pertaining to the elliptical geometry. These six parameters can be determined from two fields: elliptical symmetrical surface wind speed and surface wind vector. The former is deriveable from cross-polarized SAR images such as RadarSat-2 using the 2D symmetric hurricane estimates for the wind model developed by Zhang et al. (2017). This model yields only wind speed without direction, so it must be combined with another model (e.g., inflow angle model) to retrieve the full wind vector. If VH-polarized ScanSAR data are used, a hybrid backscattering model establishing the relationship between the cross-polarized NRCS and the radar incidence angle under different wind conditions has to be used to simulate the NRCS dependence on incidence angle. The C-band Cross-polarized Coupled-Parameters Ocean (C-3PO) model is available for retrieving hurricane winds by including the radar incidence angle, namely:

$$\sigma_0 = \left[0.2983U_{10} - 29.4708\right]\left[1 + 0.07 \cdot \frac{\theta - 34.5}{34.5}\right] \tag{9.1}$$

where σ_0 = the NRCS of radar signal in dB; U_{10} = wind speed at 10 m wave height; θ = incidence angle. Compared with C-2PO, C-3PO has a lower bias, a slightly smaller RMSE of <3 m·s^{-1} for wind speeds up to 40 m·s^{-1}, and marginally higher R^2 (Zhang et al., 2017).

9.3.2 Waves

The strong winds associated with powerful hurricanes also induce waves. Wave parameters include significant wave height and mean wave period. They are commonly retrieved from SAR images owing to their capability of cloud penetration. SAR-based sensing of ocean waves relies on their backscattering signal using a number of algorithms, such as the Max-Planck Institute (MPI) algorithm and the Parameterized First-guess Spectrum Method (PFSM). Initially developed for C-band SAR imagery, MPI is independent of radar frequency. Both MPI and PFSM require an initial "guess" wave spectrum that can be served by an empirical wave model, a numerical wave model, or a parametric wave spectrum model, to invert the wind–sea wave spectrum (Shao et al., 2015). With modification, both algorithms are also functional with TerraSAR-X imagery in a number of steps, including the calculation of HH-polarized NRCS that is sensitive to high wind speeds, and the polarization ratio to convert HH-polarized to VV-polarized NRCS based solely on incidence angle (Figure 9.5). After the ambiguous wind direction is derived from the empirical wave model, the European Centre for Medium-Range Weather Forecasts (ECMWF) wind direction data may be used to determine the real wind direction from SAR imagery. It may be necessary to resample the data to unify the spatial resolution of ECMWF wind products (spatial resolution: 0.25° × 0.25°) and SAR-X data, the spatial resolution of which is 5 m. With the assistance of XMOD2 for wind retrieval from TerraSAR-X, the VV-polarized NRCS is expressed as a function of the angle among

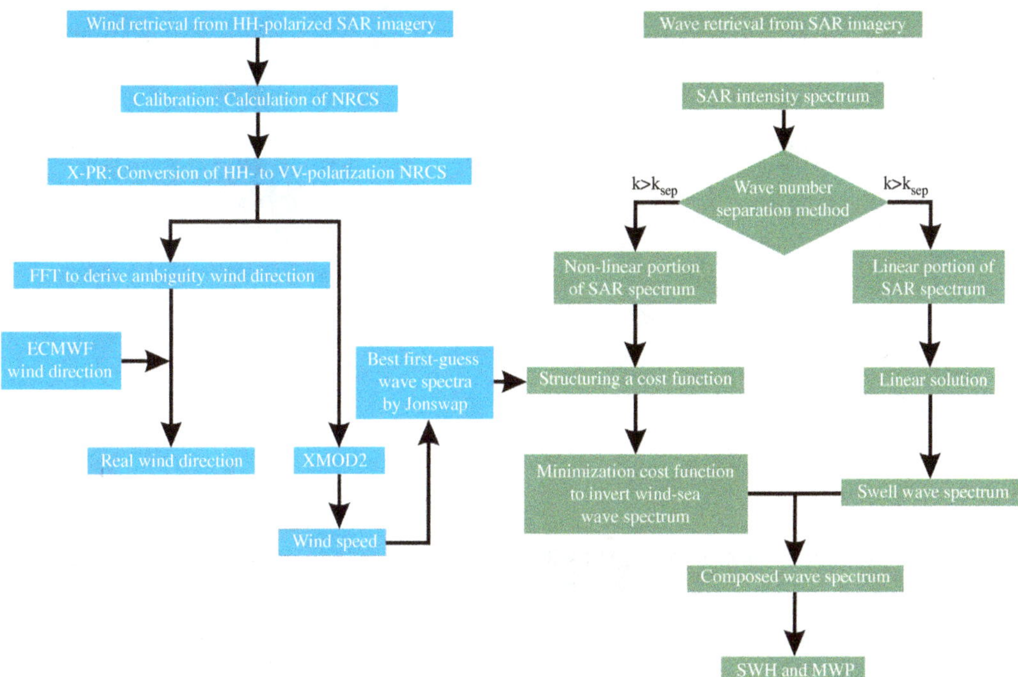

FIGURE 9.5 The procedure of retrieving wind and wave parameters from HH-polarized SAR imagery using the parameterized first-guess spectrum method. NRCS – normalized radar cross-section; X-PR – X-band polarization ratio (model); SWH – significant wave height; MWP – mean wave period. FFT – Fast Fourier transformation. XMOD2 – a model for retrieving wind from TerraSAR-X imagery. (Source: modified from Shao et al., 2015.)

the radar look direction, the wind direction, and the incidence angle, while the coefficients of the equation are functions of sea surface wind speed and radar incidence angle.

The retrieval of wave parameters requires the use of SAR intensity. In the PFSM method, the linearly mapped spectrum portion is separated from the SAR intensity spectrum via the calculated wave number threshold. The portion of the linearly mapped SAR spectrum is where the wave numbers are smaller than the separation wave number. Significant wave height (SWH) and mean wave period (MWP) are output directly from the inverted 2D wave number spectrum (Figure 9.5).

SAR imagery of ocean waves is affected by non-linear imaging mechanisms that distort short-waves, so is not perfectly suitable for retrieving short wind waves, such as those generated by a hurricane. The alternative to image-to-wave spectra inversion is to empirically estimate integral wave parameters, such as significant wave height or wave period, from SAR images using the fetch- and duration-limited wind-wave growth relationships or the H-model (Zhang et al., 2018), in which significant wave height H_s and wave period T_p are expressed as a function of dimensionless variance, dimensionless frequency, and dimensionless duration. H_s and T_p can be calculated directly from the wind field at 10 m high (U_{10}) if fetch and duration are known. The equivalent fetch and equivalent duration inside the hurricane are also obtainable from the wind-wave triplets of H_s, T_p, and U_{10}. If the triplets are measured from a scanning radar altimeter, the empirical equation in terms of fetch (unit: km) and duration (unit: hour) can be determined for the three sectors of a hurricane: left, right, and back (Zhang et al., 2018), hence the H-3Sec model, all being functions of the radial distance r (unit: km) from the hurricane center. The fetch and duration relationships may be represented in different ways, resulting in two more types of H models, H-LUT and H-Harm. H-LUT is able to simulate surface waves in the azimuthal and radial directions under storms, while H-Harm considers the influence of the radius of the maximum wind speed of storms in the fetch- and duration-limited simulations. All the three H-models can effectively quantify the significant wave heights inside the

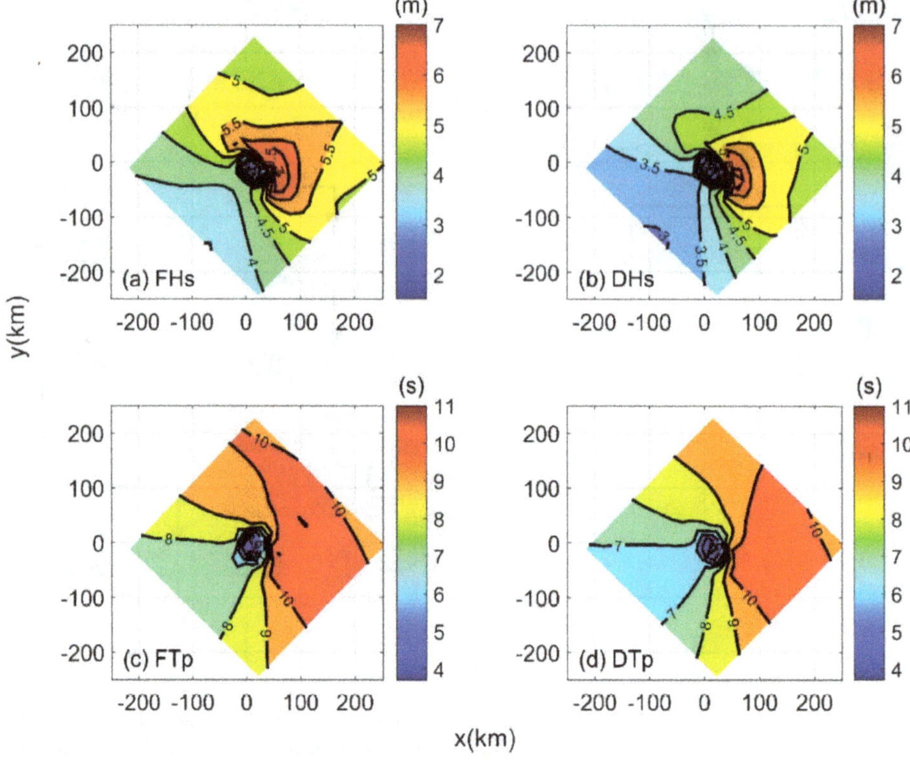

FIGURE 9.6 Wave height (top row) and period (bottom row) associated with Hurricane Gustav at 1128 UTC on 30 August 2008 retrieved from SAR images using different growth functions: (a) fetch-limited significant wave height; (b) duration-limited significant wave height, (c) fetch-limited wave period; and (d) duration-limited wave period. (Source: Zhang et al., 2018.)

hurricane based on winds observed from Sentinel-1A SAR images, except near the hurricane eye (Figure 9.6). These models can also potentially simulate H_s and T_p for wind waves inside hurricanes from RadarSat-2 ScanSAR mode images reliably.

9.3.3 Rainfall

Hurricanes are usually accompanied by intensive and heavy rainfalls. For instance, typhoon Morakot dumped 2,777 mm of rain on 7–8 August 2009 in Taiwan. Rainfall data are routinely recorded at ground rain gauges. However, they may be damaged by heavy rainfall or destroyed by hurricane-triggered landslides. Besides, rain gauges have limited spatial coverage, and their density is unable to depict the spatial variation of rainfall intensity in adequate detail. Thus, there is a need to use remote sensing to estimate hurricane-triggered rainfall. Remote sensing data acquired from ground radar and space-borne sensors allow quantitative estimation of precipitation. So far four types of rainfall data products have been derived by using various remote sensing means, including the IR-dominated PERSIANN, the PERSIANN-Cloud Classification System (PERSIANN-CCS), the microwave-dominated CMORPH product, and the TRMM Multi-satellite Precipitation Analysis (TMPA) 3B42RT and 3B42V6 products. They all face different limitations in producing accurate precipitation estimates at the hydrologically relevant scale, especially for extreme typhoon events such as Morakot.

PERSIAN-CCS is an automated system for estimating precipitation from geostationary infrared satellite images, from which local and regional cloud features are extracted to estimate rainfall distribution at a spatial resolution of 0.04° × 0.04° half-hourly. The system can retrieve widely ranging rainfall rates at a given T_B and detect variable rain/no-rain IR thresholds for different cloud types. PERSIANN-CCS products have two versions, real-time (RT) and microwave-adjusted (MW). The former is converted from real-time Geostationary Operational Environmental Satellite (GOES) images using the PERSIANN-CCS-RT algorithm. The latter is derived from microwave precipitation estimates from low earth-orbiting satellites, such as TRMM and other microwave satellites using the PERSIANN-CCS-MW algorithm. It sequentially calibrates the RT products into IR and microwave combined products.

CMORPH is a precipitation data product derived from microwave and IR remote sensing data based on half-hourly geostationary IR images that are used to extrapolate the passive microwave precipitation estimates. It has the combined desirability of microwave sensing (high quality) and infrared sensing (high spatial and temporal resolutions). It is a quasi-global (60°S–60°N), high-resolution (0.1° in latitude/longitude), half-hourly product. Therefore, it improves the estimation of multi-hour precipitation accumulation and produces better rainfall estimates than the simple averaging of available microwave-based estimates and other merged results that incorporate microwave and infrared information in the estimation. This data product comes in a few forms, including 3B42RT and 3B42V6. The former is a near-real-time product and 3B42V6 is a post-real-time product. They cover the region of 50°S–50°N and 60°S–60°N, respectively. The real-time product uses rainfall estimates from TRMM Precipitation Radar and TRMM Microwave Imager to calibrate the estimates derived from low earth orbiting passive microwave radiometers. All of these precipitation products underestimate rainfall by various amounts, such as radar by 18%, 3B42RT by 19%, PERSIANN-CCS by 28%, 3B42V6 by 36%, and CMORPH by 61% (Chen et al., 2013). Although CMORPH has the highest underestimation, it still outperforms others in terms of the spatial pattern of rainfall distribution, with a fairly good correlation coefficient of 0.70 for two-day rainfall accumulation. However, it underestimates the tropical warming cloud precipitation system the most. PERSIANN-CCS provides a better estimate of total rainfall accumulation but has a looser spatial and temporal correlation. In particular, it overestimates rainfall at the early stage but underestimates it at a later stage of a typhoon event.

9.4 MONITORING OF HURRICANE EFFECTS

Hurricanes not only dump heavy rainfalls but also cause strong storm surges that impact the coastal environment in various ways, such as triggering coastal erosion, disturbing coastal ecosystems, and degrading estuary water quality, all of which can be detected by means of remote sensing.

9.4.1 Coastal Erosion

Coastal erosion refers to the removal or depletion of sands and sediments from a beach or the collapse of coastal cliffs caused by the undercutting of their base by hurricane-triggered wave surges. The collapsed debris on the beach eventually dissipates under the slaughter of persistent waves and strong currents, thus causing the shoreline to retreat and the coastal landscape to evolve (e.g., removal of coastal vegetation, see Figure 9.7). Furthermore, hurricane-triggered waves may also re-suspend sediments that are transported elsewhere by currents. The net effect is a depleted sediment reserve in one part of the coastal system, but beach progradation and even the formation of tidal flats elsewhere. In a hurricane, the first casualty of storm surges and wave run-ups is sandy beaches of offshore islands that can shield and mitigate their effects. There exists a complex relationship between the patterns and processes of a hurricane and winter-storm sedimentation. An insight into this coastal process is imperative to comprehend the origin and evolution of sedimentary systems, especially tidal flats.

Hurricane-induced beach erosion or progradation is ideally monitored from VHR images such as IKONOS of 4 m (multispectral bands) and 1 m (panchromatic band) resolutions. Although coarse-resolution MODIS data have been used to reveal the passage of hurricanes, they can only show whether they have suspended sediments across the shallow shore. Only images of an ultra-fine

(a) Pre-hurricane image

(b) Post-hurricane photograph

FIGURE 9.7 Coastal changes around Fort Macon, North Carolina of the USA induced by Hurricane Dorian in early September 2019. (a) pre-event satellite image; (b) post-event photograph taken on 7 September 2018. (Source: NOAA.)

resolution, integrated with field observations, can paint a detailed picture of the exact effects of a hurricane on a beach system, such as whether it has suspended sediments offshore of tidal flats (Rankey et al., 2004). At least two images are vital for detecting such changes, one pre- and one post-hurricane (Figure 9.7). A comparison of the pair of images can reveal whether morphometric changes have occurred to the tidal flats or beaches under study. However, this comparison is unable to reveal the subtle change in surface elevation caused by the depletion of sands or sediments. Also unable to be studied is the underwater topography. In a word, visual comparison of the images is incapable of indicating sediment change unless it has caused noticeable horizontal progradation or erosion.

Nevertheless, the attribution of beach erosion to a specific hurricane is not a mean task as it requires a reliable counterfactual scenario: hypothetical beach conditions free of the storm effects (Pérez Valentín and Müller, 2020). Since it is almost impossible to meet these conditions, shoreline changes are detected through clouds. This detection is reduced to the separation of land and water. Owing to the tidal effects, the spectral separability between the two covers in the inter-tidal zone is not clear-cut on optical imagery. Radar imagery such as Sentinel-1 is a better option because radar backscatter signal is sensitive to surface roughness. Coastal water with a rougher surface than the nearby sandy beach can be detected via simple thresholding of a VV-polarized SAR image. This method separates water from land and thereby detects the time-averaged position of the coastline. Since the detection is commonly implemented within a zone centered at the shoreline, an elongated polygon is created to serve as a mask, within which VV and VH images are automatically classified using either the supervised or unsupervised method. The quality of image classification may be improved via spatial filtering to remove inland water bodies from land and roads mistakenly classified as water. This classification separates the radar image into two zones of water and land, each represented as a polygon, and their shared boundary represents the shoreline.

The same analysis is applicable to radar images before and after a hurricane event, resulting in two sets of shoreline positions. A spatial comparison of the two classified results reveals the difference in shoreline location or the change in shoreline position. However, the change cannot be entirely attributed to the impact of a hurricane as it is potentially subject to displacement associated with (unobserved) "normal" erosional processes in the absence of a hurricane. Or the influence of one hurricane may be compounded by the residual effect of a previous one. The solution to this dilemma is to make use of regression analysis to estimate the causal effect of a specific hurricane on beach erosion (Pérez Valentín and Müller, 2020). This can be achieved by linearly regressing the observed shoreline position against the running variables (e.g., days after a hurricane), and a dummy variable representing the presence (post) or absence (prior to) of the hurricane.

The regression coefficient represents the causal effect of the hurricane on the dependent variable or shoreline position. Theoretically, only the last image before the hurricane and the first image after the hurricane should be used. But such a small sample size is inadequate to resolve the regression coefficients. Thus, a number of pre- and post-hurricane images have to be used, which is impossible to achieve. Alternatively, a number of beaches in close proximity to each other can be used to increase the sample size. Application of this approach to detecting shoreline changes at 75 locations distributed over 24 sandy beaches of Puerto Rico, caused by Hurricane Maria on 20 September 2017, showed that the shoreline displacement remained within the range of natural variability, hence cannot be attributed to the hurricane at a high level of statistical confidence at one site (Figure 9.8A). At another site, the hurricane eroded the shoreline by 3–5 m along its path, and the erosion had a spatial variation of up to 40 m in particular beaches (Figure 9.8B). The detection is achieved at an RMSE of 24 m (spatially averaged), highly comparable to the 20 m resolution of the radar image used. This method of detection is limited in that it can be used to strictly estimate the immediate, short-term effects of a hurricane. It works only under the assumption that the observed beach conditions just before the hurricane are representative of hypothetical counterfactual conditions just after the hurricane, had it not occurred (Pérez Valentín and Müller, 2020).

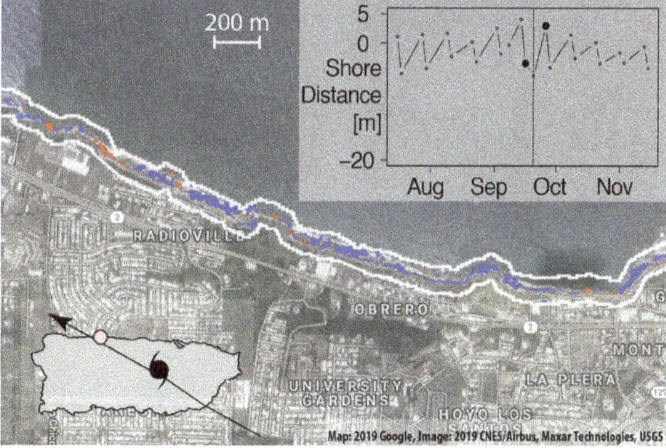

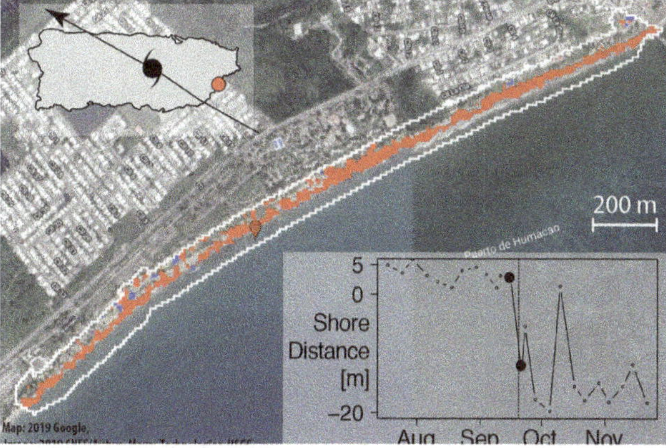

FIGURE 9.8 Changes in shoreline position caused by Hurricane Maria on 20 September 2017 in two beaches of Puerto Rico detected from Sentinel-1 radar imagery based on regression discontinuity. (A) Shoreline accretion in Arecibo (blue); (B) beach erosion in Punta Santiago (red). Arrow: Maria travel path; Insets: time-Series shoreline positions with the pre- and post-hurricane strike highlighted. (Source: Pérez Valentín and Müller, 2020.)

9.4.2 Coastal Terrestrial Ecosystem Damage

Hurricanes may not last long themselves, but they can produce immediate and lingering devastation to coastal ecosystems because of their savagely fierce velocity. The effects of hurricane strikes on coastal ecosystems can be accurately assessed by comparing the coastal ecosystem change following the hurricane landfalls. One of the most vulnerable coastal ecosystems is forests. The damage of Hurricane Katrina to the coastal forest of Louisianan is easy to detect from optical images such as Landsat TM. A composite of TM bands 4, 5, and 3 is the optimal band combination to visualize the disturbed forestlands. The disturbances are also detectable from VHR images captured immediately after the hurricane events after the images have been converted to land cover maps using supervised and unsupervised image classification methods or a combination of both (Lam et al., 2011). The classification of an IKONOS image recorded on 26 November 2006 into seven categories of water, forest, agricultural and others, developed, brackish marsh, fresh marsh, and swamp is rather accurately executed at an overall accuracy of 77% (Kappa = 0.74) with the pre-hurricane image and 81% (Kappa = 0.78) with the post-event image after Hurricanes Ivan (23 September 2004) and Katrina (26 August 2005) hit Florida (Figure 9.9). Over this period, "fresh marsh" decreased the most by

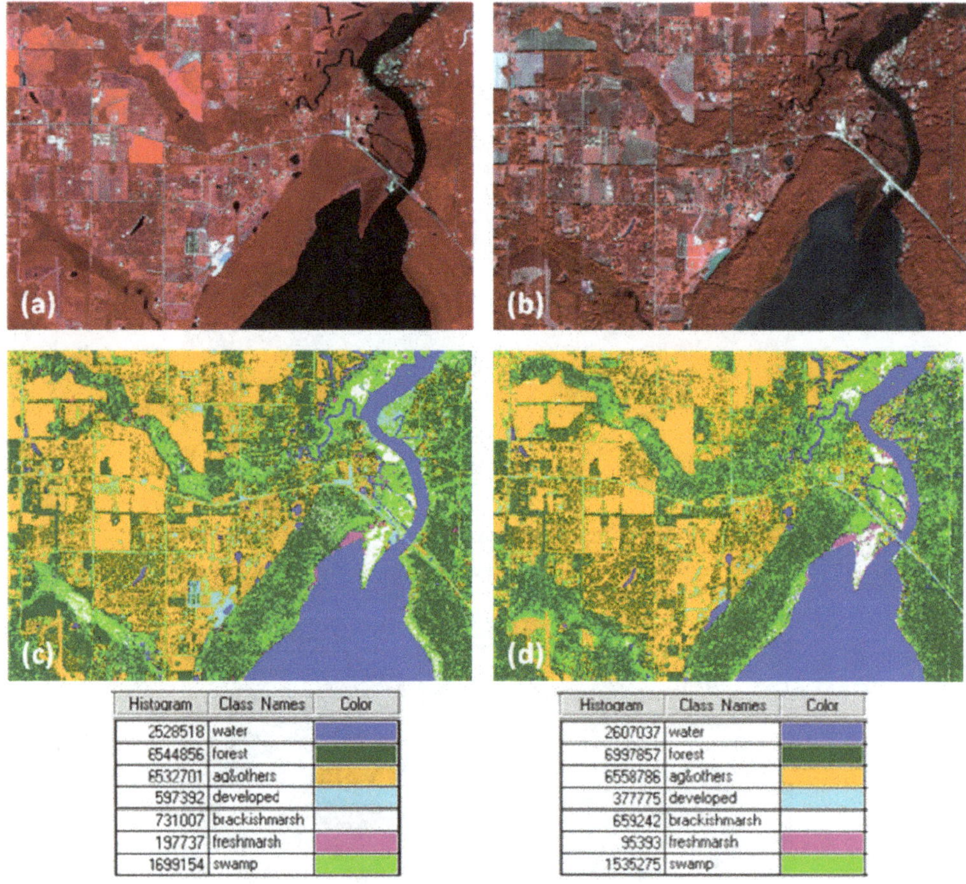

(a)
(b)
(c)
(d)
Histogram
Class Names
Color
2528518 water
6544856 forest
6532701 ag&others
597392 developed
731007 brackishmarsh
197737 freshmarsh
1699154 swamp
Histogram
Class Names
Color
2607037 water
6997857 forest
6558786 ag&others
377775 developed
659242 brackishmarsh
95393 freshmarsh
1535275 swamp

FIGURE 9.9 Changes to coastal ecosystems in Weeks Bay, Alabama induced by hurricanes Ivan and Katrina, detected from IKONOS images before and after the events. (a) False-color composite of IKONOS imagery of 18 September 2004; (b) false-color composite of IKONOS imagery of 26 November 2006; (c) land covers mapped from (a) using a hybrid of supervised and unsupervised classification; (d) land covers mapped from (b) using the same method. (Source: Lam et al., 2011. Reproduced and modified with permission from the Coastal Education and Research Foundation, Inc.)

52%, and brackish marsh and swamp decreased less by 9.8% and 9.6%, respectively. However, the largest relative decrease occurred to fresh marsh (51.76%) and developed (36.76%) relative to their pre-hurricane area. These changes demonstrate that the hurricanes rainfall inundated low-lying areas that had not fully recovered one year after the preceding strike. A comparison of the pre- and post-hurricane land covers also reveals the resilience of the ecosystems to external disturbances triggered by hurricanes.

Hurricanes can snap tree branches and damage forest canopy via exfoliation. Thus, post-hurricane assessment of the severity of forest damage may focus on defoliation, branch loss, stem loss, and other changes in forest structure, which is ideally achieved from various types of optical satellite images. Over a broad scale, MODIS imagery is commonly used because of its fine temporal resolution. It has been used to detect Hurricane Katrina-induced damage to the coastal De Soto National Forest in southern Mississippi. This detection is based on the abrupt canopy modification as reflected by changes in VIs. The loss of green leaves is directly correlated with a decreased LAI derived from optical spectral bands and indirectly correlated with a reduction in canopy-level total chlorophyll and water content (Wang et al., 2010). There exists a linear relationship between differencing normalized difference infrared index (NDII) (Eq. 9.2) and the overall damage severity, suggesting that time-series ΔNDII allows hurricane-induced forest damage severity to be quantified at a reasonable accuracy.

$$NDII = \frac{\rho_{NIR} - \rho_{SWIR}}{\rho_{NIR} + \rho_{SWIR}} \tag{9.2}$$

Post-hurricane forest damage has been assessed using different methods based on different indicators (Table 9.2). The easiest way of assessing the damage is to produce two NDVI images and then compare the pre- and post-hurricane NDVI values at the same location. For instance, the Lienhuachi

TABLE 9.2

Summary of Post-Hurricane Forest Damage Assessments Carried Out Using Different Data/Bands, Methods, and Indicators

Hurricane/ Typhoon	Time	Data	Band	Indicator	Method	Authors
Georges	09/1998	AVHRR	Red, NIR	NDVI	UID[a]	Ayala-Silva and Twumasi (2004)
Herb	07/1996	SPOT	Red, NIR	NDVI	UID	Lee et al. (2008)
Songda	09/2004	AVHRR	Red, NIR, SWIR	NDVI, NDII, LAI	UID	Aosier and Kaneko (2007)
Katrina	08/2005	Landsat TM/MODIS	NIR, SWIR, visible	Non-photosynthetic vegetation	Spectral mixture analysis	Chambers et al. (2007)
				NDII, NDVI, EVI, FPAR	Linear weight combination	Wang et al. (2010)
Fran	09/1996	SLICER	NIR laser	Canopy height		Boutet and Weishampel (2003)
Lothar	12/1999	CARABAS-II SAR	VHF-band	Backscattering amplitude	Linear regression	Fransson et al. (2002)
Saomai						Zhang et al. (2013)
					UID, PCA, CVA[b], post-classification	Wang and Xu (2010)

Source: Modified from Wang et al., 2010. Used with permission.

[a] UID – univariate image differencing

[b] CVA – change vector analysis

Experimental Forest, a low-lying, evergreen natural hardwood forest of central Taiwan (nearly half has been replaced by conifer plantation forest, mostly below 800 m), had an NDVI value of 0.89 derived from SPOT multispectral data of 20 m resolution captured four weeks before typhoon Herb (maximum wind speed = 53 $m \cdot s^{-1}$ but only 39.5 $m \cdot s^{-1}$ near the forest) (Lee et al., 2008). The NDVI value dropped to 0.82 17 days afterward. The typhoon-induced NDVI decrease is only 0.03 (0.93 – 0.90) because the forest is located more than 100 km to the south of the typhoon path. Such a negligible decrease is explained by the repetitive damage to this area by successive typhoons in the past. They averaged 0.7 per annum.

Apart from NDVI and EVI, terrestrial ecosystem damage by hurricanes has also been assessed from changes in LAI and FPAR, as well as the adjusted NDII. ΔNDVI, ΔNDII, and ΔEVI can all indicate post-hurricane forest damage to a certain degree, but ΔLAI and ΔFPAR fail to identify regions that have suffered the most damage (Wang et al., 2010). ΔNDII is the best indicator of total damage to a forest among the five indices. After proper grouping of ΔNDII values, it is possible to produce a map of categorical damage, such as no, light, moderate, and severe. The overall severity of damage is calculated from the weighted average of these four levels, adjusted by tree size or diameter at breast height and tree density measured within a plot. There is a strong linear relationship ($R^2 = 0.79$, p-value < 0.0001) between ΔNDII and in situ measured damage severity of the hurricane-impacted forest. This relationship demonstrates that the relative change of pre- and post-hurricane NDII value is able to not only quantify region-wide total damage to a forest but also indicate the damage severity (Wang et al., 2010), as illustrated in Figure 9.10.

One way of identifying disturbed forests is to classify the composite images of six indices that include NIR-to-red ratio, NDVI, Tasseled Cap index of greenness, brightness, wetness (TCW), and SAVI. Damaged forests are also detectable using univariate image differencing (UID), selective PCA, change vector analysis, and post-classification comparison (Wang and Xu, 2010). The magnitudes of forest changes in the form of continuous change imagery are computed using a quadratic-square-root transformation of the vegetation indices between the pre- and post-hurricane. Of the four change detection methods, post-classification comparison along with a composite image achieves the highest accuracy and lowest error (0.5%) in estimating the area of disturbed forest.

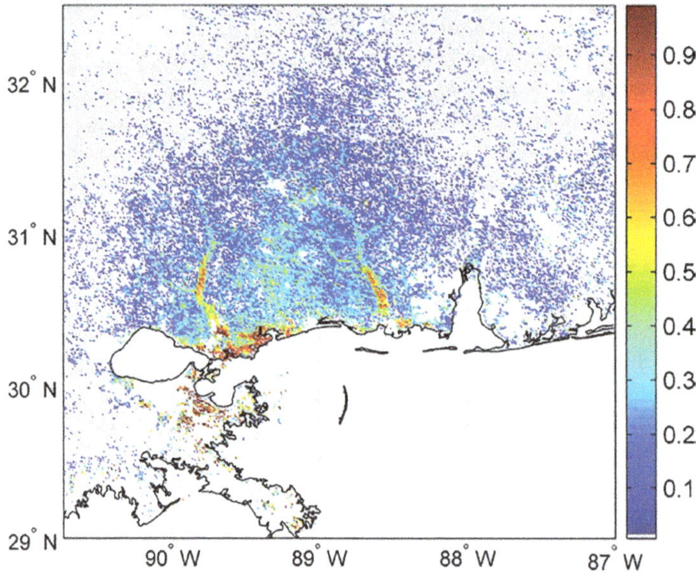

FIGURE 9.10 Regional-scale damage severity (scale: 0–1) to the coastal De Soto National Forest in southern Mississippi/Louisiana caused by hurricane Katrina on 29 August 2005, estimated from a linear weighted combination of four factors. (Source: Wang et al., 2010. Used with permission.)

Both UID and change vector analysis have a similar performance. Among the six indices, TCW outperforms the others owing to its maximum sensitivity to forest modification.

Another way of assessing the severity of forest damage by hurricanes is to use spectral mixing analysis that can quantify the fraction of each endmember in a pixel, including green vegetation, non-photosynthetic vegetation (NPV), and shade. After the shade portion is removed through pixel value standardization, the difference in NPV (ΔNPV) is considered to represent the severity of disturbance (Figure 9.11). Through regression analysis of ΔNPV against other factors, it is possible to quantify the contribution of each factor to the detected disturbance severity, such as forest structure, tree phenology characteristics, and landform. Forest damage induced by a typhoon is also detectable via NDVI differencing. The regression of ΔNDVI against a number of factors, such as tree size, canopy structure, biogeographical origin, wood density, local topography, and wind intensity, as well as environmental factors (e.g., elevation, aspect, and distance to typhoon path), is able to pinpoint the exact causes of damage and identify those at-risk forests of hurricane damage. Most of such data are derivable from satellite data products, including DEM and global forest canopy height.

Overlay of the generated damage level map with landform and forest characteristics can reveal how the observed disturbance patterns vary with environmental variables, such as wind velocity and elevation, as well as the type of forest most at risk of hurricane damage. For instance, forests located over 700 m suffered a decrease of 0.09 in NDVI value, three times that of below 700 m trees, indicating the homogenizing effect of typhoons on vegetation cover (Lee et al., 2008). The fact that the number of pixels having a reduced NDVI is lower at a higher elevation, but their percentage is higher at a lower elevation suggests that trees at the top of a mountain are more exposed and more vulnerable to damage than their low-lying counterparts in the same forest. Besides, trees growing on windward slopes are more easily damaged than those on the leeward side. Furthermore, the mean NDVI of the natural hardwood forest significantly decreased from 0.95 ± 0.04 to 0.87 ± 0.05, in contrast to the 0.89 ± 0.08 to 0.83 ± 0.05 decrease of the conifer plantation forest (paired *t*-test, both p values <0.001) because it is located at a lower elevation. The NDVI of the natural hardwood forest decreased significantly more than that of the conifer plantation forest (one-way ANOVA, $p < 0.001$). These findings have been confirmed in another study and are accounted for by the fact that more aggregated trees are harder to be damaged easily (Zhang et al., 2013). But the distance from the typhoon path shows no obvious diverse influence on tree damage within 25 km. However, if the distance is shorter, higher levels of disturbance to coastal forests are found closer to the hurricane track, and the disturbance lessened as the hurricane moved further inland (Feng et al., 2020).

9.4.3 Coastal Marine Ecosystem Change

Apart from human activities, coastal ecosystem disturbances can also be triggered by extreme meteorological events such as hurricanes. Although hurricane winds cannot penetrate the ocean water deeply, they can still exert a profound impact on coastal marine ecosystems because they induce vertical entrainment, mixing, upwelling, and heat loss to the atmosphere through evaporation, and hence cool SST sharply. In turn, SST and the mixed layer depth are significantly correlated with surface chlorophyll-a (Chla) enhancement. Through their interactions with oceanic eddies, such as intensifying pre-existing cyclonic eddies, suppressing pre-existing anti-cyclonic eddies, and inducing the formation of cyclonic eddies under non-eddy conditions, hurricanes also profoundly influence the growth and redistribution of phytoplankton. They enhance Chla in the surface layer in two ways: (i) upward entrainment of deep maximum Chla and (ii) production of new phytoplankton owing to the supplemented nutrients in the euphotic layer. Such marine ecosystem changes can be detected using remote sensing reliably and readily via sea level anomaly (SLA). Daily SLA products are available at a spatial resolution of 0.25° and daily merged Chla products have been released by the GlobColour project (http://hermes.acri.fr). They have been retrieved from widely ranging sources using different methods. Chla is retrieved using a neural network algorithm, while OC4v5 from SeaWiFS, OC4Me from MERIS, and OC3v5 from MODIS and VIIRS data. These diverse

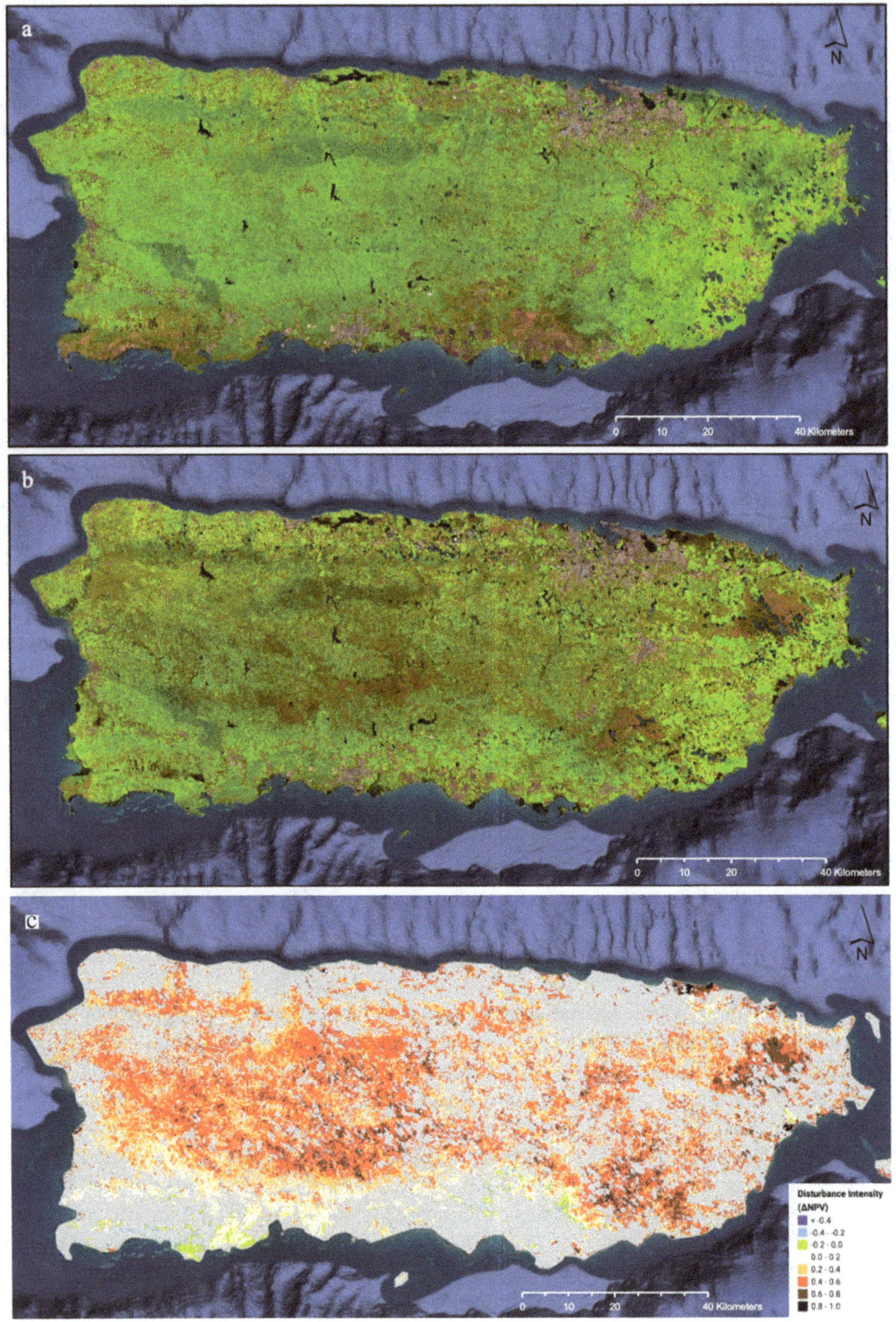

FIGURE 9.11 False-color composite (SWIR band – R, NIR band – G, and Red band – B) of Landsat 8 images of Puerto Rico before hurricane María (20 September 2017), centered at (18.20°N, 66.48°W). (a) Before hurricane; (b) after hurricane; and (c) disturbance level one month after the event. (Source: Feng et al., 2020. Used with permission.)

data products may be merged via weighted averaging. Besides, Chla can also be calculated from daily MODIS reflectance at 443, 488, 531, 555, 645, 667, and 678 nm.

One of the most important marine ecosystems is seagrass beds that are especially vulnerable to hurricane impacts due to their shallow depth. They may suffer from hurricane-triggered die-off. Shallow seagrass beds are detectable from satellite images, even in turbid bay waters. All seagrass spectra from several optical sensors, including Landsat TM/ETM+, ASTER, SPOT-4, and KOMPSAT-2, show low reflectance in the green wavelength and high reflectance in the NIR spectrum (Kim et al., 2015). The spectral pattern of underwater seagrass is markedly distorted by the water column, which makes it difficult to accurately map seagrass beds. The water column effect can be avoided or minimized by using low-tide images or images recorded in the season when the water has relatively high clarity. In this way, seagrass beds are mapped at a rather high user's accuracy of 71.4% via the automatic classification of KOMPSAT-2 multispectral bands of 4 m resolution. The same mapping may be repeated using images obtained after a hurricane. A comparison of the area of seagrass beds before and after the hurricane reveals a die-off of seagrass beds off the southern coast of South Korea in 2012. It was hit by three consecutive typhoons of Bolaven (27 August), Tembin (30 August), and Sanba (17 September), coinciding with a die-off of seagrass beds in September of the same year. However, it is impossible to attribute the observed changes to a specific typhoon event or typhoons at all. It remains to be ascertained whether typhoons can result in an immediate loss of seagrass beds. The answer to this question lies in examining the long-term trend of the seagrass beds and comparing it with the occurrence of hurricanes. If no similar drastic reduction in seagrass beds by typhoons is observed in the preceding decade, the observed decline in seagrass beds cannot be traced to hurricanes. While remote sensing is effective at detecting marine ecosystem changes over time and relating them to the timing of typhoons, it cannot supply solid evidence to prove that the changes are caused by hurricanes due to the effects of other biological processes (e.g., resilience and self-recovery). Nor can it enlighten how the ecosystem has responded to hurricane disturbances in this case.

The investigation into how phytoplankton has responded to consecutive typhoons has to rely on remote sensing in combination with numerical simulation and Argo profile data (Ma et al., 2021). Numerical simulation data pertain to water salinity and temperature, and the Argo temperature and salinity profiles are needed to determine the mixed layer depth, the depth at which the water density gradient first reaches 0.01 $kg{\cdot}m^{-3}$. Typhoon-induced variations of phytoplankton are detected from regional Chla concentration maps before and after the event. The change in Chla concentration is then linked to physical and biological parameters of the ocean such as SST and SLA. Responses of phytoplankton by size to hurricanes can be further separated by the continental shelf and the deep ocean. The former experienced a 1.5-fold increment in micro- and nano-phytoplankton as high as that of pico-phytoplankton, three times higher than the 0.5-fold increment in the deep ocean. Thus, phytoplankton of different size classes in different parts of the ocean responds to the same hurricane differentially (Ma et al., 2021). Phytoplankton in the shallower continental shelf is more responsive to hurricanes than their counterparts in deeper ocean waters because the hurricane impacts cannot penetrate deep into them.

9.4.4 Water Quality

Hurricanes affect not only the coastal marine ecosystems but also the physical and biochemical properties of nearshore seawaters, such as re-suspending sediments in estuaries. Assessment of hurricane impacts on estuarine water quality is significant to coastal management. The assessment of hurricane-induced total suspended sediments (TSS) concentration in estuaries is best carried out using cloud-free optical images with visible light bands that can penetrate water deeper than longer wavebands. This sensing is theoretically founded on the curvilinear relationship between suspended sediments and their reflectance in these spectral bands. The relationship is linear at a low TSS level below 600 $mg{\cdot}L^{-1}$ but non-linear above this level. TSS is usually estimated from multiple bands.

Their ratio suppresses the background influences (e.g., atmospheric and solar effects). For instance, the ratio of logarithmically transformed MODIS band 1 to band 2 is effective at mapping TSS concentration. The intercept of the linear regression model of in-water reflectance against TSS is regarded as representing the atmospheric effect and should be eliminated. This correction substantially improves the estimation of TSS concentration. The best regression model between the corrected MODIS bands of 250 m resolution and the observed TSS concentrations is expressed as:

$$\log(b2)/\log(b1) = -0.1356[\log(TSS)]^2 + 0.7402\log(TSS) + 0.6836 \ (R^2 = 0.85, p < 0.001) \quad (9.3)$$

Model validation reveals an RMSE of 5.5 mg·L^{-1} (n = 21). Thus, the MODIS b1/b2 reflectance–TSS curvilinear regression model can be used to map large-scale TSS concentrations during a hurricane (Chen et al., 2009). Application of this model to the pre-, concurrent-, and post-hurricane MODIS images reveals drastic changes in TSS concentrations in the Apalachicola Bay of Florida (Figure 9.12). The mean concentration of about 21.8 mg·L^{-1} peaked at 54.9 mg·L^{-1} during the hurricane, almost three times higher than the pre-hurricane level. It declined to 23.8 mg·L^{-1} five days after the hurricane had passed. This level is still twice higher than the pre-hurricane concentration, suggesting that the fierce winds of the hurricane have caused strong sediment re-suspension in shallow estuaries.

Additionally, the downpour of a hurricane may also induce a huge amount of surface runoff in urban areas that is discharged into the nearby bays or estuaries. The storm water from urban areas may cause biochemical pollution near the coast as the injected effluences into coastal water promote the growth and thriving of water column phytoplankton. High levels of Chla concentration often signal poor water quality, and a persistently high level may lead to algal blooms. The abnormally high Chla is easily detectable from optical images of a coarse spatial resolution such as MODIS data. Freely available MODIS images of 250 m resolution are the ideal data source for monitoring the impact of hurricanes on the bio-quality of nearshore water because of their extremely fine temporal resolution of 1–2 days. The influence of hurricanes on the biochemical water quality of coastal areas is usually studied by comparing pre- and post-event Chla concentrations estimated

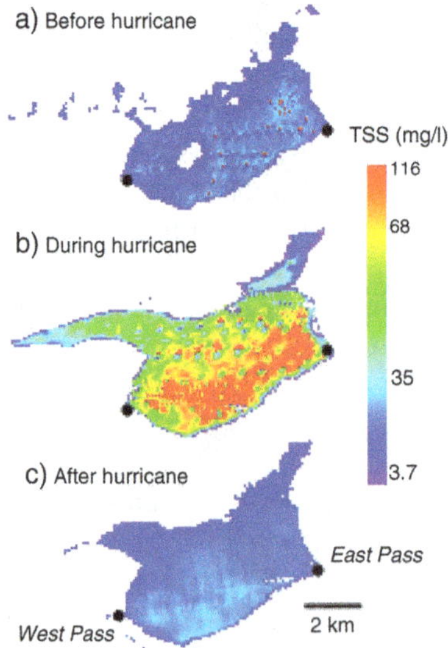

FIGURE 9.12 Comparison of total suspended sediments in the Apalachicola Bay of Florida (a) 5 days before, (b) during, and (c) 5 days after Hurricane Frances on 4 September 2004. (Source: Chen et al., 2009. Used with permission.)

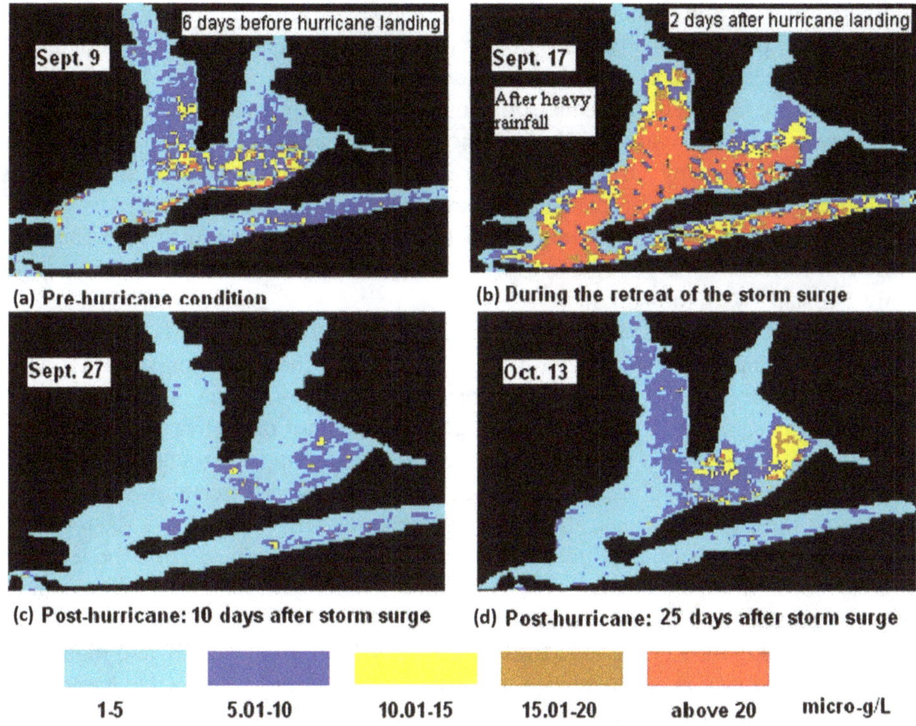

FIGURE 9.13 Concentration of Chla in the Pensacola Bay of Florida (a) before the landfall of Hurricane Ivan on 17 September, (b) during landfall, (c) 10 days, and (d) 26 days after the retreat of the storm surge retrieved from a logarithmic model based on the ratio of MODIS bands 1 and 2. (Source: Huang et al., 2011. Used with permission.)

from respective images. After logarithmic transformation, the relationship of Chla with the ratio of MODIS band 1 to band 2 is best expressed as:

$$\log(Chla) = 125121\left[\log(\text{b1})/\log(\text{b2}) + 1.5\right]^{-13.8} \quad (R^2 = 0.67) \tag{9.4}$$

The application of this model to MODIS images enables the spatial distribution of Chla concentration to be mapped and quantified (Figure 9.13). Prior to the landfall of Hurricane Ivan on 5 September 2004, the Pensacola Bay of Florida had a mean Chla concentration of 5.3 $\mu g \cdot L^{-1}$ (Huang et al., 2011). It rose substantially to 14.7$\mu g \cdot L^{-1}$, with more than a quarter of the water area having a poor water quality as Chla shooted up to > 20 $\mu g \cdot L^{-1}$. This is because the hurricane dumped 400 mm of rainfall during its 48-hour landfall near the Pensacola Bay. At the end of the storm surge on 18 September 2004, the mean Chla level bounced back to 2.0 $\mu g \cdot L^{-1}$, because of the effective flushing of the polluted water from the Bay to the Gulf of Mexico by stormwater (Figure 9.13). Heavy nutrient loads from urban stormwater runoff and storm-surge inundation stimulated the temporary Chla bloom. The river flow peaked four days after the peak storm surge but did not alter the good water quality in the Bay, suggesting that urban stormwater runoff instead of the river inflow is the major perpetrator of water pollution in the Bay during the hurricane.

9.5 SENSING OF TORNADOES

A tornado is a violently twisting column of air extending from a thunderstorm all the way to the ground. Virtually, just a wind of a tremendous speed, a tornado is invisible unless it forms a condensation funnel comprising water droplets, dust, and debris. Meteorologically, it is a cyclone particularly prevalent in mid-western US/Canada. Tornadoes are also a deadly natural hazard owing to their sheer wind speed up to 180 km·h^{-1}. In extreme cases, they can have a diameter of up to 3 km, reach a top wind speed of 480 km·h^{-1}, and touch the ground for up to more than 100 km. Dissimilar to hurricanes that are formed over oceans, tornadoes form unexceptionally over terrestrial surfaces. Moreover, they have a much shorter life expectancy than hurricanes, traveling a couple of kilometers before dissipating, leaving a trail of destruction behind. For this reason, it is almost impossible to monitor tornadoes, let alone forecast and prepare for them. So far only a handful of studies have been carried out using remote sensing, most of them focused on the assessment of tornado destruction using weather radar data. They serve as the primary data source for detecting hook signatures commonly associated with tornadoes.

Tornado damage is studied at four scales, ranging from path, neighborhood, individual buildings, to member connection (Womble et al., 2018), of which the last is least studied while the first is the most researched. Tornado paths or tracks are highly visible on false-color composites of multispectral images due to their distinctive color formed out of destroyed vegetation and/or littered debris on the damaged path, in huge contrast to the intact surroundings (Figure 9.14). The savage winds of a tornado may have uplifted furniture, snapped tree branches, blown down tree trunks, and shredded buildings, all of which contribute to the pale color of the tornado track on the composite. Analysis of satellite imagery is able to yield information on the length of the tornado touchdown, the spatial extent of affected areas, and the spatial variability of damage as judged by the quantity of debris.

As illustrated in Figure 9.14, tornado tracks are detectable from moderate- and high-resolution images. If the touchdown area is vegetated, the track is automatically detectable based on NDVI derived from Earth resources satellite images such as Landsat TM, ETM+, ASTER, and Sentinel-2. If the tracks are highly distinctive, they can be identified at least partially via visual inspection of a single NDVI image (Molthan et al., 2014), even though long-term NDVI change and image-enhancement techniques, such as PCA, facilitate visual inspection. The rate of successfully detecting tornado damage tracks from satellite images is related to the severity of damage. Areas suffering from estimated maximum damage of around EF (equivalent to T0 through T3)-2 are identified at a

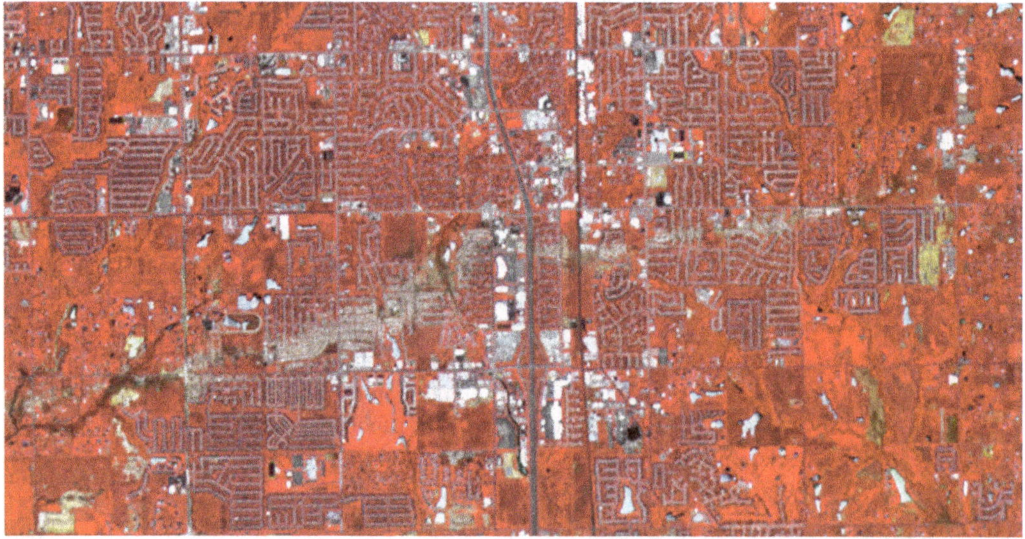

FIGURE 9.14 Track (absence of vegetation) of the EF-5 tornado that struck Moore, Oklahoma on 20 May 2013 on the false-color composite of an ASTER image acquired on 2 June 2013 (R – infrared, G – Red, and B – Green Bands). (Source: Womble et al., 2018.)

rate of 65.6%. The satellite-detected lengths of tornado tracks are favorably comparable to official survey results at an R^2 of 0.88–0.93, but the maximum widths are detected much less reliably at a lower R^2 value of 0.39–0.52.

Tornado damage tracks are directly detectable from two images acquired prior to and post the tornado touchdown after they have been mapped using supervised classification, unsupervised classification, or object-oriented classification through spatial comparison of the classified results. Of these classification methods, the object-oriented approach achieved the highest accuracy in detecting the damage track of a tornado that landed in Oklahoma City on 3 May 1999, followed by PCA and image differencing of Landsat TM imagery (Myint et al., 2008). Object-oriented classification of PCA components 3 and 4 produced the most accurate detection (98.3%) that is 15–20% more accurate than the other two methods that produced comparable outcomes. While the selected PCs can improve detection accuracy by 5–10% (Figure 9.15), unsupervised classification of PC components 2, 3, and 4 using clustering analysis into binary damaged vs non-damaged categories achieved an overall accuracy of 84.2%, or 85.0% using only components 3 and 4. Both are slightly higher than 79.2% achieved using the maximum likelihood classifier, though the accuracy rose to 87.5% if PC components 3 and 4 were classified. Naturally, the accuracy of detection will decrease in sparsely vegetated regions where spectral bands of images of a coarser spatial resolution may have difficulty discriminating building debris from intact buildings.

Neighborhood-scale building damage caused by tornados can be assessed using the same methods as those used to assess earthquake damage (tornado damage to forests can be studied in the same way as hurricane-triggered forest damage, so is not repeated here). Tornado damage assessment at the level of individual structures and cladding panels requires either LiDAR data or super-fine resolution imagery. If the affected area is rural where building density is low, it is possible to assess the damage to individual buildings using super-fine resolution photographs or low-altitude drone images. The immediate aftermath assessment is ideally achieved using an unpiloted aerial

FIGURE 9.15 Damage track of Oklahoma City Tornado outbreak on 3 May 1999, shown in PCA component 3 of a Landsat TM image captured on 12 May 1999. (Source: Myint et al., 2008.)

system (UAS), the strongest advantages of which are flexibility and fine resolution, in addition to a low operation cost. It can fly over the affected area soon after the tornado has passed or dissipated, even over unnavigable areas because of road blockage by tornado debris. A super-fine resolution image of up to 1.2 cm enables the identification of damage that is not even observable on the ground or from satellite images (Wagner et al., 2019). Acquired at a height lower than 130 m above the ground, UAS imagery is less subject to the impact of clouds and haze than its space-borne counterparts. If the imagery is multispectral or hyperspectral, it can differentiate bare ground from wind-strewn hay, so UAS is perfectly suited to assess the impacts of tornado high wind, such as erosion, scour, soil deposition, and topographic interactions, even in areas of low vegetation cover. If combined with RapidEye imagery of 5 m resolution, UAS images enable both the extent and detail of high wind damage to be studied. Furthermore, structure-from-motion and 3D products derived from UAS images allow the study of the interaction of wind damage with topography. This approach is a cost-effective alternative to airborne LiDAR without an overwhelmingly large volume of data to process. However, the acquisition of the perfect UAS data requires careful consideration and planning of flight operations so as to conform to the official airspace regulations as no flight is permitted in certain parts of a country such as airports, military bases, and national parks. It is absolutely imperative to obtain prior permission for flight from relevant authorities. The operator of the system may have to undergo training to obtain the certificate to operate a UAS. Logistically, the ground area to be covered and the operating life expectancy of the drone battery must also be factored in planning the flight route.

The estimation of tornado damage from UAS imagery can be carried out at binary (damage vs no damage) or triple levels (no damage, minor damage, and major damage), all based on various indicators, including the portion of roof cover failure, roof structure failure, the quantity of debris, and tree health (Chen et al., 2021). This method of estimation may be automated using machine learning methods, such as ANN and SVM based on OBIA. OBIA can produce detailed, spatially explicit tornado damage heatmaps for individual buildings at an accuracy between mid-30% (multi-class) and almost 60% (binary classes). Tornado damage heatmaps can be produced from the Gaussian process (also known as kriging) regression analysis by merging the binary (damage vs non-damage) object detection and image-classified heatmaps. They illustrate spatially continuous tornado damage estimates that can be used to generate EF-scale contours and better understand high-wind impacts. Such timely information is valuable for emergency response personnel and other decision-makers to gauge the damage quickly and devise proper disaster recovery strategies and plans. The produced damage map can assist insurance agents and emergency personnel to quantify the economic loss of the tornado damage.

REFERENCES

Aosier, B., and M. Kaneko. 2007. Evaluation of the forest damage by typhoon using remote sensing technique. In *IEEE Int. Conf. Geosci. and Remote Sens. Symp.*, IGARSS 2007, Barcelona, Spain, July 23–27.

Ayala-Silva, T., and Y.A. Twumasi. 2004. Hurricane Georges and vegetation change in Puerto Rico using AVHRR satellite data. *Int. J. Remote Sens.* 25(9): 1629–1640.

Boutet, J.C. Jr., and J.F. Weishampel. 2003. Spatial pattern analysis of pre- and posthurricane forest canopy structure in North Carolina, USA. *Landsc. Ecol.* 18: 553–559.

Chambers, J.Q., J.I. Fisher, H. Zeng, E.L. Chapman, D.B. Baker, and G.C. Hurtt. 2007. Hurricane Katrina's carbon footprint on US gulf coast forests. *Science* 318: 1107.

Chen, S., Y. Hong, Q. Cao, P.E. Kirstetter, J.J. Gourley, Y. Qi, J. Zhang, K. Howard, J. Hu, and J. Wang. 2013. Performance evaluation of radar and satellite rainfalls for Typhoon Morakot over Taiwan: Are remote-sensing products ready for gauge denial scenario of extreme events? *J. Hydrol.* 506: 4–13. DOI: 10.1016/j.jhydrol.2012.12.026.

Chen, S., W. Huang, H. Wang, and D. Li. 2009. Remote sensing assessment of sediment re-suspension during Hurricane Frances in Apalachicola Bay, USA. *Remote Sens. Environ.* 113(12): 2670–2681. DOI: 10.1016/j.rse.2009.08.005.

Chen, Z., M. Wagner, J. Das, R.K. Doe, and R.S. Cerveny. 2021. Data-driven approaches for tornado damage estimation with unpiloted aerial systems. *Remote Sens.* 13, 1669. DOI: 10.3390/rs13091669.

Corbley, K.P. 2006. Rapid post-disaster mapping: Airborne remote sensing and Hurricane Katrina. *GIM Int.* 20(8): 19.21.

Feng, Y., R.I. Negrón-Juárez, and J.Q. Chambers. 2020. Remote sensing and statistical analysis of the effects of hurricane María on the forests of Puerto Rico. *Remote Sens. Environ.* 247. DOI: 10.1016/j.rse.2020.111940.

Fransson, J.E., F. Walter, K. Blennow, A. Gustavsson, and L.M.H. Ulander. 2002. Detection of storm-damaged forested areas using airborne CARABAS-II VHF SAR image data. *IEEE Trans. Geosci. Remote Sens.* 40(10): 2170–2175.

Guimond, S.R., G.M. Heymsfield, P.D. Reasor, and A.C. Didlake. 2016. The rapid intensification of Hurricane Karl (2010): New remote sensing observations of convective bursts from the global hawk platform. *J. Atmos. Sci.* 73(9): 3617–3639. DOI: 10.1175/JAS-D-16-0026.1.

Horstmann, J., S. Falchetti, C. Wackerman, S. Maresca, M.J. Caruso, and H.C. Graber. 2015. Tropical cyclone winds retrieved from C-band cross-polarized synthetic aperture radar. *IEEE Trans. Geosci. Remote Sens.* 53(5): 2887–2898. DOI: 10.1109/TGRS.2014.2366433.

Huang, W., D. Mukherjee, and S. Chen. 2011. Assessment of Hurricane Ivan impact on chlorophyll-a in Pensacola Bay by MODIS 250m remote sensing. *Mar. Pollut. Bull.* 62(3): 490–498. DOI: 10.1016/j.marpolbul.2010.12.010.

Kim, K., J.-K. Choi, J.-H. Ryu, H.J. Jeong, K. Lee, M.G. Park, and K.Y. Kim. 2015. Observation of typhoon-induced seagrass die-off using remote sensing. *Estuarine, Coastal Shelf Sci.* 154: 111–121. DOI: 10.1016/j.ecss.2014.12.036.

Lam, N.S.-N., K.-B. Liu, W. Liang, T.A. Bianchette, and W.J. Platt. 2011. Effects of Hurricanes on the Gulf Coast ecosystems: A remote sensing study of land cover change around Weeks Bay, Alabama. *J. Coastal Res.* Spec. issue 64: 1707–1711.

Lee, M.-F., T.-C. Lin, M.A. Vadeboncoeur, and J.-L. Hwong. 2008. Remote sensing assessment of Forest damage in relation to the 1996 strong typhoon Herb at Lienhuachi Experimental Forest, Taiwan. *Forest Ecol. Manage.* 255: 3297–3306. DOI: 10.1016/j.foreco.2008.02.010.

Ma, C., J. Zhao, B. Ai, S. Sun, G. Zhang, W. Huang, and G. Wang. 2021. Assessing responses of phytoplankton to consecutive typhoons by combining Argo, remote sensing and numerical simulation data. *Sci. Total Environ.* 790, 148086. DOI: 10.1016/j.scitotenv.2021.148086.

Molthan, A.L., J.R. Bell, T.A. Cole, and J.E. Burks. 2014. Satellite-based identification of tornado damage tracks from the 27 April 2011 severe weather outbreak. *J. Oper. Meteorol.* 2: 191–208. DOI: 10.15191/nwajom.2014.0216.

Myint, S.W., M. Yuan, R.S. Cerveny, and C.P. Giri. 2008. Comparison of remote sensing image processing techniques to identify tornado damage areas from Landsat TM data. *Sensors* 8: 1128–1156.

Pérez Valentín, J.M., and M.F. Müller. 2020. Impact of Hurricane Maria on beach erosion in Puerto Rico: Remote sensing and causal inference. *Geophys. Res. Lett.* 47(6). DOI: 10.1029/2020GL087306.

Rankey, E.C., P. Enos, K. Steffen, and D. Druke. 2004. Lack of impact of hurricane Michelle on tidal flats, Andros Island, Bahamas: Integrated remote sensing and field observations. *J. Sedimen. Res.* 74(5): 654–661. DOI: 10.1306/021704740654.

Reul, N., J. Tenerelli, B. Chapron, D. Vandemark, Y. Quilfen, and Y. Kerr. 2011. SMOS satellite L-band radiometer: A new capability for ocean surface remote sensing in hurricanes. *J. Geophy. Res. C: Oceans* 117(2). DOI: 10.1029/2011JC007474.

Shao, J., X. Li, and J. Sun. 2015. Ocean wave parameters retrieval from TerraSAR-X images validated against buoy measurements and model results. *Remote Sens.* 7: 12815–12828. DOI: 10.3390/rs71012815.

Uhlhorn, E.W., and P.G. Black. 2003. Verification of remotely sensed sea surface winds in hurricanes. *J. Atmos. Oceanic Technol.* 20(1): 99–116. DOI: 10.1175/1520-0426(2003)020<0099:VORSSS>2.0.CO;2

Wagner, M., R.K. Doe, A. Johnson, Z. Chen, J. Das, and R.S. Cerveny. 2019. Unpiloted aerial systems (UASs) application for tornado damage surveys: Benefits and procedures. *Bull. Amer. Meteorol. Soc.* 100(12): 2405–2409. DOI: 10.1175/BAMS-D-19-0124.1

Wang, W., J.J. Qu, X. Hao, Y. Liu, and J.A. Stanturf. 2010. Post-hurricane forest damage assessment using satellite remote sensing. *Agricul. and Forest Meteorol.* 150(1): 122–132. DOI: 10.1016/j.agrformet.2009.09.009.

Wang, F., and Y.J. Xu. 2010. Comparison of remote sensing change detection techniques for assessing hurricane damage to forests. *Environ. Monit. Assess.* 162(1-4): 311–326. DOI: 10.1007/s10661-009-0798-8.

Womble, J.A., R.L. Wood, and M.E. Mohammadi. 2018. Multi-scale remote sensing of tornado effects. *Front. Built Environ.* 4: 66. DOI: 10.3389/fbuil.2018.00066.

Zhang, G., X. Li, W. Perrie, P.A. Hwang, B. Zhang, and X. Yang. 2017. A hurricane wind speed retrieval model for C-band Radarsat-2 cross-polarization ScanSAR images. *IEEE Trans. Geosci. Remote Sens.* 55(8): 4766–4774. DOI: 10.1109/TGRS.2017.2699622.

Zhang, L., G. Liu, W. Perrie, Y. He, and G. Zhang. 2018. Typhoon/hurricane-generated wind waves inferred from SAR imagery. *Remote Sens.* 10, 1605. DOI: 10.3390/rs10101605

Zhang, B., and W. Perrie. 2012. Cross-polarized synthetic aperture radar: A new potential measurement technique for hurricanes. *Bull. Amer. Meteorol. Soc.* 93(4): 531–541. DOI: 10.1175/BAMS-D-11-00001.1

Zhang, X., Y. Wang, H. Jiang, and X. Wang. 2013. Remote-sensing assessment of forest damage by Typhoon Saomai and its related factors at landscape scale. *Int. J. Remote Sens.* 34(21): 7874–7886. DOI: 10.1080/01431161.2013.827344

10 Floods

10.1 INTRODUCTION

Floods refer to the temporary immersion of the ground surface by water that is normally dry land. They are commonly associated with rivers, mountain torrents, storm surges, rising sea levels, and tsunamis in coastal areas (Table 10.1). Floods are triggered under three circumstances: excessively heavy rain, failure of reservoir dams, and sea-level rise triggered by climate change in coastal areas. The type of floods that can happen to a place is dictated by its location. In the lower reach of a river, floods are commonly caused by heavy rains in the catchment. The rainwater is not discharged timely, leading to a temporary and abrupt rise in water level. This ephemeral phenomenon lasts from hours to days, depending on how quickly the rainwater is discharged. In coastal areas, floods commonly take place in stormy weather, especially during hurricanes and tropical cyclones. Flash floods occur in mountainous areas with a parched catchment shortly after downpours. Each type of flood has its own causes and can result in different consequences and destruction. More importantly, each type requires different methods to study using different remote sensing data, with different foci.

Irrespective of their geographic location and type, floods are one of the most frequently occurring catastrophic natural hazards around the world. Each year, enormous economic losses and casualties are incurred by floods that do enormous damage to properties, infrastructure, and land (Figure 10.1). Compared with other types of natural hazards, floods take place more frequently but tend to be confined to a limited spatial extent when they do occur, especially flash floods. It is significant to identify areas vulnerable to floods, map flooded areas to assist emergency response and disaster recovery, assess flood hazards, and plan for flood risk reduction. The damage caused by flooding can be mitigated via remote sensing monitoring, preferably in real time. Flood monitoring concentrates on flood extent and changes in land cover. Flood extent is relatively easy to map from remote sensing images via classifying all surface covers categorically, at least one of them is related to water. Whether a water feature is permanent or temporary is determined by examining it on multi-temporal images captured before and after the flooding event. The water body that is absent from the pre-event image represents the flooded area. However, the environmental settings in which flooding takes place likely complicate the mapping.

This chapter first explains how to map flooded areas using remote sensing and detect water bodies, even in challenging environments, using a number of approaches and spectral indices. This explanation is followed by an assessment of flood damage (FD). How to evaluate flood vulnerability, risk, and hazard forms the content of Section 10.3. Also covered in the same section is an overview and evaluation of various flood forecasting and warning systems, and remote sensing data products useful for predicting floods. Section 10.4 expounds how to study flash floods using remote sensing. Finally, this chapter elaborates on how to sense coastal floods triggered by hurricanes. Topics explored in this section encompass assessment of coastal vulnerability to floods, and estimation of hurricane-induced flood depth. A particular type of flooding associated with tsunamis is also covered in the same section. The discussion emphasizes how to detect tsunami waves, map inundated areas, assess tsunami damage, and estimate tsunami debris volume.

10.2 INUNDATION EXTENT MAPPING

10.2.1 Mapping Flooded Areas

The severity and potential damage of a flood are related to the extent of inundation. Inundated areas are easily mapped from optical images using either unsupervised or supervised classification methods, such as the parallelepiped classifier (Bonn and Dixon, 2005). Water can be mapped easily

DOI: 10.1201/9781003354321-10

TABLE 10.1
Comparison of Causes, Consequences, and Relevant Parameters of Floods by Type

Type	Causes	Consequences	Relevant Parameters
River flooding in floodplains	Intensive rainfall and/or snowmelt Ice jam, clogging Collapse of dikes or other protective structures	Stagnant or flowing water outside the channel	Extent (according to probability) Water depth Water velocity Propagation of flood
Coastal flooding	Storm surge Tsunami High tide	Stagnant or flowing water behind the shoreline Salinization of agricultural land	Same as above
Mountain torrent activity or rapid runoff from hills	Cloud burst Lake outburst Slope instability in a watershed Debris flow	Water and sediments outside the channel on alluvial fans; erosion along channel	Same as above; Sediment deposition
Flash floods in the Mediterranean ephemeral water courses	Cloud burst	Water and sediments outside the channel on alluvial fans Erosion along a channel	Same as above
Groundwater flooding	High water level in adjacent water bodies	Stagnant water in floodplain (long-lasting flooding)	Extent (according to probability) Water depth
Lake flooding	Water level rise through inflow or wind-induced setup	Stagnant water behind the shoreline	Same as above

Source: EXCIMAP (2007).

using even a single spectral band due to its characteristically low spectral response that is highly distinctive from other non-water features (see Section 4.5). These two classes are easily distinguishable from each other owing to the large spectral contrast between them. Open water, especially those extensive lakes, can be detected accurately using a single threshold. However, the spatial co-existence of water with aquatic vegetation in small lakes and wetlands complicates the detection.

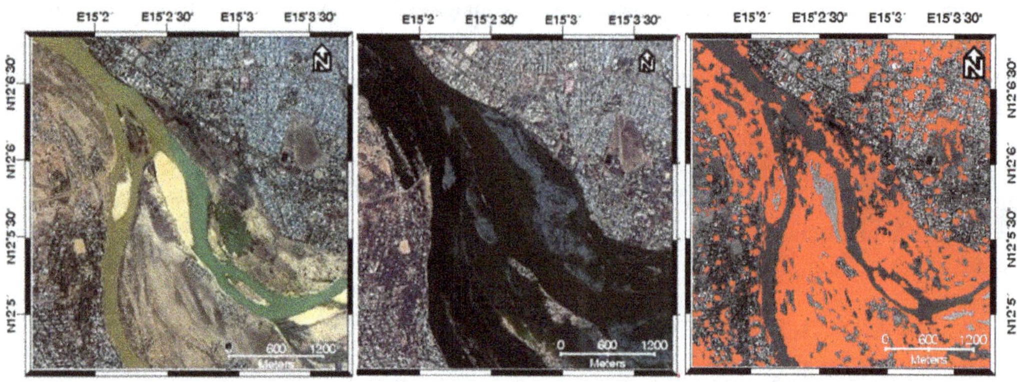

FIGURE 10.1 GSA-fused images generated from KOMPSAT-2 satellite images captured before the flood event (left) and after the flood event (middle) caused by tropical cyclones in the city of Ndjamena in Chad, and the detected flooded areas in red (right). (Source: Byun et al., 2015.)

In such challenging environments, the successful separation of flooded areas from non-water features requires the use of sophisticated methods that take advantage of multiple spectral bands and vegetation indices derived from them.

10.2.1.1 Suitable Images

Detection of flooded areas is virtually a change detection, during which areas that are not covered by water in the pre-flood image but are covered by water in the post-flood image are inundated. The pre-event image is likely to be in existence already if it is an optical Earth resources image such as Sentinel-2. However, the right post-event image may not exist. More likely, it is acquired long after the flood event because of cloud cover or the long satellite revisit period. The delay in acquiring the post-event image is avoidable by using drones that can be deployed flexibly for quick assessment. Each drone image covers a small ground area, so drone (UAS) images are excellent for mapping small-extent floods, such as inland areas near a river channel under clouds. Extensive floods are better mapped from aerial photographs or satellite images. Flooded areas can be delineated from optical images with relative ease, regardless of whether they are airborne or space-borne. All optical images face one hurdle in mapping flood extent when it is still raining heavily (after all, it is the rain that causes the flood in the first place), namely, cloud cover. If an image is contaminated by thick clouds, it does not allow the flood extent to be mapped. On the other hand, by the time the sky clears, the flood water may have receded, causing the water fraction mapped from optical imagery to be lower than the ground observations (Sun et al., 2016). This deficiency can be overcome by combining optical images with radar imagery data to timely detect flooded areas.

Compared with optical imagery, radar images that are recorded using microwave radiation can penetrate clouds (not higher frequency channels that cannot penetrate precipitating clouds). On radar images, flooded areas tend to have a smooth surface and a low backscattering coefficient. Not impacted by the weather, radar data can be used to map the changing flood extent throughout a flooding event. In particular, radar images are good at detecting surface roughness that may be smoothed by sedimentation after the floodwater has subsided. On RadarSat-1 images, there is always a very sharp contrast between land and water surfaces, especially if the images are captured at a high incidence angle where the effect of water surface waves is low (Bonn and Dixon, 2005), irrespective of the weather condition. Because of this capability, RadarSat-1 imagery has the potential to monitor and map flooded areas in near real time. Other radar images suitable for mapping floods include ALOS PALSAR, TerraSAR-X, EnviSat ASAR, and Sentinel-1A/1B (Table 1.7). These images may be combined to study floods. Multi-sensor radar images enable a flooding event to be monitored over a prolonged period. This capability is highly significant and desirable as the temporal change of flooded areas serves as a reliable indicator of future change in the flooding situation. Such information facilitates the flood control personnel to decide whether further efforts are needed to contain the flood or to mitigate its potential damages by issuing a warning for evacuation.

However, optical imagery is still preferred if the mapping takes place long after the rain has ceased, such as the assessment of FD. This assessment is eased if the floodwater is laden with silt, which is the case during downpours in rugged areas, especially if landslides have been initiated by heavy rain. If the objective of sensing is to estimate the volume of sediments or flood debris, the best data to use are LiDAR data that are good at detecting the piling up of the debris washed ashore, as in a tsunami (see Section 10.5.2.4 for more details).

10.2.1.2 Spectral Indices

There are several ways of mapping and quantifying flood extent. The first method is to make use of various spectral indices designed to map water specifically, such as NDWI (Eq. 4.17), modified NDWI (MNDWI), differencing NDWI (DNDWI) derived from two medium-resolution bands of

TABLE 10.2
Spectral Indices That Have Been Used to Map Flood Extent from VHR Imagery

Indices	Formula	Authors
VI[a] – vegetation index	NIR/R	Lillesand and Kiefer (1987)
DVI[a] – differential VI	NIR – red	Richardson and Everitt (1992)
DVW[b] – difference between vegetation and water	NDVI – NDWI	Gond et al. (2004)
IFW[a] – index of free water	NIR – green	Adell and Puech (2003)
NDWI[a] – normalized difference water index	(G – NIR)/(G + NIR)	McFeeters (1996)
NDWI-G – normalized difference water index of Gao	(NIR1 – NIR2)/(NIR1 + NIR2)	Gao (1996)
NDVI[a] – normalized difference VI	(NIR – R)/(NIR + R)	Tucker (1979)
SAVI[a] – soil-adjusted VI	1.5(NIR – R)/(NIR + R + 0.5)	Huete (1988)
OSAVI[a] – optimized SAVI	(NIR – R)/(NIR + R + 0.16)	Rondeaux et al. (1996)
SR[a] – simple ratio	R/NIR	Pearson and Miller (1972)
WI[a] – water index	NIR²/B	Davranche et al. (2013)
WII[a] – water impoundment index	NIR²/R	Caillaud et al. (1991)

Source: Malinowski et al. (2015).

[a] Two indices calculated for, respectively, NIR1 and NIR2.

[b] Two indices calculated for NDVI and NDWI derived with NIR1 and NIR2. B – blue, G – green, R – red band.

optical imagery (e.g., Landsat ETM+), NDVI, and MNDVI (Table 10.2). This table also lists the formula for calculating these indices. The exact spectral bands used to derive these indicators vary with the image used, but their wavelengths should remain roughly identical. For instance, NDWI is always calculated from the SWIR band (wavelength = 1.24 μm) and the NIR band (wavelength = 0.86 μm). They are numbered bands 10 and 8A in Sentinel-2 imagery, but bands 6 and 5, respectively, in Landsat 8 OLI imagery.

All of the indices in Table 10.2 are generic in that they do not aim at detecting floods in a particular environment or for a particular type of flood except the modified land surface water index (MLSWI) (Kwak et al., 2015). It is devised to detect monsoon-triggered river floods from MODIS data, calculated as follows:

$$MLSWI_{2,6\ and\ 2,7} = \frac{1 - \rho_{NIR} - \rho_{SWIR}}{1 - \rho_{NIR} + \rho_{SWIR}} \tag{10.1}$$

where 2 and 6 = MODIS bands 2 and 6 (SWIR: 1.628–1.652 μm), 7 = band 7 (SWIR: 2.105–2.155 μm). MLSWI is sensitive to surface wetness. An MLSWI threshold of 0.85 indicates water. MLSWI can directly detect floodwaters from the reflectance of multi-temporal MODIS bands acquired during floods and yield information on the start and end dates, the maximum extent of floods, and peak floods (Kwak et al., 2015). It is superior to NDWI (minimum value = −0.62) and LSWI (minimum value = −0.79) in distinguishing floodwater from non-flood areas near peak inundation.

If radar imagery is used for the detection, a whole different set of indices is used because radar data are mono-spectral but have two polarizations: VV and VH (Table 10.3). Thus, these indices are based only on polarization. Understandably, far fewer indices are derivable from radar data than from optical images.

All indices, however, are unable to assess ground features that lie above the surged water level but have been damaged by floods such as in a tsunami. In this case, the detection has to rely on new indices derived from UAV images and LiDAR data. Comparison of fine-print LiDAR data before and after a flood can reveal the level of FD calculated using Eq. 10.2 to buildings, such as washed

TABLE 10.3
Polarization Indices Derivable from SAR Data for Mapping Water Pixels

Index	Abbreviation	Equation	Reference
Polarization ratio	VHrVV	Y_{VH}/Y_{VV}	Brisco et al. (2011)
Normalized difference polarization index	NDPI	$(Y_{VV} - Y_{VH})/(Y_{VV} + Y_{VH})$	Mitchard et al. (2012)
Normalized VH index	NDHI	$Y_{VH}/(Y_{VV} + Y_{VH})$	McNairn and Brisco (2004)
Normalized VV index	NVVI	$Y_{VV}/(Y_{VV} + Y_{VH})$	McNairn and Brisco (2004)

Source: Huang et al. (2018).

away (bare ground height after the tsunami), completely collapsed (a huge difference in building height), and slightly damaged (e.g., no change to building height).

$$FD_{ij} = \Delta H_{ij} - \Delta DEM_{ij} \tag{10.2}$$

where i and j = the number of rows and columns, respectively; ΔH = differential water level, ΔDEM = differential elevation (Kwak et al., 2015).

10.2.1.3 Image Differencing and Thresholding

A flood manifests the change in surface cover from non-water to water. Normally, a single one-time image is unable to indicate newly inundated areas, either pre-event or post-event. The detection of flooded areas, especially those caused by a hurricane, requires two images. All non-water areas covered with water in the post-event image are flooded. This change is detectable using two broad methods of image differencing, followed by thresholding, and change detection. The former detects water based on a single threshold derived from the differencing image of the pre- and post-event images. Ideally, both images should be acquired from the same sensor, as this guarantees that the spatial resolution and number of spectral bands of the two images are identical so as not to introduce new factors that may degrade the quality of detection. Of the two, the pre-event image is regarded as the reference, with which the post-event image is co-registered. Superimposition of these two images can reveal qualitatively the areas that have been inundated and show the boundary of flood extent. Water and non-water pixels are separated via a single threshold. All pixels with a value below the threshold represent water, and above it non-water. Thresholding converts the differencing image into a binary image of flooded and non-flooded pixels. This approach may be implemented either globally or locally. Global thresholding makes use of a universal threshold estimated from the histogram of the entire image. Local thresholds are more specific as they are derived from and applicable to sub-windows. The generation of such threshold values requires the entire image to be partitioned into sub-areas, within which a threshold is estimated locally. The size and shape of all sub-areas vary widely, depending on the proportion of water and non-water features and their variability. Although thresholding is a promising and simple approach for mapping flooded areas quickly, it is highly vulnerable to image noise.

This drawback can be overcome using the universal image quality index proposed by Byun et al. (2015). This window-based index measures local spectral distortion and is easily calculated from a single band based on a moving window of a fixed size. It takes into account local spatial properties (i.e., luminance, contrast, and correlation information), different from traditional pixel-based similarity measures based solely on pixel values in isolation. Thus, it is robust against several types of image noise such as white Gaussian, salt and pepper, mean shift, and multiplicative noises. Although derived from only a single band (e.g., NIR), this index achieves an overall accuracy of 75.04% that is comparable to that of SVM-based supervised change detection because the NIR band is sensitive to flood-affected areas and produces false hits in non-flooded areas.

Apart from raw bands, the differencing image can also be produced from VI images to take advantage of multispectral bands, such as NDWI (Figure 10.1). The difference of the two-time NDWI values is known as DNDWI (Ogashawara et al., 2013). Thresholds can also be calculated from multiple bands such as MNDVI derived from radiometrically corrected optical data. While it is relatively easy to calculate DNDWI, it is rather challenging to set the exact threshold that reliably separates water pixels from non-water pixels, just as with one of the images in mapping water bodies. The determination of the optimal threshold may need repeated trials to finalize, even though this task can be facilitated by the histogram of the NNDWI layer. With a threshold of DNDWI = 0.05, flooded areas are mapped at an accuracy of 85.7%, while non-flooded areas are more accurately mapped at 92.1%.

A single threshold may enable water to be differentiated into open water from non-water but not into specific water types, a task that requires the use of multiple thresholds in different bands. Their combination is conducive to accurately differentiating open water into more specific sub-classes. For instance, a pixel is considered to represent the water-vegetation sub-class if its value is $>T_{\text{final}}$ (the final threshold) but $<T_{\text{upper}}$ (the upper threshold) and its MNDVI > T_{MNDI} (Kordelas et al., 2018). Multiple thresholds may be combined with the optimum threshold of individual image segments to derive the final threshold. The joint application of T_{final} with that based on MNDVI enables the differentiation of marshland into open-water, water-vegetation, and non-water areas at a combined kappa coefficient of 0.88 and overall accuracy of 97.7%. Both are highly comparable to 0.95 (kappa) and 99.0% (overall accuracy) achieved using supervised RF classification of a set of bands and indices (NDWI, MNDWI, NDVI, and MNDVI) extracted from Sentinel-2 data. Higher accuracies could not be attained because of the omission of areas with emergent dense vegetation completely covering water. The co-presence of water and vegetation cannot be detected in the SWIR histogram. This is a common problem in coastal floodplains where floods take place beneath vegetation and cannot be detected due to the obstruction of the water signal by the vegetation canopy.

Thresholding is simple, easy to implement, and fully automatic with minimal human intervention, but the use of a universal threshold from a histogram may not lead to a satisfactory separation of water and non-water features if they are not balanced in their spatial distribution, or their spectral properties vary spatially. Besides, the threshold may be subject to personal bias and is likely to be area- and even scene-specific. This limitation can be circumvented by producing a differencing image from classified maps, in a post-classification change detection. The classified image allows newly inundated areas to be easily determined from the difference in pixel values. For instance, if a pixel used to have a value of 3 (e.g., covered by vegetation) but has a value of 7 (e.g., covered by water) now, then a difference of 4 means that it has been converted from vegetation to water. Namely, it has been flooded in the interim of the two image acquisitions. While post-classification eliminates the necessity of setting the precise threshold of change vs no-change, it is still subject to classification inaccuracy because not all land covers, especially those mixed covers (e.g., vegetation partially submerged underneath water) can be classified accurately. Image classification inaccuracies inevitably degrade the quality of mapping, an issue to be discussed in-depth in the following section.

10.2.1.4 Image Classification

In addition to vegetation indices, inundated areas can also be mapped using image classification that can yield quantitative change detection information. Images can be classified using supervised and unsupervised methods. In unsupervised classification, the initially grouped spectral clusters may be aggregated to form two broad categories of water and non-water in a post-classification session (Figure 10.2). If another pre-event image is masked for permanent water bodies, its overlay with the classification-extracted water area reveals newly inundated areas. Unsupervised classification can be repeated twice to map water bodies into specific types. In the first round, the entire image is classified into only two categories of water and terrestrial, formed by amalgamating all spectral clusters (Hudson and Colditz, 2003). Then, the water cluster is further grouped into ten clusters in another round of unsupervised classification, during which more specific covers are mapped, such

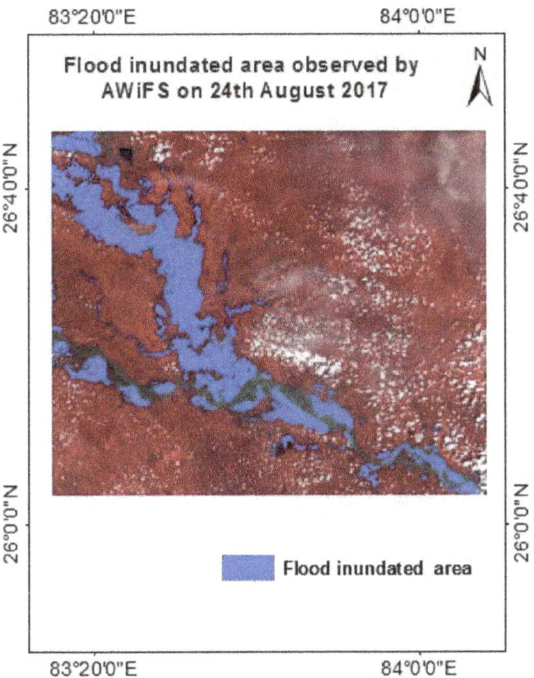

FIGURE 10.2 Areas inundated by the 24 August 2017 flood in the state of Uttar Pradesh, India, mapped from optical AWiFS images using unsupervised classification of ten clusters. (Source: Anusha and Bharathi, 2020.)

as water, formerly flooded, non-flooded, and clouds. Naturally, some of these covers pertain to flood extent and nature, such as permanent water, flooded and wet areas (e.g., formerly flooded surfaces) (Hudson and Colditz, 2003). The flood extent is derived by amalgamating inundated areas and permanent water.

Supervised classification is more demanding and complex than unsupervised classification in that samples of all the possible covers to be classified must be selected to train the computer. The representativeness of these samples dictates the quality of the final classification results. Supervised classification is able to differentiate the water class into more specific types such as marshland, rice paddies, and temporary ponds, but it takes a long time to refine the training samples to produce satisfactory results (Kordelas et al., 2018). In comparison, unsupervised image classification does not require specialist expertise to perform and can be potentially transferable to other sites of similar characteristics for rapid mapping of flood extent.

Apart from per-pixel classification, flood-affected areas can also be mapped using the object-oriented method, in which the raw images are first segmented to form patches and then classified using the described methods. OBIA is suitable for classifying VHR images, such as KOMPSAT-2 and WorldView-2 images (Table 1.5). Dissimilar to per-pixel classification, object-oriented image classification allows diverse spatial, spectral, and even geometric parameters to serve as the inputs. In general, the more inputs are used in the (object-oriented) image classification, the more accurately flooded areas are mapped. The highest overall accuracy achievable is 95% obtained using a decision-tree classifier based on objects, which is not possible with the pixel-based classification that achieves an accuracy of only 77% (Malinowski et al., 2015). Nevertheless, perfect classification is not possible because of omission errors. Namely, those flooded areas underneath dense vegetation are not well represented in the flood map due to the lack of a SWIR band. This band will undoubtedly improve the mapping accuracy as water strongly absorbs infrared radiation.

10.2.2 Detection in Challenging Environments

Although water has a spectral reflectance uniquely distinct from that of terrestrial covers, such a disparity is not always translated to high accuracy in detecting flood extent. Nor does it guarantee accurate extraction of water pixels in certain challenging environments (e.g., a high degree of spatial heterogeneity). Two of them merit an elaboration here: coastal wetlands and floodplains. In coastal marshlands, vegetation spatially co-exists with water, so it is challenging to detect the precise extent of flooding. This detection is still based on the reflectance of water vs non-water targets on a pair of images obtained before and during the flood event. If viewed from above, it is impossible to differentiate flooded forests from non-flooded areas due to canopy blockage.

The task of mapping water surfaces is not straightforward with marshland floods, even though accurate quantification of surface water extent is critical to understanding their role in ecosystem processes. This environment is rather challenging because of the presence of aquatic vegetation typical of heterogeneous or variable wetlands. In this environment, water is prone to mixing with vegetation and soil. The classification of surface water extent vs non-water requires training datasets derived from prior class masks generated from a water body dataset or composited dynamic surface water extent probabilities (Huang et al., 2018). The use of class probabilities reduces omission errors by 10% for water bodies and commission errors by 4% among non-water classes. Thus, the inclusion of prior water masks reflective of the dynamics in surface water in a wetland is helpful to the accurate mapping of water bodies from SAR data.

Coastal marshlands are subject to periodic tidal flooding that reduces spectral reflectance, especially in the NIR and SWIR wavelengths. Such flooding can be detected using the Tidal Marsh Inundation Index (TMII) derived from daily 500 m MODIS surface reflectance data (O'Connell et al., 2017). This index is based on the normalized difference of green and SWIR reflectance combined with a phenology parameter ($NDWI_{2,5}$) composed of the moving average of the normalized difference of NIR and SWIR reflectance, or

$$TMII = 1 - \frac{1}{e^{Flood\ status}} \tag{10.3}$$

$$Flood\ status = 0.25 + 16.56 NDWI_{4,6} - 25.20\overline{NDWI_{2.5}} \tag{10.4}$$

where $NDWI_{a,b}$ = NDWI calculated from MODIS bands *a* and *b*. $\overline{NDWI_{2.5}}$ refers to the moving average mean of NDWI derived from MODIS bands 2 (green) and 5 (SWIR). This index is produced from a consideration of a suite of normalized difference indices and various band combinations using a binomial (flood vs non-flood) generalized linear model. A TMII threshold is essential to demarcate the boundary of flood-affected areas and identify flooding across the annual cycle. TMII-based classification is 67–82% accurate in mapping flooding conditions, and 75–81% accurate for mapping dry conditions across training, test, and validation datasets. Overall, TMII achieved an accuracy of 77–80% in classifying marsh flooding. Moreover, TMII-filtered composites are less noisy than the existing MODIS MOD13 16-day NDVI product and fit field data better than MOD13. TMII is applicable to other coastal areas, judging from the similar findings obtained from marshlands of different species within the Gulf Coast of a narrow tidal range.

Floodplain is another challenging setting for mapping flooded areas using remote sensing data alone because of its unique geographic setting: low-lying land with a high moisture content interspersed with dense vegetation. The mapping is even harder to achieve accurately at a detailed level. If a flood is confined to localized shallow water in riverine floodplains, the difficulty of mapping its exact extent as patches of herbaceous vegetation likely arises from its spatial mixing with open water (Malinowski et al., 2015), creating a high degree of spatial heterogeneity. This mapping requires VHR images and the use of ancillary data, such as a DEM. The combination of images with the DEM effectively overcomes the limitations of remote sensing imagery in the detection. Since water has a level surface, if one exposed area of a certain height is inundated in a flood, other areas of a similar or lower height must also be inundated, irrespective of whether it is covered by vegetation or not. DEM data enhance the accuracy of mapping flood extent, especially if the coastal area happens to be a forested swamp. In particular, DEM data are essential to drastically rectify the under-detection of flooding beneath forest canopies, especially within bottomland forests and hardwood swamps (Wang et al., 2002).

The omission of inundated vegetation along lakes and rivers in a heterogeneous environment is rife with fine-resolution radar images, even though a Sentinel-1 false-color backscatter composite illustrates the strong double bounce of such features clearly (Huang et al., 2018). The reason is that such areas are too small to be included in the pure training samples. Additional omission errors can also arise from the coarse resolution (5 m × 20 m) of the image used in mapping small lakes (e.g., <500 m^2). If flooding is confined to localized shallow water in riverine floodplains, VHR images such as WorldView-2 with a spatial resolution of 2 m (0.5 m for the panchromatic band) may come to the rescue, facilitated by auxiliary data, such as spectral indices designed to map water surfaces and vegetation, PCA, DEM and slope, and pre-classified high vegetation and shadows (Figure 10.3). If the flood crest does not coincide with the acquisition of the remote sensing data, PCs are more useful for delineating flooded areas in coastal floodplains than raw bands (Hudson and Colditz, 2003), regardless of whether the images are multi- or mono-temporal. This is because wet (formerly flooded) surfaces cannot be easily distinguished from non-flooded areas. PCA is particularly effective at enhancing flood-affected areas after the pass of the flood crest.

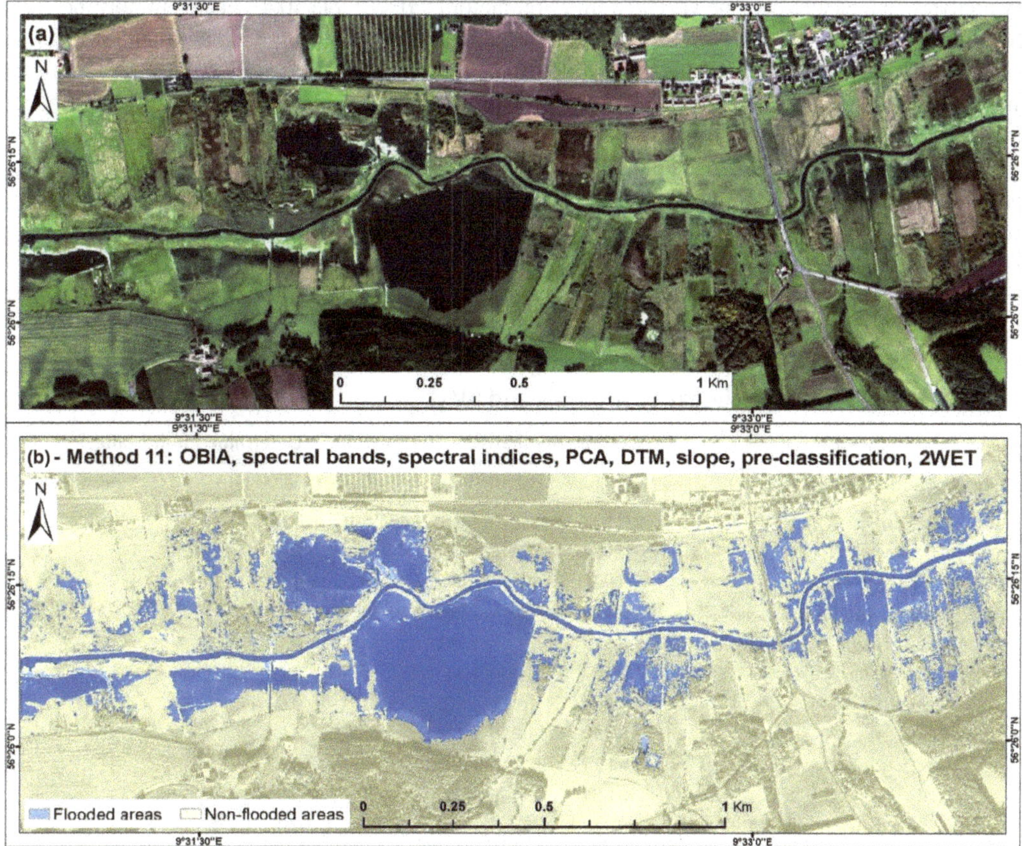

FIGURE 10.3 (a) Flooding of a riverine floodplain in Central Jutland, Denmark shown on a WorldView-2 image, and (b) the flood extent mapped using OBIA based on all available inputs. (Source: Malinowski et al., 2015.)

10.3 VULNERABILITY, HAZARD, AND RISK MAPPING

10.3.1 Vulnerability

It is impossible to map flood-susceptible areas in a low altitudinal range, subtropical floodplain directly from remotely sensed data. Instead, they have to be modeled from all the variables that play a role in causing floods, such as elevation, aspect, slope, curvature, rainfall, soil, land use/land cover, stream density, distance to stream or river, distance to road, and slope moisture load. Land use/land cover plays an important role in affecting the onset of runoff, erosion, sedimentation, and hence flood risk in many drainage basins around the world. Flood susceptibility of river channels is related to annual rainfall, soil type, stream density, distance from stream, distance from road, in addition to TWI, altitude, slope aspect, slope, curvature, land use/land cover, and geomorphology (Arora et al., 2021), all of which can be derived from a DEM or satellite images. Additional factors, such as NDVI, TPI, TRI, stream potential index, and various other proxies, have also been considered in modeling flood susceptibility. The importance of these factors to flood vulnerability varies with the geographic setting and type of floods. For instance, flash floods take place commonly in hilly regions and low-lying floodplains in humid tropical and subtropical climates. Here slope plays a relatively more significant role in flood occurrence than other factors. In low-lying floodplains, geomorphology is much more important than lithology, and its importance exceeds that of all other factors combined (Arora et al., 2019).

The independence of all flood-conditioning variables is evaluated automatically using collinearity analysis as measured by the variance inflation factor (VIF) (Table 10.4). Factor importance can be determined using boosted or linear logistic regression and weight of evidence modeling. After the most significant contributing factors are identified, the zoning of flood-susceptible areas requires the use of accurate and appropriate bivariate or multivariate models, or machine learning models such as SVM. Other advanced hybrid models for this purpose include ANFIS and metaheuristic model-based ensembles. The modeled flood susceptibility is in raster format, and the modeled detail depends on the spatial resolution adopted to represent all the input layers. The final susceptibility may be displayed in a few categories to illustrate its spatial pattern (Figure 10.4).

TABLE 10.4

Independence and Importance (Weight) of 12 Flood-Conditioning Factors Determined via Collinearity Analysis and SVM

	Collinearity Statistics		SVM-Determined Importance	
Conditioning Factor	**Tolerance**	**VIF[a]**	**Relative Weight**	**Rank**
Elevation	0.538	1.858	0.07	10
Curvature	0.924	1.082	0.10	3
Distance to roads	0.926	1.079	0.07	8
Distance to streams	0.776	1.289	0.12	2
Geomorphology	0.913	1.095	0.13	1
Land use/land cover	0.707	1.415	0.08	6
Rainfall	0.906	1.103	0.04	12
Slope	0.749	1.336	0.10	4
Slope aspect	0.874	1.145	0.07	9
Soil	0.784	1.276	0.05	11
Stream density	0.733	1.365	0.09	5
Topographic wetness index	0.623	1.605	0.07	7

Source: Modified from Arora et al. (2021). Used with permission.

[a] VIF = variance inflation factor.

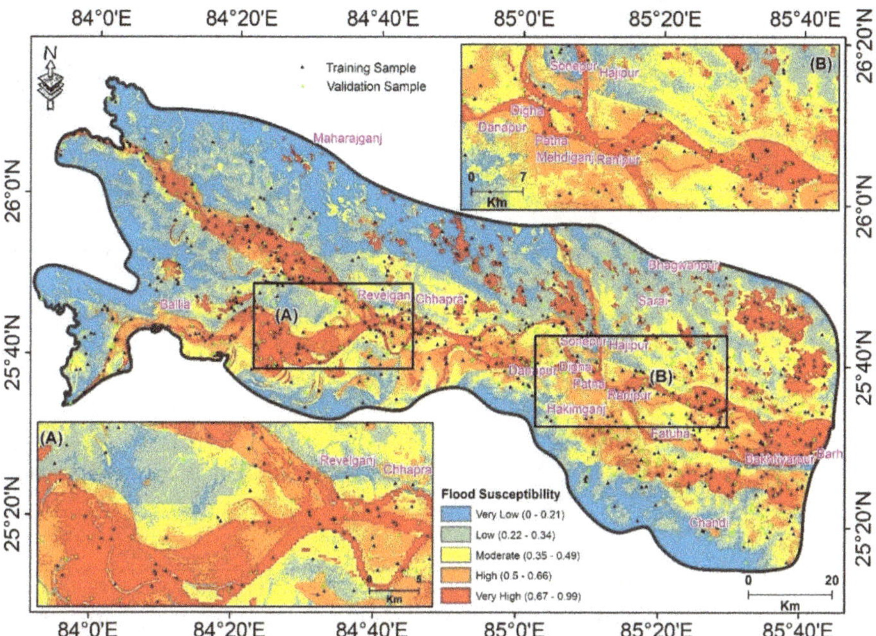

FIGURE 10.4 Flood susceptibility in the Middle Ganga plain, India modeled using an ensemble of adaptive neuro-fuzzy inference system (ANFIS) model with a generic algorithm from 12 factors listed in Table 10.4. (Source: Arora et al., 2021. Used with permission.)

Although the same modeling approach can be adopted to zone urban flood vulnerability, additional variables must be taken into account, such as density of population and drainage block sites. Since not all people can cope with a flood with the same level of composure and ease, it may be necessary to identify special groups of vulnerable cohorts, such as the elderly and children (Sowmya et al., 2014). So this kind of modeling relies less on remotely sensed data than on socioeconomic data. Compared with rural areas or watershed, urban areas have a flood vulnerability less affected by slope gradient, soil, surface wetness, and even streams as the ground is mostly paved and flat. Instead, it is the amount or proportion of impervious surfaces that is critical to surface runoff. Also important in urban flood vulnerability modeling is drainage that dictates how quickly the rainwater is discharged. Common to all environments is the factor of distance to rivers and water bodies, even though its importance to flood vulnerability varies with the exact environment.

10.3.2 Flood Hazard

Flood hazard is related to inundation extent, water depth, and flood water flow velocity (Table 10.5). Flood risk refers to the potential adverse consequences of a flooding event. It is derived from the probability of a flood event and the potential adverse consequences to human health, the environment, and the economic activity associated with a flood event (EXCIMAP, 2007). Apart from inundating surface covers, floods also trigger secondary damage (e.g., landslides in the watershed) and damage buildings when the floodwater forms torrents in the downstream channel. More damage will incur to the inundated areas if the floodwater is laden with silts and pebbles.

It is important to map the spatial distribution of flood hazards for planning watershed-level land use, mitigating FD, and rapidly responding to flood emergencies. Although large-scale global hazard maps have been produced by the Dartmouth Flood Observatory based on observations, they do not show inundation probability distributions that are essential to flood hazard mapping. Flood hazard on a smaller scale is usually modeled using hydrologic models. Of the diverse inputs to the

TABLE 10.5

Comparison of Flood Hazard Map with Flood Risk Map in Terms of Their Content and Use

	Hazard Map	Risk Map
Content	Flood extent according to probability classes • Flood depth • Water flow velocity • Flood propagation • Degree of danger	Assets at risk according to past events • Flood vulnerability • Probable damage • Probable loss (per unit time)
Purpose and use	Land use planning and land management • Watershed management • Water management planning • Hazard assessment on local level • Emergency planning and management • Planning of technical measures • Overall awareness building	Basis for policy dialogue • Priority setting for measures • Flood risk management strategy (prevention, mitigation) • Emergency management (e.g., the determination of main assets) • Overall awareness building

Source: EXCIMAP (2007).

models, remote sensing supplies only a small portion of the needed data (Figure 10.5). Despite this inadequacy, the spatial distribution of inundation probability can be produced from time-series satellite images. They are excellent at determining the inundation frequency and extent of historic floods on a scale ranging from catchment to regional, national, and even global. Although flood hazard has been modeled from readily available geomorphological parameters, this approach may suit the watershed scale only. More critically, it is unable to provide a spatial perspective into the

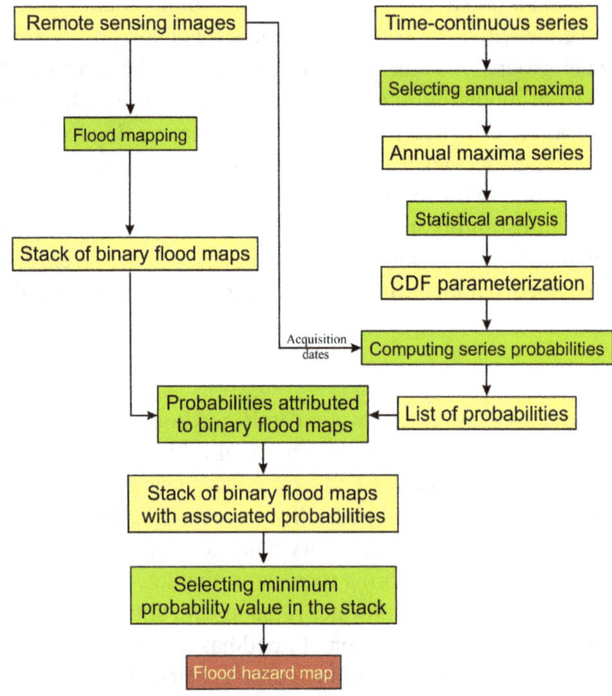

FIGURE 10.5 Flowchart of mapping flood hazard from time-continuous series remotely sensed data from which the frequency of historic flood events is detected. Yellow boxes: input data and intermediate results; green boxes: data processing. (Source: Modified from Giustarini et al., 2015.)

hazard. In contrast, satellite images offer a number of benefits in mapping flood hazards at a fine spatial resolution. The mapping is repeatable from the same data source using the same method. Thus, the produced flood hazard maps are consistent and comparable across all regions around the world, particularly those regions lacking ground observations. The mapping of an area into zones of different levels of flood risk is commonly known as flood hazard zoning.

There are two general ways of zoning flood hazards. The first approach is to combine remote sensing with process-based hydrologic–hydraulic models (Giustarini et al., 2015) because many geomorphologic variables critical to flood hazards are not directly obtainable from remotely sensed data. This approach takes advantage of the strength of each discipline, such as time and space continuity of global flood inundation models and fine spatial resolution of remote sensing imagery. The flood hazard maps produced from multi-annual remote sensing data show the flood inundation extent over long periods, even on a global scale. But the coarse resolution (1 km) of the maps may be of little use for small areas. Besides, process-based hydrographic models are a leading source of uncertainties about the modeled hazard because the required hydrological data are incomplete and insufficient. It is thus desirable to zone flood hazards by combining multi-year remote sensing data with hydrodynamic modeling. Remote sensing data are not recorded continuously, and when they are recorded, they may not coincide with the flooding events. In the zoning, time-series remote sensing data are used to map inundation frequency (Figure 10.5). It elucidates how often an area has been inundated in the past and the result is expressed as the probability of flooding. The combination of long-term records of hydrological data with time-series satellite data adds a spatial perspective on flood frequency. Table 10.6 details the type of data that have been used to zone flood hazards at various spatial scales. Remote sensing is still the leading source of data in all cases. After stacked binary flood maps are combined with the computed probabilities of series values from historic remote sensing data, the final flood hazard map is produced by selecting the minimum probability value in the stack (Figure 10.5).

The second approach is to quantify flood hazard using a number of critical parameters and express it as a flood hazard index (FHI). So far FHI has been calculated from the linear combination

TABLE 10.6
Common Methods of Mapping Flood Hazards and the Data Used

Authors	Geographic Area	Nature of Map	Data Sources
Sakamoto et al. (2007)	Cambodia and the Vietnamese Mekong Delta	Spatiotemporal variation of annual inundation extent	MODIS time-series imagery (2000–2005)
Thomas et al. (2011)	Central eastern Australia	Annual inundation	Landsat images (1979–2006)
Pulvirenti et al. (2011)	Italy	Spatiotemporal variation of inundation extent	Multi-temporal COSMO-SkyMed data
Chini et al. (2013)	Japan	Spatiotemporal variation of inundation extent	Multi-temporal COSMO-SkyMed data
Brakenridge (2022)	Global	Archived flood events	Remote sensing sources + news, governmental and instrumental sources
Skakun et al. (2014)	Namibia	Relative frequency of inundation	Landsat images (1989–2012)
Huang et al. (2014)	Southeastern Australia	Probability map of inundation	Remote sensing observations combined with gauged time-series data: MODIS time-series imagery (1983–2010) and gauged discharge data
Giustarini et al. (2015)	Seven catchments in the UK	Hazard	Remote sensing data combined with modeled time-series data: EnviSat time-series imagery (2007–2010) and daily global flood inundation maps over 30 years

Source: Modified from Giustarini et al. (2015).

of two standardized factors: flooded area and MNDWI which is considered a proxy for qualitative turbidity (Kumar et al., 2018). The use of this parameter is justified by the fact that more turbid water does potentially more damage to crops, plantations, and the aquatic habitat of marshland. Although flood frequency and extent can be easily derived from remote sensing data, they do not depict flood hazards realistically as FD can be related to more factors than just land cover, such as topography. Thus, more variables than these two parameters need to be considered, such as flood-affected areas, mean turbidity, mean flood depth, and flood duration (Kumar, 2016). Of these parameters, flooded areas are extractable from optical images such as Landsat ETM+ and OLI using MNDWI. Flood depth is derived from a DEM. Alternatively, the maximum flood inundation depth can be modeled from streamflow recorded at gauging stations. Nevertheless, the FHI derived from only two parameters is just as effective as the four-factor index. Irrespective of the exact number of factors considered, FHI is easy to compute, and its value is easily converted to a hazard level. For instance, a value between 0.9 and 1.5 is considered to represent low hazard, a value between 1.6 and 2.1 medium hazard, 2.2 and 2.8 high hazard, and 2.9 and 4.0 very high hazard (Figure 10.6).

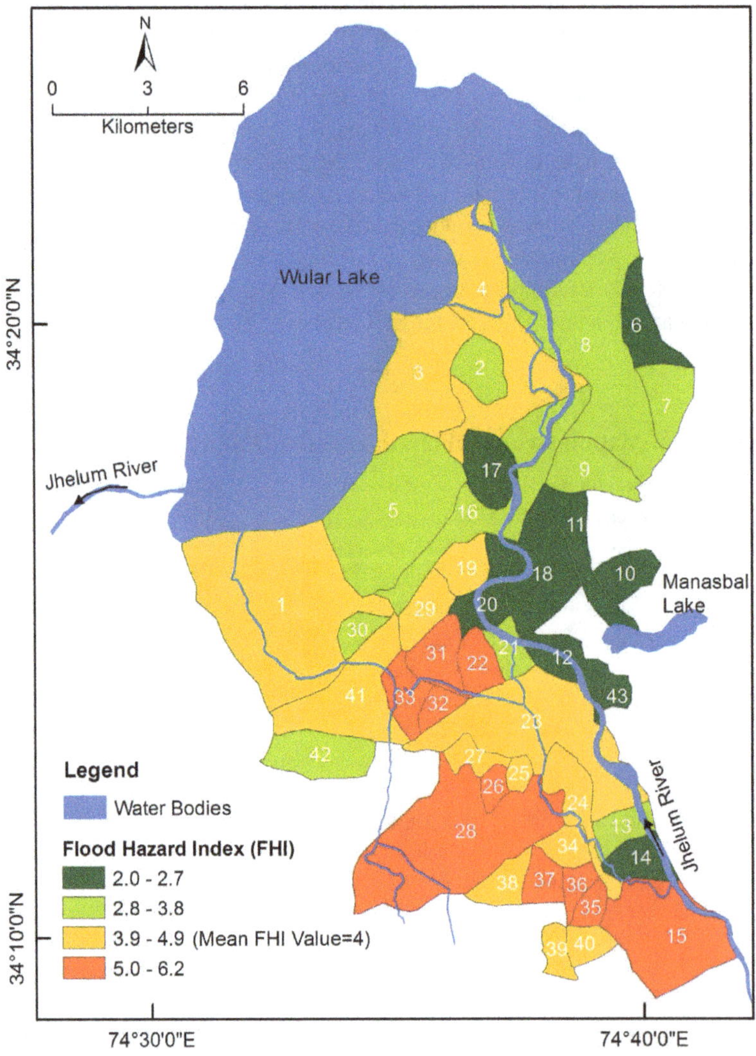

FIGURE 10.6 Village/town level flood hazard index in Sonawari sub-district of Bandipore district (Jammu and Kashmir) derived from the linear combination of four standardized parameters of flood-affected areas (%), mean turbidity, flood depth, and flood duration. (Source: Kumar, 2016. Used with permission.)

TABLE 10.7

Factors Considered in Mapping Flood Hazards, Their Weight, and Five Grades

		Scale of Hazard				
Factor	**Weight**	**5 – Very High**	**4 – High**	**3 – Moderate**	**2 – Low**	**1 – Very Low**
Slope (%)	0.2690	0–0.9	0.9–6.4	6.4–12.8	12.8–22.8	22.8–88.9
Soil drainage capacity	0.1793	Eutric vertisols	Eutric fluvisols	Haplic luvisols	Chromic luvisols	Eutric leptosols
Elevation (m)	0.0860	1,780–2,240	2,240–2,700	2,700–3,160	3,160–3,620	3,620–4,080
Land use (level of flood abstraction)	0.0562	Swamp	Agriculture/bare land	Grassland	Woodland/ shrubland	Plantation forest
Drainage density (km^{-2})	0.3712	2.8–3.5	2.1–2.8	1.4–2.1	0.7–1.4	0–0.7
Daily max rainfall (mm)	0.0383	76.9–70.5	70.5–64.0	64.0–57.5	57.5–51.1	51.1–44.6

Source: Gashaw and Legesse (2011). Used with permission.

This index is applicable to assessing floods at the village/town level as it is based on the portion of flooded areas to the total town/village area. Whether the same hazard will remain unchanged in the next flood in the future remains unknown.

Catchment-level flood hazard has been mapped from flood probability, average daily precipitation, and inundation depth (Mojaddadi et al., 2017). Instead of three factors, Gashaw and Legesse (2011) considered six factors, each graded at five hazard levels from 1 (low) to 5 (very high) (Table 10.7). The weights of these factors sum to 1 and are determined using a machine learning method. None of the factors are related to the extent of flooding. Instead, it is the elevation that is considered in the mapping. The water-related factor is the daily maximum rainfall. Virtually, this method of mapping flood hazards is scenario-based. The same area faces different levels of flood hazard, depending on the daily rainfall. Of the six factors, the most important is drainage density which indicates how quickly the accumulated rainwater is discharged. The next most important factor is slope which governs how quickly the rainwater flows on a slope to reach the river channel. Soil drainage capacity is the third most important factor that governs how quickly the rainwater is absorbed by the soil. A higher portion of absorption means less water available for flooding. Elevation, land use, and daily maximum rainfall are not so important, all receiving a weight smaller than 0.1 (10% each). Daily rainfall is not so critical if the rain spreads out evenly within 24 hours. In comparison, hourly maximum rainfall is a more reliable factor in predicting flood than daily rainfall.

Remote sensing is valuable to this method of assessment. Of the six factors in Table 10.7, both land use and drainage density can be mapped from remotely sensed data. Slope and elevation can be derived from a DEM. Soil drainage capacity is interpreted from field surveys and aerial photographs. Only daily maximum rainfall cannot be acquired from remote sensing images at a fine scale. Compared to FHI, this method of deriving flood hazards from the six factors is flexible in that it can be applied to the landscape scale rather than the village level (Figure 10.7). Since it does not require information on the flood extent, it is applicable to predict future flood hazards.

10.3.3 Flood Risk

Flood risk assessment pertains to the evaluation of the probability of flood occurrence and the degree of flood-induced damage, including the assessment of the danger of floods, the vulnerability of the flood-bearing body, and the probability of loss. In order to mitigate flood risk, it is essential

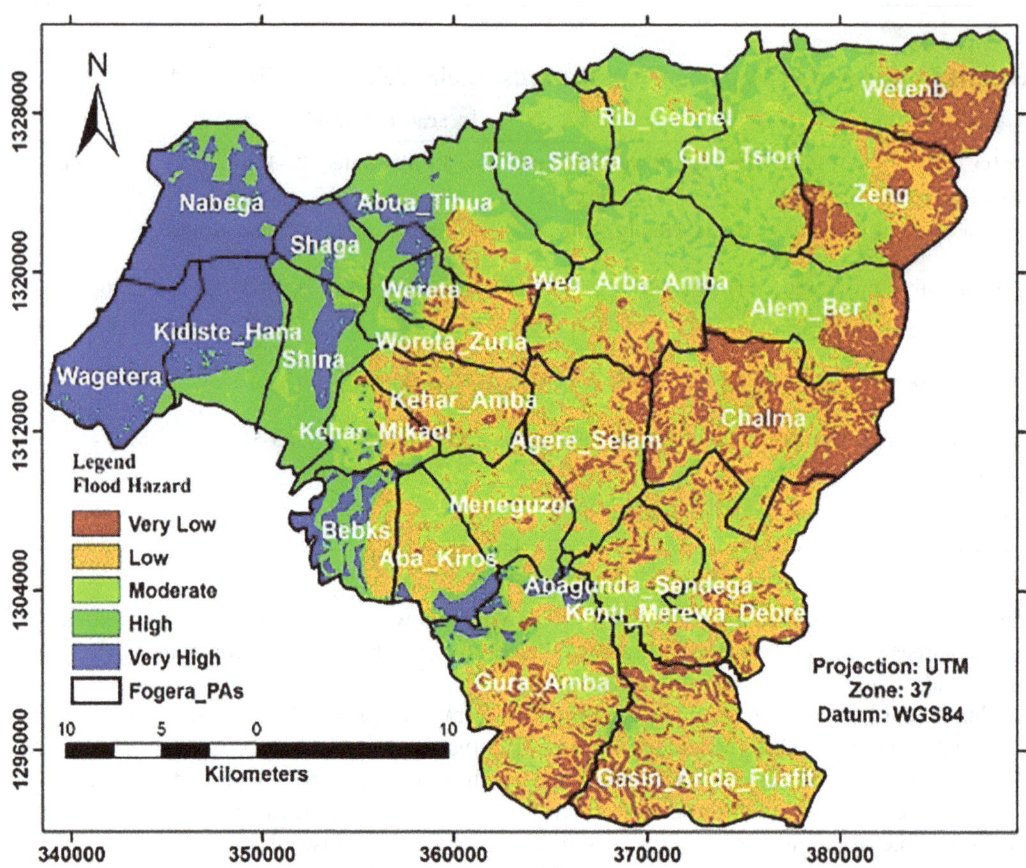

FIGURE 10.7 Distribution of regional flood hazard in Fogera Woreda, Northwest Ethiopia at five levels derived from weighted averaging of six factors shown in Table 10.7. (Source: Gashaw and Legesse, 2011. Used with permission.)

to detect flood-hazardous and risky areas as an essential component of risk management. Precise assessment of flood risk requires information on individual vulnerability parameters that may include population, economic assets, economic activities, and environmental issues. Population data include distribution by municipality, postal code or address (building), or the average number of residents per dwelling or property, and distribution of particularly vulnerable cohorts, such as homes for the elderly, location of schools, hospitals, sports facilities, and other infrastructure with a concentration of people. Under assets and economic activity, information on the distribution of buildings (as a proxy for the affected population) is also essential. The impact of damage on the economy should indicate the location of a particular activity. The loss to agriculture should be further differentiated by freshwater or saline flooding.

Most of the information needed for flood risk assessment can be acquired from remote sensing data, even though the difficulty and level of accuracy vary with the nature of the information. Land use/land cover is mapped or updated from satellite images using the per-pixel or object-oriented classification methods according to needs. Land covers are commonly classified into residential areas, farmland, forest, industrial, commercial, infrastructure, and facilities at risk. Of these features, those related to land covers of an adequate size can be mapped from medium resolution, multispectral optical images at a rather high accuracy. However, mapping of facilities requires VHR images or aerial photographs. When this proves impossible (e.g., sports facilities), GPS may be used

TABLE 10.8

Factors Considered in Estimating Flood Risk, Their Weight, and Five Scales of Grading

		Scale of Risk				
Factor	**Weight**	**5 – Very High**	**4 – High**	**3 – Moderate**	**2 – Low**	**1 – Very Low**
Flood hazard	0.3333	Very high	high	moderate	low	Very low
Population (person·km^{-2})	0.3333	2,396–400	400–225	225–200	200–175	175–149
Land use type (based on sensitivity to flooding)	0.3333	Agriculture	Grassland/bare land	Swamp	Plantation forest	Woodland/ shrubland

Source: Gashaw and Legesse (2011). Used with permission.

for the mapping. Needless to say, the process of mapping is lengthy and requires extensive ground-truthing to retain a credible accuracy.

The essential factors in mapping flood risk at the district level are flood hazards and the population at risk. In addition, land use also plays a role in flood risk. These three parameters are treated equally by Gashaw and Legesse (2011) in mapping flood risk at five levels (Table 10.8).

Instead of population and land use type, a total of 13 factors, including elevation, aspect, slope, curvature, stream power index (SPI), TWI, sediment transport index, topographic roughness index, distance from a river, geology, soil, surface runoff, and land use/cover, are used to estimate watershed-level flood risk by Mojaddadi et al. (2017), eight of them are derived from a DEM. These variables closely resemble those used to estimate flood hazards. After all, flood risk represents the expectation of the vulnerability function. The consideration of so many variables requires an objective way of weighting them, which is achieved using frequency ratio, a bivariate statistical model. This ratio is calculated by dividing the sum of flood-affected areas or the number of pixels on a remote sensing image with a specific range (for continuous variables) or of a given attribute (for discrete variables) of a flood predictor by the sum of all flood-affected areas, namely,

$$Frequency\ ratio = \frac{\dfrac{N_p\left(LX_i\right)}{\Sigma_{i=1}^{m} N_p\left(LX_i\right)}}{\dfrac{N_p\left(X_j\right)}{\Sigma_{j=1}^{n} N_p\left(X_j\right)}} \tag{10.5}$$

where $N_p(\mathrm{LX}_i)$ = the number of flood-affected pixels in class i of factor X; $N_p(X_j)$ = the number of pixels of factor X_j; m = the number of classes or possible attribute values of factor X_i; n = the total number of factors considered.

The simultaneous consideration of so many factors is handled using sophisticated analytical methods, such as LR, decision trees, adaptive network-based fuzzy inferencing, and SVM. The machine learning methods can quantify the exact importance (weight) of the considered flood-conditioning factors. In particular, the frequency ratio model is excellent at ranking multiple flood-conditioning parameters. A flood risk map is produced by multiplying the flood probability map by the hazardous triggering layer (T_{pi}) created through the integration of the maximum flood inundation depth with daily mean precipitation, or

$$Flood\ risk = f\left(P_s,\ T_{pi}\right) \tag{10.6}$$

where P_s = probability calculated from the ensemble of frequency ratio and SVM analysis. Both P_s and T_{pi} must be standardized prior to the integration to ensure that they make an equal

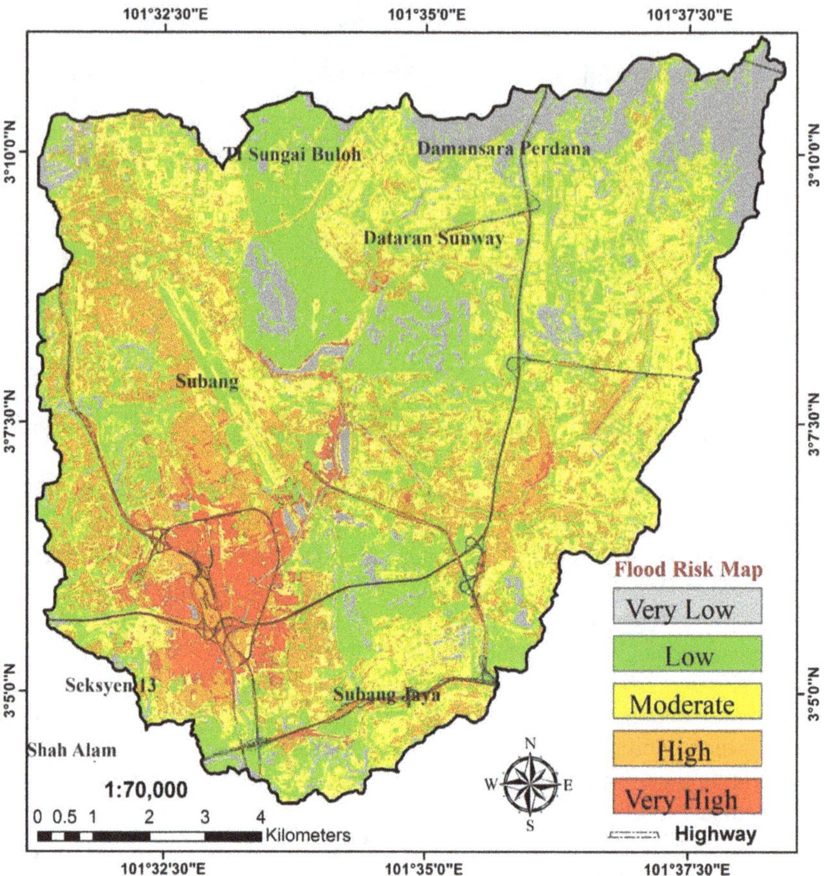

FIGURE 10.8 Distribution of flood risk at five levels in the Damansara River catchment in Selangor, Malaysia produced by integrating flood vulnerability with a flood hazard map based on three factors: flood probability determined from 13 variables using FR, daily average rainfall, and flood inundation depth. (Source: Mojaddadi et al., 2017.)

contribution to flood risk. The resultant flood risk map is enumerated qualitatively in a few categories (Figure 10.8).

10.3.4 Flood Forecast and Warning

Severe floods are commonly associated with moisture-saturated air masses or clouds. They can be predicted by monitoring the movement of such clouds, as exemplified by the tracking of hurricanes. The tracking is ideally accomplished using meteorological satellite images. Their fine temporal resolution enables the same storm to be monitored every 30 minutes or shorter. The satellite images do not need to have a fine spatial resolution as only the general position of the air masses suffices in predicting the likely area of heavy rainfall and potential floods. So far several satellite missions have been launched to forecast heavy rainfall events at a high accuracy long before they eventuate. These prediction systems play a significant role in minimizing flood risk and rapid flood response. Some of the early warning systems are operational on a global scale, such as the global flood detection system (GFDS) and the global flood awareness system (GloFAS). As an experimental system, GFDS is designed to detect and map major river floods in near real time from daily passive microwave satellite images at a spatial resolution of 0.09° × 0.09°. The GFDS flood signal is well correlated with in situ measured streamflow (Revilla-Romero et al., 2015). The GloFAS scheme forecasts medium-range daily ensemble streamflow and maps flood threshold exceedance probabilities at 0.1° resolution for all rivers with an upstream area over 4,000 km^2 around the world. The MODIS Near Real-Time Global Flood Mapping Project is another prediction system that produces 250-m resolution global daily surface and floodwater maps in 10° × 10° tiles. The MODIS Water Product is also a good data source for flood forecasting. This raster layer covers the global land areas. Each pixel of the layer has one of four possible values: 0 – invalid, 1 – non-water pixel, 2 – a reference pixel with water, and 3 – flood pixel. GFDS, GloFAS, and MODIS Water Product all produced consistent predictions of large floods (Revilla-Romero et al., 2015). However, the spatiotemporal characteristics of the monitored floods vary widely among the three systems. All of the three automated flood prediction (detection, monitoring) systems can provide early warning and useful independent information for characterizing major floods. However, the spatial resolution is too coarse to predict flash floods. Besides, optical MODIS data are subject to cloud contamination.

The USGS also releases a dynamic surface water extent product produced from Landsat images. Initially, the product covers only the US and its territories. This data product is based on independently tested algorithms for detecting mixed pixels comprising both open water and its partial mixture with vegetation and soil. Rigorous tests and evaluations over various landscapes in North America confirm that this product is a reliable data source for open-water bodies and partially inundated surfaces under cloud-free conditions.

Existing flood data products are available at various spatial resolutions. For instance, the MODIS water mask (MOD44W) data product has a resolution of 250 m. The SRTM water body dataset (SWBD) derived from the SRTM data has a spatial resolution of 90 m, finer than other similar products. Included in the dataset are permanent open-water bodies, the boundaries of which are represented in vector format. This product covers most of the terrestrial surface of the Earth from 56°S to 60°N. More information on the SWBD data product is available at https://dds.Cr.Usgs.Gov/srtm/version2_1. These datasets, however, may not be sufficiently detailed for mapping local-scale water bodies. They can serve as a reliable source for selecting samples to train the computer in automatically mapping water bodies from fine-resolution images.

10.4 FLASH FLOODS

Flash floods are one of the most deadly natural disasters when they strike because the rapid speed of occurrence leaves little time to respond and evacuate. They refer to the sudden and violent arrival of a torrent of rainwater in a terrifying manner. Torrent rainwater surges along narrow gullies or in low-lying valleys without any warning. Such floods have a short duration but a relatively high peak discharge. They pose a higher level of threat to human safety and can do more damage than regular floods because of the tremendous speed of water flow. Flash floods are triggered by heavy rain, dam break, levee failure, or rapid snowmelt. Slow-moving or multiple thunderstorms over the same area also produce flash floods when the ground is so dry that it cannot absorb the rainwater timely. Those areas full of tall mountains and steep slopes are prone to flash floods as the surface runoff is confined to narrow gullies, causing the accumulated rainwater to rise quickly at the gully bottom.

The variables that are important to flash floods include slope gradient, slope aspect, TWI, TPI, profile slope curvature, convergence index, SPI, land cover, soil, and lithology (Costache et al., 2021).

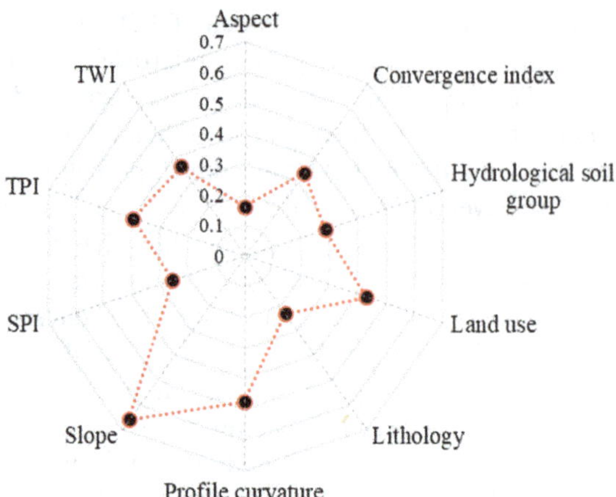

FIGURE 10.9 Importance of ten topographic and hydro-environmental variables to flash floods determined using linear SVM. (Source: Costache et al., 2021.)

Slope curvature is important to flash floods because it accelerates surface runoff. The convergence index is a morphometric variable differentiating area forming parts of river valleys from those situated along interfluvial divides. Soil affects flash floods mainly through vertical infiltration of rainwater. The variables significant to flash flood susceptibility are slightly different from those critical to general floods, although some are common to both. As illustrated in Figure 10.9, slope and profile curvature are the most important parameters to the formation of flash floods, followed by land use. The least important factor is slope aspect, while the remaining six factors have an importance that is rather similar to each other.

From these variables, the flash flood potential index value can be predicted using machine learning algorithms, such as deep-learning neural network–weight-of-evidence (Figure 10.10). This modeling approach is underpinned by an assumption that surfaces that have been affected by torrential rain in the past have a high chance of being struck by it again in the future. The areas affected by torrential rain are easily detected from remote sensing images based on the presence of torrential micro-reliefs such as ravines and gullies. These sites are located in the upper river basin where the slope is relatively steep without vegetation, both of which are conducive to flash floods.

In terms of hazard, a flash flood is no different from a general flood in the sense that it is also related to the same set of variables, and the hazard is estimated from remotely sensed parameters using hydrological modeling. Instead of deriving a single value for hazard, flooding extent, flow depth, and velocity in an arid environment may be predicted separately based on stream geometry and Manning's n values extracted from VHR images, peak discharge, and hydrographs (Mashaly and Ghoneim, 2018). Since only DEM and land use/land cover are used in the simulation, this prediction is virtually a scenario-based analysis of rainfall at a given intensity as no historic flooding events are necessary. In the prediction, the three variables (layers) are treated independently from each other. Hence, it is impossible to know the realistic level of flash flood hazard, a deficiency that can be overcome by jointly modeling flash floods in a GIS. In the modeling, basin morphometric parameters such as drainage network (stream order, basin order, stream length, and bifurcation ratio), basin geometry (length, width, perimeter, area, elongation ration, circulatory ratio, texture ratio, stream frequency, drainage density form factor ration), and basin relief (relief ratio, relative relief, and ruggedness number) can all be taken advantage of after they have been derived from a DEM. Since the topic is not related to remote sensing closely, it is not expounded further.

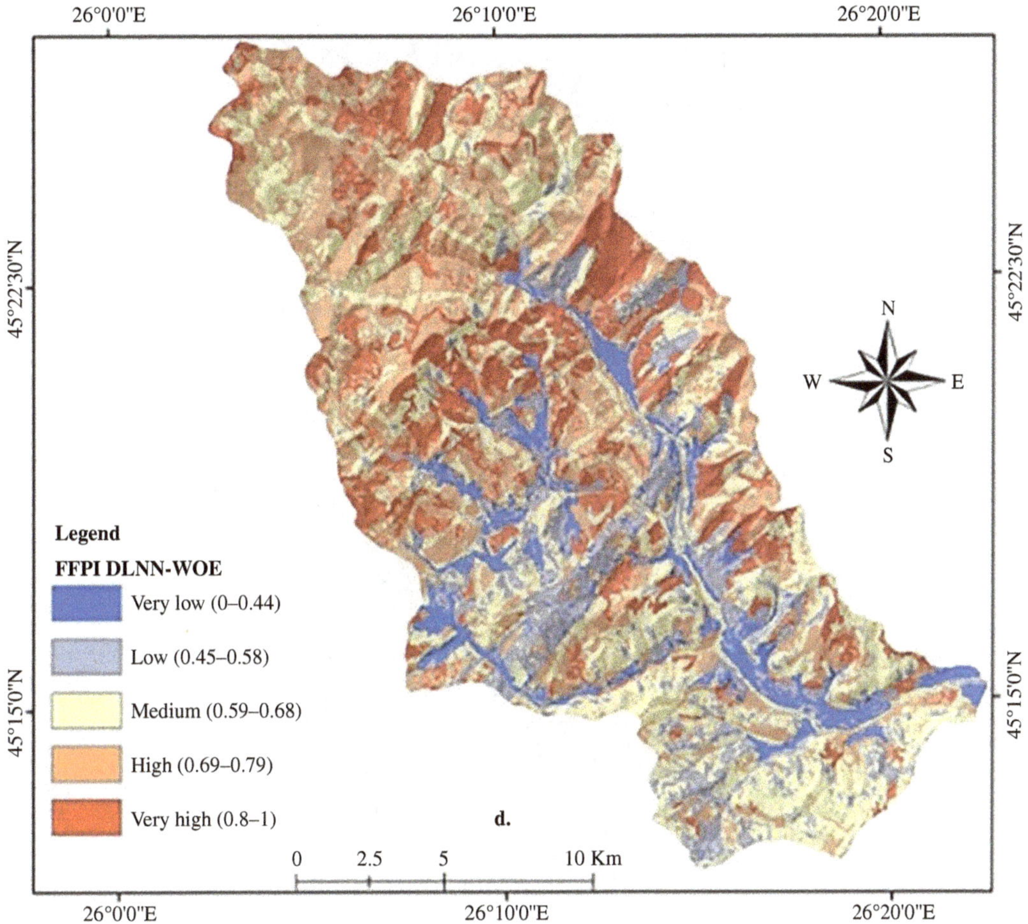

FIGURE 10.10 Watershed-level flash flood potential index at five levels estimated from ten variables using deep-learning neural network weight-of-evidence. (Source: Costache et al., 2021.)

In addition to the usually considered flood factors of land use/land cover, lithology, slope, altitude, plan curvature, relief, stream networks, stream density, and distance from streams, four new flash flood-conditioning parameters (alluvial plain width, valley width, basin gradient, and mean slope) have also been considered to map flash flood susceptibility in an arid region (Elmahdy et al., 2020). Essential in the mapping is a flash flood inventory map that serves as an excellent indicator of flash flood susceptibility. This map can be produced from remotely sensed images and allows the selection of flash flood locations to establish the relationship between the conditioning parameters and flash flood susceptibility. These locations can also be used to train the machine learning algorithms if the mapping is done automatically using the computer. Commonly used machine learning methods are boosted regression tree (BRT), CART, and NB tree models (see Section 1.4.2 for more details). Of these models, BRT is the most accurate, followed by CART and NB, achieving an $F1$ score of 0.91 and the highest AUC value of 0.92 (Elmahdy et al., 2020). The mapped susceptibility indicates that extreme flash floods commonly take place in narrow alluvial plains of the mountainous and coastal areas in the UAE, characterized by steep slopes, a large relief, surface runoff, and a high density of streams. Of all the factors, altitude and slope are the most important. The best flash flood susceptibility map is divided into seven basins, each having its flash flood susceptibility (Figure 10.11).

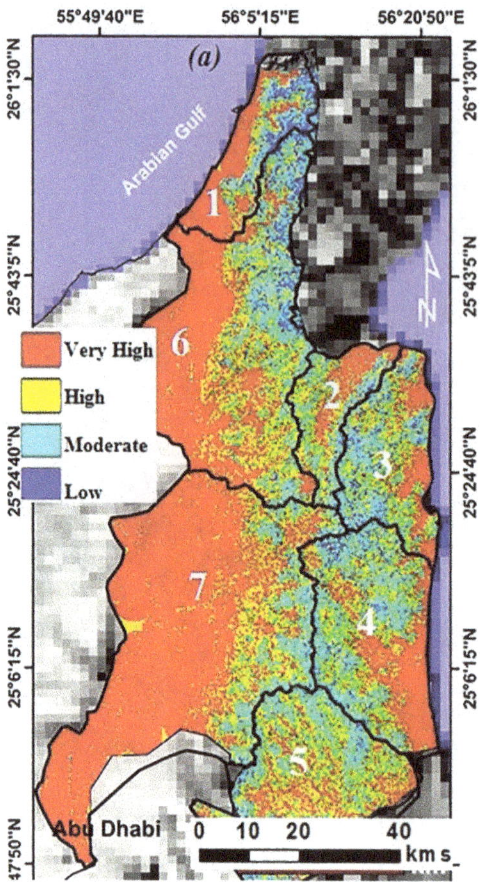

FIGURE 10.11 Flash flood susceptibility in the UAE modeled using the boosted regression trees machine learning method based on 12 conditioning parameters. (Source: Elmahdy et al., 2020.) 1–7: basins.

10.5 COASTAL FLOODS

Coastal floods are commonly caused by three means, hurricane-induced storm surge, monsoon rainfall, and global warming-triggered sea-level rise combined with savage storms. These processes may interact with each other, resulting in more severe floods than individually. Coastal areas are particularly prone to storms and floods caused by severe weather conditions, such as hurricanes owing to their geographical proximity to the sea, low-lying topography, and relatively high population density. Dissimilar to heavy rain-triggered floods that take place within a short time, coastal inundation caused by storm surges can last for days, during which ferocious waves slap the shore and can induce coastal erosion. During a hurricane, strong gales trigger a sudden surge in the seawater level. Horizontally, the surge can fan out more than 100s km of coastline. Vertically, the surge can reach a height of over 6 m near the center of a category 5 hurricane (Klemas, 2009). The surge may be amplified by king tides or accompanied by downpours. The combination of these two processes causes inundation of low-lying areas and does enormous damage to infrastructure, and even leads to loss of lives, such as the New Orleans flood by Hurricane Katrina in 2005. The storm surge breached the levees protecting the city and flooded about 80% of it to a maximum depth of 6 m. It is significant to delineate floodplains, and zone areas that need protection from storm floods, identify plans for development, and scope various kinds of flood protection measures.

TABLE 10.9
Data Needed in Mapping Coastal Inundation Areas

Geography	Hydrology	Others
DEM	Rainfall	Historical flood extent and depth
Aerial photographs/images	Discharge	Operating rules
River/drainage cross section	Water level	Regulation reports
Reservoir	Tide level	
Hydraulic structures	Wave	
Sewer system		
Coastal dyke/seawalls		
Sea bathymetry		
Land use condition		

Source: Doong et al. (2016).

10.5.1 Storm-Triggered Floods

10.5.1.1 Data Needs

A storm surge is the sudden and abrupt rise of seawater above the usual level under the action of gales in coastal areas. It is the main cause of ephemeral coastal floods that can do enormous damage to nearby residential buildings and infrastructure. With the anticipated sea-level rise caused by global climate warming, storm surges will become more common and frequent and cause more floods and damage to coastal areas. The minimization of storm surge hazards requires the mapping of coastal areas prone to the hazard. As with all other kinds of natural hazards, storm surge hazards and vulnerability cannot be sensed directly using remote sensing. So they are commonly estimated via numerical simulation models that require remote sensing data as inputs. Other data needed in mapping coastal inundation are listed in Table 10.9.

Remote sensing plays a vital role in supplying information to predict the path of a hurricane, its intensity and landfall, and storm surges and waves in coastal flood monitoring. While remote sensing is unable to predict storm surge height and waves itself, it is able to provide the data indispensable for the prediction, such as the width and slope of the continental shelf, and water depth. It can also supply information on the elevation of the coastal areas. Its relation to the expected maximum height of water enables the prediction of the area likely to be flooded in a range of storm scenarios. Any coastal areas that lie below the wave height will be inundated in a storm unless it is protected by seawalls. This assessment also needs to take into account tidal height. Only when the tide is at its maximal height will the predicted coastal inundation eventuate. These two applications are addressed differently using different methods from those adopted for flood area mapping. Essential in the mapping is the actual surface elevation that is best determined using LiDAR data. Apart from LiDAR, DEM can also be acquired from radar images, altimeters, and scatterometers to measure the amplitude of short surface waves, including surface wind velocity and roughness. The mapping is based on scenarios, plus the impact of waves (e.g., 3 or 5 m).

10.5.1.2 Vulnerability Indices

With continued climate change, the sea level will rise and induce storm surges and floods in coastal areas. Coastal susceptibility to flooding due to the foreseen sea-level rise has been assessed using the coastal vulnerability index (CVI) derived from eight parameters: (a) geomorphology, (b) slope,

(c) relative sea-level change rate, (d) mean tide range, (e) shoreline erosion and accretion, (f) population, (g) bathymetry, and (h) coastal flooding (Islam et al., 2015), or

$$CVI = \sqrt{\frac{a \cdot b \cdot c \cdot d \cdot e \cdot f \cdot g \cdot h}{8}} \tag{10.7}$$

The aforementioned parameters must be ranked on the basis of their potential contribution to physical changes on the coast, as the sea level rises. Not all coastal areas are equally vulnerable to sea-level rise-induced inundation because local geomorphological features also exert a significant control over the vulnerability, such as low tidal flats (i.e., supratidal plains and intertidal plains), depressions, low elevated dunes, cheniers, spits, and levees. The derivation of CVI is ideally implemented using remote sensing and GIS. For instance, shoreline erosion and accretion can be easily detected from multi-temporal fine-resolution images, and all the considered factors (layers) are easily integrated into a GIS. The derived CVI value is usually grouped into three to five classes to show the spatial distribution of the vulnerability pattern (Figure 10.12).

Apart from storm surges, coastal flooding or waterlogging can also be caused by monsoons in certain parts of the world, such as the south-western coastal Upazila of Bangladesh. Waterlogging hazard in the pre-monsoon and monsoon seasons has been mapped and quantified from satellite images. Change detection of maps produced from multi-temporal images reveals the inundated or waterlogged areas by monsoon water. Analysis of multi-temporal satellite images can also reveal the proportion of the waterlogged area (including floodplain) by tidal saline water and

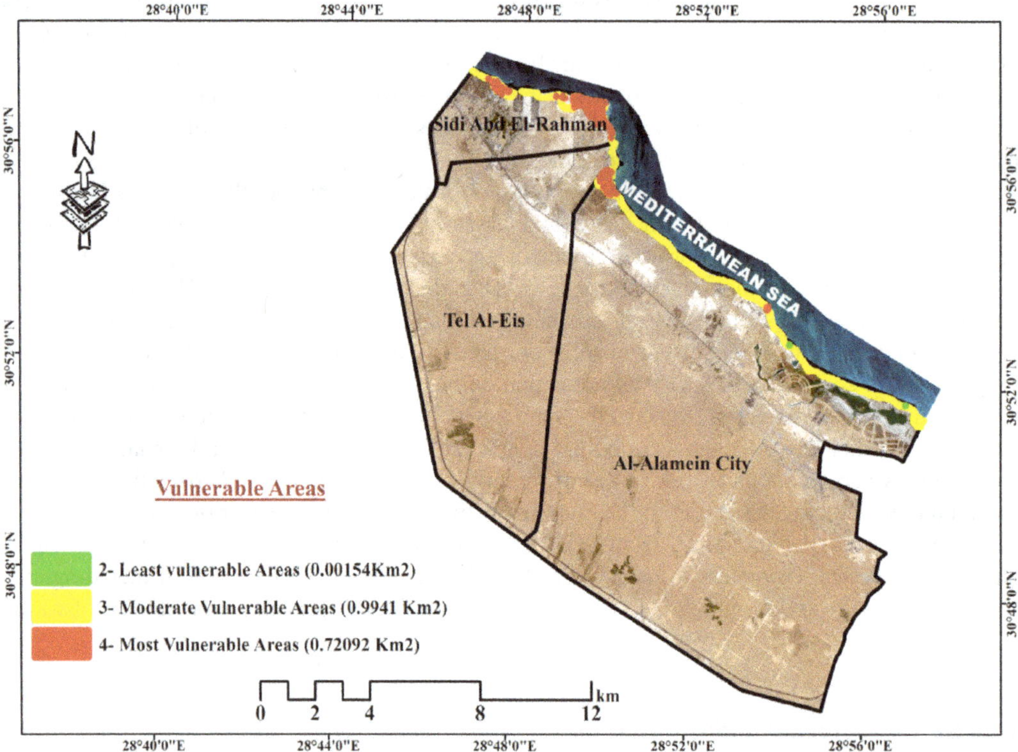

FIGURE 10.12 Coastal vulnerability to climate change impacts in Al-Alamein New City of Egypt based on remote sensing and GIS modeling. (Source: Marzouk et al., 2021. Used with permission.)

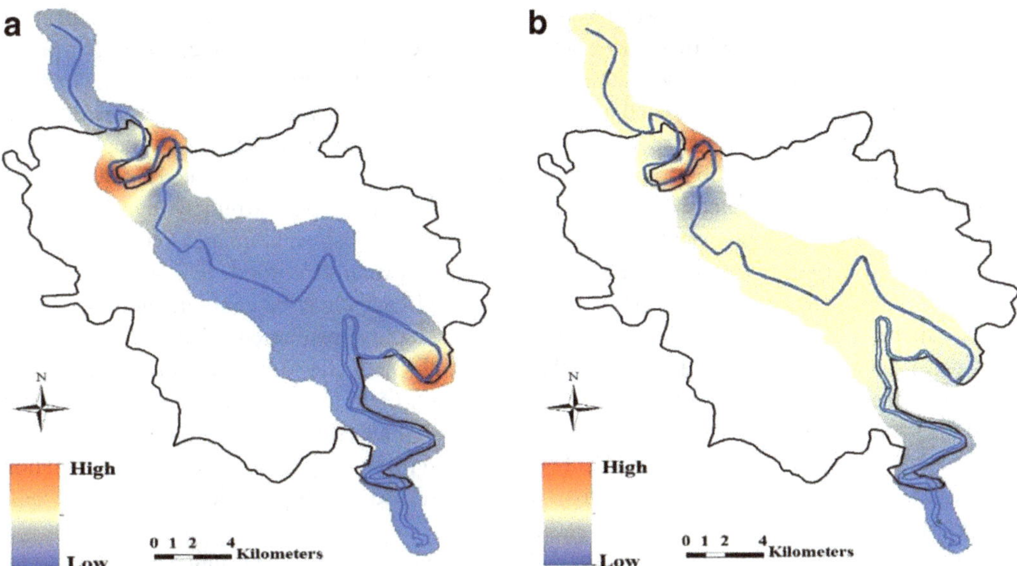

FIGURE 10.13 (a) Waterlogging hazard on the southwest coast of Bangladesh based on three factors: water depth, flow velocity, and sediment yield during the monsoon season, and (b) in the pre-monsoon season. (Source: Tareq et al., 2018. Used with permission.)

their use (e.g., having the potential for shrimp farming) in the pre-monsoon and monsoon seasons (Figure 10.13).

The change in waterlogged areas may be further analyzed against a number of environmental factors to identify their significance to coastal flooding. Coastal waterlogging hazard (CWH) is assessed from the velocity of flow, depth of flow, and inundation depth using an empirical model developed by Tareq et al. (2018) as follows:

$$CWH = \frac{S1}{V1 \cdot D1} + \frac{S2}{V2 \cdot D2} + \frac{S3}{V3 \cdot D3} \tag{10.8}$$

where CWH = waterlogging hazard based on sediment load (S), velocity (V), and inundation depth (D). The three factors of flow velocity, flow depth, and inundation depth are 1, 2, and 3, respectively. Thus, the hazard is proportional to the sediment yield, usually transported by flooding river water, and inversely proportional to the product of velocity and inundation depth of the river flow.

10.5.1.3 Flooding Extent and Depth

The mapping of sea-level rise-induced coastal inundation is very similar to that associated with storm surges and waves. Virtually, the same method of mapping storm-triggered coastal inundation is still useable without modification. The extent of large-scale coastal floods caused by hurricanes can be mapped rapidly from coarse-resolution optical and microwave images. For instance, Hurricane Floyd caused the Tar River to crest at a record height of 4.30 m above the flood stage at the river gauge station of Greenville, North Carolina on 21 September 1999. Whether the maximum flood extent can be detected by means of remote sensing depends on when the images are recorded. The delineation of this extent from optical images such as Landsat 8 OLI may rely on existing pre-flood data, near peak data, and data obtained a few days after the peak flood. Of the seven TM bands, bands 5 and 7 and bands 4 and 8 are able to identify the water–land (wet–dry) boundary. The detection has been eased by the advent of Landsat 8 imagery that possesses an improved spatial resolution of 15 m from the usual 30 m. The inundated areas may be detected by firstly classifying all the images into water and non-water, and then spatially comparing all the maps to detect changes in flooded, non-flooded areas, and water bodies. It is possible to achieve high overall classification accuracies around 82.5–99.3% owing to water's unique spectral behavior (Wang, 2004). Per-pixel-based evaluation against the ground truth indicates an agreement of 90.7% in terms of regular river channels and water bodies, flooded and non-flooded areas. The mapping accuracy is especially high in open fields that include cultivated lands, herbaceous cover, evergreen shrublands, and deciduous shrublands whereas the flooded areas include standing water, very wet or saturated soil, and damaged vegetation caused by floodwater. They are more difficult to map at extremely high accuracy.

In order to best map flood extent on a coastal floodplain, the post-event images must be acquired days after the river's flood crest to capture the maximum extent of a flood, if it is impossible to acquire remotely sensed data concurrently. However, the images may be contaminated by clouds, which can be overcome with microwave data, such as coarse-resolution advanced technology microwave sounder (ATMS) data. It has 22 channels (23.8–183.3 GHz), some of which are suitable for sensing the Earth's surface. But their spatial resolution of 15 km is too coarse to map coastal inundation areas. The detailed mapping of flooded areas over an extensive area requires images of a finer resolution, such as the NPP VIIRS data with a spatial resolution of 371 m × 387 m. Estimation of water fraction from such coarse-resolution VIIRS and ATMS data is possible via linearly decomposing mixed pixels (Sun et al., 2016).

Apparently, the 15-km spatial resolution ATMS data do not allow small inundated areas to be mapped. One way of overcoming this limitation is to use the water fraction and high-resolution flood (WFHF) method developed by Zheng et al. (2017). In the mapping, the extent of large-scale coastal floods caused by hurricanes, land and water are distinguishable from each other by differencing the brightness temperature of bands 3 and 4. The detection of pure water, pure land, and land-water mixed pixels is based on rules. The image is first classified into pure water and pure land pixels based on their thresholds. The difference in the water fraction at the basin scale before and after flooding is calculated to reduce the impacts of soil and vegetation and to avoid pixel-to-pixel errors. Based on the water fraction difference, the flood map derived from the ATMS data at 15 km is extrapolated to 100 m resolution using a DEM, based on the fact that water inundation always proceeds from the lowest to the highest elevation points in a basin. The WFHF-mapped flood area in the vicinity of New York City (Figure 10.14) is 88% accurate and is closely correlated ($r = 0.94$) with the 3 m resolution storm surge flooding products distributed by the Federal Emergency Management Agency (FEMA). The inundated area is mapped from VIIRS data using decision trees, followed by cloud/shadow removal, and calculation of water fraction in a pixel. The flood map initially derived from the coarse-resolution VIIRS and ATMS data is interpolated to a finer resolution (e.g., 30 m) with the assistance of SRTM data (Sun et al., 2016). The ATMS-derived flood map of 30 m resolution has a high correlation coefficient (0.95) with the FEMA storm surge flooding products, 95% of ATMS-mapped inundation areas overlap with the geotagged Flickr contributions reporting flood map. Overall, the combination of optical and microwave data with a 30 m DEM is able to produce high-quality and high-resolution flood maps over broad-scale coastal areas even from very coarse-resolution satellite images.

Floodwater depth across the inundated domain cannot be readily estimated from remote sensing analysis directly but can be determined using the Floodwater Depth Estimation Tool (FwDET) developed by Cohen et al. (2019). It requires a raster inundation map in conjunction with a DEM. Its new version FwDET v2.0 (freely available as Python scripts and ArcGIS and QGIS tools at https://sdml.ua.edu/models/) features a new flood boundary identification scheme that accounts for the lack of confinement of coastal flood domains at the shoreline. A new algorithm is used to calculate the local floodwater elevation for each grid cell, alleviating inaccurate local boundary assignment across permanent water bodies. While developed primarily to study coastal flooding, FwDET v2.0 is equally applicable to the estimation of riverine floodwater depth. Evaluated against physically based hydrodynamic simulations in both riverine and coastal environments, the results produced by FwDET v2.0 closely match model-simulated depth with a mean difference of 0.31 m in the riverine (10 m DEM) case and 0.18 in the coastal case (1 m DEM) (Figure 10.15). So the use of a finer resolution DEM is conducive to reaching a closer match.

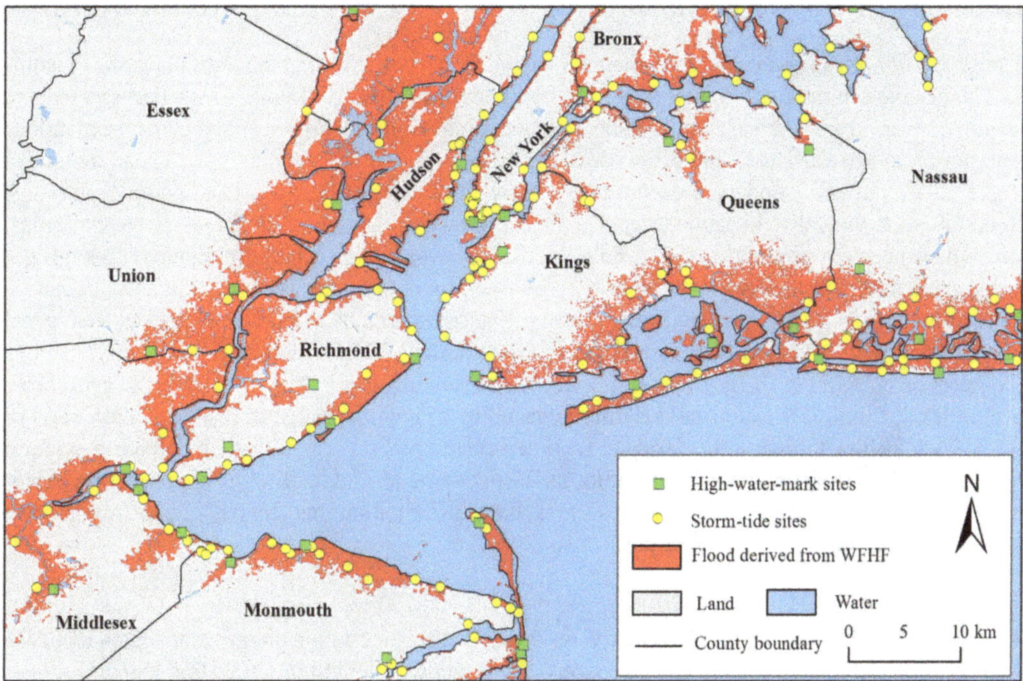

FIGURE 10.14 Extent of coastal flooding in the vicinity of New York City caused by Hurricane Sandy in late October 2012, detected from ATMS data using water fraction and a high-resolution flood mapping method. (Source: Zheng et al., 2017.)

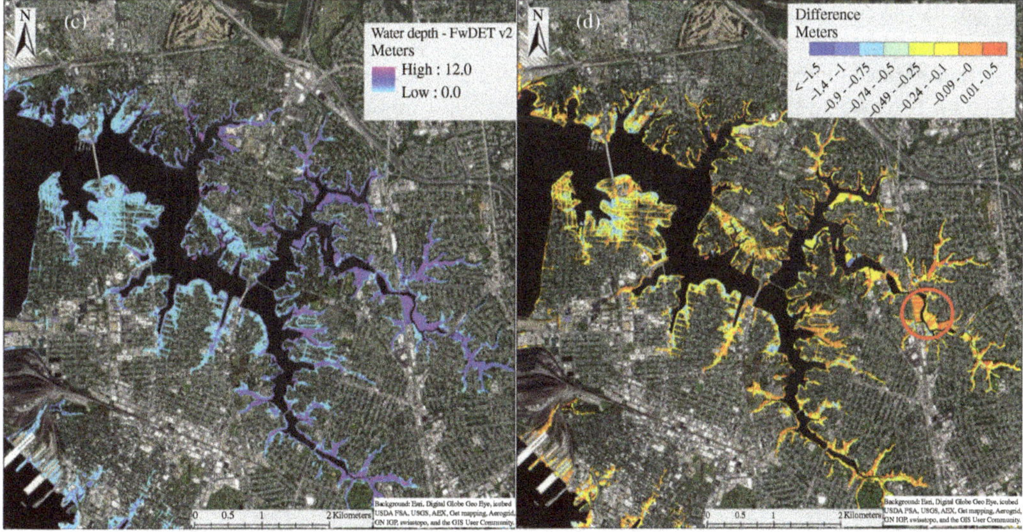

FIGURE 10.15 Depth of coastal flooding in the Norfolk–Portsmouth estuary of Eastern US caused by the August 2011 Hurricane Irene, determined using the FwDET v2.0 tool (c), and its difference from the model-simulated depth against the backdrop of a satellite image (d). (Source: Cohen et al., 2019)

10.5.2 Tsunami-Triggered Floods

The term *tsunami* originates from Japanese, meaning harbor (*tsu*) and wave (*nami*), and is commonly accepted in the English language. A tsunami depicts a series of huge waves that can reach a height of over 30 m. The surging oceanic waters encroach on coastal areas and cause tremendous devastation to infrastructure and properties, as well as heavy loss of human lives. These waves are triggered by powerful undersea earthquakes along the interface between two tectonic plates. A tsunami exerts virtually the same effects as the hurricane-triggered storm surges in coastal areas, which have been covered in the preceding sections. However, a tsunami is still unique in that it is not accompanied by heavy downpours or gales. The flooding of seawater is not permanent but can do more damage than freshwater flooding because saline seawater may kill crops and devastate coastal vegetation. After the tsunami-triggered waves have retreated, the inundated areas may still lie above the water level. Almost all the areas impacted by the tsunami-generated waves face the grim prospect of total ruin. It is the enormous damage caused by a tsunami that deserves special attention of study, such as how to assess the damage. Remote sensing plays a critical role in detecting tsunami features, such as mapping inundation limits, assessing damage, and facilitating the modeling of tsunami vulnerability. Table 10.10 illustrates the tsunami information that can be studied from remote sensing data acquired from various platforms.

10.5.2.1 Tsunami Wave Detection

The series of earthquake-generated waves on the ocean surface in a tsunami propagates from the source (epicenter) radially out in all directions. The exact wavelength of a traveling tsunami ranges from 10s to 100s km, depending on the fault rupture scale, and wave height that can range from several centimeters to several meters (Koshimura et al., 2020). Such a size far exceeds the spatial resolution of most satellite images. Hence, it is extremely difficult to track tsunamis remotely from space so as to predict their most likely impact areas. On the other hand, the sooner a tsunami is detected, the more precious time is saved to warn the nearby ships and the likely coastal areas to be impacted. The difficulty level of detecting tsunami waves varies with the earthquake magnitude and depth. Those in deep water tend to have a smaller amplitude than nearshore waves and hence are more difficult to detect, even though they may be amplified enormously as they approach the shoreline. This is because shallow water slows tsunami propagation speed and compresses its wavelength, hence amplifying its amplitude considerably. Thus, it is much easier to detect the same tsunami nearshore than in the middle of an ocean, but the detected results are less valuable to mitigate tsunami damage. In reality, it is still possible to detect tsunamis in the mid-ocean based on the detected sea surface height in the tsunami propagation direction, such as the 2004 Sumatra–Andaman earthquake-triggered tsunami from a Jason-1 satellite image two hours after its

TABLE 10.10
Tsunami Features That Are Best Studied Using Different Types of Remote Sensing Data Aboard Various Platforms

Tsunami Features	Platforms	Sensors/Sensing Methods
Mid-ocean propagation	Satellites	Altimeter (Sea surface level)
Inland penetration	Aircraft	Videos
Inundation zone	Satellites, aircraft	Optical sensors, SAR
Structural damage	Satellites, aircraft, drones	Optical sensors, SAR
Debris	Satellites, aircraft, drones	Optical sensors, SAR, LiDAR
Search and rescue	Aircraft, drones	Optical sensors, videos

Source: Koshimura et al. (2020).

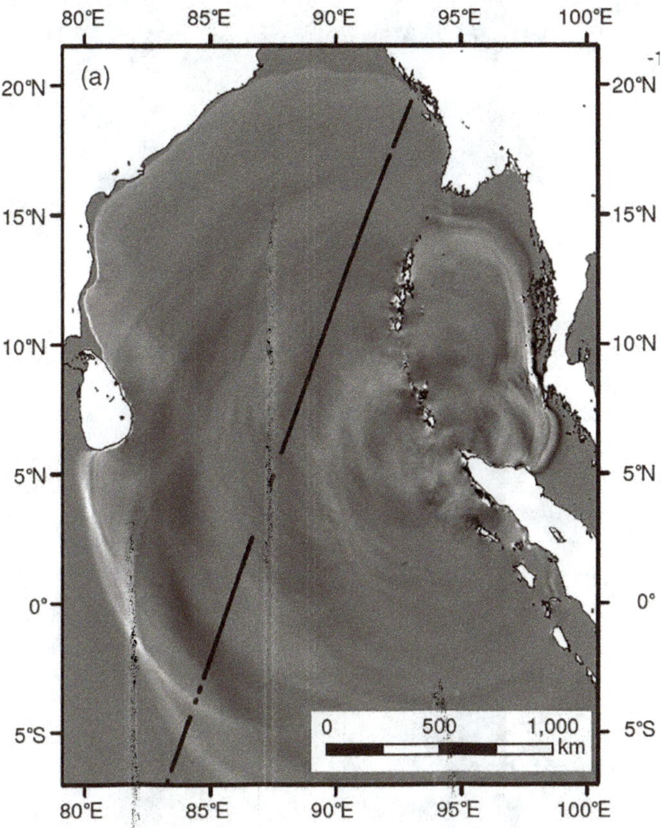

FIGURE 10.16 A mid-ocean tsunami captured on a Jason-1 satellite image two hours after the 26 December 2004 Sumatra-Andaman earthquake (thick line: Jason-1 track). (Source: Koshimura et al., 2020.)

initialization (Figure 10.16). It may be impossible to capture the entire tsunami in one image, but at least the rough location of the tsunami can be pinpointed. The direction of tsunami wave propagation cannot be detected because it requires two temporally consecutive images. They are impossible to capture because of the satellite's lengthy revisit period. The waves will likely have dissipated by the time the second image is recorded unless the sensor is tilted specifically for this purpose.

10.5.2.2 Inundation Area

The spatial extent of tsunami inundation indicates the landward encroachment distance of a tsunami upon the coastal area. Tsunami-inundated areas are easily extracted from remote sensing images because of the spectral distinction between seawater-flooded areas and those unaffected. This distinction lingers even after the seawater has receded because of the lasting damage the seawater has done to the former vegetated and built-up areas (Figure 10.17). The easiest way of detecting the inundated areas is to visually interpret the affected areas that tend to be more saturated with saline water than intact areas, and hence a darker tone or color. This disparity can also be detected with the assistance of spectral indices, such as NDWI. Other cues, such as sediment deposits and debris piled up by tsunami waves, may also be useful in determining the inundation distance and run-up, subject to coastal morphology and distance to the epicenter (Ramírez-Herrera and Pacheco, 2012). However, the use of VI in the detection is plagued by the omission of height. For instance, a tree may be perfectly healthy immediately after the inundation, but it will not be detected as being inundated from an image recorded shortly after the flooding. A better alternative is to make use of a DEM. After the inundated areas have been determined,

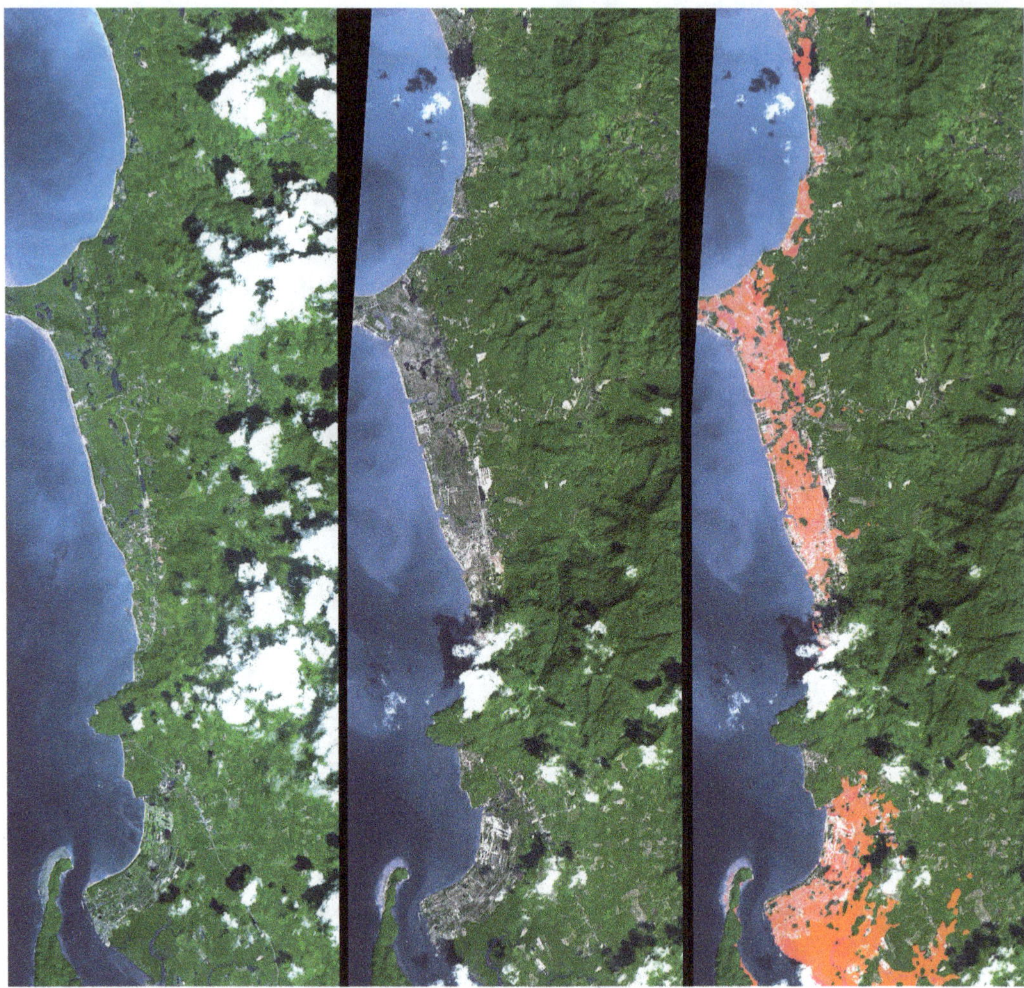

FIGURE 10.17 Multi-temporal ASTER images of Khao Lak, Thailand before 15 November 2002 (left) and after 31 December 2004 (middle), and the coastal inundation caused by the 26 December 2004 Indian Ocean tsunami (right). The area highlighted in red is 10 m above the sea level determined using the Shuttle Radar Topographic Mission data. (Source: https://photojournal.jpl.nasa.gov/tiff/PIA06671.tif.)

they can be overlaid with a DEM of the same spatial resolution to identify the maximal height of inundation. Any areas below the maximum inundation height must also be flooded, irrespective of whether their height is above the ground. This overlay can also reveal the depth of inundation. Such elevation-based detection enables inundated areas to be extracted more accurately than spectral-based detection.

The accuracy of the extracted inundation extent is subject to the spatial resolution of the satellite image and the vertical accuracy of the DEM. The vertical accuracy further varies with the ground cover, slope, aspect, and elevation (Ramírez-Herrera and Pacheco, 2012). In order to achieve the most reliable extraction, the satellite image must be captured soon after the inundation event. Otherwise, some of the affected vegetation may have regenerated if they have not been killed by the saline water, hampering the detection and lowering the reliability of the detected inundation extent. The mapped inundation extent can reveal tsunami-vulnerable areas, which is related to elevation, land use, and distance from the shoreline (Figure 10.18). The inundated areas may be confined if the affected site lies in a confined coastal plain backed by steep mountains.

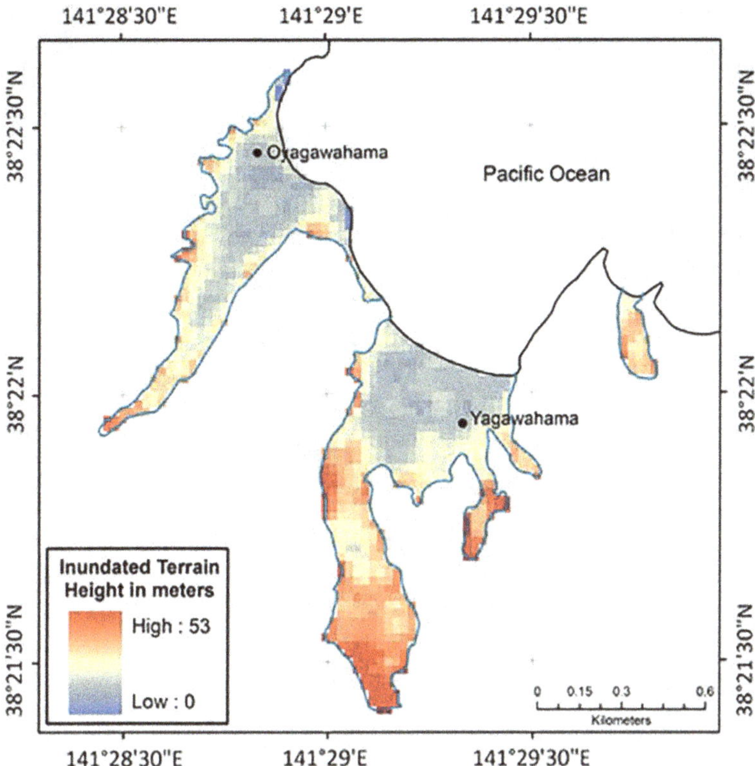

FIGURE 10.18 Areas inundated by the 2011 Tohoku tsunami at Oyagawahama and Yagawahama in the Oshika Peninsula, Miyagi Prefecture of Japan, extracted from 30 m GDEM produced from ASTER images based on maximal height. (Source: Ramírez-Herrera and Pacheco, 2012. Used with permission.)

Apart from DEMs, tsunami-inundated coastal areas can also be mapped from optical and radar satellite images, such as ASTER and EnviSat SAR data (Chini et al., 2013). Optical images allow inundated areas to be mapped into multiple covers, such as flooded, debris, damaged buildings, mud and stressed and unstressed vegetation using k-means clustering analysis. Co-registered optical and radar images enable the identification of the threshold of flooded areas from radar imagery, and the inundation between the pre- and post-event backscatterings. They can also reveal the maximum inundation width and its spatial variation across an area, together with the mean inland inundation distance (Chini et al., 2013). The accuracy of the mapped inundation may be compared with the modeled results. For instance, the Tohoku University's Numerical Analysis Model for Investigation Near-field Tsunami No. 3 (TUNAMI N3) is able to simulate the inundation area and produce inundation depth. Since it is related mostly to modeling, it is not explored further.

10.5.2.3 Damage Assessment

A tsunami does enormous damage to all features in the inundation zone, including cropland, buildings, properties, and infrastructure. The damage may be assessed using multi-temporal or mono-temporal images. If assessed from mono-temporal data, they must be acquired after the tsunami. Post-tsunami data allow the detection of building damage in built-up areas only, as unaffected buildings must always be upright. Any buildings that deviate from this are damaged. In comparison, the use of pre- and post-event images offers more flexibility (Figure 10.19). General damage can be assessed by comparing land covers before and immediately after a tsunami strikes in a change detection. This spatial comparison reveals the level of damage. Ideally, the two images should originate from the same sensor and cover the same ground area, as with all types of change detection.

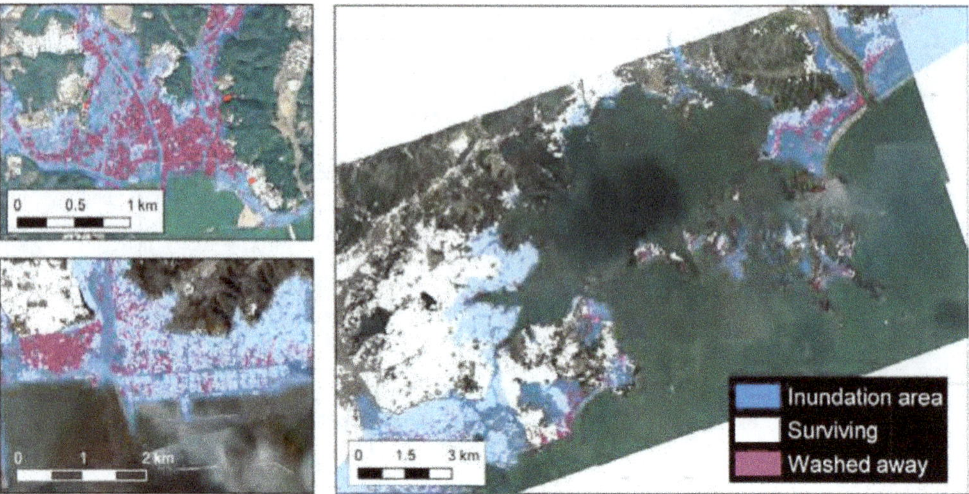

FIGURE 10.19 Building damage visually inspected from aerial photographs taken soon after the 2011 Tohoku tsunami in Minami-Sanriku town (upper left), Ishinomaki city (lower left), and Matsushima town and Shiogama city (right) of Japan. Red: washed away houses; blue: tsunami inundation areas. (Source: Suppasri et al., 2012.)

These images can be optical or microwave. The area inundated by the surging seawater can be detected automatically based on NDVI. Any areas that used to be vegetated but are devoid of vegetation after the tsunami are construed as having been affected by it. The surged seawater has likely destroyed the original vegetation. Damage to cropland is assessed from the area of inundation and the type of crops affected. Since crops are easily ravaged by the saline seawater, presumably, all inundated croplands suffer total crop failure, even if the inundation may last only days. However, whether the land is ready for immediate replanting is another type of damage not directly visible in the aftermath image. Further loss of land productivity of the inundated land should be factored in all realistic damage assessments.

The principles and method of assessing tsunami damage to buildings are almost identical to those of earthquake damage assessment. If buildings are toppled by the torrent waves of a tsunami, then they can be easily detected via subtracting the pre-event digital surface model (DSM) constructed from LiDAR data from another post-disaster DSM. Building-level damage assessment, especially if the buildings are still standing, requires fine resolution (e.g., a pixel size of 0.80 m), large-scale aerial photographs, or VHR satellite images. The areas damaged by a tsunami are easily mapped using image classification, during which buildings may be broadly classified as collapsed, washed away, or survived based on the presence of the building roof, as revealed by visual interpretation (Gokon and Koshimura, 2012). The mapping can be made more detailed by including an additional class called "slightly damaged" from pre- and post-event TerraSAR-X data acquired in the StripMap mode with a spatial resolution up to 3 m in a decision-tree classification (Gokon et al., 2015). The assessment of building damage severity is underpinned by the assumption that washed away or collapsed buildings have less intense backscattering on radar imagery than intact buildings. The overall detection accuracy is rather limited at only 67.4% (kappa = 0.47) in Sendai of Japan where the method is developed. The main reason for inaccuracy is the invalid assumption as shifted debris from collapsed buildings also causes an increase in backscattering. Accuracy rises slightly to 70.9% in classifying the damage to three types of collapsed, washed away, and survived from pre- and post-event WorldView-2 images (Bai et al., 2018).

One way of improving the mapping accuracy is to resample the remote sensing image to a finer resolution after spatial filtering. Then the filtered image is split into multiple tiles, only those tiles containing sufficient built-up areas are selected for further classification into various types of damage, such as washed away, collapsed, and slightly damaged using a modified wide residual network. The washed away regions contain most of the buildings washed away, as revealed by the ground truth data. Built-up areas are readily extracted and differentiated from non-built-up areas from a post-event TerraSAR-X image using the SqueezeNet network at an accuracy of 80.4% (Bai et al., 2018). The accuracy decreases to 74.8% (kappa = 0.60) in mapping building damage levels. Although this accuracy is slightly higher than 67.4% and 70.9% achieved in other similar studies, the mapping is done for the damaged regions, not individual buildings due probably to the low spatial resolution of the image used.

Building-level damage assessment is possible only when the building footprint is available. When this is not the case, built-up areas have to be classified from VHR images. No matter what kinds of images are used, remote sensing-based damage assessment ignores structural damage to the still standing buildings. The mapped results show nothing about whether the survived buildings are still inhabitable. Nevertheless, the devastated (e.g., washed away) buildings mapped via remote sensing can still be useful to evaluate the structural vulnerability against tsunamis at the local scale.

10.5.2.4 Debris Volume Estimation

In post-disaster response and management, it is essential to determine the volume of debris deposited by a tsunami. The presence of tsunami debris is easily detected from pre- and post-event remote sensing images. Virtually, any areas that have a higher elevation after the tsunami than before the event must be caused by newly deposited debris. The estimation of debris volume, however, is much more complex and demanding than tsunami damage assessment as it requires sub-meter resolution stereoscopic images such as airborne Pi-SAR2 (Koyama et al., 2016). These X-band images have a spatial resolution of 0.3–0.6 m in the along-track (azimuth) direction and 0.3–0.5 m in the cross-track (slant range) direction. The debris volume is computed from the area and height of the debris pile using Eq. 3.3, of which area is easily estimated from the number of mapped debris pixels multiplied by pixel size. The pile height, measured from the top of a pile to its base, is estimated either from stereoscopic images or radar images. If estimated from a pair of radar images, they must be separated by a baseline of r_d, and both sense the same pile from a distance of R_1 and R_2, respectively (Figure 10.20). Debris height z_t is calculated as follows:

$$z_t = z_r - \Delta R_1 cos\theta_1 \tag{10.9}$$

$$\Delta R1 = R_1 - R_{1r} \tag{10.10}$$

$$\theta_1 = cos^{-1}\left(R_1^2 + r_d^2 - R_2^2\right)/2R_1 r_d \tag{10.11}$$

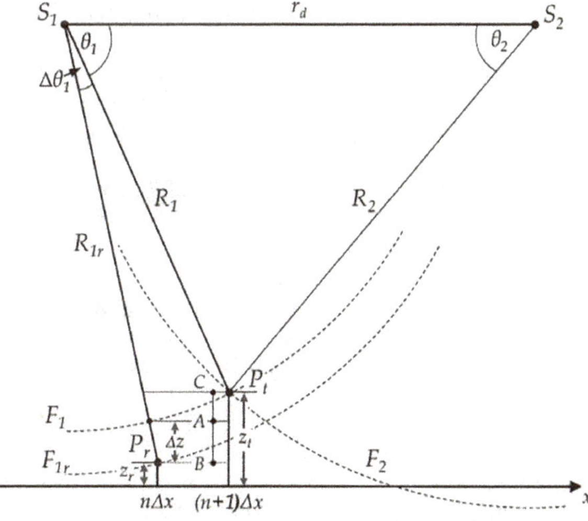

FIGURE 10.20 Geometric configuration of two radar antenna positions, S_1 and S_2, in relation to the phase difference Δz ($R_1r - R_1$). R_1, R_1r, and R_2 represent the radar range fronts. (Source: Koyama et al., 2016.)

where z_r = height of the reference object (usually set to 0); R_1 and R_{1r} are determined from the radar velocity and by timing the propagation of the radar pulses from the radar antenna to the receptor.

Radar ranges R_1 and R_2 are calculated as follows:

$$R_1 = R_{1r} + c\Delta t_1 / 2 \tag{10.12}$$

$$R_2 = R_{2r} + c\Delta t_1 / 2 \tag{10.13}$$

where c = speed of light (e.g., 2.9979246×10^8 m·s^{-1}); $\Delta t_1 = t_1 - t_1'$; $\Delta t_2 = t_2 - t_2'$. If the debris height is detected from two-time TerraSAR-X images, the threshold of debris height is calculated from the difference (d) in the average backscattering coefficient within a 13 × 13 pixel window between the pre- and post-event images (after–before) using the following empirical equations:

$$Z_{R1} = -1.21\text{d} - 4.36\text{r} \tag{10.14}$$

$$Z_{R2} = 1.21\text{d} - 4.36\text{r} \tag{10.15}$$

where r = correlation coefficient of the two SAR images calculated within the same 13 × 13 pixel window (Koyama et al., 2016).

After the base area of the debris piles is calculated via mathematical morphology analysis, the debris pile volume can be estimated on a voxel basis by totaling all the 3D pixels (Δx, Δy, Δz) that belong to the same pile. If estimated from high-resolution polarimetric stereo-SAR images, the estimated height has an RMSE smaller than 0.6 m, while the estimated debris volume has an RMSE of 1,099 m^3 (Koyama et al., 2016). Since the stereo images with low coherence are co-registered with each other using the eFolki algorithm, the gradient-based optical-flow technique does not require the use of pre-event images or any auxiliary information such as orbits or DEM data in the estimation.

REFERENCES

Adell, C., and C. Puech. 2003. Will the spatial analysis of water maps extracted by remote satellite detection allow locating the footprints of hunting activity in the Camargue? *Bull. Soc. Fr. Photogramm. Teledetec.* 2003: 76–86.

Anusha, N., and B. Bharathi. 2020. Flood detection and flood mapping using multi-temporal synthetic aperture radar and optical data. *Egyptian J. Remote Sens. Space Sci.* 23(2): 207–219. DOI: 10.1016/j.ejrs.2019.01.001

Arora, A., A. Arabameri, M. Pandey, M.A. Siddiqui, U.K. Shukla, D.T. Bui, V.N. Mishra, and A. Bhardwaj. 2021. Optimization of state-of-the-art fuzzy-metaheuristic ANFIS based machine learning models for flood susceptibility prediction mapping in the Middle Ganga Plain, India. *Sci. Total Environ.* 750: 141565. DOI: 10.1016/j.scitotenv.2020.141565

Arora, A., M. Pandey, M.A. Siddiqui, H. Hong, and V.N. Mishra. 2019. Spatial flood susceptibility prediction in Middle Ganga Plain: Comparison of frequency ratio and Shannon's entropy models. *Geocarto Int.* 32: 2085–2611. DOI: 10.1080/10106049.2019.1687594

Bai, Y., C. Gao, S. Singh, M. Koch, B. Adriano, E. Mas, and S. Koshimura. 2018. A framework of rapid regional tsunami damage recognition from post-event TerraSAR-X imagery using deep neural networks. *IEEE Geosci. Remote Sens. Lett.* 15: 43–47.

Bonn, F., and R. Dixon. 2005. Monitoring flood extend and foresting excess runoff risk with Radarsat-1 data. *Nat. Hazards* 35: 377–393.

Brakenridge, G.R. Global Active Archive of Large Flood Events. Available online: http://floodobservatory.colorado.edu/Archives/index.html (accessed April 26, 2022).

Brisco, B., M. Kapfer, T. Hirose, B. Tedford, and J. Liu. 2011. Evaluation of C-band polarization diversity and polarimetry for wetland mapping. *Can. J. Remote Sens.* 37: 82–92.

Byun, Y., Y. Han, and T. Chae. 2015. Image fusion-based change detection for flood extent extraction using bi-temporal very high-resolution satellite images. *Remote Sens.* 7: 10347–10363. DOI: 10.3390/rs70810347

Caillaud, L., B. Guillaumont, and F. Manaud. 1991. *Essai de discrimination des modes d'utilisation des marais maritimes par analyse multitemporelle d'images SPOT, application aux marais maritimes du Centre Ouest.* Brest, France: IFREMER, p. 45.

Chini, M., A. Piscini, F.R. Cinti, S. Amici, R. Nappi, and P.M. DeMartini. 2013. The 2011 Tohoku (Japan) tsunami inundation and liquefaction investigated through optical, thermal, and SAR data. *IEEE Geosci. Remote Sens. Lett.* 10: 347–351. DOI: 10.1109/LGRS.2012.2205661

Cohen, S., A. Raney, D. Munasinghe, D.J. Loftis, A. Molthan, J. Bell, L. Rogers, J. Galantowicz, R.G. Brakenridge, A. Kettner, Y.F. Huang, and Y.P. Tsang. 2019. The Floodwater Depth Estimation Tool (FwDET v2.0) for improved remote sensing analysis of coastal flooding. *Nat. Hazards Earth Syst. Sci.* 9: 2053–2065. DOI: 10.5194/nhess-19-2053-2019

Costache, R., A. Arabameri, T. Blaschke, Q.B. Pham, B.T. Pham, M. Pandey, A. Arora, N.T.T. Ling, and I. Costache. 2021. Flash-flood potential mapping using deep learning, alternating decision trees and data provided by remote sensing sensors. *Sensors* 21: 280.

Davranche, A., B. Poulin, and G. Lefebvre. 2013. Mapping flooding regimes in Camargue wetlands using seasonal multispectral data. *Remote Sens. Environ.* 138: 165–171.

Doong, D.-J., W. Lo, Z. Vojinovic, W.-L. Lee, and S.-P. Lee. 2016. Development of a new generation of flood inundation maps—A case study of the coastal city of Tainan, Taiwan. *Water* 8: 521. DOI: 10.3390/w8110521

Elmahdy, S., T. Ali, and M. Mohamed. 2020. Flash flood susceptibility modeling and magnitude index using machine learning and geohydrological models: A modified hybrid approach. *Remote Sens.* 12: 2695. DOI: 10.3390/rs12172695

EXCIMAP (European exchange circle on flood mapping). 2007. Handbook on Good Practices for Flood Mapping in Europe. 57 p.

Gao, B.-C. 1996. NDWI—A normalized difference water index for remote sensing of vegetation liquid water from space. *Remote Sens. Environ.* 58: 257–266.

Gashaw, W., and D. Legesse. 2011. Chapter 9: Flood hazard and risk assessment using GIS and remote sensing in Fogera Woreda, Northwest Ethiopia. In: *Nile River Basin*, ed. A. M. Melesse, 179–206. Dordrecht: Springer. DOI: 10.1007/978-94-007-0689-7_9

Giustarini, L., M. Chini, P. Hostache, F. Pappenberger, and P. Matgen. 2015. Flood hazard mapping combining hydrodynamic modeling and multi annual remote sensing data. *Remote Sens.* 7: 14200–14226. DOI: 10.3390/rs71014200

Gokon, H., and S. Koshimura. 2012. Mapping of building damage of the 2011 Tohoku earthquake tsunami in Miyagi Prefecture. *Coast. Eng. J.* 54: 1250006.

Gokon, H., J. Post, E. Stein, S. Martinis, A. Twele, M. Mück, C. Geiß, S. Koshimura, and M. Matsuoka. 2015. A method for detecting buildings destroyed by the 2011 Tohoku earthquake and tsunami using multitemporal TerraSAR-X data. *IEEE Geosci. Remote Sens. Lett.* 12: 1277–1281.

Gond, V., E. Bartholomé, F. Ouattara, A. Nonguierma, and L. Bado. 2004. Surveillance et cartographie des plans d'eau et des zones humides et inondables en régions arides avec l'instrument VEGETATION embarqué sur SPOT-4. *Int. J. Remote Sens.* 25: 987–1004.

Huang, C., Y. Chen, and J. Wu. 2014. Mapping spatio-temporal flood inundation dynamics at large river basin scale using time-series flow data and MODIS imagery. *Int. J. Appl. Earth Obs. Geoinf.* 26: 350–362.

Huang, W., B. DeVries, C. Huang, M.W. Lang, J.W. Jones, I.F. Creed, and M.L. Carroll. 2018. Automated extraction of surface water extent from Sentinel-1 data. *Remote Sens.* 10(5): 797. DOI: 10.3390/rs10050797

Hudson, P., and R. Colditz. 2003. Flood delineation in a large and complex alluvial valley, lower Pánuco basin, Mexico. *J. Hydrol.* 280: 229–245. DOI: 10.1016/S0022-1694(03)00227-0

Huete, A.R. 1988. A soil-adjusted vegetation index (SAVI). *Remote Sens. Environ.* 25: 295–309.

Islam, M.A., M.S. Hossain, and S. Murshed. 2015. Assessment of coastal vulnerability due to sea level change at Bhola Island, Bangladesh: Using geospatial techniques. *J. Indian Soc. Remote Sens.* 43(3): 625–637. DOI: 10.1007/s12524-014-0426-0

Klemas, V.V. 2009. The role of remote sensing in predicting and determining coastal storm impacts. *J. Coastal Res.* 25(6): 1264–1275.

Kordelas, G.A., I. Manakos, D. Aragonés, R. Díaz-Delgado, and J. Bustamante. 2018. Fast and automatic data-driven thresholding for inundation mapping with Sentinel-2 data. *Remote Sens.* 10(6), 910. DOI: 10.3390/rs10060910

Koshimura, S., L. Moya, E. Mas, and Y. Bai. 2020. Tsunami damage detection with remote sensing: A review. *Geosciences* 10(5): 177. DOI: 10.3390/geosciences10050177

Koyama, C.N., H. Gokon, M. Jimbo, S. Koshimura, and M. Sato. 2016. Disaster debris estimation using high-resolution polarimetric stereo-SAR. *ISPRS J. Photogramm.* 120: 84–98.

Kumar, R. 2016. Flood hazard assessment of 2014 floods in Sonawari sub-district of Bandipore district (Jammu & Kashmir): An application of geoinformatics. *Remote Sens. Appl. Soc. Environ.* 4: 188–203. DOI: 10.1016/j.rsase.2016.10.002

Kumar, R., R. Singh, H. Gautam, and M.K. Pandey. 2018. Flood hazard assessment of August 20, 2016 floods in Satna District, Madhya Pradesh, India. *Remote Sens. Appl. Soc. Environ.* 11: 104–118.

Kwak, Y., B. Arifuzzanman, and Y. Iwami. 2015. Prompt proxy mapping of flood damaged rice fields using MODIS-derived indices. *Remote Sens.* 7: 15969–15988, DOI: 10.3390/rs71215805

Lillesand, T.M., and R.W. Kiefer. 1987. *Remote Sensing and Image Interpretation* (2nd ed.). New York: John Wiley and Sons.

Malinowski, R., G. Groom, W. Schwanghart, and G. Heckrath. 2015. Detection and delineation of localized flooding from WorldView-2 multispectral data. *Remote Sens.* 7: 14853–14875. DOI: 10.3390/rs71114853

Marzouk, M., K. Attia, and S. Azab. 2021. Assessment of coastal vulnerability to climate change impacts using GIS and remote sensing: A case study of Al-Alamein New City. *J. Clean. Prod.* 290. DOI: 10.1016/j.jclepro.2020.125723

Mashaly, J., and E. Ghoneim. 2018. Flash flood hazard using optical, radar, and stereo-pair derived DEM: Eastern Desert, Egypt. *Remote Sens.* 10(8): 1204. DOI: 10.3390/rs10081204

McFeeters, S.K. 1996. The use of the Normalized Difference Water Index (NDWI) in the delineation of open Water features. *Int. J. Remote Sens.* 17: 1425–1432.

McNairn, H., and B. Brisco. 2004. The application of C-band polarimetric SAR for agriculture: A review. *Can. J. Remote Sens.* 30: 525–542.

Mitchard, E.T.A., S.S. Saatchi, L.J.T. White, K.A. Abernethy, K.J. Jeffery, S.L. Lewis, M. Collins, M.A. Lefsky, M.E. Leal, I.H. Woodhouse, and P. Meir. 2012. Mapping tropical forest biomass with radar and spaceborne LiDAR in Lopé National Park, Gabon: Overcoming problems of high biomass and persistent cloud. *Biogeosciences* 9: 179–191.

Mojaddadi, H., B. Pradhan, H. Nampak, N. Ahmad, and A.H. bin Ghazali. 2017. Ensemble machine-learning-based geospatial approach for flood risk assessment using multi-sensor remote-sensing data and GIS. *Geomat. Nat. Hazards Risk* 8(2): 1080–1102. DOI: 10.1080/19475705.2017.1294113

O'Connell, J.L., D.R. Mishra, D.L. Cotten, L. Wang, and M. Alber. 2017. The tidal marsh inundation index (TMII): An inundation filter to flag flooded pixels and improve MODIS tidal marsh vegetation time-series analysis. *Remote Sens. Environ.* 201: 34–46. DOI: 10.1016/j.rse.2017.08.008

Ogashawara, I., M.P. Curtarelli, and C.M. Ferreira. 2013. The use of optical remote sensing for mapping flooded areas. *Int. J. Eng. Res. Appl.* 3(5): 1–5.

Pearson, R.L., and L.D. Miller. 1972. Remote mapping of standing crop biomass for estimation of the productivity of the short-grass Prairie, Pawnee National Grasslands, Colorado. In *Proc. the 8th Int. Symp on Remote Sens. of Environ.*, Ann Arbor, MI, USA, 2–6 October 1972, pp. 1357–1381.

Pulvirenti, L., M. Chini, N. Pierdicca, L. Guerriero, and P. Ferrazzoli. 2011. Flood monitoring using multi-temporal COSMO-SkyMed data: Image segmentation and signature interpretation. *Remote Sens. Environ.* 115: 990–1002.

Ramírez-Herrera, M.T., and J.A. Pacheco. 2012. Satellite data for a rapid assessment of tsunami inundation areas after the 2011 Tohoku tsunami. *Pure Appl. Geophys.* 170. DOI: 10.1007/s00024-012-0537-x

Revilla-Romero, B., F.A. Hirpa, J.T.-D. Pozo, P. Salamon, R. Brakenridge, F. Pappenberger, and T. De Groeve. 2015. On the use of global flood forecasts and satellite-derived inundation maps for flood monitoring in data-sparse regions. *Remote Sens.* 7: 15702–15728. DOI: 10.3390/rs71115702

Richardson, A.J., and J.H. Everitt. 1992. Using spectral vegetation indices to estimate rangeland productivity. *Geocarto Int.* 7: 63–69.

Rondeaux, G., M. Steven, and F. Baret. 1996. Optimization of soil-adjusted vegetation indices. *Remote Sens. Environ.* 55: 95–107.

Sakamoto, T., N. van Nguyen, A. Kotera, H. Ohno, N. Ishitsuka, and M. Yokozawa. 2007. Detecting temporal changes in the extent of annual flooding within the Cambodia and the Vietnamese Mekong Delta from MODIS time-series imagery. *Remote Sens. Environ.* 109: 295–313.

Skakun, S., N. Kussul, A. Shelestov, and O. Kussul. 2014. Flood hazard and flood risk assessment using a time series of satellite images: A case study in Namibia. *Risk Anal.* 34: 1521–1537.

Sowmya, K., C.M. John, and N.K. Shrivasthava. 2014. Urban flood vulnerability zoning of Cochin City, southwest coast of India, using remote sensing and GIS. *Nat. Hazards* 75(2): 1271–1286. DOI 10.1007/s11069-014-1372-4

Sun, D., S. Li, W. Zheng, A. Croitoru, A. Stefanidis, and M. Goldberg. 2016. Mapping floods due to Hurricane Sandy using NPP VIIRS and ATMS data and geotagged Flickr imagery. *Int. J. Digital Earth* 9(5): 427–441. DOI: 10.1080/17538947.2015.1040474

Suppasri, A., S. Koshimura, M. Matsuoka, H. Gokon, and D. Kamthonkiat. 2012. Application of remote sensing for tsunami disaster. In: *Remote Sensing of Planet Earth*, ed. Y. Chemin, 143–168. Rijeka, Croatia: InTech.

Tareq, S.M., M.T.U. Rahman, A.Z.M.Z. Islam, A.B.M. Baddruzzaman, and M.A. Ali. 2018. Evaluation of climate-induced waterlogging hazards in the south-west coast of Bangladesh using geoinformatics. *Environ. Monit. Assess.* 190(4): 230. DOI: 10.1007/s10661-018-6591-9

Thomas, R.F., R.T. Kingsford, Y. Lu, and S.J. Hunter. 2011. Landsat mapping of annual inundation (1979–2006) of the Macquarie Marshes in semi-arid Australia. *Inter. J. Remote Sens.* 32: 4545–4569.

Tucker, C.J. 1979. Red and photographic infrared linear combinations for monitoring vegetation. *Remote Sens. Environ.* 8: 127–150.

Wang, Y. 2004. Using Landsat 7 TM data acquired days after a flood event to delineate the maximum flood extent on a coastal floodplain. *Int. J Remote Sens.* 25(5): 959–974. DOI: 10.1080/01431160310001500022

Wang, Y., J.D. Colby, and K.A. Mulcahy. 2002. An efficient method for mapping flood extent in a coastal flood plain using Landsat TM and DEM data. *Int. J. Remote Sens.* 23(18): 3681–3696.

Zheng, W., D. Sun, and S. Li. 2017. Mapping coastal floods induced by hurricane storm surge using ATMS data. *Int. J. Remote Sens.* 38(23): 6846–6864. DOI: 10.1080/01431161.2017.1365387

11 Wild and Coalmine Fires

11.1 INTRODUCTION

As a natural hazard, fires fall into two types, those associated with vegetation and those with coalmines. The former is commonly known as wildland fires or simply wildfires. They refer to the automatic combustion of woody fuels in the wilderness. This natural phenomenon usually takes place in areas where there is an abundant supply of combustible fuels, such as forested or vegetated areas. Thus, wildfires are also nicknamed bushfires. They can be triggered naturally by lightning or deliberately ignited by humans as a way of clearing the bushland for agriculture, especially in tropical areas. In this case, fire is used as a means of deforestation. In traditional slash-and-burn agriculture, artificially lit fires get out of control and spread to the wilderness, forming wildfires. Fires are also used as a way to manage grazing land in savannahs and open forests with distinct grass strata (silvopastoral system). Finally, fires are deliberately lit to deplete the accumulated litters in a controlled burn to minimize future fire hazards. They are known as prescribed fires. Some areas on the Earth are especially vulnerable to the attack of wildfires, such as California, Brazil, Spain, and Australia, due to the dry, hot weather combined with the availability of abundant fuels. If not depleted in a periodic, controlled burn, they can be ignited by lightning and cause huge damage to the natural and built environments.

It is important to study wildfires as they can cause tremendous loss in timber, soil nutrients, and the capacity of carbon sequestration from the atmosphere. They also damage the local ecosystems, destroy the habitat of highly vulnerable and prized species, reduce biodiversity and the value of the forest ecosystem, and increase the risk of soil erosion. Fires also destroy properties and cause casualties. The large quantity of smoke produced by wildfires degrades air quality, adversely affects human health, and aggravates global climate warming as a consequence of the rising trace gas emissions in the atmosphere. Remote sensing has proved to be an effective way of studying fires. In particular, it is able to provide early fire warnings to minimize fire damage and to bring wildfires under control quickly once they start. Wildfires can be studied by means of remote sensing in diverse ways, including active fire detection, fire spread monitoring, fire affected-area delineation, mapping of the spatial extent of the burned area, burn intensity, and assessment of fire impacts on ecosystem and habitat. Remote sensing plays a crucial role in mapping areas at fire risk, assessing fire intensity, and post-fire recovery. The gathered information is essential to containing the fire and putting it under control quickly to reduce environmental damage and mitigate economic and ecological losses.

This chapter first introduces the principles of thermal sensing to lay the theoretical foundation for the detection of fires from remotely sensed data. Included in the discussion are radiation laws and the handling of the atmospheric effects. It also compares and contrasts the pros and cons of optical and thermal sensing in quantifying fire temperature. Then the discussion progresses to the sensing of wildfires from diverse satellite images using different detection algorithms. How to quantify fire intensity and monitor its spread forms the content of Section 11.3. This topic is followed by an in-depth examination of how to assess burn intensity. Also explored in the same section is how to assess fire risk and hazards. Finally, this chapter elaborates on how to sense coalmine fires. Topics covered in this section include detection of surface and subsurface coalmine fires, determination of fire depths, and monitoring of fire propagation using thermal infrared sensing.

11.2 PRINCIPLES OF THERMAL SENSING

Remote sensing is ideally positioned to detect fires, monitor active fires and their spread, and assess post-fire degradation. Fires are detected from visible light, MIR, and TIR bands, based on LST anomalies on thermal bands, or a change in surface reflectance in VNIR bands. Fire detection from

DOI: 10.1201/9781003354321-11

thermal bands relies on the exceptionally intensive heat emitted by fires based on the principles described below.

11.2.1 Planck's Radiation Law

In thermal sensing, a reference object is needed to benchmark the quantity of thermal energy radiated by the target. This reference object is served by the blackbody, a hypothetical object that completely absorbs all the incident energy (mostly short wavelengths) to it and emits all the absorbed energy in long wavelengths without any loss during the conversion from shortwave to longwave radiation. According to Planck's blackbody radiation law, the spectral radiant exitance L (W m^{-2} µm^{-1}) of an object is related to its radiant temperature T_{rad} in Kelvin in the form of

$$L_\lambda = \frac{2\pi hc^2}{\lambda^5}\left(e^{\frac{hc}{\lambda kT_{rad}}} - 1\right)^{-1} = \frac{C_1}{\lambda^5}\left(e^{C_2/\lambda T_{rad}} - 1\right)^{-1} \tag{11.1}$$

where h = Planck's constant (6.626068×10^{-34} J·s^{-1}); c = speed of light (2.9979246×10^8 m·s^{-1}); k = Boltzmann's constant (1.380662×10^{-23} J·K^{-1}); and λ = central wavelength of a spectral band (µm). The radiant temperature (T_{rad}) is inversed from the Planck's equation relating spectral radiance (L_λ) to the radiant temperature as follows:

$$T_{ras}(K) = \frac{C_2}{\lambda Ln\left(\frac{\varepsilon C_1 \lambda^{-5}}{\pi L_\lambda} + 1\right)} \tag{11.2}$$

where ε = emissivity; C_1 and C_2 = the first and second radiation (calibration) constants, respectively. They are expressed as $C_1 = 2hc^2/\lambda^5$ and $C_2 = hc/(k\lambda)$. Their exact value varies with the spectral bands of satellite imagery used. For Landsat TM band 6, $C_1 = 3.74151 \times 10^8$ m^{-2} µm^4, $C_2 = 1.43879 \times 10^4$ µm K. More calibration constants are provided in Table 11.1 for all eight Landsat spectral bands, together with the sensible temperature range. For other images, the calibration parameter values specific to a scene are supplied in their header files.

The radiant temperature T_{rad} (K) of the target is calculated from spectral radiance L_λ. If the spectral radiant exitance is captured on a thermal band, then its radiant temperature T_{rad} can be inversed from Eq. 11.2 theoretically. This temperature is converted to the kinetic temperature of the target via $T_{\text{kin}} = \varepsilon^4 T_{\text{rad}}$, which is the principles of thermal sensing.

The spectral radiance received at the sensor is rendered as the image pixel value expressed in a digital number (DN) in a spectral band. For the purpose of undertaking atmospheric calibration, there is a need to convert the pixel value or DN of a spectral band to the at-sensor radiance, which is achieved using Eq. 11.3:

$$L(i) = L_{min} + \frac{L_{max} - L_{min}}{DN_{max}} DN_{obs} \tag{11.3}$$

where $L(i)_{\text{max}}$ = maximum radiance recorded in spectral band i; $L(i)_{\text{min}}$ = minimum radiance recorded in spectral band i; DN_{max} = maximum pixel value of the spectral band (255 or 2^n, n = quantization level of the spectral band); DN_{obs} = the observed value of the pixel in question. For Landsat TM band 6, Eq. 11.3 can be expressed as the following empirical relationship:

$$L_i = \text{G} \times \text{DN} + \text{O} = 0.0056322\ \text{DN} + 0.1238\ \left(\text{mW} \cdot cm^{-2} \cdot sr^{-1} \cdot \mu m^{-1}\right) \tag{11.4}$$

where G (gain) and O (offset) come with the data (saved in the image header file). This equation converts the pixel value of a spectral band to the at-sensor radiance in a given band i or L_i.

TABLE 11.1

The Calibration Constants for Typical Landsat Spectral Bands Using the Central Wavelengths Provided

Band	Calibration Constant	Landsat 4–5 TM	Landsat 7 ETM+	Landsat 8 OLI
Blue	C_1	4,438,321,096.9	4,530,975,960.1	4,530,975,960.1
	C_2	29,665.3	29,788.1	29,788.1
	Central wavelength (µm)	0.485	0.483	0.483
	Min T (°C)[a]	1,050	1,051	
	Max T (°C)	1,490	1,483	
Green	C_1	1,996,944,890.0	2,162,655,164.6	2,162,655,164.6
	C_2	25,285.9	25,692.2	25,692.2
	Central wavelength (µm)	0.569	0.560	0.560
	Min T (°C)	960	900	
	Max T (°C)	1,410	1,301	
Red	C_1	958,297,463.8	936,779,686.0	951,059,600.7
	C_2	21,832.6	21,733.6	21,799.5
	Central wavelength (µm)	0.659	0.662	0.660
	Min T (°C)	810	755	
	Max T (°C)	1,170	1,119	
Pan	C_1		679,053,366.2	1,109,245,158.4
	C_2		20,379.1	22,480.7
	Central wavelength (µm)		0.706	0.640
	Min T (°C)		702	
	Max T (°C)		1,056	
NIR	C_1	283,104,930.3	293,423,595.4	245,950,026.9
	C_2	17,107.8	17,230.7	16,633.1
	Central wavelength (µm)	0.841	0.835	0.865
	Min T (°C)	620	595	
	Max T (°C)	1,000	926	
SWIR1	C_1	2,198,908.4	2,279,816.4	2,208,831.3
	C_2	6,475.1	6,522.1	6,480.9
	Central wavelength	2.222	2.206	2.220
	Min T (°C)	120	92	
	Max T (°C)	290	258	
SWIR2	C_1	9,006,526.7	9,798,088.9	9,738,850.3
	C_2	8,584.5	8,730.4	8,719.8
	Central wavelength (µm)	1.676	1.648	1.650
	Min T (°C)	220	206	
	Max T (°C)	430	417	
TIR	C_1	607.76	666.09	Band 10: 774.9 Band 11: 480.9
	C_2	1,260.56	1,282.71	Band 10: 1,321.1 Band 11: 1,201.1

C_1 Unit: w m^{-2}·sr^{-1}·µm^{-1}; C_2 Unit: Kelvin.

Source: Nádudvari et al. (2020).

[a] Pixel-integrated temperatures that can be detected wsithout achieving saturation.

11.2.2 Atmospheric Effects

The sensing of active wildfires is impacted by the atmospheric effects, sun glint, and smoke. It is not always necessary to calibrate these effects quantitatively, depending on the type of fire information to be retrieved. If the nature of sensing is qualitative, such as mapping of burned areas, post-fire vegetation recovery (e.g., inactive fire effects), and detection of subsurface coalmine fires, the atmosphere has a negligible effect on the retrieved results, so its effect can be safely ignored. However, if the target of sensing is the temperature of a subsurface coalmine fire, the atmospheric influences must be addressed before a credible accuracy level of retrieval is possible. A prerequisite of tackling the atmospheric effects is a breakdown of the at-sensor radiance. The measured radiance on a spectral band of wavelength λ or L_λ comprises two parts, L_p (path radiance) and L_s (surface radiance), with L_s being attenuated by the atmospheric transmittance (τ) (Tetzlaff, 2004). Mathematically, it is a function of the radiance leaving the surface in the thermal spectral region, plus the path radiance and that from the atmosphere, or

$$L_\lambda = L_p + \tau L_s = L_p + \tau\varepsilon_s L_s + \tau(1-\varepsilon_s)F/\pi \tag{11.5}$$

where L_s = blackbody radiance on the ground surface; τ = ground-to-sensor transmittance; ε_s = ground surface emissivity; F = thermal downwelling flux on the ground. The second term on the right-hand side of Eq. 11.5 is related to the emitted surface radiance reaching the sensor, while the third term is the atmospheric radiance reflected at the surface and attenuated along the surface-sensor path. In the middle IR spectrum, path radiance L_p encompasses both reflected and emitted radiation during daytime. During the nighttime, only the emitted thermal component remains in TIR and MIR bands. In this case, only the emitted radiance of different air layers between the target and the sensor affects the total at-sensor signal.

The atmospheric effects are handled in various ways. The simplest way is to make use of the darkest object in an image. Since this object does not receive any incident radiation, in theory, its reflected radiation should be zero. If a pixel of the darkest object has a non-zero value, then it must represent the reflected radiation from the atmosphere. It can be eliminated by simply subtracting this pixel's value from all the pixels in all the bands. This method of calibration is highly crude and imprecise because the atmospheric transmittance is a function of wavelength. The atmospheric effects in visible light bands differ from those in longer wavelength thermal bands, so the subtraction by the darkest pixel's reflectance in a visible band cannot eliminate the atmospheric effects in other bands completely. More accurate calibration requires the use of models, such as the 6S (Second Simulation of a Satellite Signal in the Solar Spectrum) and the FLAASH (Fast Line-of-sight Atmospheric Analysis of Spectral Hypercube) atmospheric correction models. Another way is to measure L_p, τ, and F directly in the field, coincident with satellite overpasses, or derive them via a radiation transfer model under normal atmospheric conditions, such as the MODTRAN code.

11.2.3 Useful Thermal Data and Fire Products

There are two atmospheric windows for thermal sensing: 3–5 and 8–14 μm, over which all thermal bands of satellite images are recorded and useable for sensing wildfires and other thermal phenomena such as volcanoes. Virtually, all optical satellite images can be used to study wildfires so long as they have at least one TIR band. The most commonly used optical images that contain thermal bands are Landsat TM, ETM+, OLI, and ASTER. A wide range of coarse-resolution images (e.g., MODIS, GOES, and ATSR – Along Track Scanning Radiometer) also contain thermal bands. These images differ widely in their spatial and temporal resolutions and are good at detecting fires of different sizes and temperatures (Table 11.2). Those with a fine spatial resolution are the best at detecting small fires, while those with a fine temporal resolution are excellent at studying historic fires and sensing fire spread rates. Apart from affecting the nature of the detectable fires, image properties also dictate whether fires can

TABLE 11.2

Comparison of the Main Features of Public-Domain Satellite Data Commonly Used to Study Fires

Imagery	ASTER	ETM+	OLI	Sentinel-2	MODIS
No. of TIR bands	5	1	2	0	16
Spectral range (μm)	8.12–11.65	10.42–12.50	10.60–11.19, 11.50–12.51	N/A	3.66–14.385
Spatial resolution (m)	90	60	100	10, 20, 60	1,000
Swath width (km)	60	185	185	290	2,330
Revisit period (days)	16	16	16	10[a]	12 hours
Quantization level (bits)	12	8	12	12[b]	12
Saturation temperature (K)	382–419	322–347.5 (68°C)	Up to 400 (b10 or 370 for b11)		321

[a] Halved to five days with the deployment of two identical satellites.

[b] Sentinel-2 data products have a radiometric resolution of 16 bits.

be detected in the first place. Earth resource satellite data with a spatial resolution finer than 30 m are better than meteorological satellite data in studying wildfires and coalmine fires. The utility of meteorological satellite data in studying fires is compromised by their coarse spatial resolution. Nevertheless, the sensing of fires from all optical satellite images faces two disadvantages. First, they are acquired during the daytime when the solar radiation is rather strong. Both emitted and reflected energy from the target is recorded in the same band. The fire signal captured in a thermal band is overwhelmed by the strong solar radiation. Second, their temporal resolution measured by more than ten days (Table 11.2) causes most wildfires to be missed, let alone monitoring fires in real time competently. Thus, images of a fine temporal resolution are indispensable to studying fire development.

One way of shortening the revisit period is to make use of two identical satellites as with Sentinel-2. Its 2A and 2B satellites shorten the revisit period by half to five days. But it is still inadequate for monitoring fire spread in real time and in issuing fire warnings. This inadequacy is further exacerbated by the effect of smoke. Nevertheless, optical earth resources satellite images are very good at assessing the burned area and fire recovery based on the amount of regenerated vegetation. In comparison, meteorological satellite data such as MODIS are more useful for tracking fires owing to their short revisit period measured by hours instead of days (Table 11.2). MODIS thermal bands are good at detecting thermal anomalies, from which the fire ignition spot may be pinpointed upon close inspection.

So far a variety of fire products have been produced from different satellite data, all freely accessible to the public. The most comprehensive and significant fire data products are derived from MODIS data. They fall into two types: active fire and burned area. The former shows the location of current fires or burning areas, and the latter depicts the extent of burn scars over a specified period (Justice et al., 2002). Both types of products have been extensively assessed for quality assurance and validated for their accuracy prior to release to the public at levels 2 and 3 in the HDF (Hierarchical Data Format) format (Table 11.3). Level 2 fire product (MOD14) is the most basic that shows only active fires and other thermal anomalies (e.g., volcanoes). Further processing of level 2 products generates higher level products (how these products are derived will be explained in Section 11.3.1). At each level, the same product may be further differentiated into daytime and nighttime products produced from daytime or nighttime images or temporally aggregated to a fixed period, such as 15 days or a month. The date of product availability varies with its nature, but generally from mid- and late 2000 to early 2001. MOD14 Collection 6 has an overall daytime global commission error of 1.2%. Regionally, the error ranges from around 4.2% in Boreal North America and Boreal Asia,

TABLE 11.3

MODIS Fire Products and Their Available Time

Nature of Product	ESDT[a]	Release Date
Level 2 fire product	MOD14	February 2001
Rapid response	–	April 2001
Level 2G daily daytime	MOD14GD	November 2000
Level 2D daily nighttime	MOD14GN	November 2000
Level 3 daily	MOD14A1	November 2000
Level 3, 8-day summary	MOD14A2	August 2000
Global daily fire QA browse imagery	–	November 2000
1° global climate modeling grid	MOD14CMG1	Fall 2002
10-km global climate modeling grid	MOD14CMG2	Fall 2002

Source: Justice et al. (2002). Used with permission.

[a] Earth Science Data Type, a unique descriptor assigned to each MODIS product. As the rapid response and global browse fire products are not distributed from the ECS DAACs, no ESDTs are necessary for these products.

2.2% in Equatorial Asia and Central Asia, to 0.2% in Northern Hemisphere South America (Giglio et al., 2016). The main cause of detection inaccuracy stems from fire size with those smaller than image pixel size omitted.

Existing fire data products, however, may not be useful for small fires, especially coalmine fires that require fine-resolution images to study, such as UAV data. A UAV equipped with a digital camera can be deployed at a low altitude above the fire field more flexibly than satellites. At such a low altitude, the sensing is much less subject to the influence of clouds by flying on sunny days or under clouds. The spectral range of sensing is confined mostly to the visible light spectrum, which makes UAV data ill-suited to study coalmine fires directly. However, they are excellent in studying linear features and brightness of the ground fissures in a coal fire area owing to their super-fine spatial resolution (Wang et al., 2015). It is critical to map these ground openings that serve as the conduit of air to fuel subsurface coal combustion. These features are almost impossible to detect from satellite images due to their long revisit periods, coarse spatial resolution, and frequent cloud coverage. UAV images have a spatial resolution as fine as 0.04 m if they are RGB, or 0.37 m if they are TIR (Li et al., 2018). RGB images enable fire-affected areas to be detected at an accuracy of 93%. The fine spatial resolution of UAV TIR imagery helps to achieve reliable identification of weak and small coal fire areas. Their LST can be retrieved accurately at an RMSE of 1.03K. However, UAV images are limited in that each image covers only a small ground area. A number of images need to be mosaicked to cover a large ground area after each image has been individually geo-referenced.

11.3 SENSING OF WILDFIRES

In a wildfire, the combustion of dead or dry fuels produces intense heat, and most likely is accompanied by smoke and frequently by flames, even though not all vegetation is burned completely in all fires, depending on the type. Sometimes only the ground-level vegetation is burned, while the upper canopy remains intact. In terms of vertical stratification, wild forest fires fall into three types: ground, surface, and crown (Roy, 2004). Ground fires occur near the forest floor in the humus and peaty layers beneath the litter. Ground fires produce intense heat without flame. They are rather rare. Surface fires take place on or near the ground densely covered by litter, ground cover, scrub, and regenerated vegetation. Occurring at a much higher altitude, crown fires affect tree crowns and burn the foliage but are unlikely to burn trees to death (Roy, 2004).

The intensity of thermal energy released by wildfires is a function of wavelength. The emitted radiation peaks in the MIR spectrum. According to Planck's radiation law, the radiative energy emitted from an active fire at this wavelength is much more intensive than the ambient surroundings but less so in TIR wavelengths. After the ground temperature has been estimated remotely from a thermal band using Eq. 11.2, any hotspots having an abnormally high temperature are construed to represent fires. A fire is detected if a pixel's temperature is excessively higher than that of its neighbors. For instance, all pixels saturated in AVHRR band 3 (wavelength: 3.55–3.93 μm) are assumed to contain fires as saturation occurs at a brightness temperature around 321K that is higher than the temperatures of most non-fire features on the ground (Kennedy et al., 1994). This assumption is valid only when the radiation reaching the sensor stems from the emitted energy (e.g., during nighttime). Because of the intense heat, the radiant energy recorded on satellite images enables even small active fires less than 1% pixel size to be detected.

11.3.1 Fire Detection and Spread Monitoring

11.3.1.1 General Principles

Fire detection algorithms and their complexity vary with spectral bands because of their unique wavelength designation, but the general philosophy of detection remains largely universal. This detection is founded on temperature anomalies in a single band or a combination of multiple bands. They are detected using different algorithms. A good algorithm should be sensitive to small fires, but also minimize false alarms or commission errors. All fire detection algorithms are threshold-based, supplemented with either spatial or temporal contextual tests. They differ from each other in the number of thresholds used and how the thresholds are derived. In general, the joint use of multiple thresholds is conducive to achieving more reliable detections. False alarms are avoided or significantly suppressed via refining the originally detected results based on additional constraints. The thresholds may be pre-defined or fixed, or dynamic, calculated from the images used. Fixed thresholds are more likely to be image-specific, restrictive, and have limited applicability to different areas with different images or spectral bands. This restriction disappears if the thresholds are calculated from the scene dynamically. The threshold method is simple but rather imprecise, so it is not reliable in places (e.g., West Africa) where the vegetation ranges from dry Sahelian grasslands to moist tropical forests with considerable seasonal variability (Kennedy et al., 1994).

In addition, the detection outcome may turn out to be merely a fire hotspot, not an active fire, so additional criteria may be imposed to check temperatures on more than one band to confirm the initially detected results. For AVHRR data, the simplest threshold is derived from saturation in band 3. All pixels that are saturated or near-saturated in this band are assumed to manifest fires. More sophisticated and complex detection may involve several thresholds derived from multiple bands, such as those proposed by Kaufman et al. (1990). It has three thresholds derived from AVHRR bands 3 and 4:

1. $T_{b3} \geq 316K$;
2. $T_{b3} \geq T_{b4} + 10K$; and
3. $T_{b4} \geq 250K$.

It is cautioned that the use of these thresholds for large environmentally heterogeneous areas may not always produce perfect results due to high commission errors. These universal thresholds may not always yield satisfactory detections in different geographic areas (Kennedy et al., 1994), so have to be modified for an environment different from those in which they are determined as below:

1. $T_{b3} \geq 320K$;
2. $T_{b3} \geq T_{b4} + 15K$; and
3. $T_{b4} \geq 295K$ (cloud detection threshold).

A fire pixel is detected if it meets all three conditions simultaneously.

11.3.1.2 From MODIS Data

MODIS imagery has 24 SWIR-TIR bands that offer abundant choices in detecting fires. However, the most useful bands for sensing temperature are only those with a wavelength of 0.4 and 1.1 μm. For collection 6 MODIS, the detection algorithms are rather simple involving only two bands and three thresholds. A fire is detected if either of the following two conditions is met (Giglio et al., 2016):

1. $T_4 > 360K$ daytime (330K nighttime) and
2. $T_4 > 300K$ daytime (315K nighttime) and $T_4 - T_{11} > 25K$ (10K nighttime).

The first condition is imposed to detect large and intense active fires. If none of the conditions is met, then the detection focuses on the temperature of the pixel in question relative to the mean background temperature, namely

1. $T_4 - T_{11} >$ median $(\Delta T) + 3\ \delta\Delta T$
2. $\Delta T > \overline{\Delta T} + 3.5\delta_{\Delta T}$
3. $\Delta T > \overline{\Delta T} + 6$
4. $T_4 > \overline{T_4} + 3\delta_4$
5. $T_{11} > \overline{T_{11}} + \delta_{11} - 4$
6. $\delta_4' > 5$

where $\Delta T = T_4 - T_{11}$, δ = standard deviation. δ'_4 = 4-μm brightness temperature mean absolute deviation of pixels rejected as background fires, $\overline{T_4}$ and δ = the mean and standard deviation calculated from a quarter of all background pixels exclusive of clouds, water, and fires with at least six members within a window of 21 × 21 pixels after water and clouds have been identified with respective masks.

Apart from these major algorithms, additional ones have been developed to reject water pixels in coastal areas, forest clearing, and sun glint (Giglio et al., 2016). This Collection 6 active fire detection algorithm achieves lower omission errors of large fires and reduces false alarm rates in tropical ecosystems over its predecessors.

11.3.1.3 From ASTER and ETM+ Imagery

All fire detection algorithms from ASTER and Landsat ETM+ images are threshold-based, taking into account the contextual information using differential radiometric responses of SWIR band 7 and NIR band 5, so the detection is based on reflectance (ρ) and its change, not temperature. Active fires are detectable from Landsat 8 OLI data built on the ETM+ active fire algorithm (Schroeder et al., 2016). Fire-sensitive band 7 (and to a lesser extent NIR band 6) suffers from frequent saturation of fire-affected pixels. This detection is differentiated into daytime and nighttime. The daytime algorithm involves three thresholds derived from two bands:

$$\rho_7 / \rho_5 > 2.5 \text{ and } \rho_7 - \rho_5 > 0.3 \text{ and } \rho_7 > 0.5 \tag{11.6}$$

where ρ_i = TOA reflectance on Landsat 8 band i. Since exceptionally strong and extensive fires can cause saturation in band 7, saturated fire pixels have to be identified based on the following conditions:

$$\rho_6 > 0.8 \text{ and } \rho_1 < 0.2 \text{ and } (\rho_5 > 0.4 \text{ or } \rho_7 < 0.1) \tag{11.7}$$

Potential active fires must meet two conditions:

$$\rho_7 / \rho_5 > 1.8 \text{ and } \rho_7 - \rho_5 > 0.17 \tag{11.8}$$

A contextual test is carried out to narrow down potential active fire pixels, the value of which must satisfy the following constraints:

$$\rho_7 / \rho_5 > \overline{\rho_7 / \rho_5} + \max\left[3\sigma_{\rho_7/\rho_5}, 0.8\right] \textit{and } \rho_7 > \overline{\rho_7} + \max\left[3\sigma_{\rho_7}, 0.08\right] \textit{and } \rho_7 / \rho_5 > 1.6 \quad (11.9)$$

where max [] = a function that returns one of the larger values inside the bracket; $\overline{\rho_7 / \rho_5}$ and σ_{ρ_7/ρ_5} = the mean and standard deviation of the two reflectance ratios, calculated over a window of 61 × 61 pixels (pixel size: 30 m), exclusive of water and shadow pixels. They are determined using the following criteria:

1. $\rho_4 > \rho_5$ and $\rho_5 > \rho_6$ and $\rho_6 > \rho_7$ and $\rho_1 - \rho_7 < 0.2$ and
2. $\rho_3 > \rho_2$ or ($\rho_1 > \rho_2$ and $\rho_2 > \rho_3$ and $\rho_3 > \rho_4$).

These algorithms successfully detected fires that are significantly smaller than those shown in current operational satellite fire products from Landsat 8 OLI data (Schroeder et al., 2016). They enabled a 4 m^2 effective fire area combusting at 950K to be detected at an accuracy more than 50% higher than other daytime images of a coarse spatial resolution. If nighttime images are used for the detection, the detectable effective fire area decreases to 1 m^2. The commission errors of detection vary widely around the globe but average less than 0.2% owing to the additional multi-temporal analysis of co-located pixels. Thus, real-time fire detection is not feasible with these algorithms if no prior images of the study area exist.

In detecting unambiguous fires, band 4 outperforms band 5, so it is better to replace band 5 by band 4 in the above algorithms (Kumar and Roy, 2018). This substitution is justified by the fact that $\rho_7 - \rho_4$ is more sensitive to unambiguous fires as band 4 is less sensitive to fires than band 5. Unambiguous fires are detected if $\rho_4 \le 0.53\rho_7 - 0.214$. Pixels adjoining the detected unambiguous active fires are also considered to be of the same type if $\rho_4 \le 0.35\rho_6 - 0.044$. Potential fire pixels are identified if they meet the following rules:

$$\rho_4 \le 0.53\rho_7 - 0.125 \text{ or } \rho_6 \le 1.08\rho_7 - 0.048 \quad (11.10)$$

Water pixels are detected using the rule of $\rho_2 > \rho_3 > \rho_4 > \rho_5$. Commission errors over building roofs are removed through a multi-temporal test. Finally, an active fire that lasts more than six months is rejected, as detected by the following rules:

$$\rho_{7-median} > k_1 \ or \ NDVI_{fire} < NDVI_{peak} - k_2 \quad (11.11)$$

where $\rho_{7\text{-median}}$ = the median reflectance of band 7 observed over the preceding six months; $NDVI_{fire}$ = the NDVI of an active fire, $NDVI_{peak}$ = the peak NDVI observed over the preceding six months. k_1 and k_2 = constants determined from the scatterplot of any two bands among bands 4, 5, 6, and 7. These modifications reduce omission errors, while commission errors are almost the same as with the original algorithms. They enable the detection of smaller and cooler fires more accurately than the Schroeder et al. (2016) algorithms for Landsat 8 data (Kumar and Roy, 2018). The minimum simulated fire size of 1 m^2 is detected with a 50% probability when the fire has a temperature of 1,060K, lower than 1,180K detected by the Schroeder et al. (2016) algorithms.

11.3.1.4 From GOES Imagery

The R-Series GOES-R satellites are the state-of-the-art geostationary meteorological satellites currently in operation. Their major payload comprises the Advanced Baseline Imager (ABI) resembling the AHI sensor. The latest satellite, GOES-16, is able to image the Earth at a spatial resolution of 1–2 km in 16 spectral bands. Of these bands, bands 7 (3.80–3.99 μm) and 14 (10.82–11.60 μm) are suitable for detecting active fires. Both of them have a spatial resolution of 2 km, half of that of

GOES-13. ABI operates in multiple scan modes and yield continuous full-disk images of the Earth as frequently as every 5 minutes. Such a super-fine temporal resolution is ideal for studying fires and their development, especially those short-lived ones. Fires are detectable from GOES images using several algorithms, such as the Fire Thermal Anomaly (FTA) algorithm proposed by Wooster et al. (2015), the Xu et al. (2010) algorithm, and the MIR radiance method of fire radiative power (FRP) retrieval developed by Wooster et al. (2005) (Figure 11.1). The FTA algorithm was originally developed for use with the Meteosat Spinning Enhanced Visible and InfraRed Imager (SEVIRI) data. Its philosophy of detection is still threshold-based, supplemented with an additional temporal contextual constraint. Active fires detected from GOES images using the Xu et al. (2010) algorithm closely match those from MODIS data. Fire pixels, whose radiative power exceeds 30 MW, are detected from GOES images with an omission error of less than 10%. The FRP of fire clusters detected near-simultaneously by both GOES and MODIS have a bias of only 22 MW. Most of the fires having an FRP substantially lower than ~30 MW, the most common fire type, were not successfully detected from the ABI images because of their coarser spatial resolution than MODIS, resulting in a high omission error of 68% compared to MODIS center of swath data (Xu et al., 2021).

11.3.1.5 From VIIRS Imagery

The VIIRS sensor aboard the S-NPP satellite is a recent addition to the wide variety of space-borne satellites, the images of which have found applications in detecting fires. This sensor carries two sets of multispectral scanners with a revisit period of 12 hours or shorter, depending on the latitude (Schroeder et al., 2014). VIIRS images fully cover the globe in 22 spectral bands over the wavelength range of 0.412–12.01 μm (Table 8.4). Of these bands, M4 (0.545–0.565 μm) and, to some

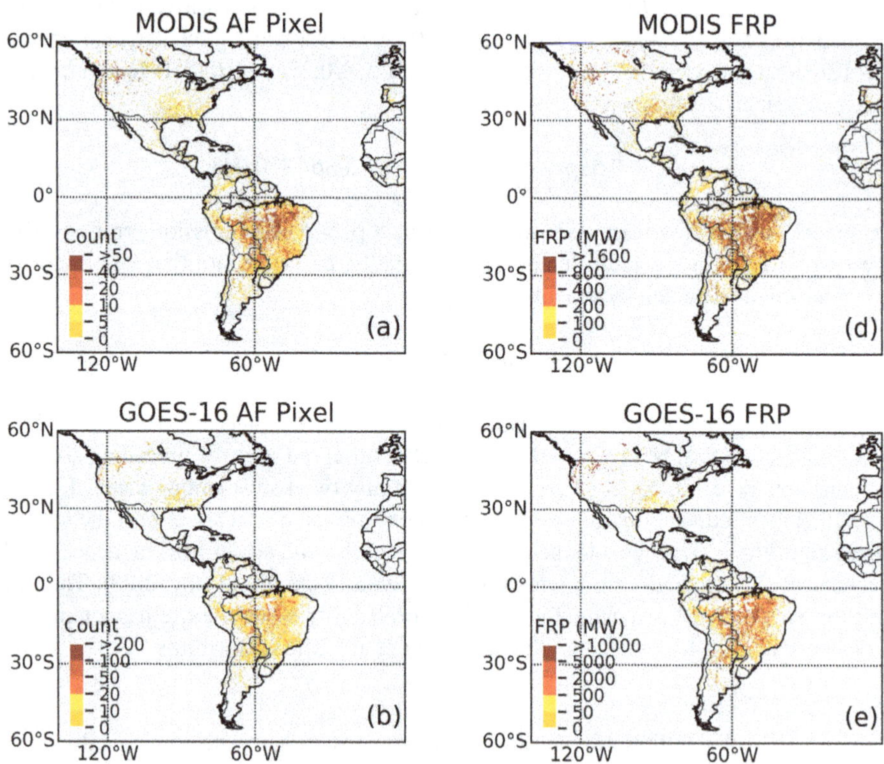

FIGURE 11.1 Total active fire (AF) pixel count and total fire radiative power (FRP) measured within 0.5° grid cells across the Americas in September 2017, detected from MODIS imagery (a and b) in comparison with those from GOES-16 imagery (c and d). (Source: Xu et al., 2021.)

degree, M5 (0.662–0.682 μm) is commonly used to detect daytime and nighttime fires of biomass burning and other thermal anomalies. The results are released in diverse types of data products, one of which is the 375-m 6-Min L2 Swath near real-time active fire product. VIIRS imagery is able to detect small fires early on and improve the mapping of large wildfires. It is particularly good for identifying lightning strikes and supporting tactical firefighting and evacuation.

Active fires are detected from VIIRS bands using algorithms built on the MODIS Fire and Thermal Anomalies product, supplemented with contextual analysis. Candidate fire pixels are identified based on the following rules:

$$BT_4 > BT_{4s} \text{ or } \Delta BT_{4-5} > 25 \text{ K } (\text{daytime image}) \tag{11.12}$$

Or

$$BT_4 > 295 \text{ K or } \Delta BT_{4-5} > 10 \text{ K } (\text{nighttime image}) \tag{11.13}$$

where BT_4s = the background reference brightness temperature of band I4 (3.550–3.930 μm) calculated over a window of 501 × 501 pixels. It has two crucial thresholds of BT_{4S} and BT_{4M}:

$$BT_{4S} = \min[330,\ BT_{4M}] \text{ K} \tag{11.14}$$

$$BT_{4M} = \max[325,\ \text{M}] \text{ K} \tag{11.15}$$

where M = median B_{T4} value of 501 × 501 pixels within a window. For edge and corner pixels, the window must encompass at least 10 valid observations. Otherwise, the temperature is set to 330 K.

The preliminarily detected fire pixels are subject to further contextual analysis of bands I4 and I5 (difference and standard deviation) from which mean brightness temperature ($\overline{BT}$) and standard deviation ($\overline{\delta}$) are calculated, as well as the difference in BT between them ($\overline{BT_{4-5}}$) based on the background sample. In addition, the mean ($\overline{BT_4'}$) and mean absolute standard deviation ($\overline{\delta_4'}$) of potential background fire pixels within the working window are also calculated from I4 BT. These parameters are jointly considered in the following rules to detect fire pixels from daytime images at a high confidence level:

$$\Delta BT_{4-5} > \Delta\overline{BT_{4-5}} + 2\delta_{4-5} \tag{11.16}$$

$$\Delta BT_{4-5} > \Delta\overline{BT_{4-5}} + 10 \tag{11.17}$$

$$BT_4 > \Delta\overline{BT_4} + 3.5\delta_4 \tag{11.18}$$

$$BT_5 > \Delta\overline{BT_5} + 5\delta_5 - 4 \textit{ or } \delta_4' > 5 \tag{11.19}$$

For nighttime images, the rules change to

$$\Delta BT_{4-5} > \Delta\overline{BT_{4-5}} + 2\delta_{45} \tag{11.20}$$

$$\Delta BT_{4-5} > \Delta\overline{BT_{4-5}} + 10 \tag{11.21}$$

$$BT_4 > \overline{BT_4} + 3\delta_4 \tag{11.22}$$

An additional constraint is imposed to reject pixels bordering desert areas on daytime images:

$$\rho_2 > 0.15 \textit{ and } \overline{BT_4}' < 345 \textit{ and } \delta_4' > 5 \textit{ and } BT_4 > \overline{BT_4}' + 6\delta_4' \tag{11.23}$$

where ρ_2 = reflectance of VIIRS band I2 (0.85–0.88 μm).

Even refined with all these conditions, daytime false alarms may still pop up over water bodies due to specular reflection (sun glint), in concrete pavements and industrial parks, all of which have high reflectance. They can be removed via spatial filtering. In general, these algorithms have robust performance, achieving a very low commission error (false alarm). On the global scale, detection accuracy varies spatially, with Eastern China having the highest commission error of 1.2% (Schroeder et al., 2014). Here low confidence daytime pixels have a false alarm rate of nearly 40% due to strong solar reflection off high-rise buildings in urban areas.

11.3.1.6 From Sentinel-2 Imagery

Sentinel-2 imagery differs from all the aforementioned satellite images in that it has no TIR bands (Table 11.2). Its 13 spectral bands span from 0.443 μm (central wavelength) to 2.202 μm in wavelength. Thus, the detection of wildfires has to rely on reflectance. VNIR bands enjoy the advantage of being insensitive to atmospheric contamination caused by smoke plumes in detecting fires. Smoke plumes associated with active fires and smouldering areas are best detected from Sentinel-2 bands 1 (0.45–0.52 μm), 2 (0.52–0.60 μm), and 3 (0.63–0.69 μm). Their spatial resolution ranges from 10 to 60 m, ideal for detecting small fires and smokes. The philosophy of detection is the same multi-criterion approach as described previously, based on the reflectance of three bands, usually B4, B11, and B12 (Hu et al., 2021). B4 (0.65–0.68 μm) is a red band not sensitive to fires, and B12 (2.10–2.28 μm) is a SWIR band insensitive to smoke plumes. These two bands are used primarily to detect unambiguous active fires, the spectral reflectance of which must meet the following condition:

$$\rho_4 \le a\rho_{12} + b \tag{11.24}$$

where ρ_{12} = reflectance of band 12; a and b = coefficients determined through regression analysis of B4 pixel values against those of B12. Depending on the specific type of biomes, additional criteria may be imposed to reduce commission errors based on the following two conditions:

$$\rho 11 \ge \mathrm{c} \text{ and } \rho 12 \ge \mathrm{d}$$

where ρ_{11} = reflectance of band 11 (1.52–1.70 μm); c and d = the 0.99 quantiles of B11 and B12 for a given type of biomes, such as tropical and subtropical moist broadleaf forest, temperate conifer forest, and boreal forest (Table 11.4). Two additional constraints are imposed for refinement:

$$\frac{\rho_{12}}{\rho_{11}} \ge 1 \; or \; \rho_{12} \ge 1 \tag{11.25}$$

TABLE 11.4

Additional Fire Detection Criteria from Sentinel-2 Imagery for Different Biome Types

Biome	C_1	C_2	C_3
Tropical and subtropical moist broadleaf forest	$\rho_{0.66} \le 1.045\rho_{2.20} - 0.071$	$\rho_{2.20}/\rho_{1.61} \ge 1$	
Tropical and subtropical dry broadleaf forest	$\rho_{0.66} \le 0.681\rho_{2.20} - 0.052$		
Tropical and subtropical grasslands, savannah, and shrubland	$\rho_{0.66} \le 0.677\rho_{2.20} - 0.052$		
Mediterranean forests, woodland, and scrub	$\rho_{0.66} \le 0.743\rho_{2.20} - 0.068$	$\rho_{2.20} \ge 0.355$	$\rho_{1.61} \ge 0.475$ or $\rho_{2.20} \ge 1.0$
Temperate conifer forest	$\rho_{0.66} \le 0.504\rho_{2.20} - 0.198$		
Boreal forest/taiga	$\rho_{0.66} \le 0.727\rho_{2.20} - 0.11$		

Source: Hu et al. (2021).

Depending on the specific types of biome, additional criteria (C_2 and C_3) are imposed to reduce the commission error with the Boolean "AND" operation for Mediterranean forest, woodland, and scrub (Table 11.4). For the biomes in different regions, the corresponding criteria are combined to detect active fires based on C_{1-3}. It should be noted that, in C_3, the logical operation of "OR" is evoked to reduce omission errors and must be met.

No matter which and what thresholds are chosen, they always represent a trade-off between commission errors and omission errors. A conservative threshold can potentially avoid omission errors of small and cool fires. On the other hand, it can also reduce potential commission errors by excluding highly reflective building roofs and soil-dominated pixels without resorting to multi-temporal data, so fires can be detected rapidly from a current image without checking its history from recent time-series images. The joint use of multi-criteria can achieve a commission error as low as 14% and an even smaller omission error of 4% (Hu et al., 2021).

11.3.2 Fire Spread Monitoring

Fire monitoring is synonymous with studying fires in the same area repeatedly to survey how they have changed spatially so that limited resources can be deployed strategically to contain them most effectively. Active fires are best monitored from multi-temporal data. In order to yield detailed information on how fire spreads, the satellite images used must have a fine temporal resolution to timely detect its spatiotemporal changes. Fire monitoring is virtually a change detection that can be implemented either qualitatively or quantitatively. Qualitative monitoring focuses on the fire extent or the nature of the surface cover in the fire-affected area, usually a change from vegetative cover to burned ground. Quantitative monitoring yields quantitative information on the fire spread, such as the flame front rate of spread (ROS) for active fires, calculated as follows:

$$ROS = \frac{L}{t_2 - t_1} \tag{11.26}$$

where L = the length of the fire front spread vector. It can be estimated from a pair of time-series images. t_1 and t_2 = time of the satellite overpass. The multi-temporal satellite images needed for determining ROS should have a short temporal resolution, such as 12 hours of MODIS images, but the coarse spatial resolution (>500 m) of MODIS TIR bands may not be sufficiently fine to allow small fires to be detected. A better alternative is to make use of Sentinel-2 imagery with a spatial resolution of 20 m and a temporal resolution of five days.

Fire monitoring may be automated by vectorizing fire pixels that have been successfully detected using the aforementioned algorithms in the preceding section to form polygons (Ruecker et al., 2021). Polygon boundaries are considered the start line of the fire front. The jagged outlines stemming from vectorization are usually smoothed to enable precise pinpointing of the fire front. The same detection is repeated for another satellite image, preferably acquired by the same sensor, to detect the endpoints of the same fire. This is achieved by clustering active fire pixels together, and by constructing a concave hull envelope around each cluster. The outmost endpoint of the vector is then connected to the start point to form the displacement vector. It is possible for a large number of start point – endpoint pairs to form, but not all of them are genuine. They are then checked against additional rules such as the exclusion of those connects either crossing burned areas, barriers, or associated with different ignition events. These rules help to remove the plausible ones. A fire front's local ROS is calculated from the displaced fire front over a duration, by which the displacement distance in the direction perpendicular to the local fire front is divided to calculate ROS in $m{\cdot}s^{-1}$ (Figure 11.2). The calculated ROS results can then be related to fire driving parameters, such as wind speed, fuel availability, and the coalescing of separate fire fronts to explore its variability across the burning plot and with major drivers.

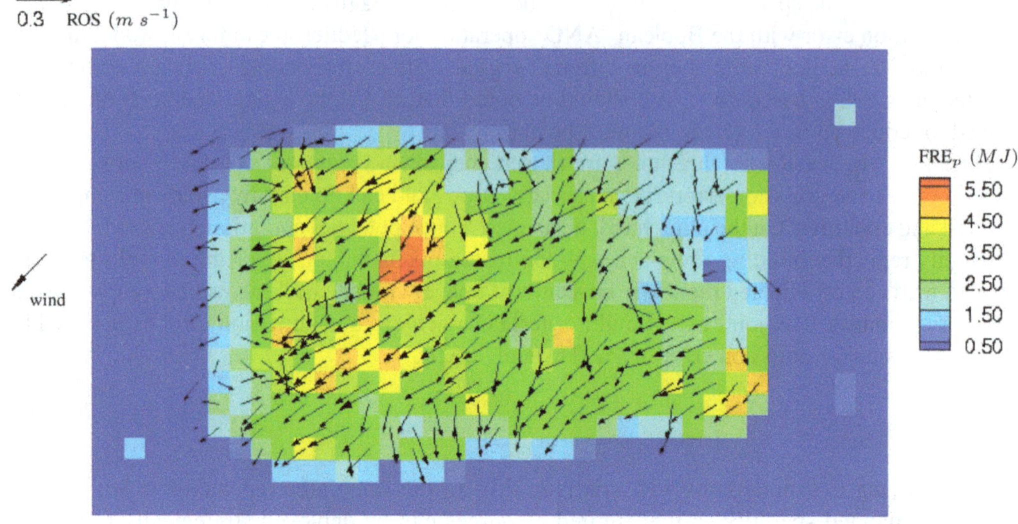

FIGURE 11.2 Spatial variation in flame front ROS detected from hand-held thermal image data overlaid on top of the fire radiative power field. (Source: Paugam et al., 2013.)

11.3.3 Sensing of Fire Burn

Mapping of burned areas is crucial to post-fire management, planning of recovery strategies and measures aiming at reducing soil erosion, and identification of the cause-effect relationship of post-fire dynamics. Fire-affected areas are readily recognizable from the false-color composite of multispectral bands based on the distinctive physical changes a fire induces to the surface cover. These changes include removal of vegetation, depletion of biomass and canopy, ground charring, deposits of charcoal and ash, and soil discoloring. They are especially enhanced immediately after the burn. On satellite images, these changes are reflected in spectral reflectance that decreases the most at 0.788 and 0.913 μm, while the reflectance at 0.21 and 2.37 μm increases the most after a fire (Van Wagtendonk et al., 2004). Owing to these spectral variations, areas affected by fires can be mapped from satellite images if they are sufficiently extensive or can register on the images (e.g., several folds of the image pixel size). Of all Landsat TM bands, band 4 (0.76–0.90 μm) is the most appropriate for discriminating fire scars from other covers in the deforested Amazon (Pereira and Setzer, 1993). The spectral discrepancy between fire scars and other covers is twice higher in this band than in other bands, of which band 5 (1.55–1.75 μm) is good at detecting aged scars, but recent scars can be mixed with water. In comparison, fire intensity is much more difficult to study via remote sensing.

11.3.3.1 Fire Intensity

Fire intensity is closely associated with the quantity of heat released from the combustion of biomass. It is a manifestation of the burning intensity of an active fire. An intense fire has a higher temperature and spreads much faster than a less intense one. Fire intensity is commonly gauged by the thermal energy released from a 1-m strip of the actively combusting area measured from the leading edge of the fire front to the rear of the flaming zone (Alexander, 1982). One way of calculating fire intensity is Byram's (1959) equation that is widely accepted to quantify wildfire behavior. Byram's fire intensity I is commonly derived from the observed FRP that can be directly measured from the radiant energy released from a fire in the field. It can also be estimated from satellite images using a number of methods without field samples, which can yield the spatiotemporal distribution of fire intensity. One of the simplest methods of determination involves the product of only three parameters:

$$I = H \times w \times ROS \tag{11.27}$$

where H = the heat yield of the combusted unit fuel ($kJ \cdot kg^{-1}$); its exact value varies with fuel type, w = the weight of the available (or burned) biomass per unit area ($kg \cdot m^{-2}$), ROS = rate of spread. Of the three parameters, H contributes the least to I. Fuel weight varies over a relatively narrow range (about 10-fold), whereas ROS varies over a much broader range (about 1,000-fold). Thus, ROS is the most significant contributor to fire intensity. Eq. 11.27 allows I of a moderate intensity to be estimated from TIR imagery for actively spreading flame fronts at a resolution up to 0.13 m (Johnston et al., 2017). The estimated results from proper time-series images are in close agreement with those obtained using the traditional methods (R^2=0.34–0.73) without ground samples or ancillary data, demonstrating that TIR sensing is a viable means of sensing fire intensity quantitatively. However, at 2.37 µm, the spectral response of fire exhibits the largest variation, suggesting that it is the best at potentially assessing fire intensity. However, no results have been produced to verify this claim.

11.3.3.2 Burn Severity

Burn severity is defined as the degree to which the fire-ravaged landscape has been burned or adversely affected by a fire event. It differs from fire intensity in timing. It pertains to the state left behind by the fire, dissimilar to fire intensity that applies to a process currently underway or actively burning fires. Burn severity is indicative of the heat impact of a wildfire on the fire-affected site, including its physical, chemical, and biological changes following a fire (Table 11.5). The fire impacts are assessed via over- and under-story vegetation and soils. In the field, burn severity is assessed based on four indicators: litter, soil, fuel, and vegetation, of which vegetation is the most reliably estimated from remotely sensed data. Naturally, fire severity has been assessed by means of remote sensing.

Remotely sensed burn severity is qualitatively enumerated at a few discrete levels of unburned, light burn (surface), severe burn (surface), and crown fire (Turner et al., 1994), based on the portion of the burned areas in the total study area. A low portion indicates a weak burn. Since images are highly sensitive to burned areas, the four levels of burn severity are easily mapped via digital classification of raw satellite images using the maximum likelihood method. While description has been supplied to specify each severity, it remains unknown how one severity is differentiated from another quantitatively from remotely sensed data. The absence of concrete remote sensing criteria for differentiating one severity level from another means that it is difficult to evaluate the mapping accuracy and identify the causes of a severe burn. So the conclusions are rather vague, such as "*pre-existing conditions (e.g., pine-beetle infestations and stand fire history) could have*

TABLE 11.5
Criteria of Grading Burn Severity in the Field Based on Four Indicators

Burn Severity	Litter	Fuel	Vegetation	Soil
Light	Not consumed	Few small fuels consumed	Some scorch, especially shrubs	Not altered
Moderate	Completely consumed	Small fuels consumed	Small-diameter trees and shrubs killed	Darkened with white ash
Severe	Completely consumed	Small/medium fuels consumed	All vegetation killed, including rhizomes	Stability lost, reddish with white ash

Source: Modified from Ryan and Noste (1985).

contributed to some bias for particular severity classes" (Turner et al., 1994). Besides, image classification is suitable to map complete burn. If a burn is partial and spatially heterogeneous (e.g., some pockets of vegetation are left behind), burn severity is more precisely quantifiable using vegetation indices.

The commonly used NDVI captures mostly the spectral information of vegetative covers. The inter-annual change of pre- and post-fire NDVI is able to characterize fire-devastated sites (White et al., 1996). A large decrease in NDVI is indicative of a more severe burn, and vice versa. A decrease in annual NDVI of about 0.4 is considered to indicate a low burn in forested areas, but this value rises to about 0.7 for riparian vegetation. The accuracy of mapping fire severity from Landsat TM band 7 is not very high at only 63.0%. The accuracy of NDVI-based identification remains unknown; presumably, it is less accurate than this level. One cause of the low accuracy is that NDVI is prone to non-fire spectral variations, such as that related to a low canopy cover. These variations across a range of vegetation and soil conditions bring down the reliability of the estimated burn severity. Such vegetation-centered assessment of burn severity suffers further complications if the post-fire image is acquired sometime after the fire, because in the interim the fire-damaged vegetation could have regenerated. In addition, this method is also subject to the rate of vegetation regeneration. Thus, the change in NDVI needs to be further differentiated by vegetation type as some types of vegetation regenerate much faster than others. In order to avoid such non-fire-caused variations, the post-fire satellite image should be captured immediately after the fire. The sooner the post-fire image is captured, the more genuine the burn severity is detected from it.

Apart from the generic NDVI, more special indices have been proposed for quantifying burn severity. The two useful ones are normalized burn ratio (NBR) and differenced NBR (dNBR). The former is founded on the fact that healthy vegetation has rather high reflectance in the NIR spectrum, in sharp contrast to burned areas, the reflectance of which is low. Thus, healthy vegetation not affected by fires has a high NBR value, while severely and recently burnt areas have a low value. This ratio is able to indicate quantitatively the magnitude of spectral changes triggered by a fire, and hence burn severity. NBR is usually derived from NIR band 4 and SWIR band 7 of Landsat TM and ETM+ images as follows:

$$NBR = \frac{\rho_{NIR} - \rho_{SWIR}}{\rho_{NIR} + \rho_{SWIR}} \tag{11.28}$$

NBR enables burn severity to be detected accurately on a repetitive basis, highlighting burnt areas in large fire zones. It is a better alternative to NDVI that can be biased by bare ground. dNBR represents the arithmetic difference between the pre- and post-fire NBR values (Eq. 11.29). It can be expressed as a few categories after the differences are grouped according to suitable thresholds of burn severity.

$$\Delta NBR = NBR_{pre-fire} - NBR_{post-fire} \tag{11.29}$$

ΔNBR likely varies spatially across the fire-impaired landscape due to the heterogeneity of fuel factors and the local environment. It can have a negative value which indicates vegetation regrowth. While it is relatively easy to calculate NBR and ΔNBR, care must be taken to interpret their physical meaning. For instance, what NBR value corresponds to what level of burn? In practice, predetermined severity thresholds are used to convert the calculated dNBR into discrete levels of burn severity (Table 11.6). A value below –0.100 indicates intact, and a value above this threshold means burned. High burn has a threshold above 0.660.

Of the two types of indices, generic vegetation indices are superior to special indices as they are able to indicate quantitatively the amount of biomass remaining on the ground after a fire. Thus, burn severity is assessed more precisely over a whole continuous range instead of a few discrete categories. Both types of indices fail to take into account the physical and chemical effects of a fire,

TABLE 11.6
Burn Severity Levels Based on Calculated ΔNBR Ranges According to the USGS

Severity Level	ΔNBR Range
Enhanced regrowth, high (post-fire)	–0.500 to –0.251
Enhanced regrowth, low (post-fire)	–0.250 to –0.101
Unburned	–0.100 to +0.999
Low burn	+0.100 to +0.269
Moderate-to-low burn	+0.270 to +0.439
Moderate-to-high burn	+0.440 to +0.659
High burn	+0.660 to +1.300

only the biological effect. No matter which type is used, the mapped burn severity is always subject to the spatial resolution of the image used. While a coarse spatial resolution image does not allow small burned areas to be mapped (e.g., a large omission error), a fine-resolution image will cause large commission errors if the area is not fully covered by natural vegetation such as small plots of tilled land (Filipponi, 2019). Such non-vegetation pixels can be removed through filtering prior to the formal mapping.

11.3.3.3 Burned Area Assessment

Burned areas or fire scars represent a change in land cover from former vegetation-covered to non-vegetated at present. Burning of the ground surface introduces a change to its spectral response with the degree of change related to burn severity and vegetation type. Burned areas can be mapped from optical images, either mono- or multi-temporal pre- and post-fire images. Fire scars can be detected from even a single band, or its combination with other bands (Van Wagtendonk et al., 2004). Burned and unburned shrub-savannah are separable maximally from each other in the shortwave mid-IR (SMIR) – longwave mid-IR (LMIR) domain of MODIS bands among all two-band space given the variable vegetation type, photosynthetic state, combustion remnants, and spectral evolution with time after burn (Trigg and Flasse, 2001). This space is sufficiently robust to withstand the perturbing effects of scattering caused by optically thick smoke plumes. The separation function is expressed by the mid-infrared burn index (MIRBI) as follows:

$$\text{MIRBI} = 10\ \text{LMIR} - 9.8\ \text{SMIR} + 2 \tag{11.30}$$

MIRBI is highly sensitive to spectral variations caused by burn but relatively insensitive to intrinsic variability triggered by vegetation change (e.g., natural growth). If mapped from Landsat 5 TM data, the radiance of TIR band 6 (10.4–12.5 μm) and the normalized difference reflectance of NIR band 4 (0.76–0.90 μm) to that of MIR band 7 (2.08–2.35 μm), namely, $(\rho_4 - \rho_7)/(\rho_4 + \rho_7)$, are the most suited to map burned areas (García and Caselles, 1991). This ratio is also effective at monitoring vegetation that has regenerated in burned areas, so the difference between burned areas and vegetated areas is reduced. Burned areas can also be mapped from Sentinel-2 time-series data that have a high revisit frequency and finer spatial and spectral resolution than Landsat TM and ETM+ images. A commission error of around 25% and an omission error of around 40% are indicative of the mapping accuracy if burned areas are mapped based on a set of empirical thresholds (including burned area index) (Filipponi, 2019), during which many small burned areas are eliminated via image filtering prior to the formal index-based thresholding.

Owing to the huge spectral difference between the two, they can be distinguished from each other with relative ease if the burn is complete. In this case, fire scars can be mapped using supervised and unsupervised classification. In the supervised classification, a category called burned area is created during training sample selection. This method is good at detecting dichotomous fire scars of burn-affected and intact that are easily identifiable from their lower reflectance than healthy vegetation on satellite images. If the burned areas are not so distinct, multi-temporal images must be used. Their differencing enhances the distinctiveness of burned areas from the unburned background. Burned areas can also be mapped using NBR and ΔNBR.

It must be noted that the spectral disparity between burned and intact vegetation is subject to a number of variations caused by the following:

i. Type of pre-fire vegetation covers: they may vary seasonally in their spectral behavior;
ii. Remnant combustion residues: if the burn is incomplete, some vegetation may still remain (e.g., stands of tall trees); and
iii. Self-regeneration and recovery: Some vegetation recovers much faster than others, and the newly regenerated vegetation may disguise the spectral signal of burned areas (Trigg and Flasse, 2001). The level of vegetation recovery depends on the space available and the type of vegetation. Riparian and forest vegetation takes longer to recover to the pre-fire levels one year after a fire than steppe vegetation, the biomass of which is uplifted by fires (White et al., 1996). The longer after the fire, the smaller the pre- and post-fire NDVI difference (García and Caselles, 1991), the less distinctive the fire scars and hence more difficult to map.

Burn area data product has been produced from MODIS data, derived based on the spectral, temporal, and structural changes to the surface. The product consists of five layers: burn date, burn date uncertainty, quality assurance, the first day (of a fire), and the last day. It also contains information on the spatial extent of recent fires, not fires that have occurred in previous years. The latest version (Collection 6) became available in 2017 (Figure 11.3). The new product (MCD64A1) derived using new algorithms supersedes the previous products (Giglio et al., 2018). The new algorithm for mapping burned areas significantly improves the ability to tackle the effect of cloud and aerosol contamination. The detectable fire size is much smaller while the uncertainty of the burn date is reduced modestly to the nearest day. The extent of unmapped areas is also minimized considerably due to the use of 268 MODIS tiles (Giglio et al., 2020). The product is freely available in both the HDF version and in the GeoTIFF format at a spatial resolution of 500 m. Validated against Landsat 8 OLI image pairs acquired 16 days apart, globally, the MCD64A1 product has an estimated commission

FIGURE 11.3 Detection of burned areas. Left: color composite of Landsat 7 imagery (path – 98; row – 71) recorded on 20 October 2002; right: MCD64 monthly burn product of October overlaid on top of the color composite. Blue – previously burned areas; red – recently burned areas. (Source: Giglio et al., 2020.)

error of 40.2%, and an omission error of 72.6%, due mostly to Landsat imagery's much finer spatial resolution (−54.1%). The coarse resolution of MODIS imagery caused small burned areas to be systematically underestimated. In using the data on burned areas in agricultural fields, caution needs to be exercised because cropland burning is much harder to map reliably, so the actual accuracy is likely to be lower than the stated accuracy.

11.3.4 Fire Risk Mapping and Modeling

Described in a variety of phrases (Table 11.7), fire risk refers to the potential or likelihood for a fire to occur, and is considered identical to fire danger or hazard by some researchers. Fire risk stems from a comprehensive range of climate, environmental, and topographic factors such as weather, fuel, terrain, and ignition parameters. It is expressed as a probability or likelihood indicative of fire potential. The assessment of fire risk may also be based on fire-related factors, such as the availability of fuel (litter and live vegetation), land use, insolation, vegetation, topography (elevation, slope, and aspect), agroclimate, and fire history (Pradhan et al., 2007). Some of them are related to the short-term variation of fire danger that is mostly temporary. Others pertain to the long-term spatial variation (e.g., they can be treated as constant within a short period). Some of these factors affect the beginning of a fire, while others govern its spread. Some factors such as trails and roads have a mixed effect on fire hazards. On the one hand, they serve as fire breaks or pathways for suppressing the fire. At the same time, they can be potential conduits for trampers and campers and hence increase forest fire hazards induced by human activity (Chuvieco and Congalton, 1989).

Fire risk is estimated from its contributors and influencers. Whatever factors are considered in mapping fire risk, they must be able to be represented as spatial layers in raster format to be manipulated mathematically in a GIS. The handling of the large number of factors considered in fire risk modeling is beyond the capability of remote sensing, but it still plays an indispensable role in the modeling. Regardless of the different philosophies of and approaches to fire risk modeling, they all

TABLE 11.7
Comparison of Common Terminologies Adopted to Describe Fire-Related Hazards and Their Determination from Various Factors

Terms	Factors Considered	Weighting of Factors	Authors
Fire risk	Land cover, land surface temperature; distance from the road, proximity to settlement; aspect, elevation, slope	Weights sum to 1	Parajuli et al. (2020)
Cumulative fire risk index	Fuel type, aspect, accessibility, slope, elevation	Integers from 1 to 4, not summing to 1	Roy (2004)
Fire hazard	Vegetation; slope, aspect, road, elevation	Weights do not sum to 1.	Chuvieco and Congalton (1989)
Fire susceptibility	Land use, soil, slope, aspect, NDVI, agroclimate	Frequency ratio of fire hotspot	Pradhan et al. (2007)
Fire potential (index)	Fuel, moisture, dead fuel moisture Moisture content of small dead fuels; live ratio	$100(1 - Fm10hr_{corrected})(1 - Lr)$[a]	Burgan et al. (1998) López et al. (2002)
Fire danger	Probability of ignition; Fuel hazard component; Human risk factor	No weighting, only mathematical combination	Chuvieco and Salas (1996)
Fire severity	NDVI	Pre- and post-fire changes	White et al. (1996)

[a] Fm = fuel moisture; Lr = live ratio, defined as the percentage of live fuel load to the total fuel load.

require land covers and topographic data that are ideally produced from images. Land covers in general, and vegetation or fuel type in particular, are obtainable via land cover classification of satellite images. Fuel factors are virtually land covers. Those covers that are critical to forest fire risk are all related to vegetation, such as forest, shrubland, and grassland, of which forest can be further broken down as deciduous and evergreen and differentiated as open vs closed canopy. High-quality land cover maps can be produced from various fine-resolution Earth resources satellite images, following the same principle as burned area mapping except that more types of covers are mapped. So the same data and data analytical procedure are applicable without modification. Topographic factors can be derived from a DEM created from LiDAR data (see Section 1.3.4 for more details).

To a large degree, the fire risk of a geographic area is a function of the past fire behavior, such as forest fire occurrence associated with different types of land cover, past events of fire, and burned area, all of which can be mapped from historic satellite images. Forest fire locations and historical fire hotspots are also identifiable from satellite images. Such information is indispensable to realistic modeling of fire risk as it is assumed that those areas that have suffered frequent fires with the largest burn area in the past are fire hotspots and face a high fire risk (Figure 11.4). In reality, this assumption is not completely valid as the fire risk is actually lowered after the highly combustible fuels have been depleted in a recent fire. Regression analysis of such data enables the establishment of the association between fire risk and a particular land cover attribute (e.g., shrubland) and determination of the strength of the association.

In addition to fire history, accurate forest fire hazard mapping must also take into consideration vegetation species, stand density, height–area ratio, and stress conditions (Chuvieco and Congalton, 1989). Other important factors to consider are topography, soil moisture, and LST. Topography mainly affects how fires are spreading. LST may be regarded as the proxy for relative humidity and moisture content of fuels. A higher surface temperature is equivalent to drier fuels that burn much faster than moist fuels. Such data can be obtained from a combination of optical, thermal, and radar satellite data directly (Leblon, 2005). In comparison to these static factors, dynamic variables of fire are much

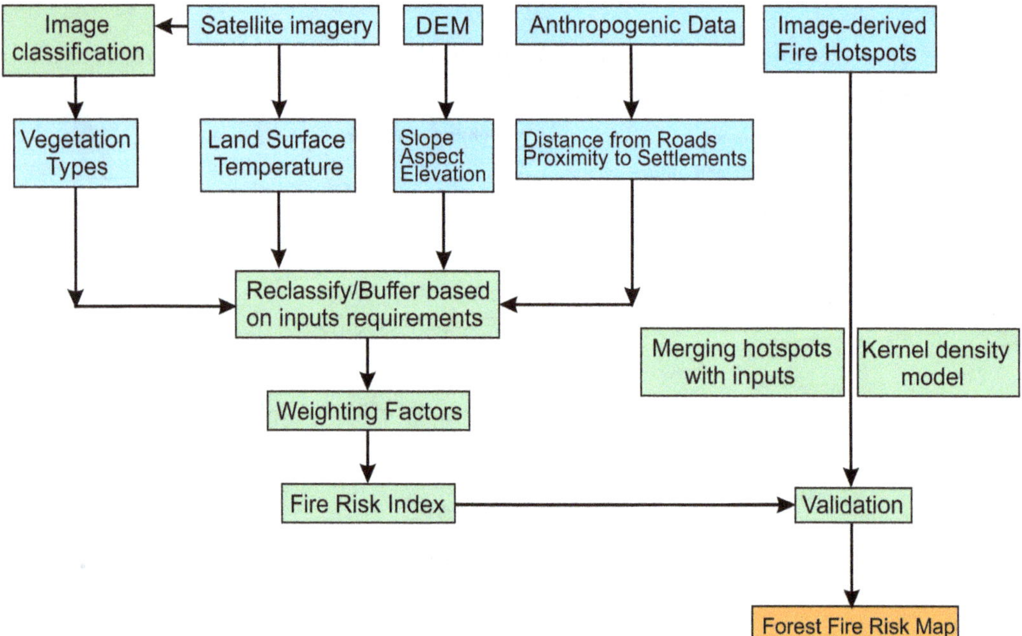

FIGURE 11.4 Procedure of assessing forest fire risk using various data inputs. Blue: data sources and variables; green: data processing; orange: results. (Source: modified from Parajuli et al., 2020).

thornier to handle in mapping fire hazards, such as wind speed and direction that directly control the ROS of a fire. They are usually ignored because of the difficulty in their accurate mapping at a local scale due to their high temporal volatility. They are considered only in scenario-based analysis.

The quantification of fire risk has been implemented in four ways. The first and the most common way is to linearly combine all the considered factors after weighting. Conceptually, forest fire risk or susceptibility is expressed as the sum of all fire contributing factors, or $\sum_{i=1}^{n} Fr_i$ (Fr_i: rating of the *i*th factor's type or range). These factors can be vegetation, slope, aspect, distance to road, and elevation (Chuvieco and Congalton, 1989), or

$$\text{Fire risk} = 1 + 100\ \text{veg} + 30\ \text{slope} + 10\ \text{aspect} + 5\ \text{road} + 2\ \text{elevation} \tag{11.31}$$

or

$$\text{Fire risk} = 40\%\text{LC} + 20\%\text{LST} + 10\%\text{S} + 10\%\text{DR} + 10\%\text{PS} + 5\%\text{A} + 5\%\text{E} \tag{11.32}$$

where LC = land cover, LST = land surface temperature; S = slope, DR = distance from the road, PS = proximity to settlement; A = slope aspect; E = elevation. In Eq. 11.32, the coefficients of all factors sum to 1 (Parajuli et al., 2020), but not in Eq. 11.31.

The second method of implementation is to zone areas of fire risk via the cumulative fire risk index (CFRISK). It integrates fuel-type index (FUI), aspect index (ASI), slope index (SLI), accessibility index (ACI), and elevation index (ELI) (Roy, 2004), or

$$\text{CFRISK} = 4\ \text{FUI} + 3\ \text{ASI} + 2\ \text{SLI} + \text{ACI} + \text{ELI} \tag{11.33}$$

The third method of quantification is to make use of vegetation variable (V=10 classes), proximity to human settlement (H=4 classes), slope (S = 4 classes), and road/fire line (R=5 classes) (Roy, 2004), or

$$\text{Fire risk} = 10V_{i=1-10} \times \left(5H_{j=1-4} + 5R_{k=1-5} + 3S_{j=1-4}\right) \tag{11.34}$$

where i, j, k, and l = the number of classes (attribute values) of each variable. For categorical factors such as land cover, they must be coded as numbers in calculating the fire risk index using Eq. 11.34, for instance, 4 (low) for needle-leaved closed forest, 3 (medium) for grassland, 2 (high) for broad-leaved open forest, and 1 (very high) for broad-leaved closed forest.

Finally, fire risk has been mapped from three fuel parameters of live ratio, moisture content, and fuel type (López et al., 2002). Live ratio is defined as the portion of live fuel load in the total fuel load. It is calculated from the NDVI of a five-year interval as follows:

$$\textit{Live ratio}\ (\%) = 0.25 + \frac{0.50NDVI_{max}}{NDVI_{absolute-max}} \tag{11.35}$$

where NDVI_{max} = the maximum NDVI of a given location in the study period (e.g., five years); $\text{NDVI}_{\text{absolute - max}}$ = the absolute maximum NDVI of any location on a continent during the same period. The proposed fire potential index can indicate the general class of fire risk. This implementation is suitable for the continental scale, while the other three implementations are more appropriate for a smaller scale.

No matter what factors are considered in mapping fire risk, all four implementations face the same two critical issues: how to avoid multicollinearity among the considered factors and how to properly weigh them. Multicollinearity means that some of the factors may be double-counted as a result of their close correlation. For instance, a higher elevation may lead to a lower temperature and less vegetation cover (lower biomass or fuel). The variation in temperature and vegetation cover is partially accounted for by topography. This issue is addressed by minimizing the correlation via a collinearity test of the VIFs. Some considered factors are easily eliminated if their VIF exceeds the pre-defined threshold in an *F*-test. The assignment of weights for a considered factor should be based on its association with historic fire events or fire hotspots by overlaying the historic fire events with each of the factor layers. The relative importance of fire factors and the location of fire ignition are judged by the closeness of the association. This association is expressed as the ratio of occurrence frequency to the total area or number of pixels (Pradhan et al., 2007), and it achieves the highest prediction accuracy among all considered factors.

After the fire risk has been mapped, this map may be overlaid with the respective input layers to identify the environmental settings vulnerable to fire. For instance, areas having a high risk of forest fire are characterized by broad-leaved closed forests; having a mean temperature of 30–35°C on southern and south-western slopes with a gradient <5°, lying below 1,000 m in close proximity to settlements (Roy, 2004).

Although fire risk and fire danger may be regarded as synonymous with each other, they are definitely derived from different factors. According to Chuvieco and Salas (1996), fire danger stems from three sources: the probability of ignition, fuel hazard, and human risk, of which fuel is directly derivable from satellite data (Figure 11.5). Human risk is related to proximity to the road network and recreational areas. The climate danger index is the most complex as it involves not only temperature and air humidity, but also topography, calculated as

$$Fire\ danger\ index = \frac{PI \times FHC}{10} + HRI \tag{11.36}$$

where PI = probability of ignition; FHC = fuel hazard component; HRI = human risk index. Missing from the consideration is fire history, namely, those areas that have experienced frequent fires in the past.

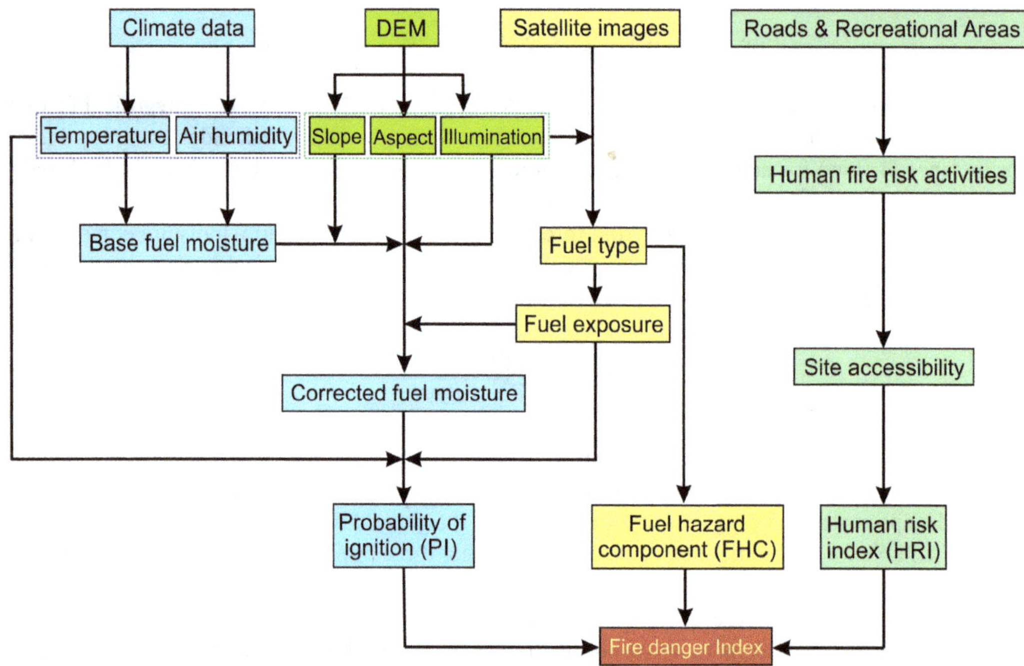

FIGURE 11.5 Derivation of forest fire risk from three considerations: the probability of ignition, fuel hazard, and the human risk index. (Source: modified from Chuvieco and Salas, 1996.)

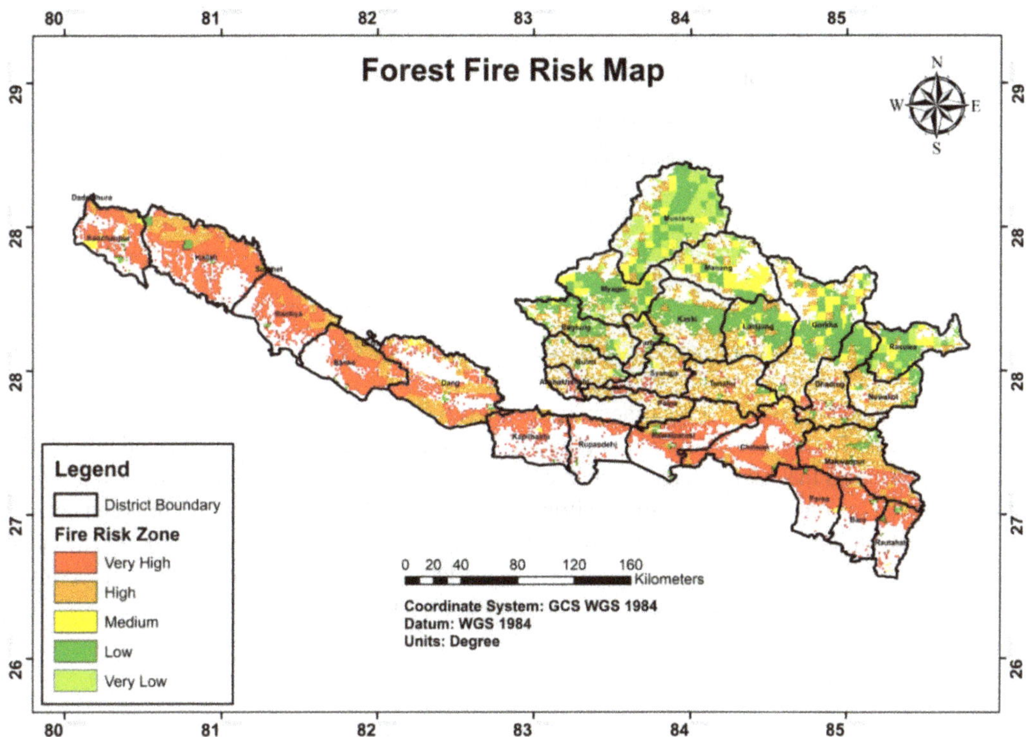

FIGURE 11.6 Spatial distribution of forest fire risk in central Nepal zoned at five levels from consideration of land cover, topography, land surface temperature, and distance to settlements and roads. (Source: Parajuli et al., 2020.)

The derived risk or susceptibility map (Figure 11.6) is crucial for haze reduction and fire prevention in forested areas and for zoning forest fire hazards. This map is usually expressed in a few discrete categories qualitatively. This zonal map represents a drastic generalization but can suppress uncertainties related to improperly assigned weights and multicollinearity among the considered factors (Parajuli et al., 2020). The mapped fire risk is commonly validated by overlaying the map with historic records of fires such as fire hotspots from satellite data. If most of the mapped historic fires fall into the very high and high categories, the map is considered highly accurate. In the comparison, historic fires may be mapped as kernel density estimation (KDE) and its comparison with the mapped risk reveals its agreement with the remotely sensed results. If high-risk areas are the same as those shown in the KDE, the mapping is deemed to have a high accuracy level. It is worth noting that the produced fire risk/hazard map just shows the susceptibility of the landscape to fire occurrence. It says nothing about the burning intensity in case a fire does eventuate.

11.4 SENSING OF COALMINE FIRES

Coal is a highly inflammable substance. In coalmines, it can be ignited spontaneously, accidentally by mining activities and external heat sources, such as nearby forest fires. Spontaneous combustion takes place as a consequence of the accumulated heat released from the interaction of oxygen in the atmosphere with coal under a favorable ambient temperature and poor thermal conductivity. The maximum alarming temperature for igniting coal is around 160°C in a number of Indian coalmines. Coal fires are associated with coal seams and piled coal heap (Table 11.8).

TABLE 11.8
Classification of Coal Fires

First Level	Second Level	Attributes
Coal seam fire	Natural coal fire	Surface/underground or subsurface
	Coalmine fire	Paleo/recent
Coal heap fire	Coal waste fire	Extinct/formant/active
	Coal stockpile fire	

Source: Zhang et al. (2004).

The former refers to the burning of coal in the minefield, either surface or subsurface. The latter refers to the combustion of mining waste and mined coal that are piled on site. Once a fire is ignited, it has the potential to burn for a long time and spread via fissures. Compared with wildfires, coal fires emit much more diverse particulate matter, SO_2, CO, NO, toxic organic compounds, and potentially toxic trace elements that may include arsenic, mercury, and selenium fumes. All of these substances are detrimental to human health and the environment. If surrounded by a forest, coal fires can destroy the natural vegetation and ecosystem and jeopardize coal mining. Thus, it is important to detect coal fires. This detection is usually carried out for the fire-affected area and focuses on fire depth, direction, intensity, and ROS. These topics will be explored in separate sections below.

11.4.1 Surface Fires

Coal fires may take place in underground mines, open coalfields, coal refuse, and coal stack during which coal seams, piles of stored coal, or spoil dumps on the surface are combusted. In terms of depth, they can be classified as surface and subsurface (Table 11.8). Surface coal fires are open and in direct contact with the atmosphere, commonly associated with spoil dumps of coal. They are usually small in scale and result from the spontaneous combustion of coal after it reacts to atmospheric oxygen. Surface fires can elevate the air temperature by up to 400°C above the normal level (Figure 11.7). This uplift in temperature is spatially variable, subject to fire depth. Coal fires, especially subsurface fires, manifest themselves on the surface as rather subtle thermal anomalies that are much more challenging to detect than forest fires. Thermal anomalies are commonly graded

FIGURE 11.7 Ground photo of a surface coal fire that produces local-scale thermal anomalies. (Source: Prakash et al., 1997.)

as low, medium, and high (Zhang et al., 2004). Low anomalies have a temperature up to 20°C above the background. An anomaly between 20°C and 120°C is considered a medium amplitude. High anomalies have a temperature between 120°C and over 300°C, but they are not as common as low-amplitude anomalies. From the thermal anomalies of a coal fire, its properties, such as the front, depth, and age, are inferred on remote sensing images through coal fire thermal models. Multi-temporal images obtained over a few days enable the determination of the rate of fire spread and the identification of newly emerged fires and stable fires.

Thermal anomalies of various amplitudes are best studied from spectral bands of varying sensitivity to temperature. Both surface temperature and temperature anomalies can be detected from individual bands with relative ease. According to Wien's displacement law, the peak radiation rises toward shorter wavelengths. Hence, surface coal fires of a very high temperature are better detected from short wavebands, such as ASTER SWIR band 9 (2.36–2.43 μm) without saturation (Raju et al., 2016). TM TIR band 6 (10.40–12.50 μm) is good at detecting low-temperature coal fires (e.g., subsurface fires) based on the equations presented in Section 11.2.1. This band saturates at 68°C, much lower than the temperature ranges of surface fires. So it is not good at detecting surface fires. Instead, TM 7 (2.08–2.35 μm) and TM 5 (1.55–1.75 μm) can sense temperatures from 160° to 420°C (Prakash et al., 1997). Nighttime SWIR bands allow the identification of the location of surface coal fires. For these reasons, Landsat TM 7 has been widely used to detect extensive surface coal fires of high-magnitude heat that cause a huge increase in radiance in SWIR bands. TM band 7 is the best at studying temperatures around 200°C, while band 5 is best used to detect temperatures around 300°C. This SWIR band is superior to TIR band 6 in detecting coal fires owing to their higher saturation temperature. Besides, their spatial resolution (30 m) is a quarter of 120 m of TM 6 and minimizes mixed pixels of the fire-affected portion of the surface, which is vital to differentiating surface fires from subsurface fires (Chatterjee, 2006). Both bands can identify fire pixels, but neither can identify individual fires. Instead, a cluster of fire pixels is spatially lumped together to form the fire zone because of the 30-m resolution.

The use of individual bands in detecting coal fires faces the challenge of how to set the temperature threshold that corresponds to a genuine fire. This threshold is commonly determined via trial and error after the image-converted temperature layer is density sliced into several (e.g., 10) DN ranges. The range corresponding to the observed coal fire is taken to indicate the threshold. A threshold of DN 137/138 on TM band 6 achieves a reasonable and overall best discrimination of fire areas from non-fire areas (Prakash et al., 1995). Additionally, the histogram of the temperature layer also facilitates the determination of the threshold (Raju et al., 2016). No matter how the threshold is determined, its value is likely location-specific. Furthermore, this threshold may vary within the entire minefield, or with the satellite image from which it is derived. Other images of the same area, if obtained under different atmospheric conditions, will likely have different thresholds. So every time a new image is used, a new threshold must be established from scratch, duplicating the efforts of mapping.

This difficulty can be circumvented by automatically searching for the local minimum that is regarded as a potential thermal anomaly. The search is repeated several times, each time with a different window size. If a pixel is deemed thermally anomalous in 70% of the cases, it is declared a thermal anomaly (Kuenzer et al., 2007). This local-minima approach can cause up to half of the detected anomalies to be false alarms. They can be easily eliminated via a comparison with the coal fire risk areas automatically delineated from multispectral satellite data. False alarms can also be minimized via additional checks against coal seams, dumps, outcrops, mining areas, and the proximity to coal outcrops and dump either mapped from high-resolution satellite images (e.g., ASTER and TM) or interpreted from geological maps (Chen et al., 2007). Another strategy for improving the reliability of the detected thermal anomalies is to create buffer zones around coal fire risk areas, such as abandoned and active mines, outcropping coal seams, coal waste spoils, coal storage piles, mining portals, or coal washery discard. Only those thermal anomalies falling inside the buffer zones are deemed to represent genuine coal fires. The joint use of both criteria effectively removes false alarms by 80% with nighttime Landsat ETM+ images, but only 50% with daytime images.

Despite the higher success rate, nighttime images are disadvantaged in that they are prone to saturation in areas of a medium- and high-amplitude anomaly. If the same analysis is carried out using multi-temporal thermal images, it is possible to detect changes in coal fire-affected areas.

A single threshold for the entire imaged area is unable to take into account the spatial variability of thermal anomalies (e.g., variable burn intensity), and that caused by the image spatial resolution (e.g., mixed pixels of fires and non-fires), and the background environment. The reliability of a single, universal threshold can be improved via two strategies. The first is to derive it within a sub-window using the moving window approach (Kuenzer et al., 2007). The sub-window histogram comprises the background temperatures and thermally anomalous components. The two are separable by the relative threshold of the first local minimum after the main histogram maximum. The central pixel in the operating window is examined thousands of times using various window sizes. It is declared an anomaly only if this is true in most cases. This method extracts thermally anomalous pixels in relation to their surrounding background and is conducive to extracting local-scale, very subtle thermal anomalies. Although it achieves a high detection rate, small fires or those too deep below the surface are omitted, which can be avoided using images of a finer spatial resolution. Other additional information conducive to minimizing false alarms is spectral reflectance or albedo of a surface cover. It helps to resolve the confusion between coal, clear water, and shadow, all of which have low reflectance in optical bands. The differentiation of shadowed areas and water from genuine coal fire pixels requires additional decision-making criteria (Kuenzer et al., 2007).

The second strategy is to take advantage of a self-adapting threshold based on the temperature gradient between adjacent pixels (Du et al., 2015). Extremely high gradient pixels are connected in two processes. The first involves segmentation of the gradient image with a pair of lower and upper bounds (e.g., 1.0–3.2 standard deviation), and the relatively high gradient values are retained. These intermediate segmentation gradient pixels form potential high gradient buffers in a ring shape centered on high-temperature anomalies. The ring buffers are then morphologically thinned to form one-pixel wide skeleton lines that are considered to represent the fire border. The second process is to segment the temperature image using 1.0δ (δ – standard deviation) so as to exclude low temperatures from cold areas along the extremely high gradient lines (e.g., false alarms removed). The temperature of those pixels along the thinned skeleton line is averaged to derive the intermediate threshold. All the immediate thresholds from multiple skeleton lines are further averaged to derive the final threshold. This method requires multiple TIR images, and the threshold is generated accurately, consistently, and uniformly, and thus ideal for detecting long-term changes in coal fires from historical images.

The reliability of remotely detected coal fires is subject to the influence of the intensity of active coal fires, the background temperature, and the quality of remote sensing data (e.g., acquisition time, spatial and temporal resolutions). Even if a suitable threshold can be determined, however, thresholding of high-temperature pixels of a moderate spatial resolution (e.g., Landsat TM 30 m) may still be futile in identifying surface fire pixels if the spectral band used has a low sensitivity range of temperature (e.g., 150–250°C) (Chatterjee, 2006). To efficiently detect surface coal fires, the band must have a temperature sensitivity range closely matching the temperature range of the coal fires. The sensitivity of TM 7 ranges from 160°C to 277°C, which means that it is able to detect coal fires, the temperature of which falls within this range. The influence of background temperature caused by solar heating can be eliminated through the analysis of topographic data in areas of an uneven surface. During daytime the coal fire signal is severely obscured by the solar radiant energy, so coal fires are best detected from nighttime images, in which the solar effect is still persistent but is much subdued. Besides, nighttime images are not subject to the topographic effect on thermal radiation or the effect is barely minimal (Chatterjee, 2006). These images are used widely (Prakash et al., 1999). Nighttime (especially pre-dawn) images are the best for detecting coal fires. A coal fire map produced from daytime satellite data suffers larger omission and commission errors than its counterpart from nighttime satellite data (Huo et al., 2014). This conclusion is still valid with high-resolution UAV TIR imagery (Li et al., 2018).

Image quality, especially image resolution, affects not only whether a coal fire is detectable, but also the type of details detectable about it. MODIS data with a fine temporal resolution measured in hours and tens of spectral bands are potentially useful for detecting temperature anomalies of a coalfield. However, the coarse spatial resolution of the thermal bands (1,000 m) prevents the derivation of quantitative information on fire outline, fire temperature, or even the separation of fires into surface and subsurface (Kuenzer et al., 2008). Only general coal fire zones and coal fire hotspot zones are separable from each other. The subtle disparity between thermal anomalies of a varying magnitude can be artificially enlarged via image enhancement, such as the ratio of MODIS band 20 to band 32. This ratio enhances sub-pixel-level hotspots or areas of a higher temperature than the background (Hecker et al., 2007). A success rate of only 42–49% is achieved in detecting areas with surface fires, suggesting that low-resolution MODIS imagery is potentially useful only in monitoring extensive areas with newly developing surface coal fires. Apart from image spatial resolution, the success rate of extracting coal fire pixels is also lowered by non-fire phenomena, such as surface relief-induced temperature anomalies and soil moisture.

11.4.2 Subsurface Fires

Subsurface fires are associated with vents through which oxygen reaches the coal seam layer (Figure 11.8). Once the coal seam is ignited, it will burn deeper. Subsurface fires cause the surface vegetation to wither or die, depending on their intensity. Very intense fires can cause pyrometamorphism of the surrounding bedrock and uplift the rock's reflectance in the 0.76–2.00 μm spectrum.

FIGURE 11.8 Burning of subsurface fires in the Wuda coalfield of Northern China. (Source: Chen et al., 2007.)

The burning of subsurface coal may further trigger land subsidence. These changes serve as valuable clues in the mapping of subsurface coal fires. They are much harder to detect than surface fires, even though the principle of detection remains unchanged: all surface temperature anomalies represent the potential manifestation of a subsurface fire. Such thermal anomalies are formed by convection of heat through fissures, cracks, abandoned mine tunnels or subsided surfaces on the ground, so they have a highly limited spatial extent. The difficulty of detecting a subsurface fire can also be traced to its subtle thermal manifestation on the surface. Even if an underground coal fire emits 400°C hot gases along a 30-m crack, the overall ambient temperature may be elevated merely by a few degrees over a pixel size of 60 m × 60 m. The exact thermal manifestation of a subsurface coal fire as surface temperature anomalies is influenced by rock and soil type, emissivity, surface cracks or fissures, and fire depth. Emissivity and surface temperature vary with surface materials of different thermal and radiative properties. It is also affected by wind and surface evaporation of moisture.

The identification of temperature anomalies may be eased by plotting the temperature profile in certain directions. This profile indicates the position at which the anomaly starts and ends and provides essential information on the delineation of fire areas from non-fire areas (Prakash et al., 1999). If the thermal anomalies are sufficiently distinctive, or the ground above the fire is heated by conduction sufficiently warm to exceed the background temperature, subsurface coal fires can be detected using the same methods as those used to detect surface coal fires described in the preceding section, such as thresholding and density slicing. The commonly used satellite imagery is still Landsat TM and ETM+ that allows subsurface coal fire zones to be detected, mapped, and quantified. Different ETM+ bands have different utilities in detecting subsurface coal fires. Through simple density slicing, SWIR bands 5 and 7 are useful in locating the position of the fire front, while TIR band 6 is strong at separating LST anomalies from the background of solar warming (Cracknell and Mansor, 1993). TM 6 allows delineation of zones affected by subsurface coal fires based on gross thermal anomalies from the background of solar warming. Subsurface coal fires are detected from this band using a threshold of 137 determined via density slicing. It enables the discrimination of subsurface coal fire pixels from non-fire pixels (Saraf et al., 1995). Areas with a DN ranging from 138 to 152 correspond to a temperature range of 25.6–31.6°C (Prakash et al., 1997). This range is construed to represent surface thermal anomalies caused by subsurface fires (Figure 11.9).

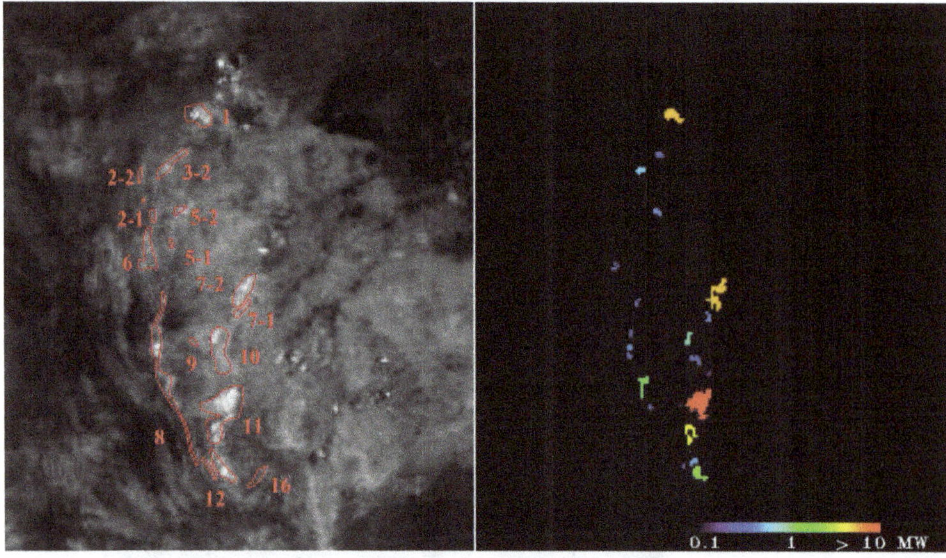

FIGURE 11.9 Left – surface thermal anomalies (red polygons) of the Wuda coalfield superimposed on a nighttime Landsat TIR band of 28 September 2002; Right – radiative energy release of coal fire pixels detected from the TIR band. (Source: Tetzlaff, 2004.)

The use of a coarse-resolution (e.g., 120 m) band poses difficulty in distinguishing subsurface coal fire pixels from their surface counterparts because pixel-integrated coal fires do not always reach the saturation temperature of around 70°C (Chatterjee, 2006). This difficulty can be circumvented by using linear unmixing of representative mixed pixels of surface fires. Spectral unmixing analysis is successful at ascertaining the approximate threshold temperature for subsurface coal fire pixels.

If the subsurface fire is intense, it has to be detected from TM bands 5 and 7 of a shorter wavelength. Their relatively fine spatial resolution of 30 m allows effective positioning of thermal anomalies caused by coalmine fires and identification of their intensity (Mansor et al., 1994). However, the location of fire fronts has to be inferred from the recorded radiant energy in the short wavelength. The success rate of extracting potential coal fire pixels is disappointedly low at only 0–17% in areas with known subsurface fires. The use of an additional fire risk layer does not boost the success rate noticeably. Subsurface coal fires are successfully detected only when the area under study is not subject to topography-induced temperature anomalies (e.g., the surface is flat and devoid of terrain-induced thermal variation), and after the exclusion of areas of abnormal geothermal fluxes (if they are present). The reliability of fire detection depends on the surface manifestation of subsurface fires. Only shallow (<10 m) underground coalmine fires can be easily detected from TIR imagery. For deeper subsurface coalmine fires, this detection is possible only for the fire zone, not individual fires, which makes it difficult to verify the remotely detected results against field data. In the field, it is almost impossible to pinpoint subsurface coal fires. As illustrated in Figure 11.9, it is the smoke emitting from vents and fissures that is relied on to infer the existence of subsurface fires in a coalfield (Prakash et al., 1995).

As mentioned previously, subsurface coal fires can cause surface subsidence or even collapse that can be detected from differential interferometric SAR (DInSAR) data. Change in surface subsidence is a potentially effective indicator of subsurface coal fire areas for detecting deeply buried small or hidden coal fires. Subsidence and coal fire areas overlap spatially to a certain degree. New cracks and fractures may emerge in the surrounding area due to the burning of coal fires. This subsidence-based detection is not always reliable as subsidence can also be caused by crust movement and coal mining (Liu et al., 2021). Thus, further constraints have to be imposed to minimize the commission errors based on the integral use of the thermal anomaly threshold, subsidence threshold, and NDVI threshold. These constraints significantly reduce the commission errors by an average of 70.4% and omission errors by 30.6% caused by non-coal fire thermal anomalies. This method is much more accurate than distributed scatterer interferometry, the second-best method. However, this method has limited performance in identifying embryonic fires and in eliminating non-fire thermal anomalies.

11.4.3 Fire Depth

Subsurface coal fires cannot take place deep down the surface because of the lack of oxygen to fuel the burning and the presence of groundwater. If a subsurface coal fire is excessively deep, its surface thermal manifestation becomes too indistinctive to detect. The exact detectable depth is complicated by the dip angle of the outcropping coal seam (Kuenzer et al., 2007). One way of estimating the depth of subsurface coal fires and their temporal propagation is to make use of the equation for linear heat flow in a semi-infinite medium, facilitated by temperature profiles drawn from a thermal scanner survey (Mukherjee et al., 1991). For instance, if a constant fire has a maximum duration of 15 years, and the temperature anomaly detected from airborne TIR remote sensing data is about 160°C produced by conduction through coal, the constant fire source is computed to be about 15 m deep.

Another method of calculating the depth of a subsurface coal fire is to make use of the location of the thermal anomaly, the location of the coal seam, and the angle of the dip of the strata (Saraf et al., 1995). Thermally anomalous pixels lie vertically above the existing fire. The surface manifestation of coal bands is studied from visible light and NIR bands. The minimum horizontal distance

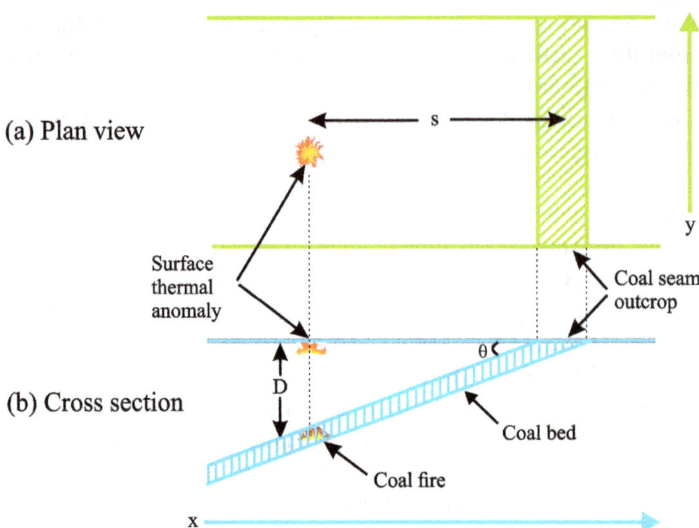

FIGURE 11.10 Schematic illustration of the principle of computing depth of subsurface coal fires from the location of the thermal anomaly, location of the coal seam, and dip angle of the strata. (Source: modified from Saraf et al., 1995.)

between the location of the outcrop and the thermal anomaly is measured from a temperature layer converted from a raw spectral band (Figure 11.10). Based on the dip (θ) of the strata that is commonly measured in the field, the depth of the subsurface coal fire is calculated based on the simple planar geometry as follows:

$$Depth = Distance_{min-horizontal} \times \tan(\theta) \tag{11.37}$$

If a fire is located 120-m south of the coal outcrop at a dip angle of 8°, its depth is calculated as $120 \times \tan 8°$ or 17 (m). This method is simple, but the accuracy of estimation is subject to the spatial resolution of the image used. If the pixel size is 30 m, the minimum horizontal distance must be at least 30 m. To achieve higher accuracy, fine-resolution images (e.g., < 5 m) should be used.

Another method of estimating the depth of subsurface coal fires is to integrate remotely sensed data with a DEM and the occurrence of the coal seams. This method requires the location of the outcrop of the coal seam and the coal fire front. In some cases, the altitude of the outcrop of the coal seam and the ground surface of the coal fire front are also essential. They are detectable from pre-dawn airborne TIR data. Such information may be acquired from a DEM stored in a GIS or determined directly from DInSAR or stereoscopic color infrared photographs. Two angles are also required: the dip angle of the coal seam, and the angle between the inclination direction of the coal seam and the spreading direction of the coal fire (Peng et al., 1997). The accuracy of detection depends on the accuracy of identifying the coal firefront.

REFERENCES

Alexander, M.E. 1982. Calculating and interpreting forest fire intensities. *Can. J. Bot.* 60: 349–357.

Burgan, R.E., R.W. Klaver, and J.M. Klaver. 1998. Fuel models and fire potential from satellite and surface observations. *Int. J. Wildland Fire* 8(3): 159–170.

Byram, G.M. 1959. Combustion of forest fuels. In *Forest Fire: Control and Use*, 61–69. New York: McGraw-Hill.

Chatterjee, R.S. 2006. Coal fire mapping from satellite thermal IR data–A case example in Jharia coalfield, Jharkhand, India. *ISPRS J. Photogra. Remote Sens.* 60(2): 113–128. DOI:10.1016/j.isprsjprs.2005.12.002

Chen, Y., L. Jing, Y. Bo, P. Shi, and S. Zhang. 2007. Detection of coal fire location and change based on multi-temporal thermal remotely sensed data and field measurements. *Int. J. Remote Sens.* 28 (15): 3173–3179. DOI: 10.1080/01431160500444889

Chuvieco, E., and J. Salas. 1996. Mapping the spatial distribution of forest fire danger using GIS. *Int. J. Geog. Info. Sci.* 10(3): 333–345. DOI: 10.1080/02693799608902082

Chuvieco, E., and R.G. Congalton. 1989. Application of remote sensing and geographic information systems to forest fire hazard mapping. *Remote Sens. Environ.* 29: 147–159.

Cracknell, A.P., and S.B. Mansor. 1993. Detection of sub-surface coal fires using Landsat Thematic Mapper data. *Int. Archives of Photogra. Remote Sens.* 29: 750–753.

Du, X., S. Bernardes, D. Cao, T.R. Jordan, Z. Yan, G. Yang, and Z. Li. 2015. Self-adaptive gradient-based thresholding method for coal fire detection based on ASTER data—Part 2, validation and sensitivity analysis. *Remote Sens.* 7 (3): 2602–2626. DOI: 10.3390/rs70302602

Filipponi, F. 2019. Exploitation of Sentinel-2 time series to map burned areas at the national level: A case study on the 2017 Italy wildfires. *Remote Sens.* 11(6): 622. DOI: 10.3390/rs11060622

García, M.J.L., and V. Caselles. 1991. Mapping burns and natural reforestation using thematic Mapper data. *Geocarto Int.* 6: 31–37.

Giglio, L., L. Boschetti, D.P. Roy, A.A. Hoffmann, M. Humber, and J.V. Hall. 2020. Collection 6 MODIS Burned Area Product User's Guide (Version 1.3). NASA, https://lpdaac.usgs.gov/documents/875/MCD64_User_Guide_V6.pdf

Giglio, L., L. Boschetti, D.P. Roy, M.L. Humber, and C.O. Justice. 2018. The Collection 6 MODIS burned area mapping algorithm and product. *Remote Sens. Environ.* 217: 72–85.

Giglio, L., W. Schroeder, and C.O. Justice 2016. The collection 6 MODIS active fire detection algorithm and fire products. *Remote Sens. Environ.* 178: 31–41.

Hecker, C., C. Kuenzer, and J. Zhang. 2007. Remote-sensing–based coal-fire detection with low-resolution MODIS data. *Rev. Eng. Geol.* 18: 229–238.

Hu, X., Y. Ban, and A. Nascetti. 2021. Sentinel-2 MSI data for active fire detection in major fire-prone biomes: A multi-criteria approach. *Int. J. Appl. Earth Obs. Geoinf.* 101, 102347. DOI: 10.1016/j.jag.2021.102347

Huo, H., X. Jiang, X. Song, Z.L. Li, Z. Ni, and C. Gao. 2014. Detection of coal fire dynamics and propagation direction from multi-temporal nighttime Landsat SWIR and TIR data: A case study on the Rujigou coalfield, Northwest (NW) China. *Remote Sens.* 6(2): 1234–1259. DOI: 10.3390/rs6021234

Johnston, J.M., M.J. Wooster, R. Paugam, X. Wang, T.J. Lynham, and L.M. Johnston. 2017. Direct estimation of Byram's fire intensity from infrared remote sensing imagery. *Int. J. Wildland Fire* 26(8): 668–684. DOI: 10.1071/WF16178

Justice, C., L. Giglio, S. Korontzi, J. Owens, J.T. Morisette, D. Roy, J. Descloitres, S. Alleaume, F. Petitcolin, and Y. Kaufman. 2002. The MODIS fire products. *Remote Sens. Environ.* 83: 244–262.

Kaufman, Y.J., A. Setzer, C. Justice, C.J. Tucker, and I. Fung. 1990. Remote sensing of biomass burning in the tropics. In: *Fires in Tropical Biota*, ed. J.G. Goldammer, 371–399. Berlin: Springer-Verlag.

Kennedy, P.J., A.S. Belward, and J.M. Gregoire. 1994. An improved approach to fire monitoring in West Africa using AVHRR data. *Int. J. Remote Sens.* 15 (11): 2235–2255. DOI: 10.1080/01431169408954240

Kuenzer, C., C. Hecker, J. Zhang, S. Wessling, and W. Wagner. 2008. The potential of multidiurnal MODIS thermal band data for coal fire detection. *Int. J. Remote Sens.* 29 (3): 923–944. DOI: 10.1080/01431160701352147

Kuenzer, C., J. Zhang, J. Li, S. Voigt, H. Mehl, and W. Wagner. 2007. Detecting unknown coal fires: Synergy of automated coal fire risk area delineation and improved thermal anomaly extraction. *Int. J. Remote Sens.* 28 (20): 4561–4585. DOI: 10.1080/01431160701250432

Kumar, S.S., and D.P. Roy. 2018. Global operational land imager Landsat-8 reflectance-based active fire detection algorithm. *Int. J. Digit. Earth* 11: 154–178.

Leblon, B. 2005. Monitoring forest fire danger with remote sensing. *Nat. Hazards* 35: 343–359. DOI: 10.1007/s11069-004-1796-3

Li, F., W. Yang, X. Liu, G. Sun, and J. Liu. 2018. Using high-resolution UAV-borne thermal infrared imagery to detect coal fires in Majiliang Mine, Datong coalfield, Northern China. *Remote Sens. Lett.* 9(1): 71–80. DOI: 10.1080/2150704X.2017.1392632

Liu, J., Y. Wang, S. Yan, F. Zhao, Y. Li, L. Dang, X. Liu, Y. Shao, and B. Peng. 2021. Underground coal fire detection and monitoring based on Landsat-8 and Sentinel-1 data sets in Miquan fire area, Xinjiang. *Remote Sens.* 13, 1141. DOI: 10.3390/rs13061141

López, A.S., J. San-Miguel-Ayanz, and R.E. Burgan. 2002. Integration of satellite sensor data, fuel type maps and meteorological observations for evaluation of forest fire risk at the pan-European scale. *Int. J. Remote Sens.* 23(13): 2713–2719. DOI: 10.1080/01431160110107761

Mansor, S.B., A.P. Cracknell, B.V. Shilin, and V.I. Gornyi. 1994. Monitoring of underground coal fires using thermal infrared data. *Int. J. Remote Sens.* 15 (8): 1675–1685. DOI: 10.1080/01431169408954199

Mukherjee, T.K., T.K. Bandyopadhyay, and S.K. Pande. 1991. Detection and delineation of depth of subsurface coalmine fires based on an airborne multispectral scanner survey in a part of the Jharia coalfield, India. *Photogr. Eng. Remote Sens.* 57(9): 1203–1207.

Nádudvari, Á, A. Abramowicz, R. Maniscalco, and M. Viccaro. 2020. The estimation of lava flow temperatures using Landsat night-time images: Case studies from eruptions of Mt. Etna and Stromboli (Sicily, Italy), Kilauea (Hawaii Island), and Eyjafjallajökull and Holuhraun (Iceland). *Remote Sens.* 12, 2537. DOI: 10.3390/rs12162537

Parajuli, A., A.P. Gautam, S.P. Sharma, K.B. Bhujel, G. Sharma, P.B. Thapa, B.S. Bist, and S. Poudel. 2020. Forest fire risk mapping using GIS and remote sensing in two major landscapes of Nepal. *Geomat. Nat. Hazards Risk* 11(1): 2569–2586. DOI: 10.1080/19475705.2020.1853251

Paugam, R., M. Wooster, and G. Roberts. 2013. Use of handheld thermal imager data for airborne mapping of fire radiative power and energy and flame front rate of spread. *IEEE Trans. Geosci. Remote Sens.* 51(6): 3385–3399. DOI: 10.1109/TGRS.2012.2220368

Peng, W.X., J.L. Van Genderen, G.F. Kang, H.Y. Guan, and Y.J. Tan. 1997. Estimating the depth of underground coal fires using data integration techniques. *Terra Nova* 9 (4): 180–183. DOI: 10.1046/j.1365-3121.1997.d01-31.x

Pereira, M.C., and A.W. Setzer. 1993. Spectral characteristics of fire scars in Landsat-5 TM images of Amazonia. *Int. J. Remote Sens.* 14 (11): 2061–2078. DOI: 10.1080/01431169308954022

Pradhan, B., M.D.H.B. Suliman, and M.A.B. Awang. 2007. Forest fire susceptibility and risk mapping using remote sensing and geographical information systems (GIS). *Disaster Prev. Manage.* 16(3): 344–352. DOI: 10.1108/09653560710758297

Prakash, A., A.K. Saraf, R.P. Gupta, M. Dutta, and R.M. Sundaram. 1995. Surface thermal anomalies associated with underground fires in Jharia coal mines, India. *Int. J. Remote Sens.* 16(12): 2105–2109. DOI: 10.1080/01431169508954544

Prakash, A., R. Gens, and Z. Vekerdy. 1999. Monitoring coal fires using multi-temporal nighttime thermal images in a coalfield in North-West China. *Int. J. Remote Sens.* 20(14): 2883–2888. DOI: 10.1080/014311699211868

Prakash, A., R.P. Gupta, and A.K. Saraf. 1997. A Landsat TM based comparative study of surface and subsurface fires in the Jharia Coalfield, India. *Int. J. Remote Sens.* 18(11): 2463–2469. DOI: 10.1080/014311697217738

Raju, A., A. Singh, S. Kumar, and P. Pati. 2016. Temporal monitoring of coal fires in Jharia coalfield, India. *Environ. Earth Sci.* 75(12): 1–15. DOI: 10.1007/s12665-016-5799-7

Roy, P.S. 2004. Forest fire and degradation assessment using satellite remote sensing and geographic information system. In: *Satellite Remote Sensing and GIS Applications in Agricultural Meteorology*, eds. M.V.K. Sivakumar, P.S. Roy, K. Harmsen and S.K. Saha, 361–400. Geneva: World Meteorological Organization.

Ruecker, G., D. Leimbach, and J. Tiemann. 2021. Estimation of Byram's fire intensity and rate of spread from space-borne remote sensing data in a savanna landscape. *Fire* 4, 65. DOI: 10.3390/fire4040065

Ryan, K.C., and N.V. Noste 1985. Evaluating prescribed fires. In: *Tec. Coord. Proc. Symposium and Workshop on Wilderness Fire*, eds. J.E. Lotan, B.M. Kilgore, W.C. Fischer and R.W. Mutch. USDA Forest Service General Tec. Rep. INT-182, 230–238. Ogden, Utah: Intermountain Forest and Range Experiment Station.

Saraf, A.K., A. Prakash, S. Sengupta, and R.P. Gupta. 1995. Landsat-TM data for estimating ground temperature and depth of subsurface coal fire in the Jharia coalfield, India. *Int. J. Remote Sens.* 16(12): 2111–2124. DOI: 10.1080/01431169508954545

Schroeder, W., P. Oliva, L. Giglio, and I.A. Csiszar. 2014. The new VIIRS 375 m active fire detection data product: Algorithm description and initial assessment. *Remote Sens. Environ.* 143: 85–96.

Schroeder, W., P. Oliva, L. Giglio, B. Quayle, E. Lorenz, and F. Morelli. 2016. Active fire detection using Landsat-8/OLI data. *Remote Sens. Environ.* 185: 210–220.

Tetzlaff, A. 2004. Coal Fire Quantification Using Aster, ETM and Bird Satellite Instrument Data. Doctoral dissertation, Ludwig-Maximilians-University, Munich, 156 p.

Trigg, S., and S. Flasse. 2001. An evaluation of different bi-spectral spaces for discriminating burned shrub-savannah. *Int. J. Remote Sens.* 22: 2641–2647.

Turner, M.G., W.W. Hargrove, R.H. Gardner, and W.H. Romme. 1994. Effects of fire on landscape heterogeneity in Yellowstone National Park, Wyoming. *J. Veg. Sci.* 5: 731–742. DOI: 10.2307/3235886

Van Wagtendonk, J.W., R.R. Root, and C.H. Key. 2004. Comparison of AVIRIS and Landsat ETM+ detection capabilities for burn severity. *Remote Sens. Environ.* 92: 397–408.

Wang, Y.-J., F. Tian, Y. Huang, J. Wang, and C.-J. Wei. 2015. Monitoring coal fires in Datong coalfield using multi-source remote sensing data. *Trans. Nonferrous Met. Soc. China* 25: 3421–3428. DOI: 10.1016/S1003-6326(15)63977-2

White, J.D., K.C. Ryan, C.C. Key, and S.W. Running. 1996. Remote sensing of forest fire severity and vegetation recovery. *Int. J. Wildland Fire* 6(3): 125–136.

Wooster, M.J., G. Roberts, G.L.W. Perry, and Y.J. Kaufman. 2005. Retrieval of biomass combustion rates and totals from fire radiative power observations: FRP derivation and calibration relationships between biomass consumption and fire radiative energy release. *J. Geophys. Res. Atmos.* 110, D24311. DOI: 10.1029/2005JD006318

Wooster, M.J., G. Roberts, P.H. Freeborn, W. Xu, Y. Govaerts, R. Beeby, J. He, A. Lattanzio, and R. Mullen. 2015. Meteosat SEVIRI fire radiative power (FRP) products from the land surface analysis satellite applications facility (LSA SAF) – Part 1: Algorithms, product contents and analysis. *Atmos. Chem. Phys. Discuss.* 15: 15831–15907.

Xu, W., M. Wooster, G. Roberts, and P. Freeborn. 2010. New GOES imager algorithms for cloud and active fire detection and fire radiative power assessment across North, South and Central America. *Remote Sens. Environ.* 114: 1876–1895.

Xu, W., M.J. Wooster, J. He, and T. Zhang. 2021. Improvements in high-temporal resolution active fire detection and FRP retrieval over the Americas using GOES-16 ABI with the geostationary fire thermal anomaly (FTA) algorithm. *Sci. Remote Sens.* 3: 100016.

Zhang, J., W. Wagner, A. Prakash, H. Mehl, and S. Voigt. 2004. Detecting coal fires using remote sensing techniques. *Int. J. Remote Sens.* 25(16): 3193–3220. DOI: 10.1080/01431160310001620812

12 Volcanoes

12.1 INTRODUCTION

A volcano is a process during which the Earth's crust at its weakest spot is ruptured to allow searing lava, volcanic ash, and gases to be ejected from a magma chamber below the ground to the surface or into the air. Volcanoes are commonly situated in places where tectonic plates are either diverging or converging, such as along the Pacific Rim, especially in Japan, Indonesia, New Zealand, and Hawaii. Here the Earth's crust is so thin that geothermal activities take place very close to the surface. Tens of volcanoes around the world erupt annually, provoking enormous damage and destruction valued at hundreds of millions of dollars. Identical to earthquakes, volcanoes are a purely natural hazard that can damage properties and infrastructure and cause loss of lives in several secondary hazards. Effusive volcanic events are considered hazardous, affecting people and their properties and disrupting their daily activities worldwide. Of all natural hazards covered in this book, volcanoes trigger the largest number of secondary hazards that do damage in the most diverse ways, some of which have already been covered, such as landslides (and avalanche debris flows). Explosive eruptions of volcanoes may also trigger pyroclastic flows of ash, volcanic gas, rocks, and lava. Volcanic gas can travel up to 15 km from the volcano vent. Volcanic ash clouds hang over the nearby land and sea, enshrouding them in a thick blanket of smoke. The dispersed volcanic plumes from the eruption vent further hampers aviation. Tiny ash particles can be suspended in the air for days. They ruin machinery, contaminate water, damage power supplies, and severely harm agricultural fields and properties. If deposited near an airport, volcanic ash prevents airplanes from taking off and landing, further thwarting rescue efforts. While tephra can deposit close to the vent, lava flows further down the volcanic cone, destroying everything on its path, and may trigger forest fires. If a volcano is located at a high elevation where the ground is frozen, then its eruption can also trigger lahars on steep volcano cone flanks. They are able to travel further distances than lava and pyroclastic flows, threatening even more communities on their way to the nearest river channel. Lahars can do more damage to infrastructure than flooding. Fast-moving, sediment-laden lahars can easily destroy bridges and houses on their path. Underwater volcanoes cause huge waves and even tsunamis that may damage the nearby seaports and built-up areas.

It is important to study volcanoes, monitor them to predict their eruption, map lava flows, and assess potential damage after an eruption. All of these tasks are ideally accomplished using remote sensing. Remote sensing of volcanoes aims at three primary targets of study: lava flows and their thermal properties, volcanic ash and plumes, and lahar flow, in addition to monitoring surface changes in land cover, topography, and morphology. Different remote sensing data have different utilities in studying volcanoes and assessing their hazards, depending on the target of sensing. For instance, the tracking of volcanic ash benefits from coarse-resolution meteorological satellite images while VHR images are essential in monitoring small lava flows. Lava volume is best estimated from non-imagery LiDAR data. TIR is the best at sensing the thermal emissions of volcanoes. The study of lava and lahar flows is best fulfilled from DEM data while surface cover changes are best detected from optical images.

This chapter first explains how to identify active volcanoes on remote sensing images and detect and map lava flows. Then it explores how to quantify lava eruption and effusion rates, followed by an in-depth examination of how to estimate lava volume and its temperature, and how to map volcanic hazards using a wide range of remotely sensed data, both graphic and non-graphic. A major section of this chapter is devoted to the exploration of how to monitor volcanoes, including monitoring

DOI: 10.1201/9781003354321-12

methods, monitoring systems, and targets of monitoring. How to retrieve the gaseous components of volcano plumes quantitatively using the best images forms the content of Section 12.5. Finally, this chapter elucidates how to map lahar flows and assess lahar hazards using various types of remote sensing data and analytical methods.

12.2 IDENTIFICATION OF ACTIVE VOLCANOES

An active volcano is defined as one that is currently erupting or is about to erupt as judged by local seismic activity. One way of determining whether a volcano is dormant or active is to check its temperature. A Crater Lake emitting hot steams is a sure sign of strong geothermal activities. Through monitoring the lake water temperature and its temporal variation, it is possible to predict whether an eruption is imminent. Active volcanoes are best identified from satellite images that are acquired periodically as the satellite orbits around the Earth all the time. These images are acquired over different portions of the spectrum. Each type of imagery has its unique strengths. Optical images are best at identifying volcanic ash and plumes ejected into the atmosphere from the vent. This kind of identification faces three limitations: (i) not all volcano eruptions are accompanied by ash and plumes. Some explosive volcanoes just eject lava and/or tephra locally. Their identification has to rely on thermal images; (ii) the spatial resolution of the images may not be sufficiently fine to resolve the volcano itself as most Earth observation satellite images have a spatial resolution ranging from a few meters to tens of meters, much broader than the volcano dimension. Even the volcano plumes may not be sufficiently extensive to be identifiable on the images unless the eruption is extremely violent; and (iii) the satellite images may have a temporal resolution that is insufficiently fine to capture the eruption. Most Earth observation satellites have a revisit period of more than ten days (Table 1.6), inadequate to capture fast-evolving volcano eruptions. Images of a finer temporal resolution are available from meteorological satellites, but their spatial resolution on the order of 100s m or coarser allows identification of only extensive volcano plumes. Hourly images have a spatial resolution even coarser, so detailed tracking of all active volcanoes at a fine time interval (e.g., hourly) from space is still not feasible at present.

An erupting volcano raises the local geothermal heat flux during volcanism. Such strong thermal activities, naturally, are best studied from thermal images. Just like the identification of wildfires, active volcanoes are identified based on their thermal anomalies. The spatiotemporal variation of the heat flux can be detected and quantified from space-borne images (Wright and Flynn, 2004). Daytime imagery is subject to the influence of solar radiation that is strongly reflected by some ground features such as snow and bare rock, causing them to have a spectral signature resembling that of an active volcano. Thus, night-time SWIR images of a medium resolution such as Landsat OLI are preferable. They enable the detection and monitoring of thermal anomalies associated with active volcanoes, but not their precise location and extent due to the scanty details exhibited on nighttime images (Bhattacharya et al., 1993). Nevertheless, thermal anomalies and the associated radiant fluxes of active volcanoes are more accurately detected from nighttime images than their daytime counterparts. In addition, ASTER imagery with a temporal resolution of 16 days also serves as a valuable data source for updating significant (>10 m) topographic measurements at active volcanoes via time-series DEMs (Stevens et al., 2004). Coarse-resolution MODIS (250 m) and GOES data may prove futile in studying volcano eruption, but they can still be useful in tracing volcanic ash plumes and their evolution owing to their hourly and half-hourly temporal resolution. So it is the medium-resolution Earth resources satellite images (Table 1.6) that have been commonly used to study volcanoes.

The ease of detecting thermal emissions associated with active volcanism depends on three factors: eruption duration, lava spatial extent, and lava temperature. A long-lasting erupting volcano is easier to identify than a transient one. If a volcano eruption lasts a few days, then it can be captured on multi-temporal images several times. The persistence of the same thermal anomalies at the same spot can confirm that indeed an active volcano has been detected. Such anomalies are easily distinguishable from a forest fire that lasts up to hours before it spreads to adjoining areas. A volcano, especially its crater, may have a dimension much smaller than the image's spatial resolution, but it is still identifiable even on coarse-resolution images because of the intense radiative heat emitted. In the worst case, an active volcano may appear just as a thermal anomaly comprising a few contiguous pixels. Thus, it is impossible to delimit the precise spatial extent of an active volcano (e.g., the lava flow path), let alone monitor its change over time, a task that is ideally accomplished from optical images. Even on optical images, a smaller, cooler volcano is much harder to detect than a larger, hotter volcano. In general, it is much easier to identify basic lava flows than hydrothermal activity.

Since a volcano can never match the pixel size of coarse-resolution images (e.g., 250 m of MODIS) in dimension, its monitoring has to rely on the detection of hotspots at the sub-pixel level. Thermal hotspots represent the location and presence of eruptive events (vents). A sub-pixel hotspot may cause only a slight increase in radiant energy in longwave bands, but a much more substantial increase in shortwave bands (Wright et al., 2002). Thus, SWIR bands are more suited to detect such hotspots than their longwave TIR (e.g., 11–12 μm) counterparts. High-temperature volcanic activities emit the most energy at 4 μm, whereas the ambient Earth's surface emits its peak radiation at 11–12 μm. These two wavelengths are most commonly used to identify hotspots. An erupting volcano can be detected automatically from MODIS bands 22 and 32 based on the normalized thermal index (NTI) calculated as

$$NTI = \frac{\rho_{22} - \rho_{32}}{\rho_{22} + \rho_{32}} \tag{12.1}$$

where ρ_{22} and ρ_{32} = reflectance in band 22 of a wavelength around 4 μm, and band 32 of a wavelength of 11–12 μm. In case band 22 is saturated, it may be replaced by low-gain band 21 (Wright et al., 2004). This index works with night-time MODIS images only. With images from other satellites, the number of the two bands will change, but not their approximate wavelengths.

The identification of active volcanoes via thermal anomalies is commonly implemented as thresholding of a single thermal band, in a manner identical to fire identification. For instance, a threshold of NTI > −0.8 is able to detect active and high-intensity eruptions (but not low-intensity ones). The use of a single threshold is disadvantaged in that it may not always yield the most reliable detection outcome. The detection can be made more accurate via consideration of contextual information, such as examination of the threshold within a local window. Additional constraints may be posed to eliminate false positive pixels from potential hotspots. For instance, the initially identified hotspot pixels based on a single threshold within a window are confirmed as hotspots if they meet the following two conditions (Ganci et al., 2011):

$$BT_{3.9} > MaxVar\left(BT_{3.9}\right) + \min\left(BT_{3.9}\right) \tag{12.2}$$

$$BT_{3.9} > Mean\left(BT_{3.9}\right) + n \cdot std\left(BT_{3.9}\right) \tag{12.3}$$

where $BT_{3.9}$ = brightness temperature at 3.9 μm, MaxVar, mean, and std = the maximum variation, average, and standard deviation of $BT_{3.9}$ within the calculation window.

The alternative to a single threshold is the use of detailed models of lava flow surface temperature, such as the dual-band model involving SWIR and TIR bands. Hotspots can be detected from these bands using the dual-band or three-member methods (Hirn et al., 2008). The former assumes that the lava field consists at most of two endmember thermal components (e.g., hot cracks within a thermally homogeneous crust) at the sub-pixel level on high-resolution satellite images. Under this assumption, the radiant values of a hotspot pixel are considered to be the weighted average of radiant energy emitted by the two distinct components, a hot component (e.g., the flowing lava with a temperature of about several hundred degrees) and a cool background. If the hot component having a temperature of T_h comprises the f_h fraction of the pixel, the crust component with a temperature of T_c has a fraction of $1 - f_h$, and three IR channels are unsaturated, the following relationships exist among T_h, f_h, and T_c:

$$R_{x,thermal} = \tau_x \left[f_h R(\lambda_x, T_h) + (1 - f_h) R(\lambda_x, T_c) \right] \quad (12.4)$$

$$R_{y,thermal} = \tau_y \left[f_h R(\lambda_y, T_h) + (1 - f_h) R(\lambda_y, T_c) \right] \quad (12.5)$$

where R_x and R_y = the radiance recorded in bands x and y, respectively; λ_x and λ_y = the wavelength of the two bands used; $R(\lambda_x, T_h)$ and $R(\lambda_y, T_c)$ = spectral radiance ($W \cdot m^{-2} \cdot sr^{-1} \cdot \mu m^{-1}$) emitted by the hot component (cool component) at wavelength λ_x (λ_y). The dual-band model is perfectly suited to medium-resolution Landsat images that may be too coarse to encompass multiple thermal components within a pixel, such as a cold background, a lava lake, or a flow consisting of cooled crust and hotter surface cracks (Pinkerton et al., 2002).

The radiant power emission (Q) or heat loss, for a presumably two-component pixel, is resolved using Stefan–Boltzmann law and derived from the contribution of each component to the overall radiant heat-flux of the pixel as

$$Q = \sigma \varepsilon A \left[f_h T_h^4 + (1 - f_h) T_c^4 \right] \quad (12.6)$$

where Q = radiant power emission (W), σ = Stefan–Boltzmann constant (5.67×10^{-8} W m^{-2} K^{-4}), ε = surface emissivity, and A = pixel area (m^2).

Of particular notice, this dual-band method involving two thermal components may represent an overly simplification and is utterly insufficient to resolve the major thermal properties of the lava field, such as the continuous variation in temperatures over some lava flow surfaces. They are better studied using the three-member model (Hirn et al., 2008). It requires a minimum of five to seven endmembers to make fractional areas sufficiently representative of the main characteristics of the temperature field (Wright and Flynn, 2003). In addition to models, lava hotspots can also be detected using an existing system such as HOTSAT. This volcano hotspot detection system can yield estimates of time-averaged lava effusion rate. As expected, such detections can reveal temperature anomalies more accurately than thresholding.

12.3 SENSING OF LAVA TEMPERATURE AND FLOW

After a volcano has erupted, lava may be ejected out of the vent near the crater and flow down the cone flank. Lava flows can be detected from various types of remote sensing data, such as space-borne radar images. In addition, InSAR data can also be used to detect surface change as with landslides. Nevertheless, it is optical images that are much better suited to study lava flows and map their surface temperature. With a temperature ranging between 200°C and 1,100°C, they radiate so significant quantities of heat that can be easily detected from space-borne images, such as ETM+, ASTER, and even MODIS data.

12.3.1 Retrieval of Lava Temperature

There are two types of temperature, kinetic and radiant. The former refers to the energy stored by an object and manifests as molecular motion. The latter is the temperature detectable by a thermal sensor remotely. From the radiant temperature T_R, the kinetic temperature (T_k) of a lava flow is calculated as

$$T_R = \epsilon^{1/4} T_k \quad (12.7)$$

where ε = emissivity. The above equation applies to the blackbody only. For real objects, there should be a coefficient. After the ambient temperature is raised by volcanism, an active volcano

TABLE 12.1

Comparison of Field Measurement and Remote Sensing Means of Sensing Lava Flow Temperature

Method	Ground Measurement	Images
Strength	Detailed, accurate	Field view of lava flow temperature
Limitations	Unsafe to be close to the target	Not so accurate
Best use	Small flow	Extensive flow, near-real-time

emits prodigious amounts of radiation that can be captured by a thermal sensor remotely, either on the ground or airborne. In the field, lava flow temperature can be sensed using either a hand-held spectroradiometer or a portable infrared radiometer at a precision of ±1°C (Pinkerton et al., 2002), or using a forward-looking infrared thermal imaging camera. Such a field method can yield the true surface temperature of lava flows precisely and reveal surface thermal complexities that are unable to be resolved on air- or space-borne images (Table 12.1). Temperatures over the range of 175°C (cooling)–1,050°C can all be measured in situ, but the measurements have to be undertaken some distance from the lava due to safety concern. This ground-based method is workable during the daytime, causing the measurement results to be subject to the effects of sunlight, and the viewing distance. It is almost impossible to delimit the exact spatial extent of each zone of the entire lava flow fields because of the obstructed view inherent with ground-based observations. This method is not safe if tephra and lava are splashed around. Nor can it generate a holistic field view of temperature distribution, especially during an active eruption. So it is not covered here. Instead, this section focuses on the retrieval of lava temperature from different spectral bands (the theoretical grounding of how to invert at-sensor radiance to obtain lava surface temperature has been covered in Section 11.2 already).

Most of the existing satellites are not designed specifically for mapping volcanic thermal activities during effusive eruptions. These thermal activities are ideally studied from TIR bands. However, TIR images recorded over the 8–14 μm wavelengths are suited to study thermal anomalies of a low temperature (e.g., coal fires) because they tend to saturate at a high temperature. For instance, TM TIR band 6 saturates at 49°C to a maximum of around 110°C, much lower than the temperature of lava flows. Hot lava flows are better studied from SWIR bands. They are able to detect pixel-integrated temperatures up to 1,000°C or even higher. Some SWIR bands are so sensitive to the radiant energy that a thermal feature can be detected from a coarse spatial resolution image, even if it makes up a small fraction of a pixel. This is because the thermal anomalies show up much larger than their actual size on an image if their temperature reaches 970–990°C. At such a high temperature, the peak radiation shifts to a shorter wavelength, as stated in Wein's displacement law, enabling the detection of high temperatures from NIR bands that saturate at 973°C. Thus, temperatures higher than this threshold need to be detected from a spectral band of a shorter wavelength, such as visible light spectral bands. If the lava flows exhibit a huge range of temperature after cooling, it may be necessary to map a specific temperature range from a single band and map the whole range from multi-bands, and then merge the results from all individual bands to form the overall temperature field of lava flows.

Lava surface temperature can be quantified from remotely sensed images using Eq. 12.7. The quantification process is complex and requires tremendous work. A general procedure of retrieving temperature from satellite images comprises four main steps:

i. Pre-processing of the thermal data to isolate the volcanogenic signal and discount the atmospheric effects;
ii. Data calibration to convert direct current to R_λ;

iii. Compensation for the absorbed surface-leaving radiance by the atmosphere. The atmosphere is composed of gaseous matter that inevitably absorbs some radiance leaving the volcanic surface. The degree of absorption is a function of wavelength. Moreover, the amount of absorption varies with the gaseous components, such as Ch_4, CO_2, water vapor, ozone, CO, N_2O, O_2, and aerosols. Absorption can be determined via the MODTRAN atmospheric model. It is able to correct the atmospheric effects that vary with wavelength, so SWIR and TIR bands are corrected separately, namely

$$R_{SWIR} = \left[L(\lambda_{SWIR}) - R_R(\lambda_{SWIR})\right] / \tau\varepsilon \tag{12.8}$$

$$R_{TIR} = \left[L(\lambda_{TIR}) - R_R(\lambda_{TIR})\right] / \tau\varepsilon \tag{12.9}$$

where R_{SWIR} and "R_{TIR}=" the corrected spectral radiance at wavelengths λ_{SWIR} and λ_{TIR}, respectively. $L(\lambda_{SWIR})$ and $L(\lambda_{TIR})$ = the spectral radiance received at the sensor, R_R = the spectral radiance contributed by the atmospheric gaseous matter, τ = atmospheric transmissivity, and ε = surface emissivity (it can be set to 0.97). The radiance value (L_i) is related to the original digital number of pixels in a spectral band.

iv. Compensation for the contaminated radiative signal by reflected sunlight if sensing is undertaken during the daytime. (This step is redundant if nighttime imagery is used.) The reflectance of adjacent "cold" lava pixels (e.g., lava not being active) is used for the pixel-wise correction. The ability to isolate the volcanogenic signal and the accuracy of isolation depend on the spectral wavelength range of the thermal band used, time of image acquisition, surface temperature, and the physical size of the volcano.

The easiest way of determining the approximate temperatures of active and cooling lava flows is to make use of inverted Planck's equation from spectral radiances captured on nighttime optical images and the calibration constants given in Table 11.1. The accuracy of temperature retrieval is affected by a number of factors, the most important being saturation of pixel values, pixel size, and the exact value of ε that is a function of surface roughness, composition, and temperature. Whatever emissivity value is adopted, it inevitably affects the accuracy of radiometer-measured temperatures. Finally, atmospheric conditions, such as haze, vapor, and cloud cover, also exert an influence.

In retrieving temperature from Landsat ETM+ bands, it must be borne in mind that thermal radiance from the active lava channel behaves differently in band 4 (0.45–0.90 μm) from band 8 (0.52–0.90 μm). In band 8, it concentrates in the upper reaches of its bandwidth (Wright et al., 2001). Caution must be exercised if a wideband such as ETM+ 8 is used to retrieve surface temperature, as the assumption that the measured radiance is evenly distributed around the central wavelength of the bandwidth is no longer valid. Consequently, ETM+ 8 produces a temperature of 922°C (28.3 $W{\cdot}m^{-2}{\cdot}sr^{-1}{\cdot}\mu m^{-1}$) in whole pixel temperature that is more than 100°C higher than 819°C (35.7 $W{\cdot}m^{-2}{\cdot}sr^{-1}{\cdot}\mu m^{-1}$) retrieved from ETM+ 4. In addition to ETM+, lava surface temperature has also been quantified from the at-sensor radiance of Landsat 8 OLI and ASTER (band 14) images. The temperature retrieved from OLI green, red, panchromatic, and NIR bands has a slightly different range among themselves (Figure 12.1).

The thermal domain within a lava flow field can be differentiated using the thermal eruption index (TEI) derived from SWIR and TIR bands (Aufaristama et al., 2018). It is based on the relativity of the SWIR spectral radiance (R_{SWIR}) to the TIR spectral radiance (R_{TIR}) on the crust ($R_{SWIR} < R_{TIR}$) and on the active lava ($R_{SWIR} > R_{TIR}$), calculated as

$$TEI = \frac{R_{SWIR} - \dfrac{R_{TIR}^2}{10R_{SWIR-MAX}}}{R_{SWIR} + \dfrac{R_{TIR}^2}{10R_{SWIR-MAX}}} \frac{R_{TIR}^2}{\left(\dfrac{R_{SWIR-MAX}}{3}\right)^2} \tag{12.10}$$

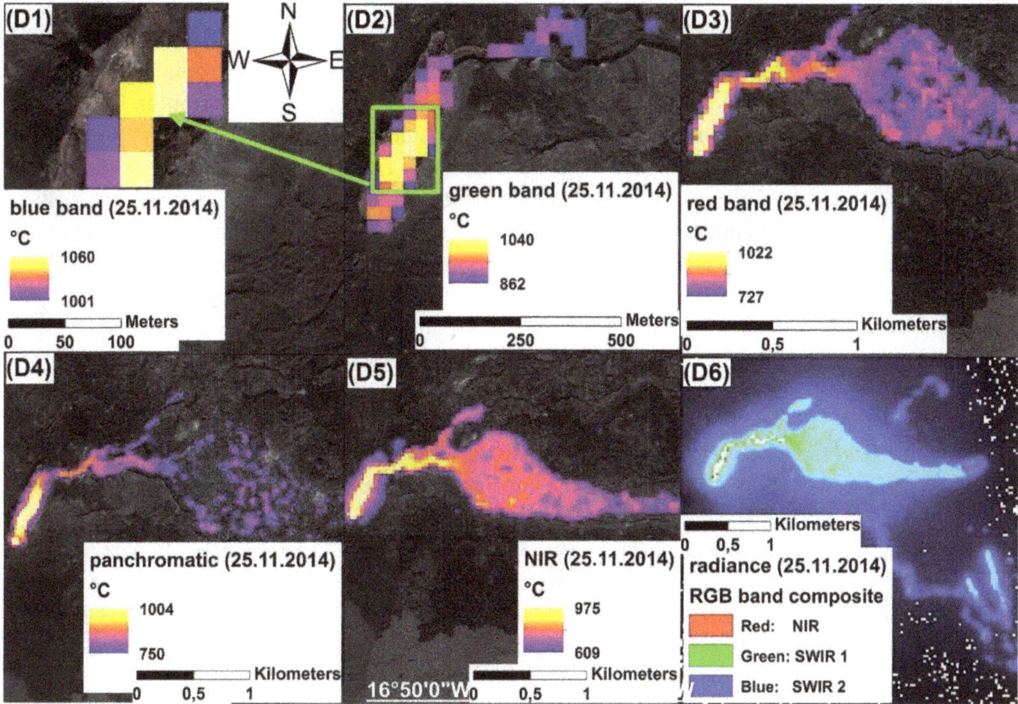

FIGURE 12.1 Distribution of surface temperature of lava flows in Holuhraum, Iceland detected from various NIR and visible light bands of Landsat 8 OLI images. (Source: Nádudvari et al., 2020.)

where R_{SWIR} and R_{TIR} = pixel-corrected spectral radiance captured in Landsat 8 OLI bands 6 and 10, respectively, and $R_{SWIR\text{-}MAX}$ = the maximum spectral radiance recorded in band 6. A TEI value >0.10 indicates hotspots and produces encouraging estimates of hotspot anomalies during an eruption. A TEI value between 0.10 and 0.51 distinguishes the active lava domain and the crust domain surrounding active lava within the lava flow field. TEI is capable of detecting hotspots, and a sub-pixel temperature of 1,096°C in an area of 3.05 m^2, about 1/30 of the pixel size.

The retrieval of sub-pixel temperature of cool and hot components is possible with the dual-band method involving two infrared bands to formulate two equations from the simultaneous solution of Planck equation in each band as shown in Eqs. 12.4 and 12.5. This dual-band method works only if the SWIR and TIR bands are available. This method enables the retrieval of the crust and the hot crack temperature for active lava flows. Besides, this method is also able to determine crust thickness (Δh), convective flux, and radian flux, all from respective models.

12.3.2 Mapping of Lava Flows

If a volcano eruption is accompanied by lava flows, it is much more hazardous. Thus, it is important to map lava flows to forecast the areas to be inundated. Lava flows may be mapped via in situ surveys, but this method is unable to delimit their spatial extent wholly with either ground instruments or even with drone-based aerial surveys if the volcano is violently spewing out ash around an erupting volcano. On the other hand, the airspace may be closed for emergency response. This leaves space-borne sensing as the only viable option. Because of their low spatial resolution, satellite images may have to be combined with the ground-based method using a very long-range (e.g., >3.5 km) terrestrial laser scanner to generate detailed results, though. Lava flows can be mapped from a wide range of satellite data using manual and automatic classification methods. Active lava flows are easily mapped from various medium-resolution images due to their high temperature, but cooled lava flows may not be so easily differentiated from previous lava deposits. These images include optical Sentinel-2 that has 13 spectral bands at three resolutions of 10, 20, and 60 m. Of these bands, those with the finest resolution (10 m) are the most appropriate for mapping lava flows. If multiple eruptions have taken place in the past, the most recent lava flow fields can still be discriminated from older ones from multi-temporal images. At least one of them must be captured before and another after the eruption.

Lava flows are mapped from remote sensing imagery based on their spectral properties. After the lava has been deposited for some time, its spectral behavior changes slightly. The glassy crusts of lava flows are degraded chemically and physically in the process of weathering. The physical breakdown of surface chill coats, the emergence and erosion of silica-rich coats, the oxidation of magma minerals, and the advent of vegetative cover can all be detected from the changed reflectance spectra. Such changes in lava spectral behavior offer the possibility of mapping lava flows and inferring their age from visible and reflective infrared bands (Abrams et al., 1991). Precise determination of age, though, may require calibration of image color.

The detection of lava flows from non-imagery LiDAR data differs drastically from image-based detection. It relies solely on surface elevation. A comparison of the elevation before and after an eruption reveals recent lava flows from the old ones through the overlay analysis of terrestrial LiDAR-derived elevation data (James et al., 2009). Although this method is rather straightforward and fast, it does face a number of issues, one of which is the low point density at long ranges. In particular, the high oblique view and non-uniform point density make it difficult to properly interpret lava features unless the LiDAR-yielded DEM data are fused with thermal images. Even so, only recent flows are differentiated from old ones without the ability to ascertain the age of flows. This method may be practical for studying flows confined to the flank of a volcanic cone due to the limited ground view from the scanning position, or multiple flows situated on a steep slope visible from a strategic position (see Figure 3.15). Naturally, this method should not be used when the lava flows are active.

Given that lava flows are detected based on surface height change, the consideration of topography via a DEM is conducive to achieving accurate mapping. If the DEM resolution is rather fine (e.g., 2 m produced from Pléiades tri-stereo images), the subtraction of one pre-event DEM from another post-event DEM reveals the spatial extent, thickness, and volume of change (Ganci et al., 2019). In the subtraction, any difference in elevation between the two DEMs is assumed to be caused by lava flows. Whenever this assumption is violated, the mapped lava flows suffer from inaccuracy. This topography-based comparison can only map those flows that have occurred in the interim. It is unable to map those flows that already exist in the pre-event DEM.

The philosophy, strengths, and limitations of each type of remotely sensed data in mapping lava flows are summarized and compared in Table 12.2. DEM comparison is the simplest but can only

TABLE 12.2

Comparison of Four Major Means of Mapping Lava Flows

Method	DEM Comparison	Terrestrial LiDAR	Optical Image + DEM	Time-Series SAR
Philosophy	Change in elevation is caused by lava flows	Construction of 3D lava field	Flows of different ages have different spectral and topographic properties	Flows of different ages have different backscattering properties
Strength	Simple, yielding information on flow extent, thickness, volume of lava	3D surface, fine-resolution results similar to DEM comparison	Able to differentiate flows by age	Able to identify lava flow age
Limitations	Only old and new lava flows differentiated	Same as DEM comparison, blind spots rife	Only the most recent flows are fully identifiable	Absence of correlation due to steep terrain
Best use	Single event eruption	Single lava flow over a steep terrain	Multiple flows deposited over long time	Multi-deposit over a gentle terrain devoid of vegetation

detect the volume, spatial extent, and thickness of lava deposits associated with a single eruption. The same is true with LiDAR data except that their spatial resolution can be much finer than DEM grid size, allowing minor deposits to be detected. Terrestrial LiDAR data have a variable point density and may contain extensive blind spots where no data are recorded (see Figure 3.15). Optical imagery is good at detecting the spatial extent of multi-deposits of past eruptions based on the mineral content and thus image color. If combined with a DEM, it is able to yield volumetric information on lava flows, as well. SAR imagery is the most versatile in that it can reveal the age of different deposits based on the changed backscattering properties of lava flows. It is best used to detect multi-deposits over a gentle terrain devoid of vegetation as it can interfere with the radar backscattering properties.

The quickest method of mapping lava flows is to automatically classify imagery and topographic data using machine learning methods that are good at fusing multi-source data (see Section 1.4.2 for more details). Better accuracy is expected by using two images, one before the eruption and another after (Figure 12.2). Consequently, the evidence of decision-making is augmented to as high as 13, including pixel values in four spectral bands of one pre- and two post-event Sentinel-2 images, plus a DEM (Corradino et al., 2019). The mapping generally follows four major steps: pre-processing, unsupervised classification, Bayesian neural network (BNN) classification, and post-processing. Pre-processing aims to filter out plumes and clouds in preparation for the classification. Unsupervised clustering analysis identifies pixels that belong to recent lava flows and non-flow pixels, such as snow, clouds, shaded and bright areas. BNN is able to differentiate older flows from newer ones. As illustrated in Figure 12.2, the mapped lava flow closely matches the outline of the observed lava flow (white line) at an accuracy of 88%, with a positive predictive rate of 88%, and a true positive rate of 86%. One important source of misclassifications is identified as topographic shadow cast by the steep flanks of a volcano on which lava flows. Such misclassifications can be removed through a DEM screening. Another source of misclassifications is the medium resolution (10 m) of the images used, which can be minimized by using a DEM of a much smaller grid size (e.g., 2 m).

Apart from the age of lava flow, the mapping can also concentrate on lava composition, which is best studied from hyperspectral images. Their rich spectral information enables the subtle spectral response of minerals in the lava field to be resolved via linear spectral mixing. Airborne hyperspectral data allow the identification of 15 endmember from a volcano field over the range of 0.40–2.50 μm (622 channels) through linear spectral mixing analysis (Aufaristama et al., 2019). They fall into six groups of surface cover, including basalt, oxidized surface, sulfate mineral, and hot material with a

FIGURE 12.2 Lava flow of the 2017 Etna eruption mapped from two post- and one pre-event Sentinel-2 images and a DEM using the neural network method, in comparison with the contour of the observed lava flow (white line). (Source: Corradino et al., 2019.)

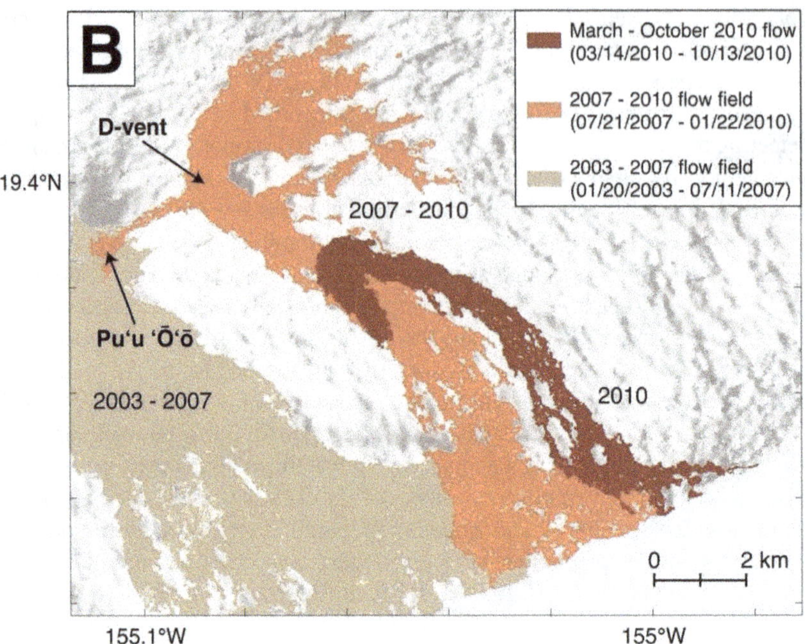

FIGURE 12.3 Lava flow field activity on the east rift zone of Kīlauea, Hawai'i from July 2003 to October 2010 over three periods mapped from InSAR coherence. (Source: Dietterich et al., 2012.)

mean overall mapping accuracy of 79% (Kappa = 0.73). The accuracy of individual classes ranges from a minimum of 70% for basalt to a maximum of 93% for sulfate. Spectral mixing analysis is unable to reveal the age of lava flows, though.

Mapping of lava flows into age groups is possible using time-series SAR images or a single optical image in conjunction with topographic data. Given that recent lava flows radically change the scattering properties of the lava surface, and decorrelate its radar signal in EnviSat SAR coherence images, lava flows of different age groups are best differentiated from InSAR coherence images independent of look angle or satellite path (Dietterich et al., 2012). The SAR coherence-based lava flow mapping results bear striking resemblance to the field maps but show more details on the internal surface structure of lava flows (Figure 12.3). From these maps, further information on lava flows can be derived, such as flow lengths, widths, and areas, all of which are useful to calculate the flow advance rate. These maps also indicate the locations of active flow paths. It is worth noting that the mapped flow paths may not be fully continuous due to the decorrelation caused by steep slopes. As with the use of all InSAR data, the coherence-based mapping of lava flows is vulnerable to the presence of vegetation where it is less successful than with bare lava flows. The achievement of better mapping requires the exclusion of vegetation-covered areas from analysis using a vegetation mask, together with non-volcano activity-caused decorrelation. Better mapping of flow age is possible only with the use of multiple overlapping images to create the coherence images as they allow noises to be filtered out.

Although SAR images are not subject to cloud contamination, the mapping of lava flows from time-series SAR images requires a huge amount of time spent on processing a large number of images. So they are less preferable to optical images. If an area is full of large lava flows spanning over a long time (e.g., a century) and they have contrasting ages, these flows can be mapped from a single optical image like Landsat 8, plus topographic features derived from a DEM using machine learning-based OBIA (Li et al., 2017). The inputs permissible in the classification are diverse, ranging from topography, geometry, to texture (Table 12.3), in addition to image properties. These inputs enable individual flows (a total of 21) to be mapped at an accuracy ranging from 76.4% to 87.7%, depending on whether topographic features are incorporated into the mapping (Figure 12.4).

TABLE 12.3

Features Commonly Utilized in OBIA of Lava Flows from Optical Multispectral Images

Imagery (e.g., Landsat 8)	Topography	Geometry	Texture[a]
Mean.b1	Mean.Elevation; Std.Elevation	Area	Homogeneity for b1, b3–b7
Mean.b4; Std.b4	Mean.Slope	Compactness	Contrast for b3–b7
Std.b5	Mean.Aspect (Categorical)	Density	Angular second moment for b5
Mean.b6			Mean for b3–b7
Mean.b7; Std.b7			Std for b3–b7
Mean.NDVI, Std.NDVI			Angular second moment (GLDV) for b2
Mean.BT; Std.BT			
Max.Diff			

Note: Std – standard deviation, b – spectral band

[a] All based on GLCM unless specified

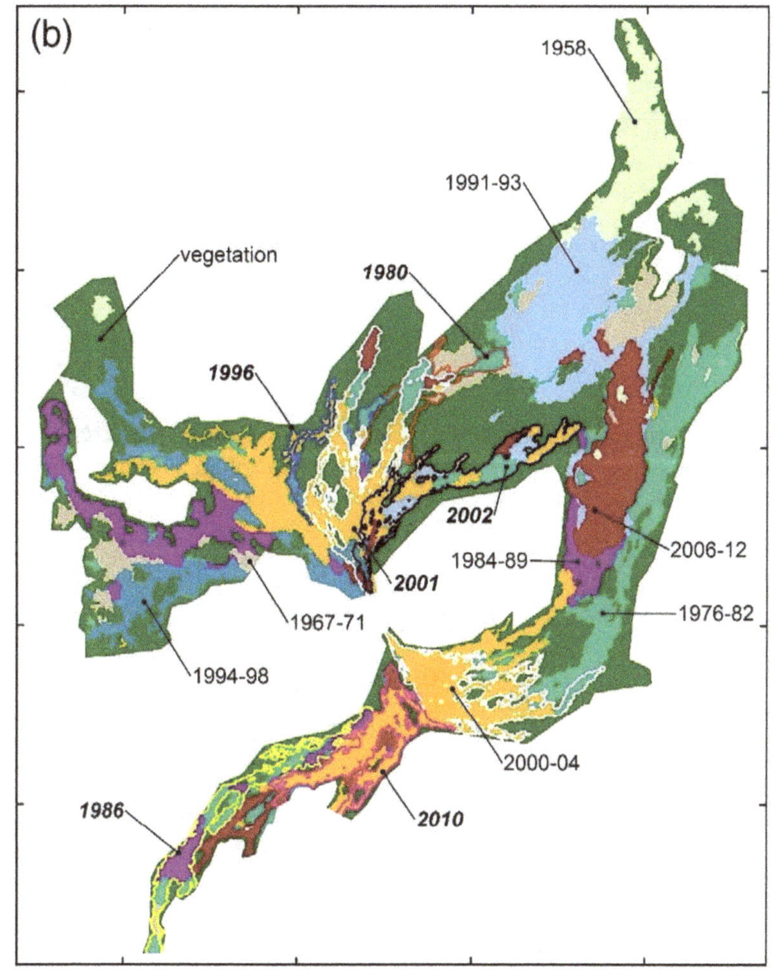

FIGURE 12.4 Distribution of lava flows by age group in Nyamuragira, Congo where more than 20 overlapping lava flows have been deposited over the last century. The map is produced from OBIA of one Landsat 8 image and 39 topographic features using Random Forest. (Source: Li et al., 2017. Used with permission.)

The accuracy rises from 84.6% to 91.3% if the individual flows are classified into only eight age groups. In general, the incorporation of topographic features in the classification improves mapping accuracy markedly.

12.3.3 Estimation of Lava Eruption and Effusion Rates

After lava flows have been mapped from multi-temporal images, lava eruption rates can be estimated from them. If the images have a short revisit period, the lava eruption rate can be estimated many times during one episode of eruption. With such information, it is possible to simulate the eruption. Realistic lava flow simulations require accurate information on total erupted volume and flow thickness. Such information is also indispensable to map lava inundation hazards realistically and refine physical models that depend on these lava flow characteristics. After lava flow parameters have been calculated from SRTM and TanDEM-X DEMs, they can be input into a volume-limited lava flow emplacement model. TanDEM-X DEMs provide precise lava flow measurements to constrain input parameters for numerical modeling of lava flows (Kubanek et al., 2015). Model performance (e.g., underestimation of flow runout distance) is highly dependent on lava thickness.

The estimation of lava effusion rate requires the determination of T_c and T_h based on their relationship with the radiance value (R) of pixels in an IR band. The radiance is calculated as the fraction-weighted (f) average of the sub-pixel radiance emitted by the three distinct temperature components of hot (h), cool (c), and background (b), or

$$R_{\lambda,thermal} = \tau\left[f_h R(\lambda, T_h) + f_c R(\lambda, T_c) + (1 - f_h - f_c) R(\lambda, T_b)\right] \tag{12.11}$$

After T_h and T_c have been resolved using the above equation, the remotely sensed radiant flux Q_R is computed as

$$Q_R = A\epsilon\sigma \sum_k T_e^4 \tag{12.12}$$

where σ = Stefan–Boltzmann constant, ε = emissivity, A = pixel surface area. k = the total number of thermal anomalous pixels. T_e = the effective radiant temperature, calculated from T_h and T_c as $\sqrt[4]{fT_h^4 + (1-f)T_c^4}$. The lava effusion rate E_R is calculated from Q_R as

$$E_R = \frac{Q_R}{D_{lava}\left(Cp_{lava}\Delta T_{stop} + C_L \Delta\phi\right)} \tag{12.13}$$

where $\Delta T_{stop} = T_h - T_{stop}$ (T_{stop} = the temperature at which the lava stops flowing, D_{lava} = lava density; Cp_{lava} = the specific heat capacity, $\Delta\phi$ = the mass fraction of crystals grown, and C_L = the latent heat of crystallization of involved lavas).

If multi-temporal images are used, the same analysis can be performed on them. Analysis of time-series images is able to reveal the temporal variation of the effusion rate. Instant lava effusion rates are commonly retrieved from cloud-free optical images using the computing routines of MyVOL and MyMOD (Hirn et al., 2008). MyVOL is devised to jointly exploit multi-payload observations of thermal features for operationally monitoring eruption rates of effusive volcanoes rapidly and robustly. MyMOD is founded on the quantitative sub-pixel IR remote sensing theory, and the dual-band and three-endmember methods. The eruption rates are derived from remotely sensed SO_2 (gas) and electromagnetic potential (very low frequency). They are highly consistent with volcanological ground truth from approximately 1 $m^3 \cdot s^{-1}$ up to the peak effusion rates of 1.2–1.3 million $m^3 \cdot day^{-1}$. Such accurate and timely information about lava effusion rates and their changes with time is crucial to predicting the distance of lava flows and the time it takes to reach it. It is possible to estimate the lava cooling rate based on the spatial distribution of temperature within the lava field, as well.

12.3.4 Estimation of Lava Volume

Accurate information on the total volume of erupted lava and lava thickness is significant to the realistic simulation of lava flows. Lava volume can be estimated using two methods, planimetric and topographic. In the former method, the lava flow area is measured from remote sensing images, and multiplied by lava thickness determined from ground surveys. The accuracy of the estimated volume is subject to uncertainties inherent in both area mapping and thickness measurement (Albino et al., 2015). In situ measured thickness at lava edges cannot capture the overall thickness and its spatial variability. Since lava thickness naturally varies spatially over the underlying terrain, this method is rather imprecise and unreliable. In comparison, the topographic method is more accurate as it is based on the surface elevation before and after the eruption, but it is also more complex methodologically.

Topography-based estimation of lava volume is almost identical to that of landslide debris volume using Eq. 3.3 except that lava always elevates the original surface. Height differences are detected by subtracting the pre-eruption DEM from the post-eruption DEM (and even co-eruptive DEM), and this difference is always positive. If the two DEMs do not have the same grid size, they have to be unified through resampling. The detected topographic changes are assumed to be caused solely by lava deposits. This differencing reveals not only lava flow fields but also the thickness of lava deposits and the lava emplacement area (Figure 12.5), from which the total volume of the

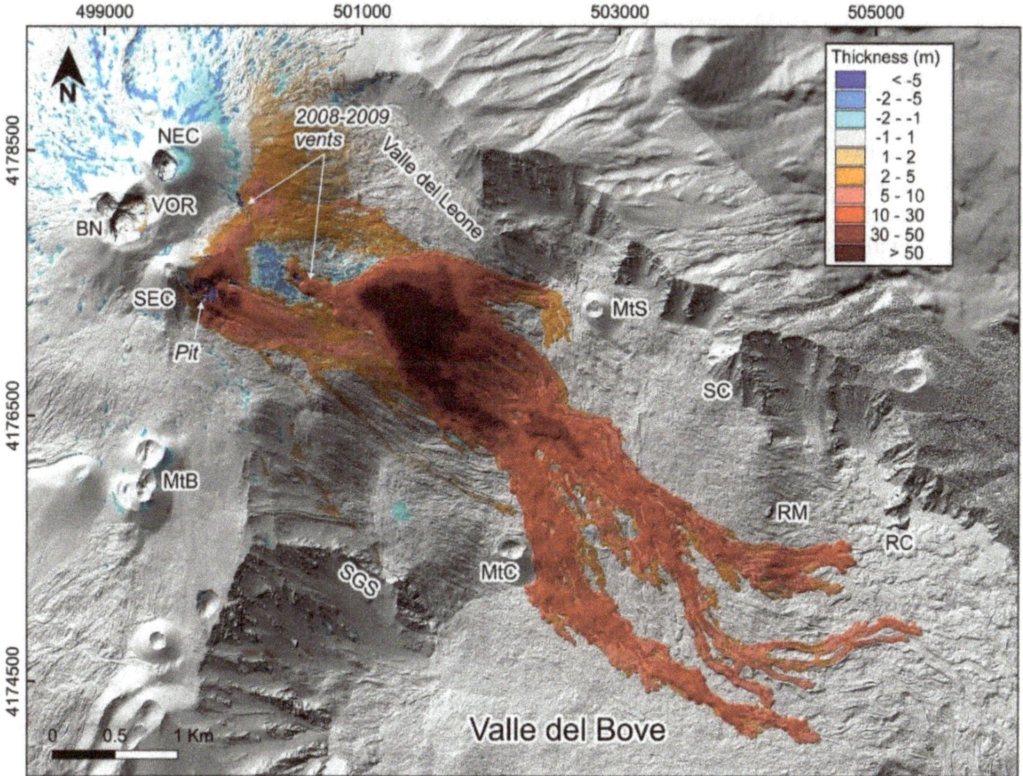

FIGURE 12.5 The thickness of lava flow of the Etna volcano detected from differencing of DEMs created from multi-temporal LiDAR data. Negative differences mean ablation (e.g., melting of snow and ice). Positive differences represent lava deposits over the previous terrain. (Source: Behncke et al., 2016.)

lava flow is calculated, and the rate of flow estimated if the eruption duration is known. Existing DEMs can be sourced from SRTM data and ASTER GDEM of a grid size of 30 m globally. An ASTER GDEM achieved an RMSE of 10 m over a volcanic landscape in the absence of significant atmospheric water vapor (Stevens et al., 2004). This accuracy is not sufficiently high to warrant the generation of a reputable estimate of lava volume, so DEMs from other sources are necessary. Precise quantification of lava volume requires fine-resolution DEMs because lava flows are confined to deeply incised gullies. A finer resolution DEM is conducive to achieving a more reliable estimate, such as ArcticDEM for estimating lava flow volume in Arctic regions. The repeated coverage and high spatial resolution (2 m) of ArcticDEM make it vital in estimating lava volume in remote regions (Dai and Howat, 2017).

Outside the Arctic region, DEMs have to be produced from contours on a topographic map, LiDAR and DInSAR data, and space-borne Earth Observation satellite images (see Section 1.3.4). Topographic maps are readily accessible and easily converted to a DEM by tracing the contours digitally. However, contours are drawn at a coarse elevational interval of 10 or even 20 m. Besides, topographic maps are second-generation data that are less accurate than the original aerial photographs or VHR images from which they are produced. They are competent at estimating the volume and shape of thick lava flows (>10 m mean thickness) based on a contour interval of 25 m (Stevens et al., 1999). This threshold may be lowered if the contour interval is smaller. In tracing contours, a sampling distance of 2 mm on a 1:25,000 topographic map is translated into a horizontal increment of 25 m on the ground, a resolution too coarse to map lava flows. Better estimations can be achieved from original remote sensing data directly.

Fine-resolution DEMs may be produced from LiDAR data directly. However, pre-eruption LiDAR data are unlikely to be in existence unless an active volcano has been repeatedly monitored already. In this case, DEMs have to be produced from aerial photographs and even satellite images. Photo-produced DEMs have a spatial resolution ranging from 0.13 m from 1:5,000 photos to 0.2 m from 1:8,000 photos (Baldi et al., 2008). Their mean vertical inaccuracies vary between 0.27 m with large-scale photos and 0.44 m with small-scale photos. These inaccuracies drop to 0.05 and 0.09 m, respectively, under the assumption of 1/5 of the minimum size after correlation algorithms. Validated against kinematic GPS data, 1 m DEM constructed from 1:5,000 stereoscopic aerial photographs taken with a WILD RC20 film camera (ground pixel size = 0.12 m) has a discrepancy of ±0.50 m with a standard deviation of 0.18 m (Baldi et al., 2002). The DEM from high-resolution stereo camera data has a much larger standard deviation of 0.71 m due probably to its much coarser spatial resolution of 0.50 m in the stereo mode. Despite the high accuracy, the photogrammetric method of producing DEMs is rather complex and time-consuming so does not allow volcanoes to be monitored timely.

In addition to aerial photographs, DEMs have also been produced from VHR tri-stereo images of the Pléiades-1 satellites launched in December 2011 (1A) and a year later (1B) (Bagnardi et al., 2016). These images have a spatial resolution ranging from 0.5 m (panchromatic band) to 2 m (multispectral bands). The stereoscopic triplet is a novel and innovative mode of image acquisition for 3D measurements not available from previous VHR satellite missions such as WorldView and Quickbird. The triplet comprises three nearly simultaneously acquired images, one backward viewing, one forward viewing, and one near-nadir viewing. These images increase point cloud density by a factor of 6.5 over conventional stereo images to over 90 million points per stereo pair while the number of pixels without height measurements (e.g., pixels falling inside topographic shadow) is reduced by 43%.

The accuracy of the estimated lava volume is affected primarily by the DEM resolution, and secondarily by its positioning accuracy in DEM differencing. The smaller the DEM grid size, the more authentically its value can capture the elevation of the pixel-covered area. DEM accuracies vary widely, depending on their sources (Table 12.4). Heights in the Pleiades-1 DEM have a standard deviation of 0.51 m in comparison with GPS data. This value is less than half (1.12 m) achieved

TABLE 12.4
Mean Height Difference and Standard Deviation between DEMs from Various Sources and GPS Heights at 19 Ground Control Points

DEM (Pixel Size)	Mean Height Difference (m)	Standard Deviation (m)
Pleiades-1 (1 m)	−2.84	0.51
TanDEM-X[a] (5 m)	−0.10	1.12
SRTM (30 m)	−3.50	3.64
ASTER GDEM (30 m)	−8.56	5.74

Source: Bagnardi et al. (2016).

[a] TanDEM-X DEM has been corrected using a linear polynomial function calculated using height differences with GPS at the ground control points

from 5 m TanDEM-X, at least seven times smaller than that from 30 m SRTM and ASTER GDEM. The height of vegetation also exerts an impact on the estimated lava volume, especially in vegetated areas. Naturally, the estimated lava volume is less reliable in vegetated areas than in bare lava fields. The estimation accuracy may be improved by excluding all the areas covered by vegetation from the estimation.

12.3.5 Mapping of Volcano Hazards

Volcano hazards stem from ash, lava flows, and possibly lahars. Volcanic ash from explosive eruptions can be monitored from coarse-resolution meteorological satellite images. A reliable assessment of eruption hazards requires the separation of volcanic ash from clouds. Defined as the lava flow inundation scenario, lava flow hazard is evaluated from a number of parameters, including the discharge (effusion) rate indicative of the current intensity and potential magnitude (volume). Thus, it is judged by hotspot location and density, lava thermal flux, and effusion rate, all of which can be output from the HOTSAT volcano hotspot detection system based on space-borne thermal infrared data (Del Negro et al., 2016). Developed by Ganci et al. (2011), HOTSAT is a multi-platform system for automatically monitoring volcanic thermal anomalies. It comprises three sub-components: data pre-processing, detection of volcano hotspots, and radiative power estimation based on real-time MODIS data in combination with SEVIRI data. Both types of imagery data are complementary to each other in their spatial and temporal resolutions. All satellite images are geo-located and calibrated in pre-processing. Smaller and less intense thermal anomalies are detected from MODIS data. The detection is updated every 15 minutes using the SEVIRI data. Volcano hotspots are identified using variable thresholds based on contextural information within a window. The output from the HOTSAT system can be input into the MAGFLOW model to simulate the lava flow path.

Lava hazard is ideally simulated via the Internet-based GIS framework called LAV@HAZARD. This user-friendly framework is based on the Google Maps application programming interface. It comprises four modules for simulating the spatiotemporal evolution of hotspots, radiant flux, effusion rate, and visualization (Vicari et al., 2011). The interactive environment visualizes volcanic hazards with a database of ca. 30,000 lava flow simulations and performs real-time scenario forecasting. This information system has a high degree of interactivity, easily readable maps, and allows fast exploration of alternative scenarios.

Table 12.5 compares field-observed lava flow features with the simulated lava flow characteristics and those derived from DEMs and satellite images (Del Negro et al., 2016). In general, the field method underestimates the lava flow thickness, and grossly underestimates the volume of lava

TABLE 12.5
Comparison of Lava Flow Features Derived from Field Observation, DEM Differencing, MODIS, and SEVIRI Images with Numerical Simulations

Lava Flow	Field	DEM[a]	Images	Simulation
Length (m)	4,400	4,100	4,336	3,560
Front altitude (m a.s.l.)	1,460	1,500	1,500	1,600
Maximum thickness (m)	50	88	n.a.	82
Average thickness (m)	16	22.1	n.a.	24.6
Area (m^2)	2,496,000	2,822,086	2,886,839	2,591,725
Volume (m^3)	40,000,000	62,374,307	63,828,240	63,828,240
Eruption rate ($m^3 \cdot s^{-1}$)	2.5	3.9	4.0	4.0

Source: Del Negro et al. (2016).
Note: n.a. – not applicable
[a] DEM is produced using the photogrammetric method at 5 m

flows, even though the area of lava flows is almost the same as remotely sensed results regardless of the method used. The estimated volume of lava flows and eruption rate hardly vary among DEM, imagery, and simulation, but the field method underestimates the volume by more than 1/3 due likely to the edge of the lava flow fields being thinner than the interior lava.

12.4 MONITORING OF VOLCANIC ACTIVITIES

12.4.1 Monitoring Methods

Volcano monitoring refers to imaging the same volcano(s) repeatedly so as to detect how it evolves over time. To some degree, it represents multi-temporal sensing of volcanism. This monitoring is ideally carried out using satellite imagery owing to its synoptic view of the event from space. Monitoring of active volcanoes focuses on those that have erupted in the recent past. The target of monitoring is variable and diverse, ranging from an eruption event, the spatial extent of lava emplacement, level of volcanic activity, lava flow rate, lava morphological change, to effusion rate. If an eruption is accompanied by ash and gases, they can be monitored easily from optical imagery, only at a coarse spatial resolution. The frequency of monitoring at a high resolution is achievable by merging multi-sensor data, but they are seldom real-time if space-borne.

Real-time monitoring is achievable only from field observations, located far or near the volcanic vent, and from airborne images remotely. Terrestrial monitoring is usually carried out using GPS, while airborne monitoring involves the use of a helicopter or a drone. GPS data are now routinely used to monitor crustal deformation in volcanic fields (Anzidei et al., 1998). Volcanoes can also be monitored using semi-permanent GPS (SPGPS). SPGPS stations are flexibly and quickly deployed in response to changing volcanic activities, adjusted based on the observed activities (Dzurisin et al., 2017). GPS-based monitoring of volcanoes is safe and feasible only when they are dormant. This monitoring is implemented by logging 3D coordinates at strategic positions and comparing the coordinates logged in different survey epochs. This comparison reveals both subsidence and deformation of the lava surface. This method is practical only at a certain number of points. Such a limitation can be overcome with the use of a helicopter or drone.

Helicopter-based monitoring can be carried out flexibly, frequently, and repeatedly. Helicopters are good at acquiring oblique thermal images that allow the creation of high-resolution (~1 m) geo-referenced thermal maps of an active lava flow field. Usually, the images are converted to thermal maps to display the distribution of and characterize lava breakouts through time, from

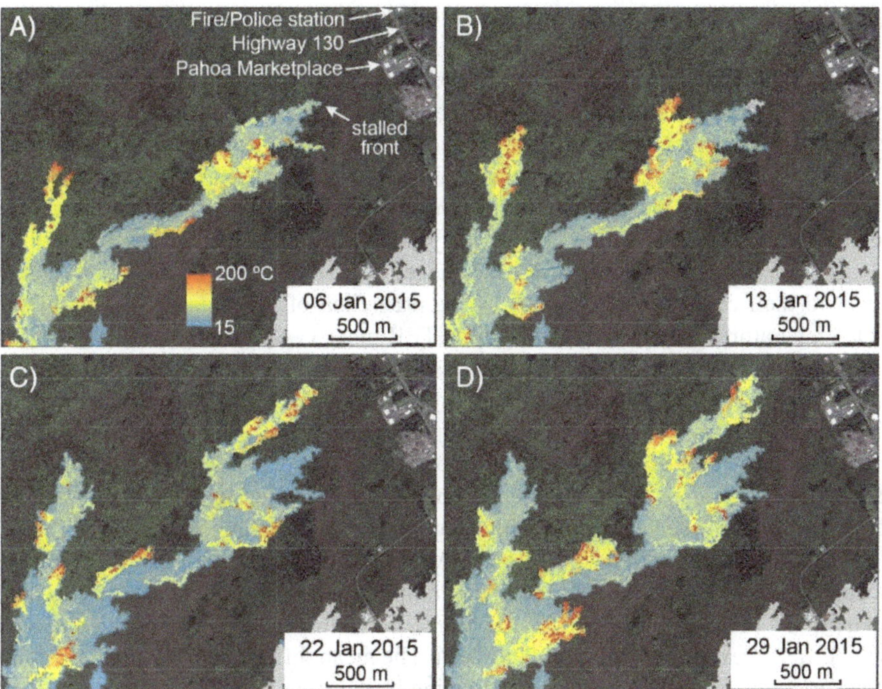

FIGURE 12.6 Weekly monitoring of the Kīlauea Volcano lava flow, located in the Big Island of Hawai'i, (A) against the background of a very high-resolution image over 6 Jan 2015, (B) 13 Jan 2015, (C) 22 Jan 2015, and (D) 29 Jan 2015. (Copyright: 2014 DigitalGlobe NextView license) using photos taken with a hand-held FLIR Systems SC620 camera. The thermal maps illustrate the changing distribution of surface breakouts during January. Gray areas: portions of the 27 June flow not covered by the thermal map. (Source: Patrick et al., 2017. Used with permission).

which subsurface lava tubes may be delineated (Figure 12.6). Owing to the fine details exhibited on thermal images, they can reveal the components of the lava field, such as the vent, master tube, distributary tubes, and surface breakouts (Patrick et al., 2017). If the monitoring period is long enough and synoptic, these thermal maps can be used to explore the pace of changes (gradual vs abrupt) and study lava flow behavior and process. Nevertheless, the thermal images may not have sufficient resolving power to estimate lava effusion rates competently.

Alternatively, monitoring may target the active front of slowly advancing lava flows solely using a low-altitude UAV quickly and repeatedly (Turner et al., 2017). UAV is a promising technology for improving response efforts following a devastating eruption. It can serve as an effective tool for mapping short-term and long-term morphological changes in a lava flow field and assessing its hazards based on flow parameters such as size, mass flux, and surrounding topography. The airborne images taken with a 16.1 megapixel Canon IXUS 127 HS camera enable bare earth DEMs to be produced at 1 m. They also allow more lava sheds to be detected than coarse-resolution DEMs and show the flow direction and flow accumulation, as well as the steepest terrain along which future lava flows are likely to follow. This method of monitoring is the best for studying changes in topography and deformation. The monitoring of the front position (length) and extension of flows (areas) has to rely on fine-resolution satellite images from which lava flow inundation scenarios (hazard) can be forecast via numerical simulation (Del Negro et al., 2016).

Imagery-based monitoring is the most flexible and can cover a huge ground area owing to the synoptic view. The full lava field is observed from one image simultaneously, yielding critical quantitative information on the volcano that no other methods can provide. The temporal interval of monitoring depends on the satellite revisit period. MODIS images fully cover the globe at a temporal resolution of one to two days at spatial resolutions of 250–1,000 m. Two MODIS spectral bands, one short-wavelength at 4 μm and another long-wavelength at 12 μm, are optimal for detecting emitted radiation. These traits make them an excellent source of data for automatically detecting and monitoring high-temperature volcanic thermal anomalies.

If carried out from space-borne images, the monitoring can be quantitative. Quantitative monitoring refers to the derivation of time-series information on changes in the volcano properties, including effusion rates. This estimation, however, is possible only with synergetic multi-sensor data, such as ASTER and MODIS. This synergy takes advantage of high-resolution SWIR-TIR ASTER and low-resolution MIR–TIR MODIS images (Hirn et al., 2008). The monitored lava effusion rates and their changes with time allow the prediction of the distance of lava flows. High-resolution temperatures may be mapped from ASTER imagery that is also an ideal source of data for sensing the accurate location of active portions of lava flows and determining the monthly effusion rate. In comparison, MODIS day- and night-time data can shorten the monitoring intervals to up to six hours.

Monitoring of volcanic activities may focus on volcano-triggered surface change or deformation that may be a precursor to an imminent eruption. This monitoring is ideally implemented via time-series DEMs in the same way as monitoring landslide deformation, except that the deformation takes place on the flanks of a volcano cone. Alternatively, deformation may be monitored from stereoscopic aerial photographs using the photogrammetric method, GPS data, or InSAR images. A comparison of the coordinates of photogrammetric control points measured on a stereo pair reveals the differences in coordinates and position. The mean difference in planimetric coordinates shows the horizontal displacement. The differencing of elevation at the same spot in the two DEMs reveals the vertical deformation. Such information is then used to infer the process of emplacement of lava flows and erosion and reveal if the volcano is in equilibrium. All the detection methods based on DInSAR and its variants in detecting landslide deformation apply to volcanoes except the type of deformation for volcanoes being more diverse, such as tube breakout.

12.4.2 Monitoring Systems

Since a volcano eruption may be an ongoing process, remote sensing-based systems have been developed to monitor and even forecast eruptions. A good monitoring system should be able to detect active volcanism at the hexameter scale repeatedly as eruption intensity can fluctuate at a temporal scale shorter than an hour. It should be reliable (false positive minimized), and the results are delivered to the public quickly or in near real-time. Several systems have been developed to monitor volcanoes from space (Table 12.6). The monitoring programs are operated by national governments or universities. Some of them are designed primarily to monitor volcano hazards in certain parts of the world or a certain number of volcanoes in particular. For instance, the Volcano Hazards Program (https://volcanoes.usgs.gov) monitors volcanoes in the US territories only. More monitoring systems are operated and maintained by research institutions. For instance, MODVOLC (http://modis.higp.hawaii.edu) is maintained by the Hawai'i Institute of Geophysics and Planetology. This open-access satellite-based system is able to monitor and catalog volcanic thermal activities around the world in near-real-time. Elevated levels of thermal emission are detected at the pixel/sub-pixel scale. The system displays the accurate location of volcano hotspots.

Developed by the University of Torino and the University of Firenze, the Middle InfraRed Observation of Volcanic Activity or MIROVA (www.mirovaweb.it) is another system for detecting near-real-time volcanic hotspots from MODIS MIR images. Currently, it monitors 215 volcanoes. The third monitoring system is HOTVOLC (http://hotvolc.opgc.fr) designed to monitor volcanic thermal activity, lava effusion rates, volcano plumes of SO_2, and ash based on IR and UV spectroradiometer data from geostationary satellite images (e.g., SEVIRI). Developed at the Observatoire de Physique du Globe de Clermont-Ferrand (France), it monitors ~50 volcanoes worldwide. Dissimilar to the aforementioned systems, SARVIEWS (http://sarviews-hazards.alaska.edu)

TABLE 12.6

Common Operational Space-Based Volcano Monitoring Systems with a Global Focus.

System	Sensor Used				Characteristics of Monitored Target			
	Data	**Bands**	**Revisit**	**Resolution**	**Algorithm**	**Key Products**	**Availability after Sensing**	**No. of Volcanoes**
MODVOLC	MODIS	MIR–TIR	6 hours	1 km	Spectral	N hotspot (pixels) Radiant flux	<12 hour	>1,000
MIROVA	MODIS	MIR–TIR	6 hours	1 km	Contextural	Thermal – VRP, NTI Lava – TADR	<4 hour	215
HOTVOLC	MSG-0 (SEVIRI)	MIR–TIR	15 minute	3 km (nadir)	Contextural	Lava – time series TSR, N hotspots, VFR Ash – BTD-5, altitude, mass, concentration, plume, area SO_2 – BTD map, area	<30 minute	~50
NASA Global SO_2	Aura, Sentinel-5P	UV	1 day	13 × 24 km OMI 50 × 50 km OMPS 3.5 km × 7 km TROPOMI		SO_2 – concentration, mass	<2 day OMI-OMPS >7 day TROPOMI	>1,000
SARVIEW	Sentinel-1	C-band	6–12 day	3.1 × 14.1 m (IW)	DInSAR, change detection	Deformation, reflectivity (RTC, change map)	≤24 hour	Variable
LICS	Sentinel-1	C-band	6–12 day	3.1 × 14.1 m	DInSAR	Deformation (wrapped and unwrapped interferogram)	<2 weeks	>1,000
Univ. Miami InSAR Viewer	Sentinel-1	C-band	6–12 day	3.1 × 14.1 m	InSAR	Ground displacement and deformation	2–3 day	~10
MOUNTS	Sentinel-1	C-band	6–12 day	3.1 × 14.1 m	DInSAR, AI	Deformation/reflectivity, wrapped and unwrapped interferogram, coherence	≤24 hour	~20
	Sentinel-2	VIS-SWIR	5 day	10 m VIS, 20 m SWIR	Contextural	SWIR image B12-B11-B8A, N hot pixels	≤12 hour	
	Sentinel-5P	UV	1 day	3.5 km × 7 km		SO_2 mass and concentration map	≤6 hour	

Source: Valade et al. (2019).

Note: Some of them may have incorporated ground-based components, only the satellite-based data are shown here

provides a monitoring service run by the University of Alaska Fairbanks based on the analysis of SAR images. If an external alert system informs the system of an ongoing hazardous event, such as volcanic eruptions, the system will start processing data and produce 6- and 12-day wrapped interferograms and radiometrically corrected terrain images at a swath width of ~250 km. On-demand higher-level analysis such as change detection can be posted to the system via the Hybrid Pluggable Processing Pipeline (HyP3).

Monitoring Unrest from Space (MOUNTS) is the first operational platform for monitoring global volcano activity using multi-sensor, freely available space-borne images, including Sentinel-1 SAR, Sentinel-2 SWIR, and Sentinel-5P TROPOMI, ground-based seismic data from Global Earthquake Catalogues GEOFON and USGS global earthquake catalogs. It makes use of open-source toolboxes and has a modular architecture for data query, downloading, processing, and result dissemination. Convolutional neural network is plugged into the workflow to automatically detect large ground deformation via differencing interferograms derived from Sentinel-1 SAR images. It offers near-real-time information on surface deformation, heat anomalies, SO_2 gas flux from Sentinel-5P, and local seismicity essential for assessing volcanic risk at an interval of 6 –12 days. Results are visualized in geocoded images, together with time-series parameters, openly accessible on the Web (www.mounts-project.com). At present, they enable a comprehensive understanding of the temporal evolution of volcanic activities of 17 volcanoes in various volcanic and climatic settings across the globe. The system has a flexible design that allows it to be adapted to monitor new targets (Valade et al., 2019).

12.5 SENSING OF VOLCANIC PLUMES

Before (and during) an active volcano erupts, it may emit gases and eject a huge quantity of ash and gases (mostly SO_2 and methane) into the atmosphere. These toxic gaseous substances are suspended in the air for days, traveling a long distance and reaching the upper troposphere and stratosphere. As the fundamental drives of volcanic activity in the magmatic system, the gases can be used to gauge the intensity of an imminent eruption. It is imperative to monitor volcanic plumes and detect their route of travel so that flights can be diverted to safer airspace and their impacts on climate assessed. The sensing of high plumes requires the use of space-borne images that have a broad spatial coverage. In this way no matter in which direction the plumes are oriented, they can always be captured in one frame of imagery. Satellite-based surveillance is a cost-effective way of monitoring how the plumes are dispersed from their source of origin. Since the plumes may change their orientation with the wind direction, fine temporal resolution images are essential to study their evolution.

Volcanic plumes are studied in two ways, time-sequential monitoring of change over broad scales and quantification of gaseous constituents. These tasks are implemented quite differently using different types of sensors (Figure 12.7). Because time-sequential monitoring is not really remote sensing, and cannot yield information on the spatial distribution of the monitored substances, it is not discussed further. Instead, attention is focused on image-based quantitative retrieval of plume components.

12.5.1 Suitable Images

Dissimilar to the sensing of solid targets that relies on the radiative energy reflected and emitted by the targets, the sensing of gaseous components of a volcanic plume is based on the measured direct absorption spectra of solar radiation. This sensing is commonly carried out using UV and IR radiations. UV sensing is not subject to the influence of water vapor in the atmosphere. Although IR sensors have been used, they are much less common than UV sensors. Sensing of volcanic gases is ideally achieved using high spectral resolution satellite images acquired from three tailored sensors: Global Ozone Monitoring Experiment-2 (GOME-2), the Ozone Monitoring Instrument (OMI), and Infrared Atmospheric Sounding Interferometer (IASI), in addition to generic multispectral satellite

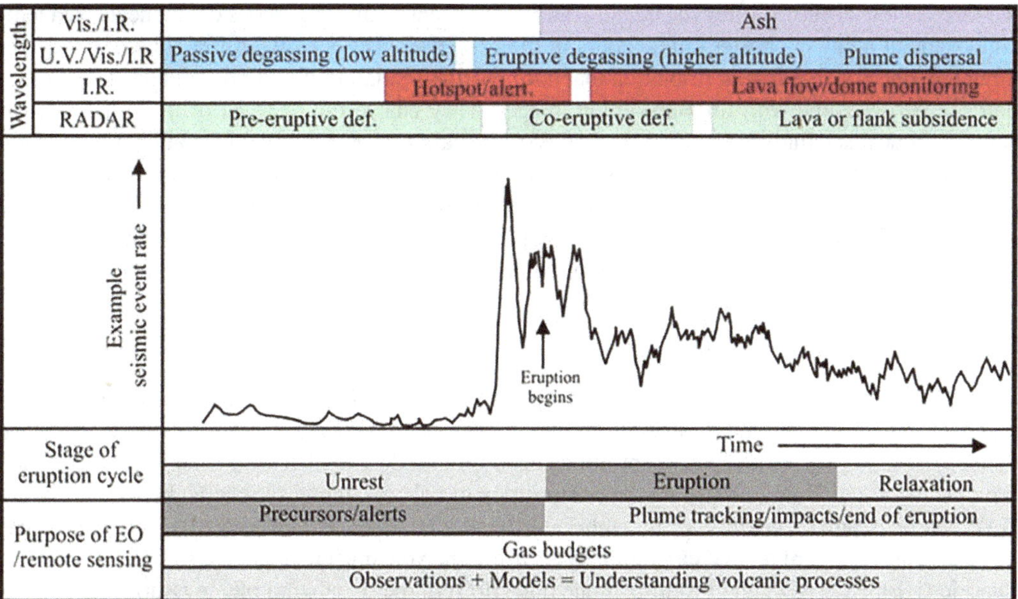

FIGURE 12.7 Best spectrum for sensing different volcanic activities and the eruptive cycle of best sensing them. The seismic signal shown is based on the seismic events seen at Mt. St. Helens (March–May 1980). (Source: Pyle et al., 2013.)

images such as the Terra-Aqua MODIS. GOME-2 is a UV–vis spectrometer aboard the MetOp-A satellite, operating over the spectrum of 0.24–0.79 μm with a spectral resolution of 0.2–0.5 μm. It senses the solar radiation backscattered by the atmosphere and reflected from the Earth's surface in nadir viewing. GOME-2 allows the retrieval of a variety of trace gases (e.g., O_3, NO_2, SO_2, H_2CO, CHOCHO, H_2O, BrO), and aerosol parameters. As a payload of the NASA's Aura satellite, the OMI (UV–vis) sensor is designed to study plume composition. Launched in 2004, this nadir-viewing imaging spectrograph measures atmosphere-backscattered radiation over the UV–vis spectrum of 0.27–0.50 μm at a spectral resolution of 0.5 μm. It is equipped with charge-coupled device detectors that acquire images at a 2,600 km swath width and a spatial resolution varying from 13 km × 24 km at nadir to 13 km × 128 km at the edges of the swath. Owing to the use of multiple wavelengths, OMI allows low-level SO_2 to be retrieved in a footprint of 13 km × 24 km. OMI is a valuable data source for monitoring volcanoes to complement airborne monitoring. It is able to capture explosive SO_2 emissions at a high temporal resolution that can be used to study volcanic processes (Lopez et al., 2013).

The hyperspectral IASI sensor aboard MetOp-A is primarily a meteorological sensor, acquiring global nadir measurements twice a day at 09:30 am and 09:30 pm with a small to medium footprint from a 12 km diameter circle at nadir to an ellipse of 20 km × 39 km at the maximal swath. This state-of-the-art sensor offers an uninterrupted spectral coverage from 645 to 2,760 cm^{-1} at a spectral resolution of 0.5 cm^{-1} with noises mostly smaller than 0.2K below 2,000 cm^{-1}. It is rather versatile, allowing the global retrieval of a host of trace gases, such as CO_2, H_2O, CO, O_3, HCOOH, CH_3OH, NH_3, and CH_4 on a daily basis, and HONO, C_2H_4, SO_2, and H_2S on a more sporadic basis (Taylor, 2019). This sensor is able to determine the height of volcanic ash. The sensing of SO_2 is achieved from IASI images in three absorption bands, the v1 band around the 8.5 μm atmospheric window, the strong v3 band around 7.3 μm, and a combination band of v1 and v3 at around 4 μm.

All satellite observations provide discrete time-series column gaseous mass present in the atmosphere, with a temporal resolution varying with the orbiting parameters and swath width of the sensor.

All three tailored sensors allow the quantification of concentrations of gaseous components on a temporal scale of a day or shorter (a few hours in the best case) (Theys et al., 2013). However, they are limited by their coarse spatial resolution that does not allow small plumes to be studied. This limitation can be overcome with finer resolution images. They enable the retrieval of fluxes from small plumes. Optical satellite images such as ASTER TIR bands of 90 m spatial resolution are excellent for studying small-scale passively degassing volcanoes (Henney, 2012). ASTER imagery enables the detection of volcanic plumes with an SO_2 column concentration of 0.5–10 g·m^{-2}. This imagery is more sensitive to SO_2 than MODIS bands. No matter which type of remote sensing imagery is used, only the strongest sources are detected due to the limited image spatial resolution and/or sensitivity at present.

12.5.2 Sensing of Plume Composition

In studying plume composition, the satellite images acquired from the aforementioned tailored sensors are limited by their coarse spatial resolution, even though they are suitable to study extensive plumes. Besides, the retrieved concentrations need ground data to check for their accuracy, which prompts the use of UAV in studying the composition of volcanic plumes (Schellenberg et al., 2019). UAV can be deployed flexibly, or fly right into the plume. Portable and rugged spectroscopic gas and ash sensors aboard a UAV can remotely measure the volcanic enhancements of CO_2, hydrogen fluoride (HF), HCl, SO_2, and bromine monoxide (BrO) in the downwind plume (Butz et al., 2017). The sensing of CO_2, HF, and HCl is based on SWIR imagery captured using a Fourier transform spectrometer EM27/SUN, while SO_2 and BrO are sensed using the UV-based Differential Optical Absorption Spectroscopy spectrometer. The spectral retrieval windows for these two gases are 0.312–0.327 μm, and 0.331–0.352 μm, respectively. The two spectrometers allow volcanic CO_2 enhancements to be detected with good confidence and the concentration of volcanic HF, HCl, SO_2, and BrO to be quantified remotely at high precision.

Of all volcanic gases, SO_2 is arguably the most readily sensed constituent via absorption spectroscopy due to its low background concentration in the atmosphere (e.g., 1 ppbv in clean air), and the strong and distinctive structures in its absorption spectrum at both UV and TIR wavelengths. In contrast, CO_2 concentration is tricky to retrieve accurately because it has a large and well-mixed atmospheric background. Remote sensing retrieval of volcanic CO_2 vertical column densities enhancements (ΔCO_2) requires compensating for the changing altitude of the sensor and background concentration variability. Thus, it is necessary to simultaneously measure the overhead oxygen columns and assume covariation of O_2 and CO_2 with altitude. The atmospheric CO_2 background can be determined by identifying background soundings via the other co-emitted volcanic gases.

Since spectroscopy requires sunlight to function, this method is operational only during the daytime when the sky is cloud-free. In spite of a large number of gaseous components sensed, it must be emphasized that this method fails to yield a spatial perspective on the concentration distribution of these gases as the sensors are not imaging. They can help to sense the content of a plume but are unable to show the spatial variations of plume coconcetrations. This deficiency can be overcome with data from imaging sensors.

The retrieval of gaseous constituents of a volcanic plume from satellite imagery may be implemented using various methods, including box, traverse, Delta-M, and modeling (Table 12.7). The box method derives a volcanic flux by treating the SO_2 mass as contained within a circle or a box whose dimensions are the total distance traveled by the plume in one day (Lopez et al., 2013). This distance is determined using a trajectory model or radiosonde wind profile. The loss of mass is taken into account by correcting the age dependency ($e^{t/\tau}$, τ = e-folding time). The plume age (t) is defined as the ratio of the distance between a given pixel and the volcano (d) and the wind speed (v) or $t = d/v$. The daily flux is then simply calculated by dividing the mass within the box by the number of days. This method is suitable for high-altitude plumes because low plumes' gaseous components have fast kinetics and quickly disperse below the detectable threshold. It is limited to cases where the age correction is well defined and accurate.

TABLE 12.7
Comparison of Four Methods Used to Retrieve Gaseous Flux in a Volcanic Plume from Satellite Imagery

Method	Box	Traverse	Delta-M	Inverse Modeling
Nature of plume	Young, high resolution	Young, high resolution	Long-lasting, high-altitude	All with complex vertical structures, mostly long-lasting
Wind field data	Needed	Needed	Not needed	Needed
Wind velocity	Constant	Not constant	n.a.	Not constant
Wind direction	Constant	Constant	n.a.	Not constant
Image overpasses	Single	Single	Multiple	Single
Plume height	Needed	Needed	Needed	Not needed
Plume lifetime knowledge	Assumed	Assumed/negligible if gained close to the volcano	Calculated	Calculated/online estimated
Lifetime correction	High resolution	High resolution	Resolution of the overpass	High resolution
Flux time resolution	One flux/overpass	High/user defined/pixel constrained	One flux/overpass	High/user-defined
Best for	First evaluation	Near real-time, comparison with ground data	Quick conversion from total mass time series	A posteriori analysis
Limitations	Heavy reliance on wind speed, flux lifetime, plume coverage	Heavy reliance on wind speed	Heavy reliance on plume lifetime, complete plume coverage	Time-consuming, strong dependence on accurate wind field data, possible misfits
References	Lopez et al. (2013)	Merucci et al. (2011)	Krueger et al. (1995)	Eckhardt et al. (2008)

Source: Theys et al. (2013).

In the traverse method, the mass of the gaseous component is integrated over the vertical column. Under a constant wind field, the flux through a traverse is expressed as

$$F(d) = \left(\sum_i VC_i l_i\right) v \cdot cos\theta \cdot e^{\frac{d}{v \cdot \tau}} \tag{12.14}$$

where VC_i = the column quantity of the *i*th pixel, l_i = horizontal length of the *i*th pixel of the profile, v = wind speed, θ = the angle between the profile direction and the transport direction. It can be determined automatically by seeking the pixels around the volcano that have the maximal column amount for simple wedge-shaped plumes. An automatic traverse definition can also be easily implemented based on the interpolation between the atmospheric wind direction profile, collected in the area of interest, and the plume altitude. Multiple traverses at increasing distances from the volcano can be used to reconstruct the flux history up to several hours before the satellite overpass on the basis of the simple relationship among transport speed, distance, and duration (Merucci et al., 2011). Eq. 12.14 assumes a single wind speed over the whole plume area and passive advection of the gas in a constant wind field, which can introduce significant uncertainties, especially if the plume has a large dimension.

The delta-M method relies on time-series SO_2 mass obtained from successive satellite overpasses based on the mass conservation equation:

$$\frac{dM(t)}{dt} = F(t) - k \cdot M(t) \tag{12.15}$$

where M = total mass contained in the plume, F = the volcanic SO_2 flux, and $k{\cdot}M$ = the SO_2 loss term that takes into account actual chemical processes (gas-phase oxidation, dry-and-wet deposition), loss from transport, but also the dilution of the outer part of the plume below the detection limit of the satellite imagery, presumably independent of position. Eq. 12.15 is solved analytically in two ways. The first method assumes a constant flux over the time interval (t) between two consecutive mass estimates, M_i and M_{i-1}. This solution requires the whole plume to be covered by one image fully, which may be problematic for very large plumes. The use of a constant k might introduce a systematic error to the retrieved total mass as it varies with a few intrinsically varying parameters (Krueger et al., 1995). The second approach is to fit an analytic function to the observed time-series mass. The time-dependent flux is then straightforwardly obtained by applying Eq. 12.15 to the fitted curve. This method has the advantage of being completely independent of the wind field and is thus applicable even if the plume is sheared in a complex wind field or if the plume is stagnant around the volcano in a low wind environment. This method is limited in that it can yield only first-order estimates of the flux as time-series fluxes are either too smooth or may contain spikes (Theys et al., 2013).

All the above three methods are functional with plumes of a single altitude so that the errors related to plume height are reasonably small to be ignored. Besides, the plume must have a simple geometry and is fully covered by one satellite image. For multi-layered plumes that do not meet these requirements, the inversion modeling approach must be used to estimate the vertical profile of SO_2 emission (Eckhardt et al., 2008). This method couples satellite-observed total SO_2 columns with an atmospheric transport model such as FLEXPART to yield the position and shape of the volcanic plume that is affected by changing winds with altitude. The method finds the vertical emission distribution that minimizes the total difference between the simulated and observed SO_2 columns while also considering *a priori* information. Mathematically, the modeled concentration **y** from the corresponding **x** observations is expressed as

$$\mathbf{y} = \mathbf{M} \cdot \mathbf{x} \tag{12.16}$$

where **M** = forward matrix of m (observed values) × n (number of unknowns or source elements). It expresses the source–receptor relationships that are calculated using the FLEXPART Lagrangian dispersion model (http://transport.nilu.no/flexpart). **M** is determined via FLEXPART-based simulation of the transport of air masses, in which several emission scenarios (e.g., release intervals in altitude and time) can be specified, but the manner of ejection is ignored, even though it is easy to identify the composition of plumes from explosive eruptions but harder at a lower altitude as they become thinner using decorrelation stretch from ASTER images (Henney, 2012). Simulation output can be specified at the predefined altitude grid and latitude/longitude grid. For each emission cell, the simulation is performed with a unit mass source assuming no chemical processes (tracer run). It may be necessary to interpolate the model output in space and time to closely match the satellite observations and then convert the results to an atmospheric column by integrating the modeled profiles. Eq. 12.16 is solved via MLR. But if the number of observations available is insufficient, they will not provide a sound solution. Instead, it is better to solve it using the Bayesian method based on measurements, the expected measurement errors, best guesses of the target state vector prior to measurement, and the associated covariance matrix. It may be necessary to apply a plume age-dependent correction for each emission altitude to account for SO_2 loss. FLEXPART-simulated plume pattern and SO_2 concentration highly resemble those derived from IASI images (Figure 12.8).

All four methods are underpinned by mass conservation in that a satellite image obtained on a given day reflects the budget of gas emissions and losses through oxidation and/or deposition since the start of an eruption. Hence, the retrieval of SO_2 fluxes from a sequence of consecutive images generally requires a sufficiently good estimate of the loss term, although in some instances the loss of SO_2 may be considered negligible.

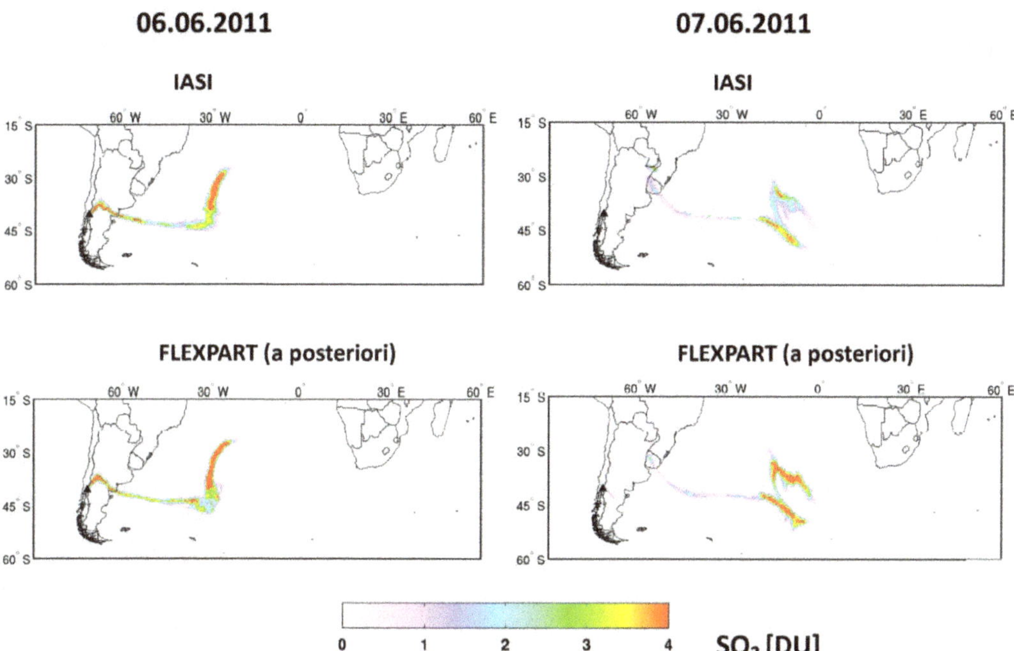

FIGURE 12.8 SO_2 columns in a volcano plume retrieved from an IASI image (evening overpasses) and their comparison with the FLEXPART-simulated columns using the *a posteriori* emissions from the inversion. The measured columns are averaged over the 0.5° × 0.5° FLEXPART output. (Source: Theys et al., 2013.)

12.5.3 Plume Tracking and Forecasting

The tracking of volcanic plumes requires coarse-resolution images that can cover a huge area per scene. Near real-time monitoring of volcanic plumes of SO_2 ash and gases is possible using Support to Aviation Control Service (SACS, http://sacs.aeronomie.be), a free online system run by the European Space Agency (Brenot et al., 2014). This monitoring is based on a combination of three UV–vis and three IR spectrometers (e.g., the OMI and the GOME-2 sensors). The IR sensors are the Atmospheric InfraRed Sounder (AIRS) and the IASI (Figure 12.9). The main products of SACS are SO_2 present in the free troposphere to lower stratosphere and aerosol/ash. Although such SO_2 is often associated with volcanic activity, there are exceptions in that not all eruptions are accompanied by detectable amounts of SO_2. Besides, SO_2 may follow low trajectories different from those of ash because of their differential altitude. The detection of volcanic ash is much more complex than SO_2, especially for dispersed plumes. If the absorption by ash is not differentiated from that by a dust storm, false detection can emerge. This confusion is easily resolved by using hyperspectral data. With optimization, it is possible to avoid false alarms and limit the number of notifications for large plumes at a success rate of 95% (Brenot et al., 2014). It is worth noting that the detected SO_2 is confined to the atmospheric layer above 3–5 km, depending on the humidity profile. However, the SACS warning system cannot forecast the direction in which the plume will travel. This inability is overcome with the use of radar dendrometers that can detect ash dispersal and fallout (Scollo et al., 2009). If combined with multispectral infrared measurements, visual and thermal cameras, this system can monitor and forecast regional volcanic plumes, but not globally.

A very critical consideration in the monitoring is the temporal resolution of the images used. Precise monitoring requires images of a fine temporal resolution as with MODIS and SEVIRI data (Marchese et al., 2019). Their temporal resolution is measured by hours or shorter (e.g., 15 minutes). If estimated from multi-annual time-series cloud-free optical satellite images, plume height and

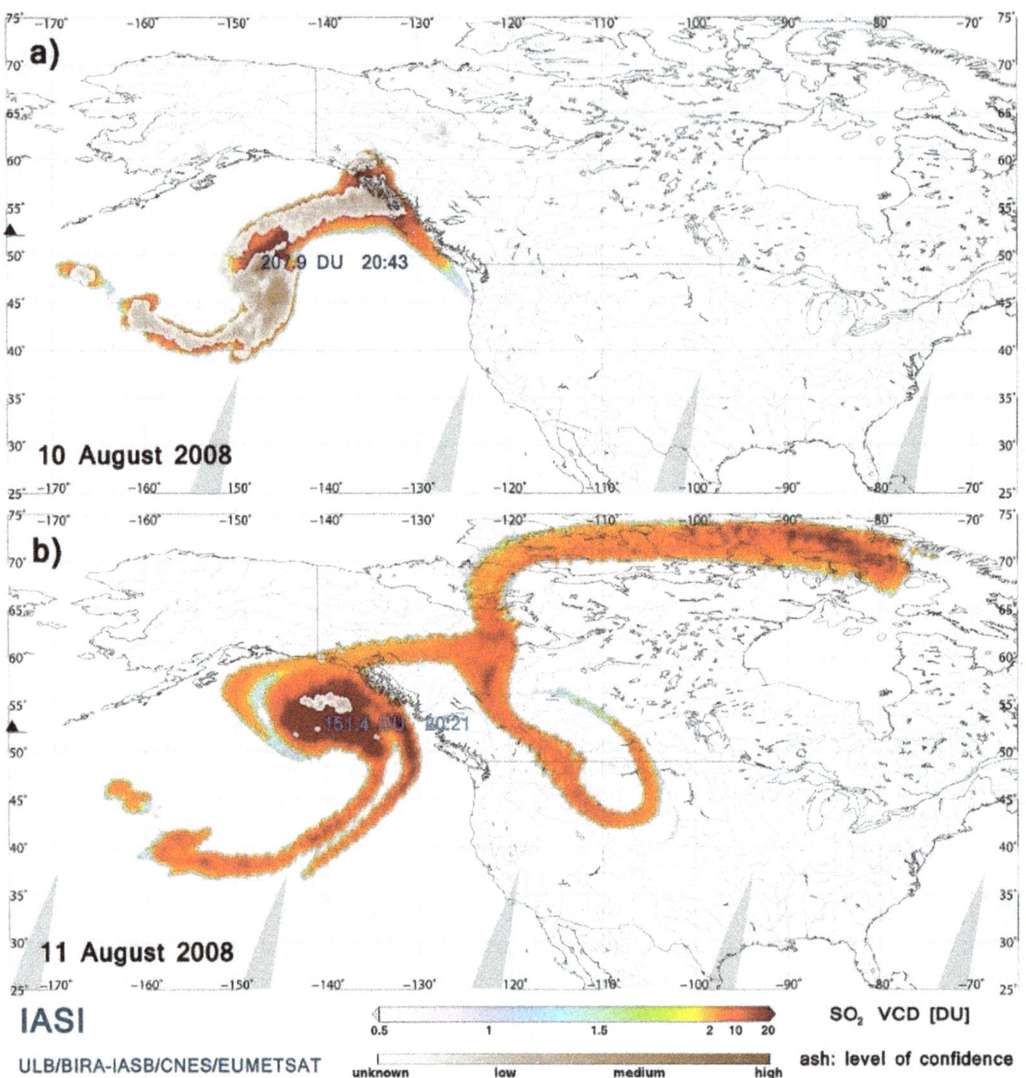

FIGURE 12.9 (a) SO_2 column concentration and ash index emitted by the Kasatochi Volcano on 10 August 2008, and (b) 11 August 2008 derived from IASI-A. (Source: Brenot et al., 2014.)

mass eruption rate may be inversed using the robust satellite techniques (RST) approach based on the RST_{ASH} indices. Ash pixels are identified based on two indices:

$$\otimes_{BT_{11}-BT_{12}}(x,y,t)=\frac{\left[BT_{11}(x,y,t)-BT_{12}(x,y,t)\right]-u_{BT_{11}-BT_{12}(x,y)}}{\sigma_{BT_{11}-BT_{12}}(x,y)} \tag{12.17}$$

$$\otimes_{BT_{3.7}-BT_{11}}(x,y,t)=\frac{\left[BT_{3.7}(x,y,t)-BT_{11}(x,y,t)\right]-u_{BT_{3.7}-BT_{11}(x,y)}}{\sigma_{BT_{3.7}-BT_{11}}(x,y)} \tag{12.18}$$

where u = temporal mean, σ = temporal standard deviation of the brightness temperature differences, and t = time. Subscription = the central wavelength of TIR and MIR bands. Negative values of the $\otimes_{BT_{11}-BT_{12}}(x,y,t)$ index generally characterize ash clouds. The RST_{ASH} product offers a better alternative for monitoring ash plumes emitted by volcanoes because the SEVIRI data have large spatial coverage and a temporal resolution of 15 minutes. The only limitation is image contamination by clouds that disables the monitoring.

For the safety of aviation, it is important to know in which direction the plume will disperse. While remote sensing itself is unable to predict the direction, it supplies vital inputs to a model that can produce the prediction. Forecasting of volcanic plumes requires timely monitoring of plumes. Accurate monitoring is possible only after volcanic ash is differentiated from meteorological clouds. This differentiation can be based on the eruption cloud height as ash is confined to the surface while clouds have a much higher altitude. The direction of dispersion is known via modeling the dispersion of volcanic plumes using TEPHRA, an advection–diffusion model based on several theories (Scollo et al., 2009). It requires inputs about erupted mass, column height, total grain-size distribution, and particle shape, as well as wind direction. In addition to showing the spatial dispersion extent, the modeled outcome also indicates the concentration level of major constituents at different distances from the vent (Figure 12.10). The shape of the plume and its direction of dispersion are both closely related to the predominant wind direction.

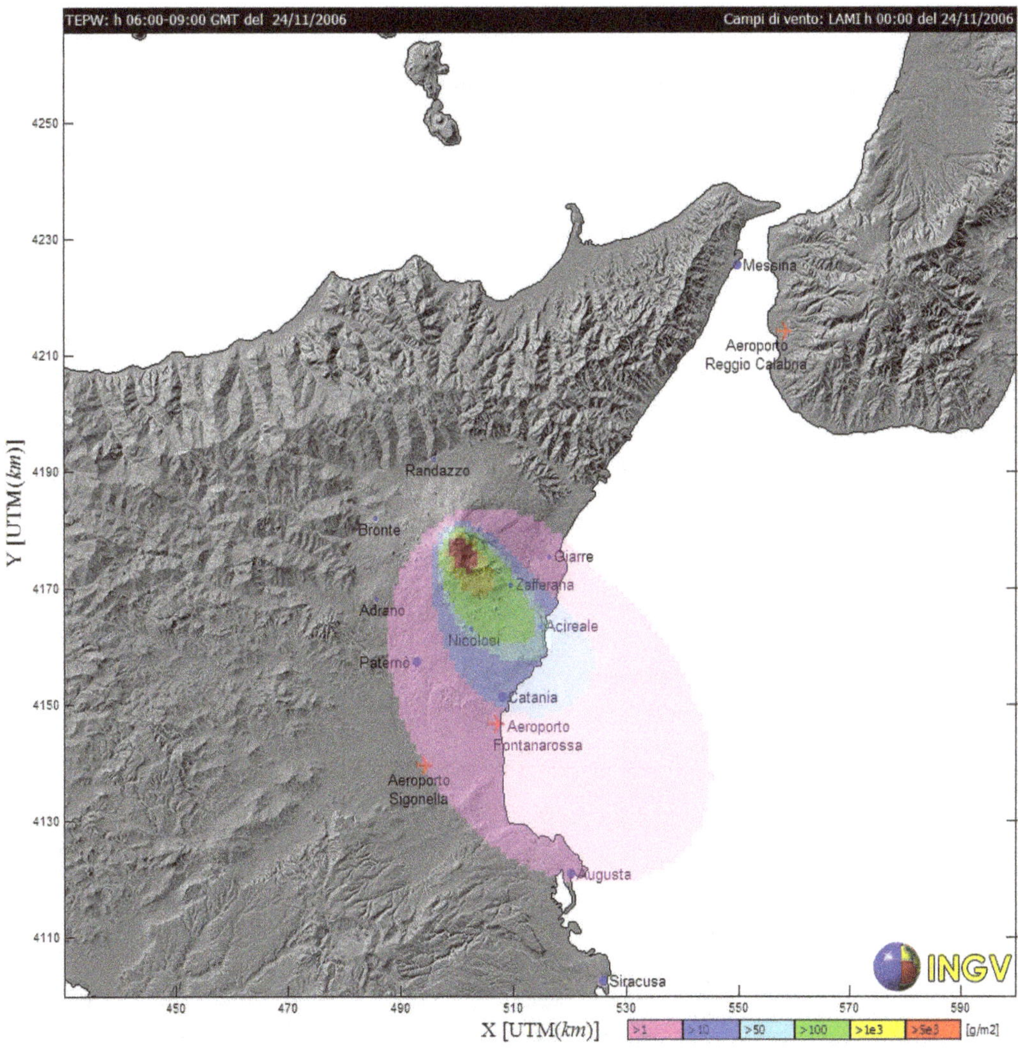

FIGURE 12.10 Direction of volcanic plume dispersion modeled using TEPHRA, reflecting the predominant wind direction. (Source: Scollo et al., 2009.)

12.6 SENSING OF LAHAR FLOWS

Originating from Indonesia, lahar is a term used to describe mudflow or debris flow comprising a mixture of pyroclastic material, rocky debris, and water on the flanks of a volcano cone. Lahars are triggered by volcanic eruptions, during which the frozen ice and snow of glaciers are quickly melted by the searing lava, or by water from the Crater Lake during an eruption. Lahars can also be formed from debris avalanches caused by large volcanic edifices failures indirectly. Thus, lahars are classified as primary and secondary. Primary lahars are formed directly by a volcanic eruption. They are bulky with a volume of around 10^8 m^3 and travel at a high speed of over 20 $m{\cdot}s^{-1}$ (Cando-Jácome and Martínez-Graña, 2019), reaching a maximum rate of 10^5 $m^3{\cdot}s^{-1}$. Secondary lahars are caused by rains. They tend to be small in volume and travel at a slower pace than primary lahars. Secondary lahars are common during heavy rains. They can be distinguished from primary lahars using morphometric parameters. Lahars can be spatially extensive or confined to a narrow channel, depending on the local topography and lahar volume. In high elevation areas, they surge tens of kilometers initially along deeply incised gullies but tend to spread out laterally on low-lying fans.

Lahars are best mapped and monitored from satellite images that enjoy a number of advantages, such as repeated coverage over a long time span. Lahars are detectable from various types of remote sensing data, including space-borne radar images. By default, lahars are cold features, so these images do not have to be TIR. Visible light and NIR images suffice. On such optical images, lahars are easily identifiable from the lack of vegetation, which can be detected from NDVI images. However, optical images are frequently contaminated by topographic shadows in high-relief areas where lahars are commonly distributed. The best way of studying lahars is to monitor surface change just like with landslides. But lahars are different from landslides in that they always travel along the lowest elevation and on the steepest terrain, so topographic data are also crucial to the monitoring.

12.6.1 Lahar Mapping

Different types of remotely sensed data have different capabilities of mapping lahar deposits using either visual interpretation or digital classification. Classification is much faster and can make use of multi-spectral bands. Visual interpretation is slow but can produce reliable maps that are commonly used to assess the accuracy of digitally classified lahar maps. Compared with the lahar deposits manually digitized from aerial photographs, unsupervised classification of a SPOT-5 image of green, red, and NDVI bands (layer) acquired one year after a volcanic eruption achieved an accuracy of 77% owing to the image's superior spatial resolution of 10 m × 10 m (Figure 12.11).

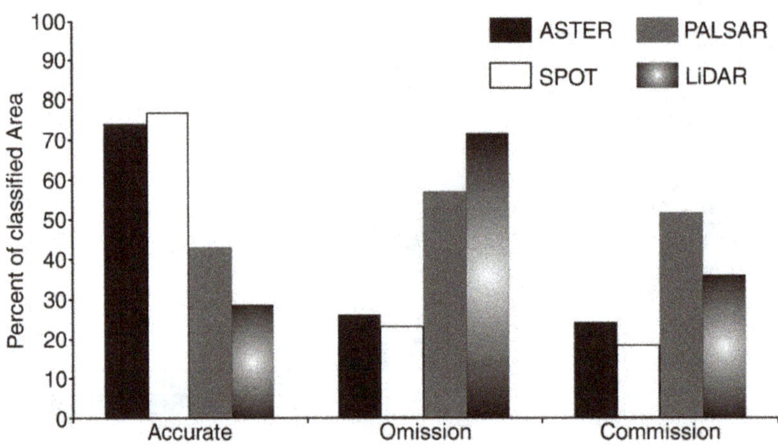

FIGURE 12.11 Comparison of accuracy of mapping lahar deposits from four types of remote sensing data. (Source: Joyce et al., 2009. Used with permission.)

The commission errors are rather high (over 24%) because most skinny deposits close to the crater fall below the detectable threshold (Figure 12.12A, B). They are better mapped from VHR images. At a spatial resolution of 15 m, ASTER imagery is not ideal, achieving a slightly lower accuracy of 74% than SPOT-5 (Figure 12.11). In comparison, ALOS-PALSAR images (resolution: 15 m) acquired two months before and nine months after the lahar deposits mapped them at only 43% correct based on coherence, much less accurate than the optical images, even though they can penetrate the clouds that frequently enshroud the volcanic mountain tops. The commission errors are rather large at a higher elevation (Figure 12.12C). Thus, optical images are much better at the mapping owing to their spectral richness. The mapping accuracy drops to only 28% with LiDAR data (Figure 12.11). They miss the lahar deposits at low elevations in the fan area where the lahar deposits are thinner than upslope ones. The limited vertical accuracy of LiDAR data causes the highest commission errors among all the four types of remote sensing data. Spatially, the commission errors are the highest at high elevations near the crater (Figure 12.12D) probably because the change in surface elevation results from seasonally changed snow thickness, not caused by lahar displacements. The low accuracy is attributed to the fact that the mapping is based on geomorphic changes caused by lahar deposits. Toward their tail, lahar deposits spread out very thinly (Figure 12.12D). As mentioned in Section 1.3.2, LiDAR data have a vertical inaccuracy of 0.15 m and a horizontal accuracy of 0.5 m, so they are unable to detect any changes smaller than 0.2 m reliably. In the mapping, a threshold of 0.5 m was adopted to screen the mapped results. This value overwhelms the magnitude of surface morphological change in the tail of the deposits. It is the main cause of the high omission errors near the tail.

Overall, no single data can achieve perfect mapping under all circumstances (e.g., cloud covers and in different seasons). Perfect accuracy is not possible probably because of the skinny path of lahars near the crater due to confinement by the steep and incised topography. The accuracy

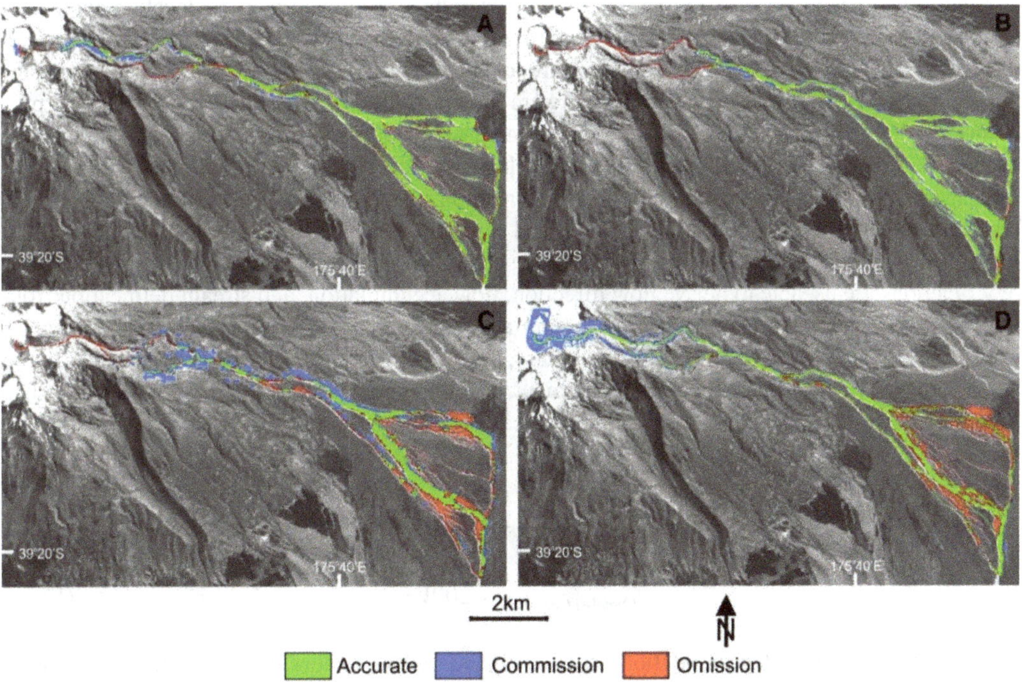

FIGURE 12.12 Comparison of lahar deposits mapped from four types of remote sensing data showing the spatial distribution of commission and omission errors, against the backdrop of ASTER green band. (A) ASTER; (B) SPOT-5; (C) ALOS-PALSAR coherence; (D) LiDAR. (Source: Joyce et al., 2009.)

indicators in Figure 12.12 show that spectral information is more important to the accurate mapping of lahar deposits than the topographic change they induce. Thus, images of a fine spatial resolution (e.g., <5 m) with more spectral bands are crucial to the achievement of higher mapping accuracy. Naturally, it is necessary and advantageous to combine different types of data for the mapping. The use of multi-sensor data increases the evidence about lahar deposits. Not surprisingly, lahars are the most accurately classified (92%) from satellite imagery combined with LiDAR data (Joyce et al., 2009).

12.6.2 Hazard Assessment

Lahar hazards refer to the potential extent of lahar impact and damage. It is related to the volume of lahar debris and the topography it is situated in. A larger lahar deposit lying on a steep slope is able to travel further downslope, impacting more areas than a smaller one sitting on a gentle slope. Remote sensing itself does not allow lahar hazards to be zoned directly, just like with landslide hazards, but it is indispensable in lahar hazard zoning. This is because topography critical to lahar hazards is ideally studied from DEMs that are routinely produced from various types of remote sensing data, such as LiDAR, radar, and satellite images (see Section 1.3.4 for more information). DEMs of different grid sizes allow lahars of different magnitudes to be studied. The level-1 digital terrain elevation dataset (DTED-1) released by NASA's Earth Observing System at a resolution around 90 m can be used to map ancient lahar deposits with a minimum volume of about 2.8×10^6 m^3, and a height <5 m above the modern floodplain (Hubbard et al., 2007). This dataset has a high inaccuracy that causes an order of magnitude error in the estimated volume of small lahar debris flows. Because of their coarse (≥30 m) resolution, ASTER-derived GDEM and SRTM DEMs are problematic in resolving stream valley hydrography critical to lahar flow paths and stream valley morphology that governs lahar filling capacity. Even so, both of them are superior to DTED-1 in resolving fine details of these features.

Similar to landslide hazard zoning, lahar hazard zoning also aims to map the location and extent of areas to be inundated by lahar flows in the future. This zoning is based on the relationship between the extent of inundation and the location and chronology of lahar deposits in the past. It can be accomplished using LAHARZ, a menu-driven software system running within ArcGIS. These GIS programs are designed to automate the mapping of lahar-inundation hazard zones. The primary inputs to the system are a DEM and the user-specified lahar volume. From the volume and user-selected drainages, LAHARZ delineates a set of nested inundation zones from the DEM (Figure 12.13). Thus, lahar hazards are zoned rapidly, objectively, and reproducibly (Schilling, 1998). In the modeling, lahar flow is treated just like water flow on a slope, so the existing ArcGIS GRID surface hydrology functions are useable to derive flow direction, flow accumulation, and stream delineation from the DEM. The major steps of successfully running LAHARZ are to (i) generate a flow direction grid, (ii) create a proximal-hazard zone boundary, (iii) select a stream, and (iv) specify one or more lahar volume (V) based on experience, judgment, and knowledge of the volcano under study. The output of LAHARZ includes cross-sectional area and total planimetric area calculated from the user-specified lahar volume (V) as

$$A_{cross-section} = 0.05V^{2/3} \tag{12.19}$$

$$A_{total-planimetric} = 200V^{2/3} \tag{12.20}$$

The LAHARZ-modeled inundation area is affected by two factors, DEM resolution and lahar flow volume. The input of GDEM, SRTM DEM, and DTED-1 to the LAHARZ software produces a similar lahar flow inundation, but the 90 m resolution of SRTM DEMs proves too coarse to indicate valley-filling cross-sectional area for small deposits due to high errors (Hubbard et al., 2007).

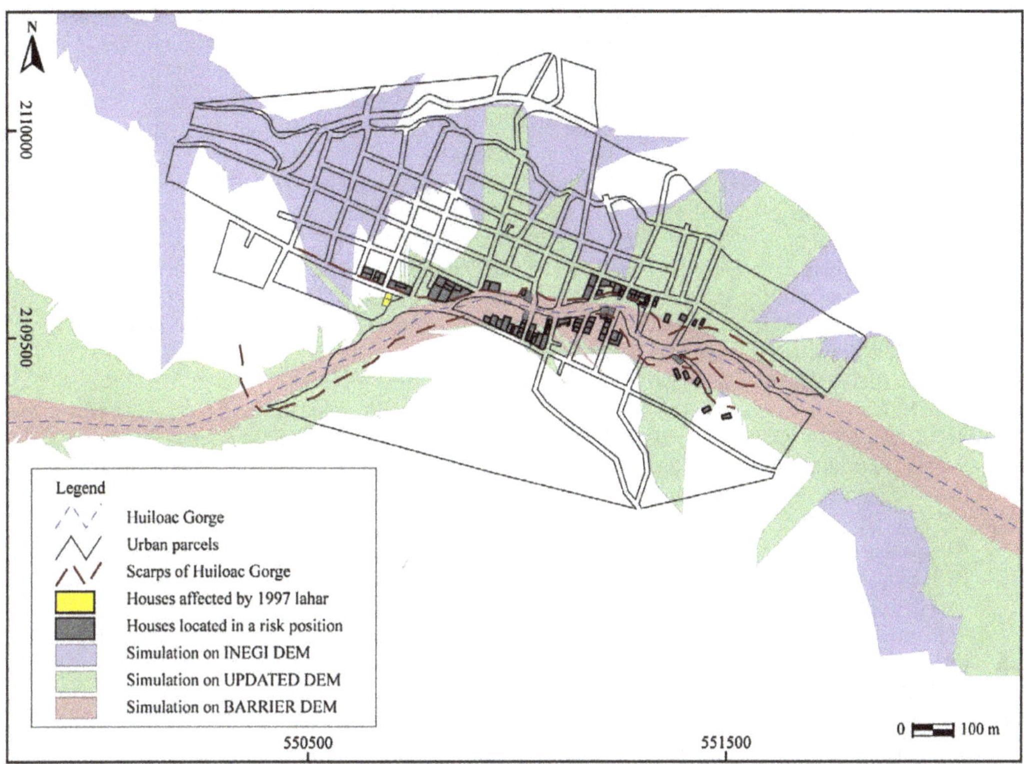

FIGURE 12.13 Spatial inundation extent of Santiago Xalitzintla by lahar flows associated with the Popocatépetl volcano of Mexico, simulated with 5-m DEMs using LAHARZ based on a large hypothetical lahar volume of 1.5×10^6 m^3. (Source: Muñoz-Salinas et al., 2009. Used with permission.)

These small but frequent flows of a magnitude <7 have to be studied from finer resolution DEMs generated from airborne LIDAR data. Naturally, the reliability of the modeled flow extent depends completely on the authenticity and accuracy of the specified lahar volume. Since it is difficult or even impossible to derive lahar volume at a high accuracy using remote sensing, this method is not as practical as it sounds. Instead, statistical analysis of multiple lahar paths is less restrictive and allows the inundation areas to be predicted mathematically.

REFERENCES

Abrams, M., E. Abbott, and A. Kahle. 1991. Combined use of visible, reflected infrared, and thermal infrared images for mapping Hawaiian lava flows. *J. Geophys. Res.* 96: 475–484.

Albino, F., B. Smets, N. d'Oreye, and F. Kervyn. 2015. High-resolution TanDEM-X DEM: An accurate method to estimate lava flow volumes at Nyamulagira volcano (D. R. Congo). *J. Geophys. Res. Solid Earth* 120: 4189–4207.

Anzidei, M., P. Baldi, G. Casula, A. Galvani, F. Riguzzi, and A. Zanutta. 1998. Evidence of active crustal deformation of the Colli Albani volcanic area (central Italy) by GPS surveys. *J. Volcan. Geother. Res.* 80: 55–65.

Aufaristama, M., A. Hoskuldsson, I. Jonsdottir, M. Ulfarsson, and T. Thordarson. 2018. New insights for detecting and deriving thermal properties of lava flow using infrared satellite during 2014–2015 effusive eruption at Holuhraun, Iceland. *Remote Sens.* 10: 151.

Aufaristama, M., A. Hoskuldsson, M.O. Ulfarsson, I. Jonsdottir, and T. Thordarson. 2019. The 2014–2015 lava flow field at Holuhraun, Iceland: Using airborne hyperspectral remote sensing for discriminating the lava surface. *Remote Sens.* 11: 476. DOI: 10.3390/rs11040476

Bagnardi, M., P.J. González, and A. Hooper. 2016. High-resolution digital elevation model from tri-stereo Pléiades-1 satellite imagery for lava flow volume estimates at Fogo Volcano. *Geophys. Res. Lett.* 43: 6267–6275.

Baldi, P., S. Bonvalot, P. Briole, M. Coltelli, K. Gwinner, M. Marsella, G. Puglisi, and D. Remy. 2002. Validation and comparison of different techniques for the derivation of digital elevation models and volcanic monitoring (Vulcano Island, Italy). *Int. J. Remote Sens.* 22: 4783–4800.

Baldi, P., M. Coltelli, M. Fabris, M. Marsella, and P. Tommasi. 2008. High precision photogrammetry for monitoring the evolution of the NW flank of Stromboli volcano during and after the 2002–2003 eruption. *Bull. Volcanol.* 70: 703–715. DOI: 10.1007/s00445-007-0162-1

Behncke, B., A. Fornaciai, M. Neri, M. Favalli, G. Ganci, and F. Mazzarini. 2016. Lidar surveys reveal eruptive volumes and rates at Etna, 2007–2010. *Geophys. Res. Lett.* 43: 4270–4278.

Bhattacharya, A., and S.K. Srivastava. 1993. Nighttime TM short wavelength infrared data analysis of Barren Island volcano, South Andaman, India. *Int. J. Remote Sens.* 14: 783–787. DOI: 10.1080/01431169308904376

Brenot, H., N. Theys, L. Clarisse, J. van Geffen, J. van Gent, M. Van Roozendael, R. van der A, D. Hurtmans, P.-F. Coheur, C. Clerbaux, P. Valks, P. Hedelt, F. Prata, O. Rasson, K. Sievers, and C. Zehner. 2014. Support to aviation control service (SACS): An online service for near-real-time satellite monitoring of volcanic plumes. *Nat. Hazards Earth Syst. Sci.* 14: 1099–1123. DOI: 10.5194/nhess-14-1099-2014

Butz, A., A.S. Dinger, N. Bobrowski, J. Kostinek, L. Fieber, C. Fischerkeller, G.B. Giuffrida, F. Hase, F. Klappenbach, J. Kuhn, P. Lübcke, L. Tirpitz, and Q. Tu. 2017. Remote sensing of volcanic CO_2, HF, HCl, SO_2, and BrO in the downwind plume of Mt. Etna. *Atmos. Meas. Tech.* 10: 1–14. DOI: 10.5194/amt-10-1-2017.

Cando-Jácome, M., and A. Martínez-Graña. 2019. Determination of primary and secondary lahar flow paths of the Fuego volcano (Guatemala) using morphometric parameters. *Remote Sens.* 11(6): 727. DOI: 10.3390/rs11060727

Corradino, C., G. Ganci, A. Cappello, G. Bilotta, A. Hérault, and C. Del Negro. 2019. Mapping recent lava flows at Mount Etna using multispectral Sentinel-2 images and machine learning techniques. *Remote Sens.* 11(16): 1916. DOI: 10.3390/rs11161916

Dai, C., and I.M. Howat. 2017. Measuring lava flows with ArcticDEM: Application to the 2012–2013 eruption of Tolbachik, Kamchatka. *Geophys. Res. Lett.* 44(24): 12133–12140. DOI: 10.1002/2017GL075920

Del Negro, C., A. Cappello, and G. Ganci. 2016. Quantifying lava flow hazards in response to effusive eruption. *Geol. Soc. Am. Bull.* 128: 752–763.

Dietterich, H.R., M.P. Poland, D.A. Schmidt, K.V. Cashman, D.R. Sherrod, and A.T. Espinosa. 2012. Tracking lava flow emplacement on the east rift zone of Kīlauea, Hawai'i, with synthetic aperture radar coherence. *Geochem. Geoph. Geosy.* 13. DOI: 10.1029/2011GC004016

Dzurisin, D., M. Lisowski, and C.W. Wicks. 2017. Semipermanent GPS (SPGPS) as a volcano monitoring tool: Rationale, method, and applications. *J. Volcan. Geother. Res.* 344: 40–51. DOI: 10.1016/j.jvolgeores.2017.03.007

Eckhardt, S., A.J. Prata, P. Seibert, K. Stebel, and A. Stohl. 2008. Estimation of the vertical profile of sulfur dioxide injection into the atmosphere by a volcanic eruption using satellite column measurements and inverse transport modelling. *Atmos. Chem. Phys.* 8: 3881–3897. DOI: 10.5194/acp-8-3881-2008

Ganci, G., A. Cappello, G. Bilotta, A. Hérault, V. Zago, and C. Del Negro. 2019. 3D Lava flow mapping at Etna volcano from Pléiades-derived DEM differences. *PANGAEA*, https://doi.org/10.1594/PANGAEA.899176, Supplement to: Ganci, G et al. (2019): 3D Lava flow mapping of the 17-25 May 2016 Etna eruption using tri-stereo optical satellite data. *Ann. Geophys.* 62(2): 1–6. DOI: 10.4401/ag-7875

Ganci, G., A. Vicari, L. Fortuna, and C. Del Negro. 2011. The HOTSAT volcano monitoring system based on combined use of SEVIRI and MODIS multispectral data. *Ann. Geophys.* 54(5): 544–550. DOI: 10.4401/ag-5338

Henney, L.A. 2012. Remote Sensing of Volcanic Plumes Using the Advanced Spaceborne Thermal Emission and Reflection Radiometer (ASTER). PhD thesis, Michigan Technological University, 297 p.

Hirn, B., C. Di Bartola, and F. Ferrucci. 2008. Spaceborne monitoring 2000-2005 of the Pu'u 'O'o-Kupaianaha (Hawaii) eruption by synergetic merge of multispectral payloads ASTER and MODIS. *IEEE Trans. Geosci. Remote Sens.* 46(10): 2848–2856.

Hubbard, B.E., M.F. Sheridan, G. Carrasco-Nunez, R. Diaz-Castellon, and S.R. Rodriguez. 2007. Comparative lahar hazard mapping at Volcan Citlaltepetl, Mexico using SRTM, ASTER and DTED-1 digital topographic data. *J. Volcanol. Geotherm. Res.* 160: 99–124.

James, M.R., H. Pinkerton, and L.J. Applegarth. 2009. Detecting the development of active lava flow fields with a very-long-range terrestrial laser scanner and thermal imagery. *Geophys. Res. Lett.* 36: L22305.

Joyce, K., S. Samsonov, V. Manville, R. Jongens, A. Graettinger, and S. Cronin. 2009. Remote sensing data types and techniques for lahar path detection: A case study at Mt Ruapehu, New Zealand. *Remote Sens. Environ.* 113: 1778–1786. DOI: 10.1016/j.rse.2009.04.001

Krueger, A., L. Walter, P. Bhartia, C. Schnetzler, N. Krotkov, I. Sprod, and G. Bluth. 1995. Volcanic sulfur dioxide measurements from the total ozone mapping spectrometer (TOMS) instruments. *J. Geophys. Res.* 100: 14057–14076.

Kubanek, J., J.A. Richardson, S.J. Charbonnier, and L.J. Connor. 2015. Lava flow mapping and volume calculations for the 2012–2013 Tolbachik, Kamchatka, fissure eruption using bistatic TanDEM-X InSAR. *Bull. Volcanol.* 77: 106.

Li, L., C. Solana, F. Canters, and M. Kervyn. 2017. Testing random forest classification for identifying lava flows and mapping age groups on a single Landsat 8 image. *J. Volcanol. Geotherm. Res.* 345: 109–124.

Lopez, T.M., S.A. Carn, C. Werner, P. Kelly, M. Doukas, M. Pfeffer, D. Fee, P. Webley, C. Cahill, and D.J. Schneider. 2013. Evaluation of Redoubt volcano's sulfur dioxide emissions by the ozone monitoring instrument. *J. Volcanol. Geotherm. Res.* 259: 290–307. DOI: 10.1016/j.jvolgeores.2012.03.002

Marchese, F., A. Falconieri, C. Filizzola, N. Pergola, and V. Tramutoli. 2019. Investigating volcanic plumes from Mt. Etna eruptions of December 2015 by means of AVHRR and SEVIRI data. *Sensors* 19(5): 1174. DOI: 10.3390/s19051174

Merucci, L., M.R. Burton, S. Corradini, and G.G. Salerno. 2011. Reconstruction of SO_2 flux emission chronology from space-based measurements. *J. Volcanol. Geotherm. Res.* 206: 80–87.

Muñoz-Salinas, E., M. Castillo-Rodríguez, V. Manea, M. Manea, and D. Palacios. 2009. Lahar flow simulations using LAHARZ program: Application for the Popocatépetl volcano, Mexico. *J. Volcanol. Geotherm. Res.* 182(1–2): 13–22. DOI: 10.1016/j.jvolgeores.2009.01.030

Nádudvari, Á, A. Abramowicz, R. Maniscalco, and M. Viccaro. 2020. The estimation of lava flow temperatures using Landsat night-time images: Case studies from eruptions of Mt. Etna and Stromboli (Sicily, Italy), Kilauea (Hawaii Island), and Eyjafjallajökull and Holuhraun (Iceland). *Remote Sens.* 12: 2537. DOI: 10.3390/rs12162537

Patrick, M., T. Orr, G. Fisher, F. Trusdell, and J. Kauahikaua. 2017. Thermal mapping of a pahoehoe lava flow, Kilauea volcano. *J. Volcanol. Geotherm. Res.* 332: 71–87.

Pinkerton, H., M. James, and A. Jones. 2002. Surface temperature measurements of active lava flows on Kilauea volcano, Hawai'i. *J. Volcanol. Geotherm. Res.* 113: 159–176.

Pyle, D.M., T.A. Mather, and J. Biggs. 2013. Remote sensing of volcanoes and volcanic processes: Integrating observation and modelling – Introduction. *Geol. Soc. London Spec. Publ.* 380(1): 1–13.

Schellenberg, B., T. Richardson, M. Watson, C. Greatwood, R. Clarke, R. Thomas, K. Wood, J. Freer, H. Thomas, E. Liu, F. Salama, and G. Chigna. 2019. Remote sensing and identification of volcanic plumes using fixed-wing UAVs over Volcán de Fuego, Guatemala. *J. Field Robot.* 36: 1192–1211.

Schilling, S.P. 1998. LAHARZ: GIS Programs for Automated Mapping of Lahar-Inundation Hazard Zones. USGS Open-File Report 98-638, Washington, 80 p.

Scollo, S., M. Prestifilippo, G. Spata, M. D'Agostino, and M. Coltelli. 2009. Monitoring and forecasting Etna volcanic plume. *Nat. Hazards Earth Syst. Sci.* 9: 1573–1585.

Stevens, N.F., H. Garbeil, and P.J. Mouginis-Mark. 2004. NASA EOS Terra ASTER: Volcanic topographic mapping and capability. *Remote Sens. Environ.* 90: 405–414.

Stevens, N.F., G. Wadge, and J.B. Murray. 1999. Lava flow volume and morphology from digitised contour maps: A case study at Mount Etna, Sicily. *Geomorphology* 28: 251–261.

Taylor, E.A. 2019. Satellite Remote Sensing of Volcanic Plumes. PhD thesis, University of Oxford, 258 p.

Theys, N., R. Campion, L. Clarisse, H. Brenot, J. van Gent, B. Dils, S. Corradini, L. Merucci, P.F. Coheur, M. van Roozendael, D. Hurtmans, C. Clerbaux, S. Tait, and F. Ferrucci. 2013. Volcanic SO_2 fluxes derived from satellite data: A survey using OMI, GOME-2, IASI and MODIS. *Atmos. Chem. Phys.* 13: 5945–5968. DOI: 10.5194/acp-13-5945-2013

Turner, N.R., R.L. Perroy, and K. Hon. 2017. Lava flow hazard prediction and monitoring with UAS: A case study from the 2014–2015 Pāhoa lava flow crisis, Hawai'i. *J Appl. Volcanol.* 6: 17. DOI: 10.1186/s13617-017-0068-3

Valade, S., A. Ley, F. Massimetti, O. D'Hondt, M. Laiolo, D. Coppola, D. Loibl, O. Hellwich, and T.R. Walter. 2019. Towards global volcano monitoring using multisensor Sentinel missions and artificial intelligence: The MOUNTS monitoring system. *Remote Sens.* 11: 1528.

Vicari, A., G. Bilotta, S. Bonfiglio, A. Cappello, G. Ganci, A. Hérault, E. Rustico, G. Gallo, and C. Del Negro. 2011. LAV@HAZARD: A web-GIS interface for volcanic hazard assessment. *Ann. Geophys.* 53(5). DOI: 10.4401/ag-5347

Wright, R., and L.P. Flynn. 2003. On the retrieval of lava-flow surface temperatures from infrared satellite data. *Geology* 31(10): 893–896.

Wright, R., and L.P. Flynn. 2004. Space-based estimate of the volcanic heat flux into the atmosphere during 2001 and 2002. *Geology* 32. 10.1130/G20239.1

Wright, R., L. Flynn, H. Garbeil, A. Harris, and E. Pilger. 2002. Automated volcanic eruption detection using MODIS. *Remote Sens. Environ.* 82: 135–155.

Wright, R., L.P. Flynn, H. Garbeil, A.J.L. Harris, and E. Pilger. 2004. MODVOLC: Near-real-time thermal monitoring of global volcanism. *J. Volcanol. Geotherm. Res.* 135: 29–49.

Wright, R., L.P. Flynn, and A.J.L. Harris. 2001. Evolution of lava flow-fields at Mount Etna, 27–28 October 1999, observed by Landsat 7 ETM+. *Bull Volcanol.* 63: 1–7. DOI: 10.1007/s004450100124

13 Avalanches

13.1 INTRODUCTION

Avalanches are a type of mass movement in which a huge quantity of snow, ice, or rocks falls abruptly and rapidly down a steep mountain slope or a cliff, usually without any warning. This process takes place so quickly (e.g., within a matter of seconds) that it is extremely difficult to foretell or avoid this kind of hazard. So far snow is the most common material involved in an avalanche, while ice or rock avalanches are much less common. Similar to landslide, a typical snow avalanche path is composed of three sections: the zone of origin or source, the zone of transition, and the zone of deposition where the snow debris accumulates (Figure 13.1). The three zones are separable by their surface morphology and topography. The first zone is an erosional zone, so its elevation is lower than the surrounding area after the event. In contrast, the last zone is where the avalanche debris accumulates, which uplifts the original surface elevation above the intact snow. It is this zone that has been intensively studied by means of remote sensing. The zone of transition has a unique texture formed by the downslope motion of compacted snow different from that of the other two zones. However, it is not always possible to demarcate these three zones on remote sensing images, or even on oblique ground photos.

Snow avalanches are considered a natural hazard in snow-covered steep mountainous regions around the world, especially in early spring, because of the large volume of debris involved and the sheer velocity of its downslope movement. In addition to being hazardous to human safety, snow avalanches can also temporarily block critical transportation routes (e.g., interstate highways in the US) and do damage to trees and vegetation on their path. Remote sensing plays a valuable role in studying snow avalanches with both advantages and limitations (Table 13.1). Advantages include objective and repeated coverage independent of the weather over long periods, and the possibility of merging multi-platform data. With the assistance of remote sensing imagery, point-based results can be upscaled to the whole study area. Disadvantages are also obvious, such as a high data cost, limited coverage per scene, and untested mapping accuracy.

Snow avalanches can be studied using terrestrial, airborne, and space-borne remote sensing data. This chapter first reviews all kinds of sensing platforms and remote sensing data that have found applications in studying avalanches, with their strengths and limitations compared and evaluated in Section 13.2. Next, this chapter comprehensively assesses the methods of mapping snow avalanches from remotely sensed images, including visual interpretation, change detection, object-oriented image analysis, and machine learning methods. After introducing each of them in detail, this chapter then compares them. The third topic of this chapter is concerned with the monitoring of individual avalanches and all avalanches in a region using different remote sensing approaches. Finally, this chapter explains how avalanche hazards are modeled from its conditioning factors following an overview of all the factors considered important to avalanche hazards. Also included in the discussion are two methods of implementing the modeling: linear weighting and machine learning.

13.2 SENSING PLATFORMS AND DATA

Avalanches have been detected and monitored using terrestrial, airborne, and space-borne sensors, each having its unique strengths and limitations (Table 13.2). Which platform-sensor combination should be deployed depends largely on the spatiotemporal scale of study. This section first introduces ground-based sensing, followed by air- and space-borne sensing. A comparison of the three platforms is made toward the end of the section, together with a summary of the pros and cons of each platform.

DOI: 10.1201/9781003354321-13

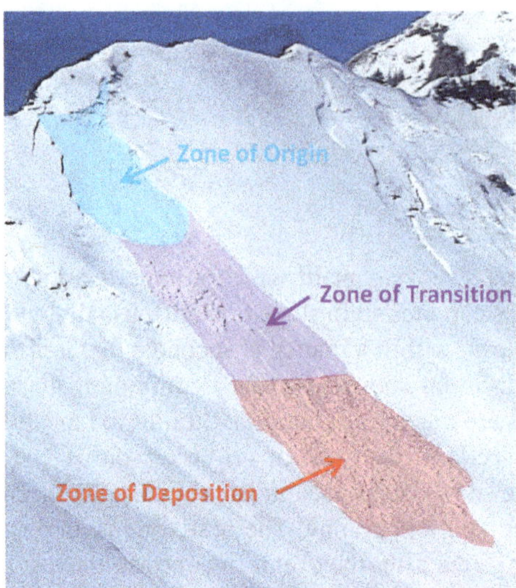

FIGURE 13.1 Three typical zones of origin, transition, and deposition of a slab avalanche that may be differentiated from each other on remote sensing imagery. (Source: Leinss et al., 2020.)

13.2.1 Terrestrial Sensing

Ground-based monitoring makes use of photographs or non-imaging LiDAR data automatically acquired at a fixed position some distance from the accumulated snow repeatedly at different times. The acquired time-lapse data reveal temporal changes in the sensed area, such as snow cover depth and snow distribution. If the photographs are taken continuously at a short interval, then they are able to indicate the glide crack dynamics and cornice fall activity and dynamics (Eckerstorfer et al., 2015). For ground sensing to function well, it is important to select a strategic vantage point to deploy the sensor. This point should maximize the coverage of the sensed area without the chance of being hit by the avalanches that may eventuate during the sensing. The distance between the sensor and the target can range from a few meters to hundreds of meters. A longer distance enables more ground to be sensed, and is less affected by the avalanches should they eventuate. Even with a long distance of sensing, terrestrial sensing has a limited view field, and not all the sensed area is fully visible from a single point. This method is inexpensive and easy to implement, and the monitoring

TABLE 13.1
Pros and Cons of Remote Sensing in Detecting Snow Avalanches

Advantages	Disadvantages
Covering otherwise inaccessible terrain;	Comparably few studies have proven the concept so far;
An unbiased, safe observation vantage point without direct contact with avalanches. Thus, their physical properties not altered in any way;	Mapped results not vigorously tested for accuracy;
Possibility to upscale point-measurements to larger scales (slope, basin, valley, mountain range);	VHR remote sensing data are expensive, have a small swath, and cover small areas;
Weather independent if radar data are used;	Temporal resolution is, currently, insufficient with most sensing systems;
Possibility to collect a complete, continuous data record over a given time from a given place;	Large data needed for operational sensing not fully automated in avalanche detection. Avalanche detection algorithms work well with optical data, but studies using radar data are very scarce so far;
Possibility to use multi-sensor data to retrieve different physical properties;	Limited availability of field validation data in inaccessible areas.
Decades of archived data excellent for long-term monitoring.	

Source: Adapted from Hafner et al. (2021).

TABLE 13.2

Comparison of Three Types of Remote Sensing-Based Detection of Snow Avalanches

Method	Strengths	Limitations
Terrestrial	Easily an flexibly deployed, even in real time; Repeated coverage, ideal to study change, even the avalanche process; Quantification of displaced snow mass possible with LiDAR data.	Unfavorable working conditions; Limited spatial scope; Presence of "dead area" or gap; Suitable for small areas of study or studying a few avalanches.
Airborne	A long history of archived data; Detailed to allow the mapping of the smallest avalanche; Flexibly deployable.	Subject to weather conditions; Limited spectral range; Narrow swath width and small ground coverage.
Space-borne	Multispectral bands allowing indices to be used in the detection of avalanches; Broad coverage, ideal for regional mapping; Data periodically available, excellent for ongoing monitoring.	Coarse image resolution not allowing small avalanches to be detected; A relatively short history of existence; A long return period ill-suited to study avalanche process or real-time monitoring.

can be real time. If two cameras are deployed to take stereoscopic photographs of the same target or with the use of the structure-from-motion technique, it is possible to construct a 3D model of the covered area. This model may be explored to identify the interactions between snow distribution and terrain and calculate the volume of an avalanche and the deposition area. The disadvantages of terrestrial sensing are the frigid environment that demands short maintenance intervals and unfavorable working conditions such as short daylight and frequent cloud cover.

Dissimilar to ground photographs, non-graphic terrestrial LiDAR data are excellent for mapping snow depth. Since LiDAR data show surface height at the time of sensing, two scans, one before, and one after the avalanche event, are essential to accomplish this task. Snow depth can be detected to an accuracy of 10 cm from a terrestrial laser scanner at a distance of 500 m (Prokop et al., 2013a). Although unable to detect snow avalanche events just like with aerial photography, LiDAR enables the detection of mass balance in different zones and slide paths via the subtraction of two DEMs constructed from LiDAR data. As illustrated in Figure 13.2, scanning of a slope under both

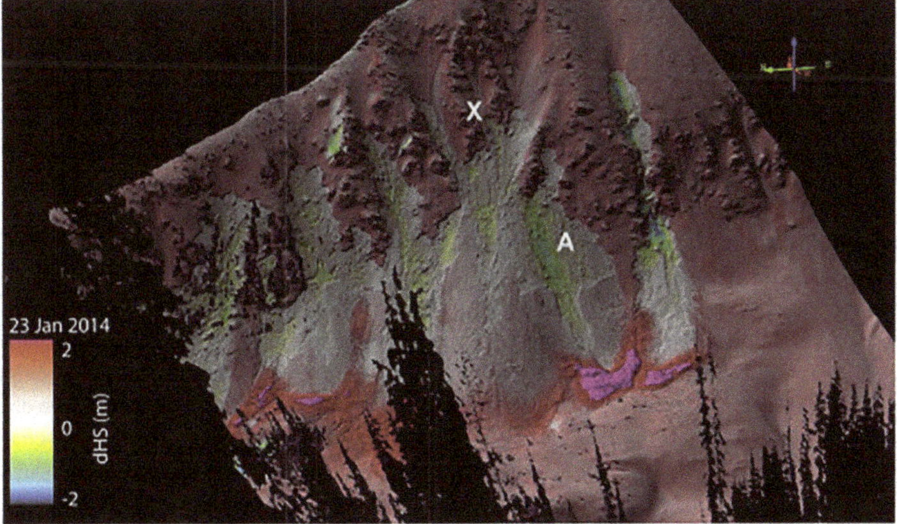

FIGURE 13.2 Change in surface snow height between 23 January and 1 February of 2014 induced by a single avalanche at point X, detected from two terrestrial laser scanner data, illustrating the mass gain (reddish) and mass loss (greenish) areas. (Source: Deems et al., 2015. Used with permission.)

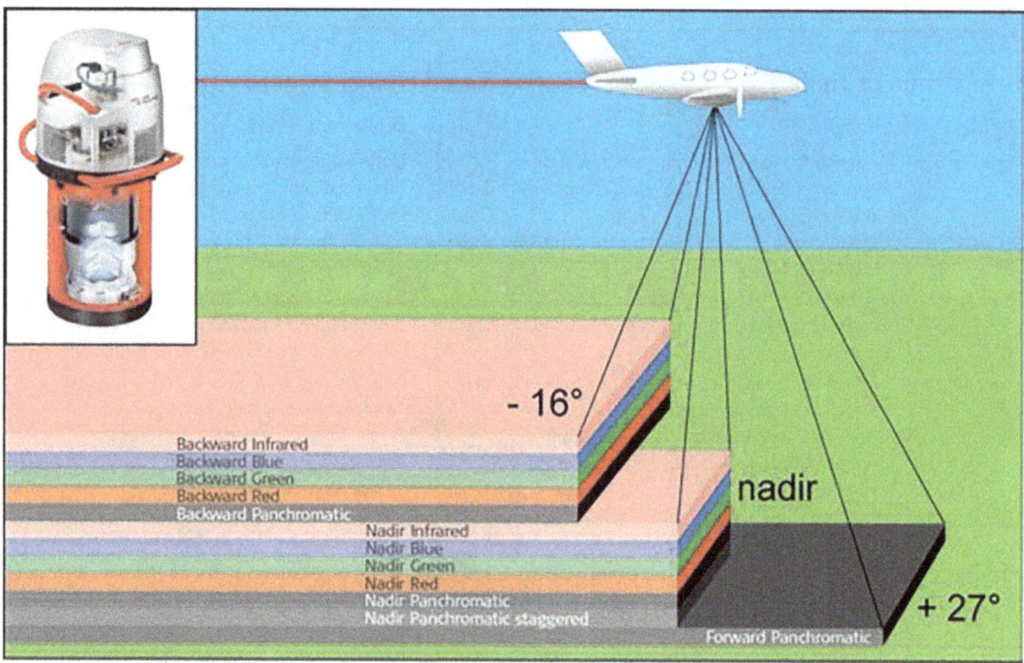

FIGURE 13.3 Three directions of acquiring airborne digital images using the Leica ADS40-SH52 scanner (inset) and the spectral bands available. (Source: Bühler et al., 2009. Used with permission.)

snow-free and snow-covered conditions can quantify mass loss in the zone of origin or the source zone and slide paths (zone of transition) of released avalanches, mass gain in the runout zone or the debris areas, and calculate avalanche debris volume based on the changed snow cover mass balance (Deems et al., 2015).

13.2.2 Airborne Sensing

Airborne sensing makes use of a camera aboard an airplane in one of three directions, nadir, backward up to 16°, and forward up to 27° (Figure 13.3). Aerial photographs taken in the nadir view cover a much smaller ground than in the other two directions, but most of the surface is exposed. Since it is impossible to predict when an avalanche is going to take place, photography can seldom be synchronized with avalanching. Instead, photographs are likely taken before and after an avalanche event. They are good at mapping the start and runout zones and inventorying the damages of an avalanche to the natural environment. Avalanches are mapped from aerial photographs via visual interpretation based on their distinctive features, such as an elongated, tongue-resembling shape comprising disturbed snow of a rough surface, a higher snow density, and a larger snow depth than the surrounding, undisturbed snow (Eckerstorfer and Malnes, 2015).

Airborne VHR optical images are obtainable using the Leica ADS40-SH52 digital push-broom scanner. It can acquire high-resolution (0.5–0.05 m, subject to flight height) images with a dynamic range of 12 bits in five spectral bands spanning from blue, green, red to infrared wavelengths, plus one panchromatic band (Figure 13.3). Such high spatial and radiometric resolutions enable the detection of subtle spectral variations within snow-covered fields, even if they are shadowed. Leica airborne data can be used to potentially map avalanches accurately in detail over widespread areas. In particular, such multi-angle data provide a new avenue for mapping avalanches into various components based on their distinctive spectral behavior. Namely, smooth snow strongly scatters the incident solar energy forward, while a rough surface of avalanche debris strongly reflects energy

FIGURE 13.4 Comparison of the Salzer avalanche deposit in Davos, Switzerland mapped from ADS40 aerial imagery using OBIA of NDAI and texture after excluding certain areas not prone to avalanche (left) with the in situ observed avalanche on an oblique ground photograph taken one day before the data acquisition. (Source: Bühler et al., 2009.)

in all directions (Figure 13.4, right). Thus, the returned radiance from the ground is a function of the illumination and observation angles that can be described by the BRDF. The multi-angular information of avalanche reflection is best quantified using the normalized difference angle index (NDAI) proposed by Bühler et al. (2009).

$$NDAI = \frac{NIR_{backward} - NIR_{nadir}}{NIR_{backward} + NIR_{nadir}} \tag{13.1}$$

This index has been used to map the Salzer avalanche deposit in Davos, Switzerland, from ADS40 aerial images using OBIA, together with texture after certain areas not prone to avalanches are excluded (Figure 13.4, left).

For sizeable avalanches, the mapping accuracy can be as high as 95% from ortho-imagery (Lato et al., 2012). For small avalanches, the accuracy is lower due to confusion with ski lifts. Aerial photographs have a limited spectral resolution, so the mapping has to rely heavily on texture information. Nevertheless, texture may be obliterated by fresh snow. In order for airborne sensing to be effective, it is rather imperative to fly over the study area soon after the avalanche event, which is not a mean task as the weather conditions can be unpredictable and blizzards may hamper the visibility of the target.

The use of VHR airborne imagery in mapping snow avalanches is both advantageous with benefits and fraught with restrictions. On the one hand, it enjoys the unparalleled strength of the longest history of data existence dating back to the 1960s or even earlier. Such long time-series data are crucial to constructing the history of avalanches so as to predict their future behavior (Huggel et al., 2007). Furthermore, the fine resolution (e.g., 0.25 m) of NIR aerial imagery allows the detection of avalanches and classifying them into release zone, track, and run-out zone. This differentiation of an avalanche into multiple zones, especially the release zone, is critically important to assessing avalanche hazards and modeling the run-out distance of avalanche debris but is rather complex and involves many thresholds. Of the three zones, the run-out zone is the most reliably detected, whereas the release zone is detected the least reliably. On the other hand, the sheer data volume restricts one image (10,000 rows by 10,000 columns) to cover a ground area of only 6.25 km^2 in one classification (Korzeniowska et al., 2017). So the same avalanche is inevitably split into neighboring tiles, introducing a source of mapping inaccuracy. Nevertheless, airborne images enable 1,648 out of 2,200 avalanches distributed in an area of 226.3 km^2 to be correctly identified at a rate of 75%, with the producer's and user's accuracies being 61% and 78%, respectively. This proportion drops to 61% in terms of the total avalanche area detected, though. It is much more challenging to delineate the precise extent of an avalanche due to its spectral similarity to the intact surroundings as they are both composed virtually of the same substance of snow. The detection accuracy is rather low for aged avalanches or snow debris whose deposits lack distinction with smooth snow or whose path is strewn with vegetation, causing the assumptions in the OBIA to be violated, but higher for fresh avalanches. Another source of misclassifications is traced to image tiling, which is redundant with space-borne images.

13.2.3 Space-Borne Sensing

Space-borne data are acquired from scanners carried onboard a satellite. They cover a much broader ground per scene than their airborne counterparts, a trait making them suitable for region-wide monitoring and detection of avalanches. The use of scanners instead of cameras also extends the spectral range of sensing to SWIR and even microwave wavelengths. Both optical and radar data have been applied to detecting avalanches, but they are processed differently, so are the image-derived parameters used in the detection.

13.2.3.1 Optical Imagery

On optical satellite images, snow avalanches resemble the snow field spectrally and in texture, so image classification is out of the question. On Landsat 8 OLI imagery of 15 m resolution, it is possible to identify avalanches via visual interpretation (Eckerstorfer et al., 2015). Given that a typical path of small avalanches is only 10–100 m long, satellite images' medium resolution (e.g., 30 m) is not sufficiently fine for detection, even though rock avalanche deposits on a volcanic cone are discernible on ASTER imagery of 15 m resolution (Huggel et al., 2007). This leaves VHR images such as SPOT 6 of 1.5 m (panchromatic band) and 4 m (multispectral bands) resolutions as the only viable space-borne data source for studying avalanches from space, even though they may not have a global coverage.

In spite of the relatively fine-resolution, it is still difficult to map avalanches from satellite images as both avalanches and the intact snow in their vicinity have highly resembling spectral reflectance in the VNIR spectrum (Figure 13.5). The exact reflectance varies slightly with snow age and purity.

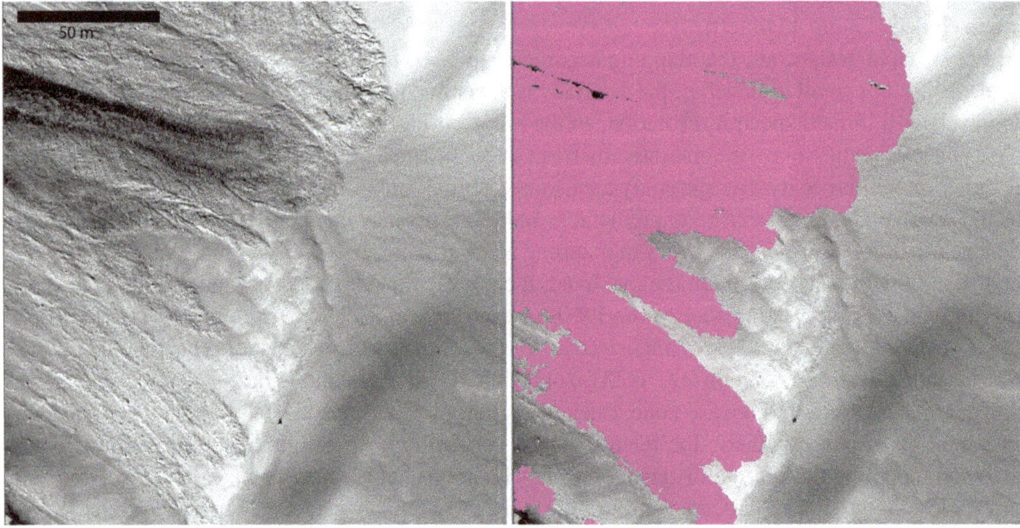

FIGURE 13.5 The appearance of avalanche debris on a panchromatic aerial photograph taken with a camera (left) and the avalanche debris mapped from the photograph (right). (Source: Lato et al., 2012.)

Dry snow appears bright in the 0.35–0.75 μm wavelengths because of the high and constant reflectance that decreases with aging or dirty snow, or in the snow of a large grain size. In the NIR and SWIR regions of the spectrum, however, reflectance decreases sharply due to strong absorption by moisture, especially over SWIR wavelengths. Because the incident radiation is absorbed by water, wet snow tends to have lower reflectance than dry snow. In microwave wavelengths, the degree of radar scattering and absorption is a function of grain size and snow density. A larger grain size tends to scatter more radiation, while denser snow absorbs more radiation. However, the relationship between scattering and density is not linear. It increases with density up to 150 kg·m^{-3} and then drops slightly beyond this density (Picard et al., 2013).

13.2.3.2 Radar Imagery

Radar imagery is captured at a few typical wavelengths from several sensors, such as the X-band (9.6 GHz) imagery from high-resolution (3 m) TerraSAR, RadarSat-2 (3–5 m), and medium-resolution Sentinel-1 (10 m) (Table 1.7). In the standard StripMap mode, TerraSAR-X imagery has a nominal SLC resolution as fine as 2.3 m × 3.3 m (rg × az). Such fine-resolution images enable more avalanche features to be studied at a higher accuracy level. Sentinel-1 imagery with a spatial resolution of 10 m has a broad spatial coverage and a fine temporal resolution (e.g., every six days). The backscatter signal of an avalanche on radar imagery depends on its snow properties. The grand backscatter intensity of a snowpack (σ^0_{snow}) comprises the scattering signal from the snow surface (σ^0_{surf}), snow volume (σ^0_{vol}), the ground below the snowpack (σ^0_{ground}), and higher order interactions between different structures in the snowpack (σ^0_{inter}) (Leinss et al., 2020), or

$$\sigma^0_{snow}(\theta) = \sigma^0_{surf}(\theta) + \sigma^0_{vol}(\theta) + \sigma^0_{ground}(\theta) + \sigma^0_{inter}(\theta) \tag{13.2}$$

Not all the terms in Eq. 13.2 are applicable to a given avalanche (e.g., the ground term disappears with very deep snow), nor are all terms equally important to all avalanches. If radar scattering takes place over a rough surface in a bi-continuous media, then σ^0_{snow} is positively correlated with ice grains and surface roughness, but inversely with incidence angle θ. The backscatter signal of dry snow depends largely on ground roughness, while smooth ground mostly scatters the incident energy forward. Of the three avalanche zones of origin, transition, and deposition, it is the depositional zone that has a very rough surface, causing a high contribution of σ^0_{vol} and σ^0_{surf} to the total backscatter intensity. Plain wet snow has weak backscattering at the air-snow interface because of the strong absorption of the incident radiation. Thus, the zone of origin is very difficult to detect from radar imagery owing to little change in surface roughness. The transitional zone should be more visible but its visibility depends on the deposition of avalanche debris. Naturally, it is the depositinoal zone that is the most easily detected because of its strongest backscatter signal and elongated tongue shape (Leinss et al., 2020).

Of all the types of radar imagery listed in Table 1.7, TerraSAR-X and RadarSat-2 Ultra-fine imagery of 3 m resolution are the most suited to detect avalanche debris to the smallest size of a cross-slope width of 45 m (Eckerstorfer et al., 2016). TerraSAR-X can clearly distinguish fresh avalanches from existing ones (Figure 13.6a). However, such data are costly, with erratic availability and a small swath width of 50 km. As with all radar images, it suffers from layover and shadow effects that are prevalent in avalanche-prone steep terrain. Coarse-resolution radar imagery is still freely available, as exemplified by Sentinel-1 SAR. It allows visual detection of avalanche debris as it lifts the local backscattering (Figure 13.6b). SAR images have a higher localized backscatter due to locally increased backscatter in the avalanche runout zone than that of the undisturbed snowpack nearby (Eckerstorfer and Malnes, 2015). Sentinel-1 imagery enables small avalanches up to 0.09 km^2 to be consistently identified more accurately with descending images than ascending images (Yang et al., 2020). By default, radar imagery has a poor spectral resolution as it is unexceptionally

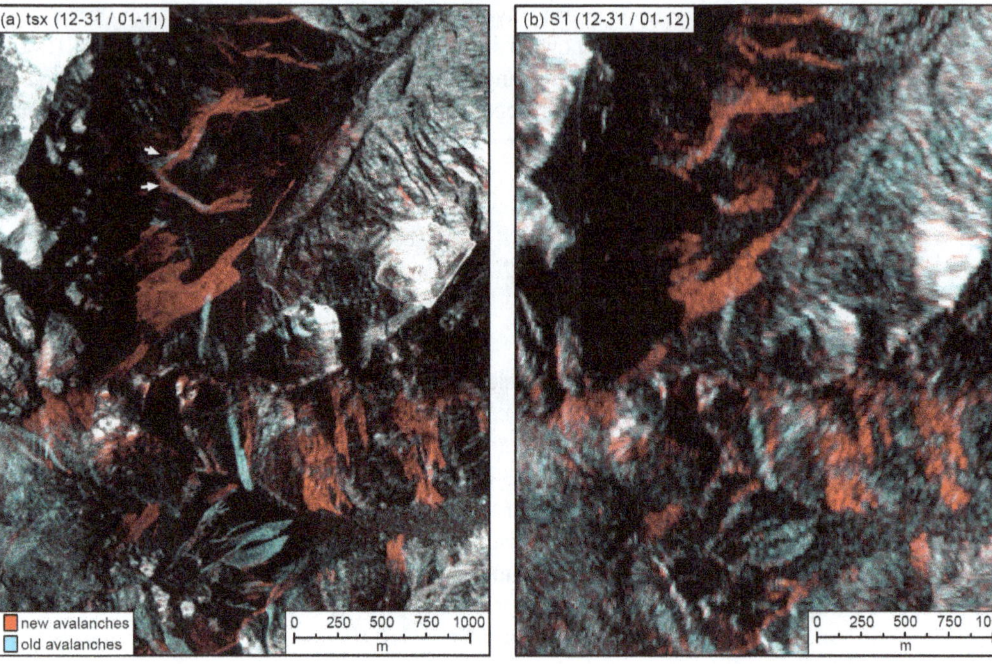

FIGURE 13.6 Appearance of new and old avalanches on a color composite of multi-temporal TerraSAR images (a) in comparison with the Sentinel-1 image pair (b). Due to the coarser (10 m) resolution, the avalanches have a less defined boundary. (Source: Leinss et al., 2020.)

mono-spectral. This deficiency is partially compensated for by its polarization that falls into two modes: vertical transmission and vertical reception (VV) and vertical transmission and horizontal reception (VH). Cross-polarized images contain subtle information about avalanche debris and may facilitate its accurate detection.

The feasibility of detecting avalanche debris from Sentinel-1A images has been demonstrated by Malnes et al. (2015) using change detection to enhance avalanche debris zones. A comparison of repeat-pass images before and after avalanches reveals an increased backscatter from avalanche debris that enables them to be detected manually. Increased backscattering can also be caused by increased snow volume, liquid water content, snow density, and surface roughness in avalanche debris, though. Sentinel-1A images acquired in the interferometric wide swath mode at a spatial resolution of 20 m × 20 m allow avalanche debris with a minimum size of roughly 80 m × 80 m to be manually detectable. In the 250 km × 150 km large ground swath mode, Sentinel-1A imagery permits the detection of 100s avalanches from a wet-snow avalanche cycle owing to the increased backscattering from avalanche debris with increased snow depth, snow water equivalent, and most importantly, surface roughness. The detection is facilitated by the tongue shape of avalanche deposits, and a sharp backscatter contrast to the surrounding undisturbed snowpack. The detection is significantly eased on an RGB composite image produced from multiple channels (e.g., one before the avalanche and another post the event), multi-sensors, or of different polarizations, as illustrated in Figure 13.6. On the composite, avalanche debris has a color quite distinct from that of non-avalanche features. The enhanced backscatter contrast between avalanche debris and surrounding snowpack facilitates detection and interpretation, even for those not experienced with interpreting avalanches on radar imagery. Nevertheless, it is not feasible to demarcate the full extent of avalanches from

TABLE 13.3
Mutual Correspondence between New Avalanches Manually Interpreted from Sentinel-1 (12-31/01-12) and TerraSAR-X Image Pairs of (12-31/01-11)

Total Detected	Common	New/Unsure	Missing
TerraSAR compared to Sentinel-1			
89	83	76/7	6
Sentinel-1 compared to TerraSAR			
164	104	100/4	60

Source: Leinss et al. (2020).

Sentinel-1 SAR images, because only the deposited avalanche debris is detectable. Nor is it possible to detect the zone of the origin or the entire avalanche path (Figure 13.6b). In fact, Sentinel-1 imagery fails to identify most small avalanches due to its limited (10 m) spatial resolution (Leinss et al., 2020). This imagery may cause overestimation of avalanche number and underestimation of avalanche size. However, if various Sentinel-1 images from multiple orbits and polarizations are combined, jointly, they notably enhance the interpretation results and reduce speckles, both of which enable the mapping results to be almost comparable to those detected from single-orbit TerraSAR-X data. Fine-resolution SAR images are crucial to mapping small avalanches. The strength of high-resolution TerraSAR-X images over medium-resolution Sentinel-1 images in mapping avalanches is amply demonstrated by the mutual correspondence of results obtained from them (Table 13.3).

As shown in the table, a total of 89 new avalanches are detected from the Sentinel-1 image pair of 12-31/01-12, plus 13 unsure, and 16 old avalanches. These images fail to identify small avalanches likely because their spatial resolution is significantly coarser than that of TerraSAR-X. Of the 89 new avalanches, 83 were also detected from TerraSAR-X, 76 of which were new and 7 unsure avalanches, while 6 avalanches were missed out. Compared to the TerraSAR-X pair of 12-31/01-11, the TerraSAR-X pair (12-31/01-11) detected 164 avalanches, of which 104 or two-thirds corresponded to 83 avalanches that were also successfully detected from the Sentinel-1 image pair. One-third (60 of 164) were missing. This number is ten times that of the previous comparison and is attributed to the 10 m spatial resolution of Sentinel-1 images that is twice coarser than that of TerraSAR-X. Consequently, a lot of small avalanches are omitted. This claim is easily confirmed by the fact that the smallest avalanches detectable from Sentinel-1 have a size of around 2,000 m^2, four times larger than the minimum area of 500 m^2 avalanches detected from TerraSAR-X. The spatial distribution of avalanches mapped from both TerraSAR-X and Sentinel-1 is highly consistent between the two types of imagery (Figure 13.7), but the spatial outline of avalanches mapped from TerraSAR-X tends to be more restrictive than those from Sentinel-1 images (Leinss et al., 2020).

13.2.3.3 A Comparison

Both optical and radar data have their strengths and limitations in mapping snow avalanches. Optical satellite data are highly suitable for avalanche detection and activity monitoring, as all parts of an avalanche, from the starting zone to the depositional area, are fully visible. However, they are also subject to cloud cover, shadow effects, and polar night darkness that prevent continuous and operational monitoring throughout the winter in high-polar regions. Radar data, on the other hand, are not detailed enough. Although optical data can be detailed, they have a narrow swath width,

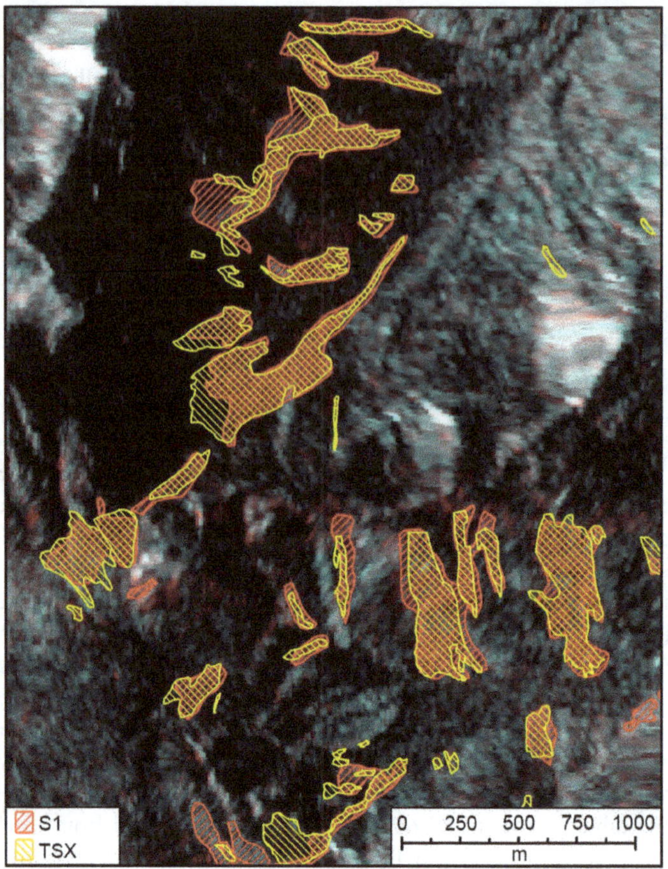

FIGURE 13.7 Avalanches manually mapped from Sentinel-1 image pairs of 12-31/01-12 in comparison with those from TerraSAR-X (12-31/01-11). (Source: Leinss et al., 2020.)

and such data are rather expensive (Table 13.4). Despite these differences, they are commonly used synergetically in practice. For instance, the SAR-derived results may be validated against results visually interpreted from optical VHR images.

If visually interpreted from VHR images, similar avalanches can be mapped from both radar and optical images. For instance, TerraSAR-X change detection mapped 316 avalanches in a study area, very close to 286 avalanches mapped from optical SPOT 6 images acquired on 24 January between the TerraSAR-X image pairs of 11 January and 2 February 2018. During the nine-day interval, additional avalanches could have been triggered by about 20 cm of fresh snow falling on 1 February. However, the mapped avalanche outlines differ significantly between them, and the outlines are sometimes split up into discrete patches that should form one polygon representing one avalanche. Optical SPOT 6 imagery enables avalanches to be mapped more completely, including differentiation of a path into three zones of origin, path, and deposition than radar data. Optical imagery is better at differentiating adjacent avalanches into multiple classes of new, old, and unsure, which is impossible with radar imagery. It allows detection of mostly the avalanche deposition zone, whereas the release zone is highly visible already on SPOT 6 images (Leinss et al., 2020). However, side-looking SAR images are inevitably prone to and prolong radar shadow and layover, within which avalanches cannot be detected. Besides, the coarse spatial resolution makes it impossible to detect small (e.g., <100 m^3) avalanches or avalanches thinner than sub-pixel size, even though this drawback disappears with VHR RadarSat-2 images.

TABLE 13.4
Comparison of the Strengths and Limitations of Three Types of Imagery in Mapping Avalanches Based on Visual Interpretation

Imagery	Strength	Weakness
SPOT 6	Daily revisit capability due to constellation of two satellites (SPOT 6 and 7); Coverage of a very large area (i.e., the entire Swiss Alps in one day) is possible upon request; Fine (1.5 m) spatial resolution well suited for avalanche detection; Wet-snow avalanches highly visible in NIR band; No radiometric saturation on snow	Cloud-free conditions essential for sensing; Data availability subject to request at a high price (~\$100,000 for covering 12,500 km²); Images obtained in far-off nadir viewing suffer geometric distortions in steep terrain; 1.5 m resolution restricts the detection of medium and small-sized avalanches
Sentinel-2 (10 m) optical	Short revisit time of every five days; Large regional coverage in several overpasses; Images acquired with relatively small incidence angles (<10°); Data freely available	Cloud-free conditions essential for sensing; 10 m resolution too coarse to map even very large avalanches
Sentinel-1 (10 m) radar	Short revisit time of six days or shorter if combined with data from other orbits; Operational in all weather and light conditions; Data freely available; Minimal "blind spots" in layover and radar shade by combining ascending and descending images; Sensitive to surface roughness changes caused by avalanche debris	Expensive pre-processing computation; Avalanches in radar shadow and layover missed out; Suitable for detection of medium avalanches (50–200 m long); Overestimation of avalanche number and underestimation of avalanche size due to smooth surfaces in the release or track area; Mapping complicated by strongly variable and changing snow conditions from pre- to post-image

Source: Hafner et al. (2021).

13.3 AVALANCHE MAPPING

Avalanches or parts thereof have been mapped from optical imagery using unsupervised OBIA, semi-automated object-based approaches (Korzeniowska et al., 2017; Lato et al., 2012), and automated change detection approaches (Eckerstorfer et al., 2019), in addition to manual mapping (Eckerstorfer et al., 2015) either solely or in combination with automatic mapping (Leinss et al., 2020). Each of these methods has its strengths and limitations, as well as best uses.

13.3.1 Visual Interpretation

Visual interpretation or manual analysis relies on the appearance of avalanches on imagery, such as shape, location, size, texture, and association. All avalanches have an elongated shape, but the depositional zone can be tongue-shaped. In terms of location, avalanches are initiated on a steep terrain on the upper slope of a valley where snow accumulates. The size of an avalanche varies enormously, depending on the topography and the accumulated snow volume. Some may have the full components of the source zone, transitional zone, and depositional zone as illustrated in Figure 13.1, while other smaller ones may have the first and last zones. Texture is a rather unique feature of avalanches, especially in the depositional zone. It is much coarser than the smooth, intact snow surface in the surroundings. The depositional zone is commonly associated with a gentler terrain where the mobilized debris comes to a stop. On the post-event photographs, medium-sized avalanche debris can be differentiated from intact snow cover even in shaded areas, based on directional, textural, and spectral information at an accuracy of 94% in terms of the fraction of correctly detected avalanche deposits (producer's accuracy = 87%) (Bühler et al., 2009). Dissimilar to other land cover features, avalanches do not benefit much from color as both the debris and the intact snow are white. In general, shadow hampers the interpretation as it blurs the image differences between avalanche and non-avalanche features and should be avoided by using radar images of a smaller incidence angle.

It is difficult to recognize avalanches from mono-spectral radar imagery based on tone. The identification is much improved from an RGB composite image that can be created by merging multi-temporal or multi-orbital radar images (Figure 13.6). Ideally, one of them should be taken before the event, and another post the event. The composite illustrates whether an avalanche is new, old, or certain as shown in Figure 13.6a (Leinss et al., 2020). Although a composite may be created by merging multi-orbital images, this is not preferred as it requires the spatial resolution of the respective images to be unified. This problem disappears if the multi-temporal images are acquired from the same sensor at the same incidence angle. However, the use of multi-temporal images may not always succeed in detecting avalanches if they have a return period of 25 days or longer, due to the obliteration of the avalanche debris signature by new snow and/or wind in the interim of image acquisition (Eckerstorfer et al., 2014). This problem can be partially circumvented using multi-orbital and multi-sensor images. The plus side of multi-orbital images is their differential incidence angles which may further enhance the appearance of an avalanche on the composite. However, the different resolutions of multi-orbital images may cause avalanches to be mapped with differential details (e.g., full outline vs only the depositional zone).

Visual interpretation of avalanches from radar imagery such as Sentinel-1 relies on the increased backscattering caused by a rough surface. Visual analysis of SAR imagery such as the high-resolution RadarSat-2 Ultra-fine imagery can detect only avalanche debris-resembling features based on the sharp backscatter contrast between avalanche debris and the undisturbed snowpack (Eckerstorfer and Malnes, 2015). However, some of the preliminarily detected avalanches may represent radar shadows and layovers or other natural features, which can be easily eliminated by performing a check against the local topography in a GIS. Shadows and layovers are commonly associated with steep terrain. Although fine-resolution (3 m × 3 m) imagery allows small avalanches to be detected, it is not practical to map avalanches at the county scale using such data. Besides, only 37% of the detected avalanches could be validated using optical imagery. If the image resolution is sufficiently fine, avalanches can be mapped as polygons, as illustrated in Figure 13.7. Even if the resolution is coarse such as 10 m of Sentinel-1 imagery, it is still possible to demarcate the avalanche outline if the avalanche has a sufficiently large size.

Visual interpretation of Spot 6/7 images delineated more than 18,000 avalanches at high accuracy in comparison with ground truth data gathered from helicopter reconnaissance (Bühler et al., 2019). Of SPOT 6, Sentinel-2, and Sentinel-1 (radar) images, visual interpretation of SPOT 6 achieved the best results at a POD of 0.74, and avalanches of all sizes are detectable (Hafner et al., 2021). Successful cases have been achieved with multispectral, panchromatic, and NIR bands. On high-resolution SPOT images (1.5 m panchromatic band), the recognition accuracy reaches 71% with a low false-negative rate of 16%, and a false-positive rate of 11%. Apart from image resolution, image shadow also degrades the accuracy of detection. It drops POD from 0.86 (127 out of 147 avalanches detected) on fully illuminated slopes to 0.79 (88 out of 112 avalanches detected) to 0.15 (6 of 33 avalanches detected) on shaded slopes, all of which are manually interpreted, while the accuracy decreases to only 64% in shadow areas. The 10 m resolution Sentinel-1 image successfully identifies only very large avalanches, and the overall POD is significantly lower (0.27) than that of the SPOT image mainly because most of the avalanches have a medium and large size, which represent the largest number of all avalanches missed. The PPV of Sentinel-1 (0.87) has a range similar to that of SPOT 6. Sentinel-2 imagery of 10 m spatial resolution detects only

TABLE 13.5
Comparison of Three Types of Remote Sensing Data (SPOT 6, Sentinel-1, and Sentinel-2) in Mapping Avalanches via Visual Interpretation in Terms of Probability of Detection (POD) and Positive Predictive Value (PPV)

		SPOT 6	Sentinel-1	Sentinel-2	Avalanche Map
POD	2018	0.74	0.17	0.07	0.46
	2019	0.76	0.52	0.04	0.82
	All	0.74	0.27	0.06	0.56
PPV	2018	0.87	0.90	0.88	0.90
	2019	0.90	0.84	0.60	0.99
	All	0.88	0.87	0.81	0.93

Source: Hafner et al. (2021).

1 in 17 avalanches and is unsuitable for the mapping of most avalanches due to its coarse spatial resolution (POD = 0.06, and PPV = 0.81) (Table 13.5).

Visual interpretation, assisted by the GLCM texture measure and NDAI, identified 106 out of 113 deposits correctly after exclusion of areas not prone to avalanches, resulting in an accuracy of 94% (Bühler et al., 2009). The accuracy is noticeably lower for (very) small avalanche deposits due to the presence of artificially piled snow near ski fields, ski lifts, and sparsely vegetated snow cover that could not be effectively excluded. It is worth noting that the assessment of the mapping accuracy of avalanches is tricky or not even feasible as the validation dataset could show avalanches as points, while fine-resolution imagery-derived avalanches can be mapped as polygons. This discrepancy creates three possible types of match: 1 to 1, 1 to many, and many to 1. In all assessments so long as one of the three possibilities is met, the identification is deemed correct.

The validation of automatically detected avalanche debris against manually identified avalanches is not as straightforward as it seems because the field validation data may not contain sufficient samples. The alternative is to compare the automatically mapped results with the manually delineated avalanche debris. This comparison can be problematic in that image-generated results are pixel-based, in drastic contrast with manually interpreted results that are in vector (e.g., polygon) format. So the pixel-based results have to be vectorized. The vectorized avalanches may not match the spatial extent of the manually interpreted avalanche debris outline (Figure 13.8). The manually and automatically detected results are considered to reach a match if the same avalanche debris shows up at a similar location in both results irrespective of its spatial extent.

In spite of the high accuracy reported so far, visual interpretation is rather slow and tedious and lacks consistency, especially over large areas on the regional scale. The alternative is to make use of OBIA, especially with 0.2 m aerial images. Visual interpretation may be slow and the results subjective, but it is still indispensable in remote sensing of avalanches. This is because visually interpreted results are treated as the ground truth, against which automatically detected results are assessed for their accuracy. Besides, image classifiers or algorithms cannot be properly trained without visually selected samples.

13.3.2 Change Detection

Given that a snow avalanche is a process involving the downslope movement of accumulated snow, the best way of detecting an avalanche path is via change detection performed on the differencing image of pre- and post-event images pair. Avalanches are detectable via the use of multiple thresholds empirically determined from training samples selected from the differencing image. Statistical parameters of these samples reveal the critical thresholds at which avalanches and non-avalanche features are separable. Thresholds can also be used to effectively mask non-avalanche pixels to minimize

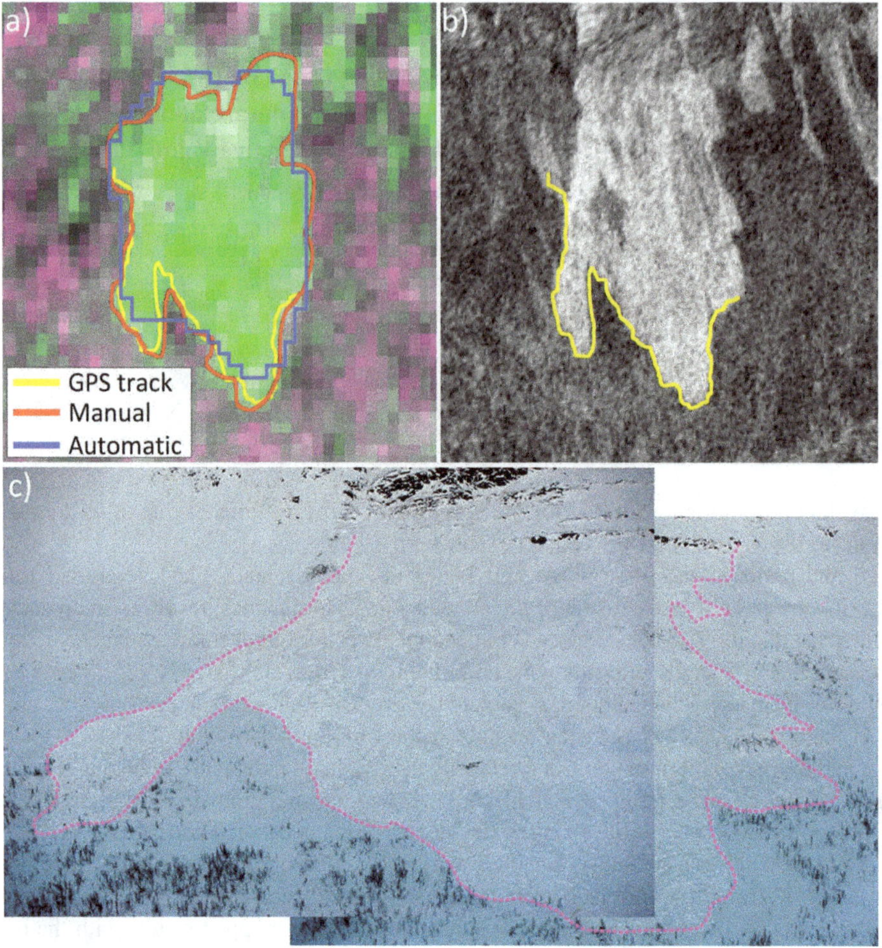

FIGURE 13.8 (a) Comparison between manually interpreted and automatically mapped outline of avalanche debris on an RGB composite of multi-temporal images (6 January 2015 and 13 December 2014) of VV polarization. (b) RadarSat-2 Ultra-fine mode single backscatter image (3 m × 3 m spatial resolution) of 3 January 2015. (c) Oblique field photograph of the avalanche debris with the interpreted outline. (Source: Vickers et al., 2016.)

confusion and enhance detection reliability. If the images are SAR, additional masks may be created to remove radar layover, shadow, and steep terrain using slope gradient (Wesselink et al., 2017). These areas have to be excluded from analysis because avalanches cannot be detected or are unlikely to be present in them. Thresholding can be applied to the post-detection results via spatially filtering out small clusters of pixels or spatially isolated pixels. For instance, any clusters smaller than the minimum avalanche size of 300 m^2 are unlikely to be genuine and can be safely eliminated. Thresholds are usually determined using two broad approaches: dominance or absolute value. The former works with window-based operations and is particularly suitable for SAR images. The latter can be set with the assistance of the differencing image's histogram. After the difference in backscatter is calculated from the two images, the portion of pixels having a change value above the threshold is determined, such as 1% above 6 dB to determine the minimum avalanche size (e.g., 100 × 100 m^2). Then the image is spatially classified using K-means clustering into two groups, avalanche and non-avalanche. The former may be overlaid with a DEM and spatially filtered to remove noise (Eckerstorfer et al., 2016).

A thorny hurdle with this automatic method of change detection is how to set the proper threshold. Whatever threshold is adopted, it always affects whether a particular sub-window of the image contains potential avalanche debris and the mapping outcome. Chances are the results are not always

TABLE 13.6

Impact of Threshold Values on the Accuracy of Automatically Detecting Avalanche Debris in Comparison with Manually Identified Avalanche Debris

Threshold (dB)	Correctly Classified (%)	Commission Error (%)	Omission Error (%)	FAR	HSS/NSS
5.0	67.5	87.3	32.5	0.563	0.527/0.458
6.0	62.3	53.9	37.7	0.464	0.574/0.493
7.0	54.8	20.8	45.2	0.275	0.622/0.528

Source: Eckerstorfer et al. (2016) and Vickers et al. (2016).

FAR – false alarm ratio; HSS – Heidke skill score; NSS – normalized skill score.

perfect no matter what threshold is adopted (Table 13.6). A conservative threshold corresponds to a higher classification accuracy or a higher fraction of correctly classified pixels and a lower omission error (e.g., fewer avalanche debris pixels are missed out). However, it is also associated with a higher commission error or FAR (Vickers et al., 2016). This is because snow-covered trees can be erroneously detected as avalanche debris if the image resolution is sufficiently fine. As the threshold rises from 5.0 to 7.0 dB, a larger fraction of the manually identified avalanche pixels are missed out, with the FAR more than halved. However, the normalized skill score does not appear to vary much with the threshold. In the end, the determined threshold always represents a trade-off between a higher classification accuracy and a smaller omission error.

This thresholding method is simple and fast as it can be implemented automatically, but the determination of the appropriate threshold is a time-demanding process, and the determined threshold could be scene-specific. Thus, a new threshold must be determined from scratch every time a new image is processed, even for the same study area or even for the same avalanche to be detected from images recorded at different times. This automatic detection method may be able to deliver promising results; however, they are obtained from the two images used for change detection only when the snow conditions are similar, such as dry–dry or wet–wet, but not dry-wet or wet-dry (Eckerstorfer et al., 2016). The thresholding method is limited in that the mapping of avalanches relies solely on their spectral signature on the image, while other useful clues are ignored, a deficiency that can be remedied using OBIA.

13.3.3 OBIA

OBIA functions most effectively with VHR images, from which a variety of spatial and contextural parameters are derivable and used in image classification (Table 13.7). OBIA allows the consideration of more properties of an avalanche than image spectral features in the decision-making, such

TABLE 13.7

Parameters Derived from VHR Panchromatic Imagery and Served As Inputs to OBIA for Mapping Avalanches

Parameter	Objective
GLCM entropy	Eliminate regions with a high level of order
GLCM dissimilarity	Eliminate regions with minimal contrast to surrounding pixels (e.g., fresh snow)
Brightness	Eliminate dark pixels (e.g., trees, rocks)
Contrast	Eliminate sharp boundaries (snow beside a rock)
Similarity	Eliminate groups of pixels that are too small to represent an avalanche
Neighbor distance	Fill in small gaps surrounded by known avalanche pixels

Source: Lato et al. (2012).

as brightness, contrast, similarity, neighbor distance, and texture (Lato et al., 2012). They are used with different objectives in mind (Table 13.7) because of their varying utilities in the mapping. Brightness helps eliminate dark pixels associated with trees and rocks. Snow avalanche debris that spectrally resembles intact snow is still distinguishable based on its signature in shape, tone, and texture. Texture refers to the spatial variability of tone and is quantifiable from the simultaneous consideration of a group of spatially adjacent pixels and is expressed in various metrics, two of which are GLCM entropy and dissimilarity. GLCM represents the distribution of pixel values at a given offset. The spatial signature of an avalanche such as texture is more distinct and useful than the spectral signature in automatic detection. The use of more image features other than spectral information in the detection creates a potential for avalanches to be mapped reliably. In order to produce reliable detections, avalanche pixels must be examined within a local neighborhood, instead of being treated individually in isolation.

In addition to texture, pixel brightness of an NIR band, indices such as NDVI, NDWI and its standard deviation (SDNDWI), and NDAI have also been considered in detecting snow avalanches in an OBIA (Korzeniowska et al., 2017). NDAI, evaluated from the nadir and backward NIR bands, is intended to separate rough avalanche debris from smooth and undisturbed snow in the surroundings. The use of such normalized indices means that any methods based on them are applicable to multiple images covering a broad area, so they are suitable for regional avalanche detection. After objects are formed via image segmentation, they are handled in two ways: thresholding and classification. For instance, vegetation and dark objects are identified via an NDVI threshold of >127, brightness <4,000, and removal of any objects <12.5 m^2 in area. Then those bright, non-vegetation segments of a high snow roughness are considered to represent rough snow (e.g., NDWI > 127, NDVI < 140, SD_{NDWI} > 0.7, and brightness >2,500). The thresholding of an entropy measure evaluated via GLCM effectively separates rough and smooth snow (Bühler et al., 2009). Thresholds can also be used to eliminate pixels not representative of avalanches, such as those with a minimal contrast to surrounding pixels (e.g., fresh snow), dark pixels (e.g., trees, rocks), and sharp boundaries (snow beside a rock) using the parameters listed in Table 13.7. Areas, where avalanches are unlikely to occur (e.g., not covered by snow), can be spectrally excluded from the analysis. Exclusion can be based on slope gradient (e.g., >35°), runout distance, and even NDAI (Bühler et al., 2019). Virtually, the classification represents a combination of OBIA with thresholding, in which thresholding is applied to eliminate non-avalanche pixels or non-avalanche containing areas.

In OBIA, the formed objects are classified into two categories of avalanche and non-avalanche, during which the considered properties of avalanche objects are compared to those of the remaining pixels in the vicinity. The classification method used to be dominated by nearest neighbor in which the object in question is assigned to a group to which the distance is the shortest. This paradigm of image classification has been gradually replaced by machine learning methods (see Section 13.3.4). After avalanches have been mapped, the classified results may be overlaid with a DEM of the same spatial resolution to perfect the mapping with the assistance of a numerical simulation tool called Rapid Mass Movement Simulation. Depending on the uniqueness of an avalanche, it may have been classified into multiple patches. During post-classification processing, the disconnected patches are merged to form one avalanche and to make the avalanche objects less fragmented.

OBIA can yield highly accurate detections. Both OBIA results and manually interpreted avalanches in the town of Dabos, East Switzerland, hardly differ from each other in the number of mapped avalanches, but some of them are mapped at a larger extent in the OBIA results than in the manual results (Figure 13.9). OBIA detected snow avalanches at a user's accuracy of over 90% (kappa = 0.79–0.85) based on the consideration of pixel brightness in the NIR band, NDVI, NDWI, and its standard deviation (SD_{NDWI}) (Korzeniowska et al., 2017). Of the 35 manually identified avalanches, 33 are correctly mapped by the digital method (Lato et al., 2012). The two omitted ones are rather small and have suffered partial melting. Besides, the automatic method correctly

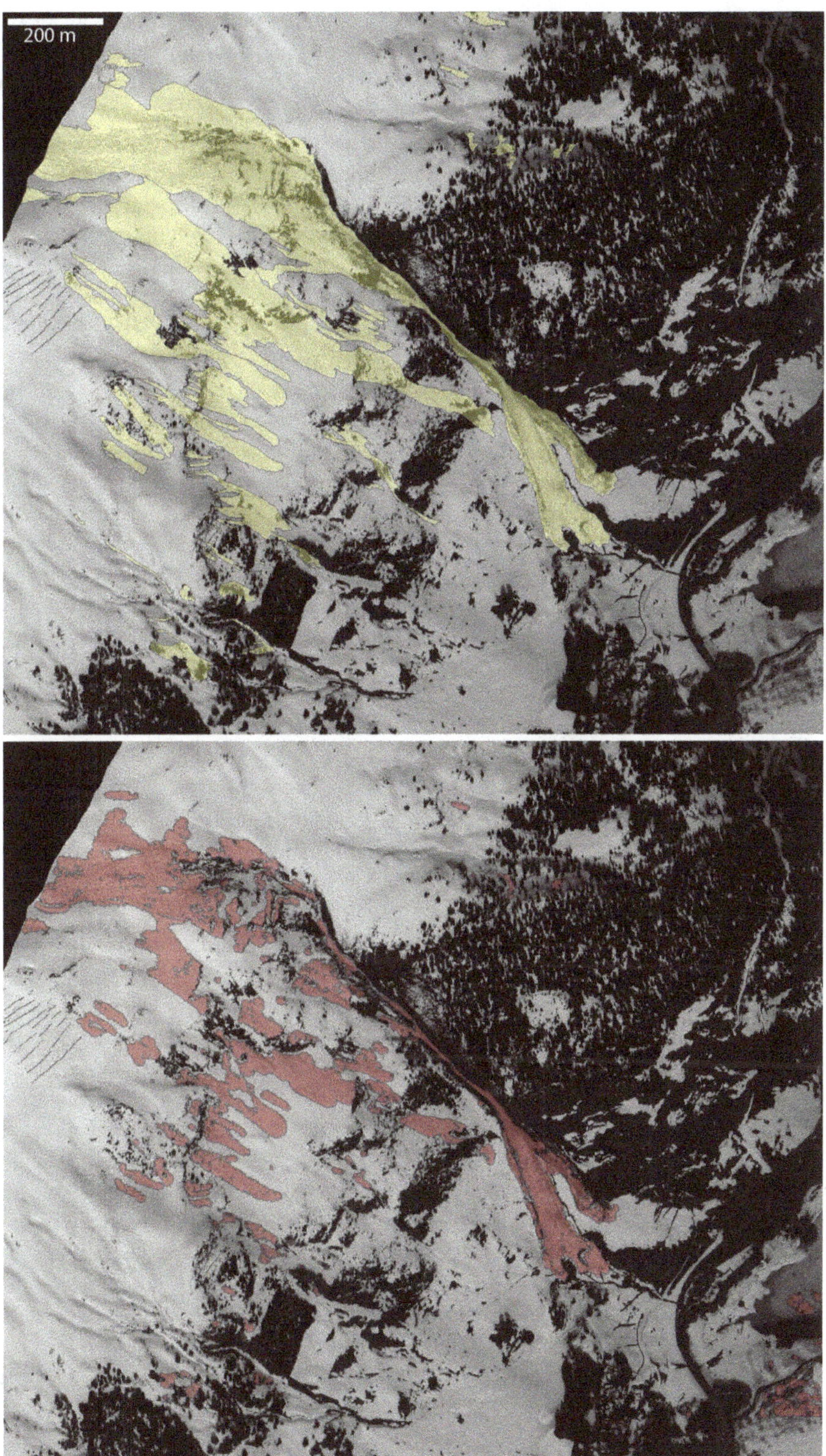

FIGURE 13.9 Comparison of snow avalanches in Davos, Eastern Switzerland mapped using visual interpretation (above) with that using OBIA based on the six inputs listed in Table 13.7, overlaid on top of a Leica panchromatic image. (Source: Lato et al., 2012.)

identified a small avalanche that was overlooked by the human interpreter. In terms of the spatial agreement, 89% of the 201,667 m^2 of the automatically mapped snow avalanche deposits overlap with 184,432 m^2 of manually mapped avalanches. If a 5 m buffer is applied to account for digitizing and translation errors, the spatial agreement increases to 95% at a size of 192,803 m^2 (Lato et al., 2012). Both the user's and producer's accuracies are higher than 90% with the errors of commission lower than 5%. These accuracy indicators are much improved over those in Table 13.6 because of the consideration of more input variables than mere pixel values. Besides, the used VHR airborne images (resolution = 0.25 m) and QuickBird multispectral bands (resolution: 2.44–2.88 m) have more spectral bands than the mono-spectral SAR image used. It must be noted that the mapping accuracy is sensitive to noise in the input VHR images. Special attention needs to be devoted to their quality, especially image resolution and image exposure, the two most important factors affecting the mapping accuracy.

OBIA is more versatile than thresholding as this methodology is applicable to any images covering any areas. It is flexible and can be adapted to images acquired by different sensors. In particular, OBIA is good at removing ski lifts and other non-avalanche features characterized by a similar entropy (Bühler et al., 2009). Besides, it is also advantageous over visual interpretation in three aspects: (i) a greater ability to detect avalanches in the shade, (ii) a higher efficiency of mapping, and (iii) the ability to analyze multiple images based on consideration of more input parameters in parallel. However, OBIA is restrictive in the sense that it allows the use of inputs derived from the image itself only. There is no room to incorporate non-imagery auxiliary data in the analysis, such as topographic data or data from other sources. This deficiency can be overcome with machine learning image classification.

13.3.4 Machine Learning Classification

Machine learning classification enjoys the advantages of fast processing of multi-dimensional data, full automation, extraordinary generalization, and reliability. Among various machine learning methods, the SVM classifier is especially strong at detecting avalanches as it is primarily a binary classifier good at classifying pixels as avalanche and non-avalanche. Its utility in mapping regional avalanches has been demonstrated by Yang et al. (2020) using Sentinel-1 data, from which six descriptors related to differences in polarization entropy (ΔH), alpha ($\Delta\alpha$), backscatter in two polarizations of VV and VH ($\Delta\sigma_{VV}$ and $\Delta\sigma_{VH}$), are used to characterize avalanches. All of them are calculated from differencing the master (e.g., the post-event) image and the slave (e.g., the pre-event) image of the same area. In addition, the coherence of phase and intensity similarity of radar echo on a pair of Sentinel-1 images (γ_{VV} and γ_{VH}) are also used in the classification. The suitability of each descriptor image is tested and judged based on the test metrics (e.g., ≥ 0.8 suitable and very suitable; ≤0.6 unsuitable). After standardization of all the metrics images to a mean of 0 and a standard deviation of 1, the eigenvalues, eigenvector, contribution rate, and cumulative contribution rate of each descriptor image are calculated from the covariance matrix of all metrics. Only those descriptors with an eigen value >1 and a cumulative contribution rate above 80% are deemed significant and retained, on which PCA is performed. The output loading matrix is considered the comprehensive model and classified using SVM that has been trained using ground-selected samples of avalanche and non-disturbed snow cover.

The results (Figure 13.10) illustrate the distribution of snow avalanches in the Aktep valley, western TianShan Mountains of China, classified from Sentinel-1A images based on six indices. In total, SVM detected 104 and 114 avalanches out of a total of 124 from the ascending images (top) and the descending images (bottom), respectively, corresponding to a POD of 84% (AUC = 0.807) and 92% (AUC = 0.938). Thus, the descending images are better at achieving a higher rate of detection than the ascending images. The main cause of misidentifications is identified as the presence of flowing water, the alpha value of which is higher. Besides, natural changes of seasonal snow also play a role. Wind-blown snow has a rough surface and hence backscatter resembling that of an avalanche.

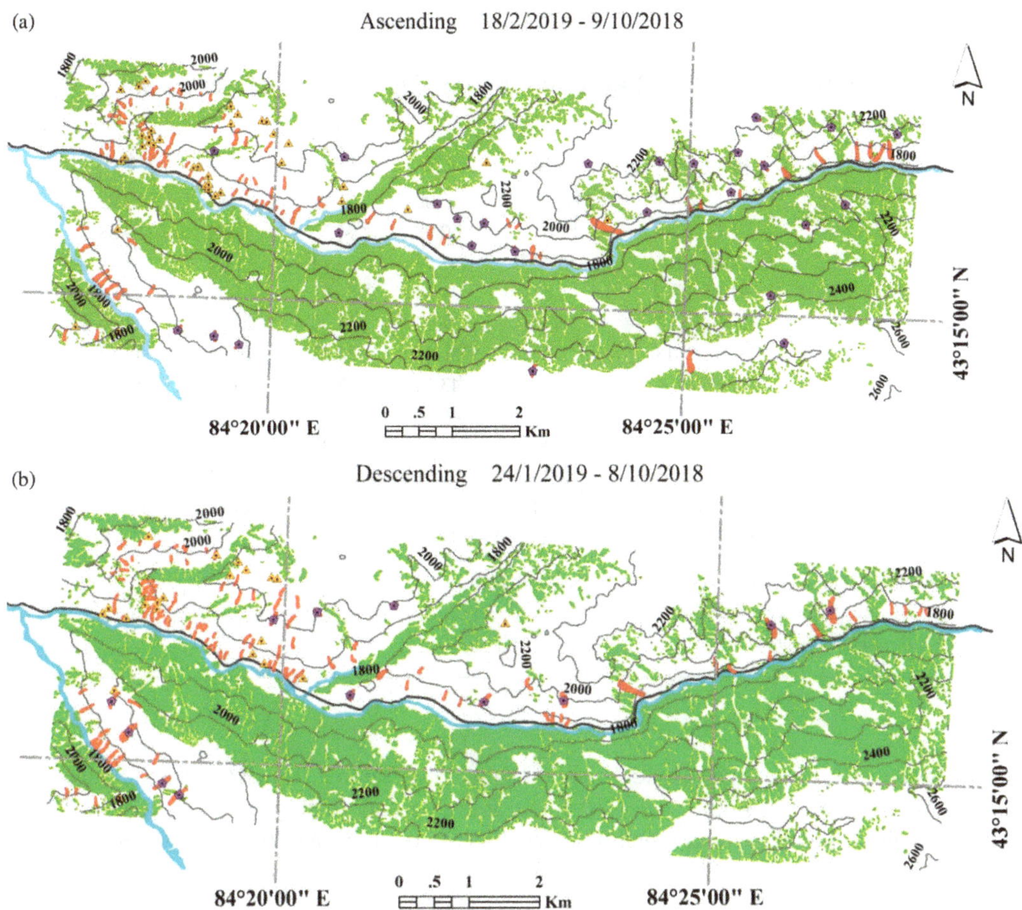

FIGURE 13.10 Distribution of snow avalanches (red patches) in the Aktep valley, western TianShan Mountains of China, detected from Sentinel-1A images based on six indices using SVM. (a) Results from the ascending image pairs; (b) Results from the descending image pairs. (Source: Yang et al., 2020.)

With the use of the six indices, SVM classification proves to be an objective, accurate, and robust method widely applicable for region-wide auto-detection of avalanches based on their scattering and interference characteristics on radar imagery.

In an effort to further increase the mapping accuracy, the number of temporal metrics is uplifted to 10, 5 of them are related to scattering variations, 2 related to polarization, plus scatter entropy, alpha, and NDVI by Liu et al. (2021). Also considered are eight avalanche conditioning factors of slope, TWI, vector ruggedness measure (VRM), TPI, aspect, profile and plan curvature, and convergence index derived from a DEM. Their importance to the occurrence of avalanches is analyzed via PCA. All the input data are classified using both SVM and LR. In spite of the large number of factors considered, the mapping accuracy is disappointedly low, with a POD <63.72% for SVM, and <64.62% for LR from either ascending or descending images due to false and missed detections of avalanche debris pixels (Table 13.8). These accuracies demonstrate that SAR images are limited in mapping avalanches due to an incorrect signal from the massive, deep frost caused by thick snow. Ascending images tend to produce better detections than their descending counterparts. However, the accuracy is significantly improved to 95.37% for SVM and 98.90% for LR after the descending and ascending images are amalgamated in the detection, suggesting that ascending and descending images offer complementary signatures of avalanches. Of the two mapping methods, LR achieves slightly higher accuracy than SVM due probably to the consideration of numerous input variables.

TABLE 13.8

Comparison of SVM and LR Accuracy in Mapping Avalanche Debris from Sentinel-1 Images

Orbit Direction	Image Pair	Algorithm	CSI[a]	POD	FAR	FB[b]
Ascending	08-12-2018/	SVM	60.00	63.72	6.20	69.92
	01-01-2019	LR	60.98	64.62	5.96	70.58
	08-12-2018/	SVM	47.54	50.73	11.68	57.44
	02-03-2019	LR	47.38	50.69	12.09	57.66
Descending	07-12-2018/	SVM	29.97	32.60	21.24	41.39
	31-12-2018	LR	32.61	35.45	19.75	44.18
	07-12-2018/	SVM	43.14	47.19	16.61	56.59
	31-12-2018	LR	46.50	50.97	15.88	60.59
Ascending	08-12/01-01/	SVM	83.06	95.37	13.44	110.19
and	07-12/31-12	LR	86.38	98.90	12.73	113.39
Descending	08-12/02-03/	SVM	81.53	94.15	14.13	109.64
	07-12/01-03	LR	84.90	98.56	14.03	114.64

Source: Modified from Liu et al. (2021). Used with permission.

[a] CSI – critical success index.

[b] FB – frequency bias.

Since Sentinel-1 imagery has a spatial resolution much coarser than TerraSAR-X and RadarSat-2, avalanches are mapped as small clusters of pixels or even groups of singular pixels (Figure 13.10), instead of polygons as shown in Figures 13.8 and 13.9. A comparison of the two mapped results in Figure 13.10 does not reveal obvious visual differences between SVM and LR.

13.3.5 Comparisons

The manual and automatic mapping methods have different capabilities for mapping avalanche debris that are further complicated by the properties of the used images. With the TerraSAR-X image pair (12-31/01-11), the manual method mapped 164 new avalanches, 110 of which were also detected by the automatic method, with 54 missing (Table 13.9), resulting in a POD of 67%, and a false-negative ratio of 33%. As illustrated in Figure 13.11, these "missed" avalanches are often small and removed during post-classification filtering. On the other hand, the automatic method detected 138 avalanches, of which 21 were absent from the manual results (Table 13.9). Of the avalanches common to both manual interpretation and automatic detection, those from the automatic detection have a smaller extent, with the upper part of the avalanches mostly missing (Figure 13.11). This is because the automatic method relies almost exclusively on the spectral properties of avalanches. Thus, the source zone or the zone of origin is mostly left out, while some of the transitional zone are also omitted. In contrast, manual interpretation makes use of more clues, such as the shape of the source zone and slope continuity (e.g., discontinuous scar after an avalanche) in delineating avalanche outlines. In terms of avalanche areas, the automatic method-produced results have a lower value owing to the omission of the zone of origin.

Different satellite images have different capabilities in detecting avalanches because of their variable spatial resolutions and wavelengths of sensing. In general, fine-resolution images allow smaller avalanches to be detected. With the Sentinel-1 image pair indicated, 68 of the 89 manually interpreted new avalanches are also automatically detected, with 21 missing, leading to a POD and false-negative ratio of 76% and 24%, respectively (Table 13.9). In the opposite

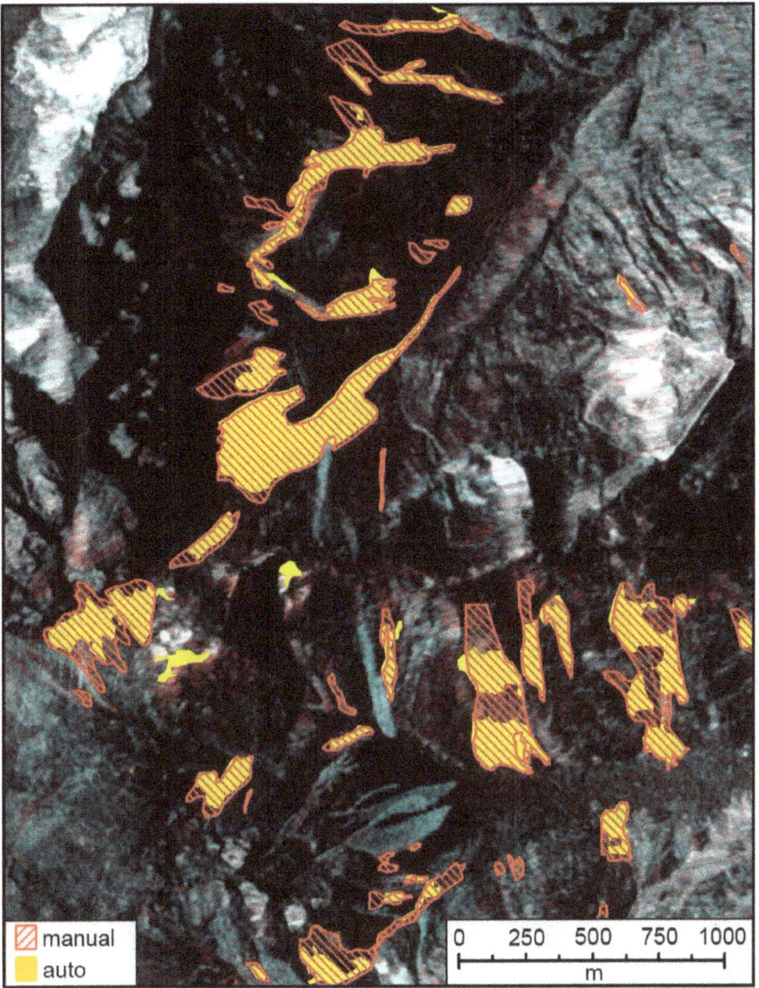

FIGURE 13.11 Comparison of new avalanche outlines (red) with automatically mapped avalanche outlines (yellow) manually interpreted from the TerraSAR-X pair of 12-31/01-11. (Source: Leinss et al., 2020.)

TABLE 13.9

Agreement (Accuracy) of Manually Interpreted and Automatically Detected New Avalanches from TerraSAR-X and C-Band Sentinel-1 Data

Image	Total	Common	POD (%)	Missing	FNR/FDR[a] (%)
	Automatic compared to manual				
TerraSAR-X	164	110	67	54	33
	Manual compared to automatic				
	138	117		21	16[a]
	Automatic compared to manual				
C-band Sentinel-1	89	68	76	21	24
	Manual compared to automatic				
	92	72		20	23[a]

Source: Leinss et al. (2020).

[a] FNR - false negative rate; FDR - false detection rate [false positive/(false positive + true positive)]

TABLE 13.10
Comparison of the Number of Avalanches Detected from TerraSAR-X Pair (01-11/02-02) with that from Optical SPOT-6 (01-24) Data

	Total	Found	Not Found	(In/Not in Cast Shadow)
(a) Of new/unsure in TerraSAR-X compared to SPOT-6	316	215	101	(84 of 17)
(b) Of SPOT-6 compared to new/unsure in TerraSAR-X	286	125	161	(57 of 104)

Source: Leinss et al. (2020).

direction, 72 out of the 92 automatically mapped avalanches are also interpreted manually with 20 missing. The 7% higher POD and lower false-negative ratio of Sentinel-1 than TerraSAR-X simply mean that the automatic method detects a higher portion of the manually detected avalanches from this image than from TerraSAR-X owing to its significantly finer spatial resolution, with the minimum size of detection being 2,000 m^2. It does not suggest that the detected results from Sentinel-1 are better than TerraSAR-X data because of the different optimal dates of data acquisition (Leinss et al., 2020). In general, an avalanche can be detected more accurately from images acquired more immediately after the event. A longer delay in data acquisition means a lower accuracy as the fresh debris loses its spectral uniqueness. The timeliness of image acquisition seems to be more important than the type of SAR images having a similar spatial resolution.

Optical and radar images have drastically differing spectral and spatial resolutions that affect their ability to detect avalanches, so it is important to compare them in avalanche detection (Table 13.10). The TerraSAR-X image pair (01-11/02-02) detected 215 of 316 avalanches as new or unsure (at a success rate of 68%) that were also detected on the SPOT-6 image. Of the remaining 101 missed ones, 84 avalanches are located in the cast shadow. The optical SPOT-6 images detected 125 of 286 avalanches (at a rate of 44%) that were also contained in the results from the TerraSAR-X images, but more than half of the optically detected avalanches were missing. A total of 20% (57 of 286) could not be mapped successfully because they were located in the radar shadow or layover, and 36% (104 of 286) had a too low backscatter contrast to be visible on the radar images. Those avalanches in shadow are not detected from SPOT-6 images but are detected from TerraSAR-X images. There are no significant differences for the lower detection limits as both TerrsSAR-X and SPOT-6 images have the same minimal detectable avalanche size of 500 m^2.

A total of 281 and 311 avalanches were detected from ascending and descending Sentinel-1A SLC images at an accuracy of 83.1% and 94.0% (measured by AUC), respectively, in one study area, and 104 and 114 avalanches (accuracy = 80.7% and 93.8%) in another, with a minimum detectable size of 0.09 km^2 for single avalanches (Yang et al., 2020). Overall, regional avalanches are mapped at POD >0.75, FAR <0.34, FOM <0.13, and TSS >0.75. These accuracy indicators demonstrate a mixed performance in comparison with that achieved by Liu et al. (2021). Their POD is never higher than 64.62%, much lower than 75%. On the other hand, their FAR value is never higher than 21.24%, markedly lower than 34% (FAR rate is even lower with the ascending images). The higher accuracy is achieved at the expense of a higher commission error. These accuracies are much better than those obtained by Vickers et al. (2016) and Wesselink et al. (2017), even though the same Sentinel-1 data are used. The improvement in mapping accuracy is attributed to the consideration of more (six) inputs based on entropy, alpha (indicative of the mechanism of scattering), backscatter, and coherence, and sophisticated PCA analysis that projects most of the information to the first few component images.

13.4 AVALANCHE MONITORING

It is essential to monitor avalanche activity in order to forecast avalanche occurrence and minimize avalanche hazards. Monitoring is critical to early warming and disaster prevention. Avalanche monitoring can be short term or long term. Short-term monitoring lasts only days or seasonally, usually carried out for individual avalanches. Long-term monitoring refers to studying the same avalanches or avalanche-inflicted areas over a period of years or decades. The target of monitoring can be individual avalanches or all avalanches in a region, depending upon the monitoring duration. Individual avalanches are monitored in the short term based on continuously recorded seismic signals that can reliably indicate the timing of the release of avalanches over an area. Long-term monitoring is usually carried out to all avalanches in a region and can be achieved using event documentation and analysis of ice-rock avalanches (Huggel et al., 2007). Apart from seismic signal analysis, long-term monitoring can also be undertaken using space- and airborne observations. Satellite and airborne data, together with seismic records and a DEM derived from SRTM and ASTER, allow the reconstruction of avalanche dynamics and identification of similar failure and propagation mechanisms (Figure 13.12). Such monitoring can provide early warning in a lead time up to hours and the avalanche event is reconstructed and analyzed within days to years afterward.

13.4.1 Individual Avalanches

Monitoring of individual avalanches is performed on large ice-rock avalanches, such as those on the Iliamna Volcano of Alaska. They have a volume on the order of 1×10^6 to 3×10^7 m^3 and take place frequently with a recent return period of two to four years. With a coarse temporal resolution, satellite images can be used to study the extent of hydrothermally altered rocks and surface thermal anomalies around the summit of the volcano where rock avalanches are likely to take place. But they cannot monitor avalanches precisely. In comparison, ground-based monitoring provides more details about avalanches. Ground monitoring may be implemented using infra-sound, in which avalanches are detected based on low-frequency sound waves (below 20 Hz) generated by avalanches using microphones. Alternatively, avalanches may be monitored by detecting ground motion generated by the downslope movement of avalanches using geophones buried in the avalanche start zone. Seismic signals help locate the avalanche source and guide airborne monitoring so that monitoring efforts can be directed at the affected area of interest. Substantially smaller avalanche activities (even those loose snow avalanches) can be detected remotely using this method.

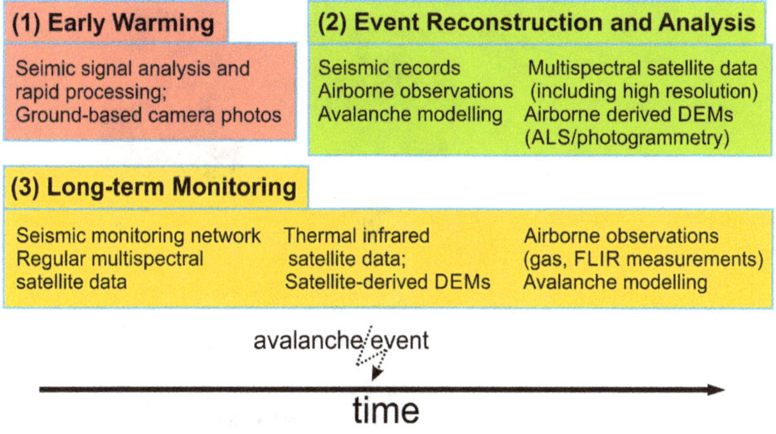

FIGURE 13.12 Schematic representation of (1) early warning, (2) event reconstruction and analysis, and (3) long-term avalanche monitoring on a time axis of a monitored avalanche event. Note: The time axis is more indicative than quantitative. The monitoring methods within each category are aligned without referring to the time axis. (Source: Modified from Huggel et al., 2007. Used with permission.)

Ground monitoring of individual avalanche activity may be implemented using the distributed fiber optic system (Prokop et al., 2013b). This time-domain reflectometer system detects seismic vibrations and acoustic signals on a fiber optic cable. Repeated tests show that it successfully detected all the 52 artificially triggered avalanches with a runout distance ranging from a few meters to approximately 250 m, and 59 unsuccessful attempts of artificial triggering. This distributed acoustic fiber optic sensing is a precise tool for monitoring avalanche activity and runout distances. However, this monitoring may not yield prognostic information on avalanche forecasting and is thus unable to minimize avalanche hazards. Since infra-sounding and seismic monitoring bears little relevance to remote sensing, both will not be explored further.

Seismic monitoring is turned into remote sensing-based monitoring by replacing geophones with digital cameras positioned strategically to face the slope where avalanches are likely to occur. Oblique ground images are periodically taken every few minutes automatically and manually downloaded. They are able to reveal even small dry loose snow avalanches that are not typically detected using the seismic method (Figure 13.13). Nevertheless, it must be acknowledged that ground images do not allow all avalanche activities to be monitored owing to the limited visibility and field of view. Depending on the interval of image recording, timing of monitoring may not be precise enough to coincide with the avalanche event itself, so the acquired image shows only the end product of it.

Monitoring may focus on different aspects of an avalanche, one of which is the volume of snow debris displaced during an avalanche. It can be estimated from a pair of stereoscopic photographs taken from two fixed positions at a similar distance to the target. They allow the construction of a 3D model of the surface with the assistance of several ground control points common to the pair of photographs using the photogrammetric method. The volume of the snow debris is estimated and monitored by studying the same avalanche at two specific times, one before and one after the event. This method is rather demanding, slow, but accurate. In comparison, the estimation is much more quickly realized with terrestrial LiDAR data at greater ease. Multi-scan LiDAR data acquired from the same fixed position are needed, one before the event and one after. After the LiDAR data are converted to a DEM, the two DEMs are subtracted from each other. The difference in height multiplied by the DEM grid cell size produces the volume of the mobilized debris (see Eq. 3.3). Both the photogrammetric and LiDAR methods yield reliable avalanche debris estimates only if the entire avalanche is fully exposed and sensed.

Compared with ground monitoring, airborne monitoring is good for assessing the avalanche failure mechanism (e.g., ramp vs wedge failure), identifying the material involved, characterizing the flow path, and broadly estimating the thickness of the failure and deposits (Huggel et al., 2007).

FIGURE 13.13 Loose snow avalanches in the Eastern Swiss Alps near Davos, Switzerland, during the winter of 2010, monitored from ground images taken every 5 minutes automatically using a digital camera deployed in the start zone. (Source: van Herwijnen and Schweizer, 2011. Used with permission.)

In comparison, satellite images such as Landsat ETM+ and ASTER allow the tracing of the avalanche trajectory because rock entrained within the avalanche causes its spectral reflectance to stand out from the surrounding ice and snow. They also allow the estimation of avalanche dimensions (e.g., width and runout length), and even the slope of the initiation zone or the ratio of vertical descent to horizontal runout if combined with a DEM. However, it is not feasible to visually estimate the amount of rock entrained along the avalanche trajectory with either airborne or satellite-based observations. Derivation of such quantitative information requires stereoscopic images using the sophisticated photogrammetric method or LiDAR data, as with ground monitoring.

13.4.2 Regional Monitoring

Regional monitoring refers to repeated sensing of all avalanches located in a specific district or province, the area of which varies enormously in size across the globe. So far regional avalanches have been detected and mapped from various data sources using diverse methods and inputs, but monitoring of avalanche activity on an ongoing basis is rare. Short-term regional monitoring may be fulfilled using a sensing system known as LISA (Linear SAR) based on the ground SAR instrument in the topographic mode. This C-band sensor has a typical penetration depth of several meters of dry snow. It has been used to monitor 50 natural and 4 artificially triggered avalanches in the Sion valley of Switzerland over a period of 50 days (Martinez-Vazquez and Fortuny-Guasch, 2006). Images recorded every 12 minutes are used to produce InSAR. The interferometric phase around an avalanche event is then unwrapped with the single-pass interferometric mode of the instrument. The computed DEMs of the focused zone just before and after the avalanche may be calibrated with a reference DEM of a much finer spatial resolution (e.g., 2.5 m × 2.5 m). The subtraction of the two-time DEMs reveals the change in snow height caused by avalanches, even though the results could be rather noisy. If multiplied by the DEM grid cell size, the volume of snow displaced in the avalanche is known. Given that the phase can have an error up to 0.2 rad, the resultant errors of the topographic maps from LISA range from 3 to 11 m. Thus, LISA has shown the ability to monitor localized changes in snow cover. However, the height resolution of the topographic map generated from these images is inadequate to yield a quantitative estimate of the volume of snow displaced in an avalanche.

The limitation of ground-based monitoring can be redressed with space-borne monitoring using optical images. In the best case scenarios, they are highly suitable for avalanche detection and activity monitoring because the entire avalanche chute, from the start zone to the depositional area, is fully visible. However, cloud cover, shade, and polar night darkness prevent continuous monitoring throughout a winter. Preferably, radar images should be used as they are particularly suited to long-term monitoring over large regions. Near-real-time (10 minutes after image downloading), region-wide monitoring is possible thanks to the automatic processing of Sentinel-1 data (Figure 13.14). Apart from imagery,

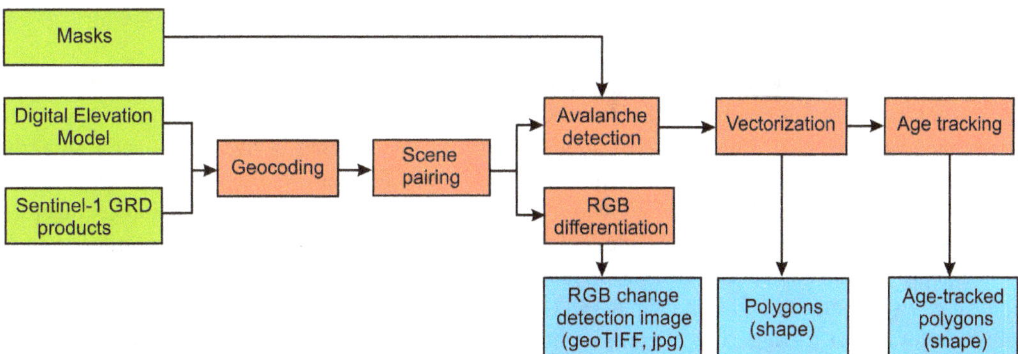

FIGURE 13.14 The workflow of a regional system for near-real-time monitoring of avalanche activity using Sentinel-1 data. Green boxes: input; red boxes: data processing; blue boxes: output. (Source: Modified from Eckerstorfer et al., 2019.)

other data needed may include a DEM and an avalanche mask determined from the results of TauDEM, together with water bodies, forested areas, agricultural areas, and glaciated areas. After geocoding, the images are paired to detect relative backscatter changes induced by avalanches (e.g., positive backscatter change) via polarization differencing (ΔVV and ΔVH) between the actual and reference images and differentiate RGB to color-code the detected backscatter change. After segmented into several classes and filtered, the segmented differencing images are used to detect class change (e.g., classified activity image – classified reference image). Potential avalanche pixels are identified via thresholds based on mean and standard deviation values automatically in both the ΔVV and ΔVH channels, resulting in a binary map of avalanches vs non-avalanches. The two sets of results may be combined to produce the final binary avalanche map. If the same procedure of detection is repeated to time-series Sentinel-1 data, the same avalanche is detected repeatedly in the detection results, creating an opportunity to determine the age of the avalanche. Compared with manually detected results, this automatic method achieved a true-positive rate (sensitivity) of 73% (actual percentage of manual detections that were correctly detected automatically) and a negative predictive value (proportion of positive and negative results) of 52%, resulting in an overall accuracy of 79% (Eckerstorfer et al., 2019). The automatic detection algorithm is quite capable of detecting avalanches that are manually identifiable.

Operational Sentinel-1 data-based long-term monitoring using the developed automatic detection method is possible only when the satellite data are in existence. Globally, avalanches can be monitored in all geographic areas covered by satellite data. If 20 m spatial resolution Sentinel-1 data are used, the smallest size of avalanches detectable is 100 m^3, or avalanches wider than sub-pixel resolution with a cutoff minimum size of 10 pixels. The major source of misidentification (e.g., a high false alarm rate) is the change in snow conditions from wet to dry in the images, which can be minimized by removing the days with high false alarms due to the transition from wet to dry snow under the assumption that dry snowpack is typically stable and consolidated, thus unlikely to lead to widespread avalanche activity. However, this manual intervention weakens the automatic detection process. Admittedly, the monitoring of avalanches from Sentinel-1 images is plagued by radar layover and shadow, as with the use of all types of radar data. The masking of these areas from the monitoring reduces the detectable areas in topographically rugged regions.

13.5 AVALANCHE HAZARD MODELING

Avalanche hazards refer to the potential or likelihood of avalanche occurrence. It is estimated via modeling. Simple avalanche flow modeling is able to reasonably replicate avalanches and is useful for avalanche hazard assessment. The zoning of avalanche hazard is significant to mitigating avalanche damage and minimizing avalanche impacts on humans and their properties.

13.5.1 Avalanche Triggers

Avalanche triggering mechanisms are complex and controlled predominantly by the relationships among terrain, weather conditions, snow structure, and metamorphism. Snow avalanches are triggered by a number of factors, the most significant being the thawing of the frozen snow in early spring as air temperature warms. In addition, strong winds also induce avalanches. Theoretically, avalanches are triggered directly by snow thawing, wind, and ground shaking. The indirect factors that may activate avalanches are diverse and their influence varies with snow characteristics, such as the depth of old snow and its character, depth of new snowfall and its moisture content, snowfall intensity, wind, and temperature (Table 13.11). Except for settlement of snow that stabilizes it, all other factors exert a triggering effect. For instance, wind drifts redistribute snow and influence energy exchange, particularly the melting and freezing of snow and ice. It is impossible to rank the importance of these factors. Nor is it possible to quantify their importance. Some of these factors can never be quantified, either via field surveys or by means of remote sensing.

TABLE 13.11

Factors Playing a Significant Role in Triggering Avalanches

Factor	Contribution
Depth of old snow	i. New snow easily slides over old snow of ≥0.6 m thick; ii. The deeper the snow, the more mass it supplies to an avalanche.
Character of old snow surface	i. A loose surface promotes cohesion with a fresh fall; ii. A crusted or wind-packed surface has poor cohesion with new snow.
Depth of new snow	i. ≥61 cm needed to produce an avalanche of dangerous proportions.
Free moisture content	i. Cementing and improving cohesion, within limits. Good cohesion may be dangerous as well as helpful.
Water–snow ratio	i. A ratio over 10% increases snow weight more than its cohesion; ii. An avalanche cycle of extraordinary violence under favorable conditions.
Snowfall (cm·hour^{-1})	i. At an intensity of >2.54 cm·hour^{-1}, snow pack grows faster than the stabilizing forces; ii. Sudden increase in load may fracture a slab beneath.
Precipitation (cm·hour^{-1})	i. Most promising single guide to avalanche hazard yet discovered; ii. Avalanche hazard critical when total water precipitation = 2.54 cm.
Wind	i. Redistribute deposited snow; grinds snow crystals to simpler and less cohesive forms; constructs stable crust and essential to the formation of fragile slab; ii. Warm wind thaws snow; sudden changes in wind direction and velocity shear snow; iii. For effective wind action, 24 km·hour^{-1} is minimal.
Temperature	i. Dry snows fall at ≤−3.9°C; ii. >28° promotes rapid snow settlement and metamorphosis; iii. Sudden rise causes cohesion loss to trigger avalanches; iv. Gradual spring warming cumulatively deteriorates snow and lead to avalanches.
Snow settlement	i. Mostly stabilizing effect; ii. In new snow, a ratio <15% = little consolidation; >80% = stabilization.

Source: Modified and adapted from American Avalanche Association.

In spite of the identified comprehensive range of factors that play a role in avalanche initiation, not all of them are commonly considered in modeling avalanche hazards because of the difficulty in acquiring the data. In practice, avalanche hazards are modeled only from general factors that can be mapped, including topographic and climatic factors, as well as surface cover. The exact variables considered include slope, aspect, curvature, elevation, terrain roughness, and ground cover. Some of them are easily derivable from remotely sensed data, such as land cover and DEM, from which more topographic variables are calculated, including aspect, gradient, curvature, and TPI. Climatic data, such as air temperature, snow depth, and relative humidity, are also obtainable from meteorological satellite data, but their spatial resolution may not be appropriate for modeling regional avalanche hazards. Thus, data commonly obtained from ground weather stations are widely used. These point data are converted to spatial distribution maps via spatial interpolation. Also critical in the modeling are historic avalanche events which can also be mapped from remote sensing data using the previously described methods. An alternative is to make use of the published information on avalanche sites. However, it is not recommended to acquire the data in the field using a GPS due to the potential avalanche hazards.

A significant variable to avalanche risk is snow volume that can be estimated from snow cover and snow depth. Snow cover information is obtainable from existing data products, such as the MODIS 8-day L3 Global 500 m Grid Snow Cover product. Snow depth, however, is the most thorny to retrieve, especially at a high accuracy level. One retrieval method is to make use of SAR data or the monthly L3 Global Snow Water Equivalent product created from the AMSR-E/Aqua L2A Global Swath Spatially Resampled Brightness Temperatures data. Another method is to make use of low frequency (<89 GHz) samples via snow map masking and brightness temperature classification. After the retrieval, snow volume is estimated by multiplying snow depth by the snow cover area

TABLE 13.12

Common Variables Considered in Modeling Watershed-Level Avalanche Hazards and Their Importance Determined by SVM

Factor	Importance[a]	Rank	Factor	Importance	Rank
LS	63.2[b]	1	Elevation	19.8	8
Lithology	54.6	2	VRM	17.5	9
RSP	41.3	3	Land use	14.7	10
TRI	36.5	4	Aspect	13.2	11
Slope	32.4	5	Distance from stream	7.3	12
TPI	28.2	6	TWI	5.1	13
Profile curvature	22.1	7			

Source: Rahmati et al. (2019).

[a] Relative importance as measured by the increase (%) in mean square error

[b] The exact value may vary with the watershed, but the importance rank remains unchanged.

derived from images. The estimated snow volume is proportional to the level of avalanche hazard but faces a huge uncertainty, especially in mountainous areas where avalanches are likely to occur.

13.5.2 Linear Weighting Methods

Snow avalanche hazards have been estimated using multi-criteria decision analysis, simple data-mining methods, probabilistic occurrence ratio (POR) (also known as the frequency ratio), and snow avalanche inventory (Kumar et al., 2017). Multi-criteria decision analysis may be based on some of the factors listed in Table 13.12. As with all types of modeling, the derivation of avalanche hazards requires proper weighting of all the considered variables. This weighting has been based on educated guesses or the expert's opinions. For instance, Abake et al. (2014) arbitrarily assigned a weight of 0.4, 0.3, and 0.3 to snow volume, altitude, and slope, respectively, in zoning snow avalanche hazards. Further scores are assigned to specific attribute values of each variable, such as 4 to slopes <30°, 7 to 30–40° slopes, and 10 to >40° slopes. In all cases, the weights are not backed by any evidence.

The deficiency of subjective weighting is avoided if avalanches are forecast from various meteorological variables using logistic models (Gauthier et al., 2017). Avalanche occurrence is best predicted from two days of accrued snowfall, daily rainfall, and wind speed near the coast, but there is no information on where it will occur as no topographic variables are considered. Besides, the assignment of weights to all the considered factors may involve a degree of subjectivity. A more objective and scientific weighting method is the analytical hierarchical process (AHP) in which all the considered variables are paired between any two of them, and the two variables in a pair are weighted relatively to each other according to the pre-determined weighting scheme, usually an odd number between 1 and 9. For instance, if a weight of 7 is assigned to slope gradient, then a weight of 1/7 is allocated to slope aspect when these two parameters are paired and compared with each other for their importance. The result of all pair-wise comparisons is represented as a square matrix. The weight of each considered variable is calculated as the row sum of the matrix, divided by the total number of variables considered. The weights of all variables sum to 1. Through AHP, the avalanche susceptibility index (ASI) is calculated as follows:

$$\text{ASI} = 0.41\text{Slope} + 0.28\text{Curvature} + 0.14\text{Aspect} + 0.09\text{TR} + 0.05\text{Elevation} + 0.03\text{LC} \quad (13.3)$$

where TR = terrain roughness; LC = land cover. The coefficients of the model represent weights. In Eq. 13.3, slope is the most important, followed by curvature and aspect. In this way, avalanche

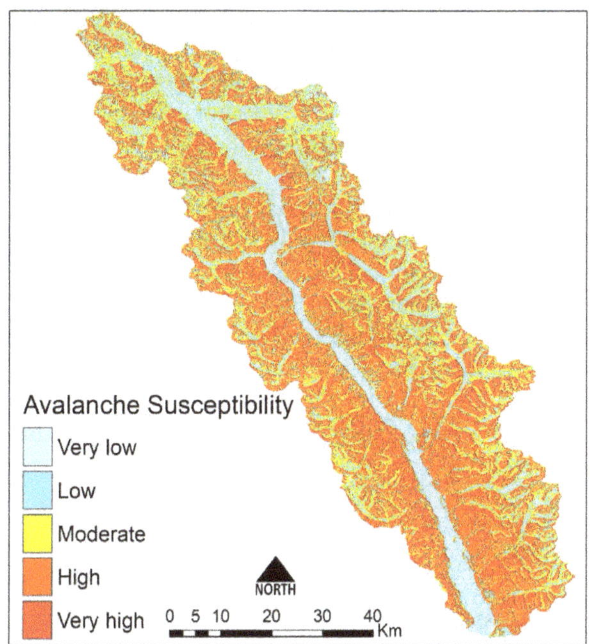

FIGURE 13.15 Distribution of snow avalanche susceptibility in the Nubra valley of Indian Himalayas modeled from remote sensing using AHP. (Source: Kumar et al., 2017.)

hazards are then treated as an arithmetic combination of all the considered variables multiplied by their weights, which can be easily implemented in a GIS (Kumar et al., 2017).

The AHP method has been used for multilevel decision analysis to demarcate avalanche-prone areas in the Indian Himalayas based on the consideration of eight parameters (slope, elevation, aspect, curvature, ground cover, snow depth, air temperature, and relative humidity) in three general categories of topography, ground cover, and meteorology (Singh et al., 2019). The three most important variables are identified as slope (0.315), elevation (0.251), and snow depth (0.186), with a combined influence over 75%. These weights differ from the coefficients in Eq. 13.3 because different variables are considered and the importance of the same variable varies geographically and with the scale of study, such as slope. The weighted averaging of the considered parameters results in an avalanche susceptibility map (Figure 13.15). The susceptibility level initially expressed as a continuous value may be classed to identify its spatial pattern. The accuracy of the modeled avalanche hazard map may be evaluated by overlaying it with the original avalanche events, and the evaluation results are expressed as AUC. Another method of validating the snow avalanche risks analyzed using physical exposure and vulnerability index is to compare them with avalanche inventory data. If the predicted high-risk areas coincide with the observed avalanches in the past, the prediction is deemed accurate. This comparison indicates an accuracy of 91% (prediction rate) in terms of AUC (Kumar et al., 2017). A slightly higher validation accuracy of 93.2% is achieved by Singh et al. (2019) probably because they considered more input variables, particularly snow depth. The obtained results indicate that this simple model performed well in identifying snow avalanche-prone areas. However, the results are unable to indicate when this hazardous potential will be turned into reality.

13.5.3 Machine-Learning Methods

The aforementioned linear weighting method has been gradually replaced with more powerful machine learning methods, such as RF, SVM, ANN, NB, and GAM (Rahmati et al., 2019). These algorithms are the best at handling multiple variables of a diverse nature, including TPI, topographic ruggedness index (TRI), TWI, VRM, wind exposition index (WEI), length-slope (LS), and relative slope position (RSP). Dissimilar to the weighted linear combination method described in the preceding section, these variables do not need to be weighted by the human analyst. Instead, the weighing of all the considered variables is handled automatically by the computer based on their contribution to the observed avalanche occurrence in the past. Such avalanche inventory data can be produced using visual interpretation or automatic mapping as described in Section 1.4.1. Also, they can be mapped using a GPS in the field, which is not recommended because of the risk of avalanches during the fieldwork. It must be acknowledged that the exact importance of the considered variables may vary with the watershed, but the ranked importance of all the considered factors remains unchanged across multiple watersheds.

Apart from individual models, avalanche hazards can also be predicted using ensemble modeling. This method predicts avalanche hazards by using different modeling algorithms or training datasets and then aggregating the prediction of each component model. An ensemble model (EM) may be created from the weighted integration of multiple models to take advantage of the component models, or

$$EM = \frac{\sum_{i=1}^{n} \left(AUROC_i \times M_i \right)}{\sum_{i=1}^{n} AUROC_i} \tag{13.4}$$

where EM = the resultant ensemble model, $AUROC_i$ = the area under the receiver operating characteristic curve (AUROC) value of the ith component model (M_i).

The effectiveness of five machine learning models (RF, SVM, ANN, NB, and GAM) in modeling avalanche hazards is compared in Table 13.13. Of these models, RF is the best in terms of AUROC in both watersheds, followed by EM and SVM, but the difference between them is rather small. This relativity still holds if judged by TSS (TPR + TNR − 1) and Matthews correlation coefficient (MCC) with the training dataset. Overall, RF is regarded as the best regardless of the study site.

TABLE 13.13

Comparative Performance of Four Machine Learning Models and the Ensemble Model in Zoning Avalanche Hazards Using the Training and Validation Datasets

		Training Dataset			Validation Dataset		
Watershed	**Model**	**AUROC[a]**	**TSS[b]**	**MCC[c]**	**AUROC**	**TSS**	**MCC**
	RF	0.981	0.893	0.884	0.964	0.862	0.865
	SVM	0.972	0.882	0.863	0.955	0.844	0.858
Darvan	NB	0.932	0.791	0.855	0.914	0.727	0.847
	GAM	0.924	0.756	0.812	0.896	0.714	0.796
	Ensemble	0.977	0.886	0.871	0.966	0.865	0.861
	RF	0.973	0.882	0.866	0.956	0.881	0.854
	SVM	0.965	0.863	0.859	0.948	0.875	0.832
Zarrinehroud	NB	0.941	0.835	0..825	0.922	0.824	0.804
	GAM	0.934	0.835	0.808	0.905	0.816	0.793
	Ensemble	0.968	0.877	0.862	0.958	0.877	0.841

Source: Rahmati et al. (2019).

[a] AUROC – area under the receiver operating characteristics curve, a cutoff-independent metric

[b] TSS – true skill statistics, a cutoff-dependent metric derived from the sum of true-positive ratio and true-negative ratio minus 1. TSS value ranges from −1 to 1, with 1 representing perfect performance

[c] MCC – matthews correlation coefficient

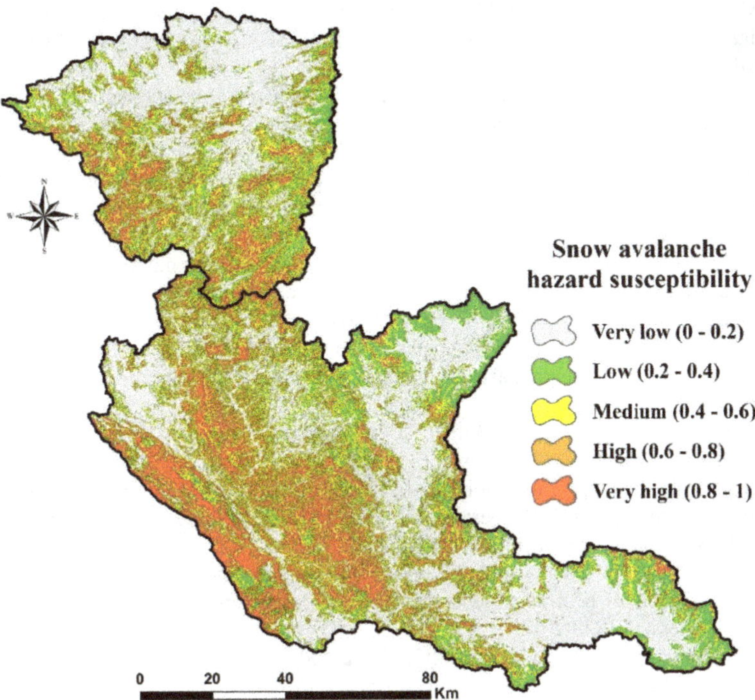

FIGURE 13.16 Snow avalanche hazard of the Zarrinehroud (938 km^2) and the Darvan (4,514 km^2) watersheds, in Kurdistan province of Iran at 5 levels modeled from 13 variables using an ensemble model. (Source: Rahmati et al., 2019.)

EM is the second best, but only marginally better than SVM. GAM is always the worst, followed by NB for the training dataset. For the validation dataset, EM and RF are almost the same accurate, followed by RF and SVM in terms of AUROC, but RF becomes the best if judged by TSS and MCC in both watersheds. Only GAM and NB are noticeably less accurate. So either RF or EM should be used. The best model is then applied to modeling snow avalanche hazards expressed on a scale of 0–1, and the result is shown in five classes, usually generated using an equal class interval of 0.2 (Figure 13.16).

REFERENCES

Abake, G., A. Al-Hanbali, B. Alsaaideh, and R. Tateishi. 2014. Potential hazard map for snow disaster prevention using GIS-based weighted linear combination analysis and remote sensing techniques: A case study in northern Xinjiang, China. *Adv. Remote Sens.* 3: 260–271. DOI: 10.4236/ars.2014.34018

American Avalanche Association. Avalanche library, https://avalanche.org/accessed on 24 April 2022.

Bühler, Y., E.D. Hafner, B. Zweifel, M. Zesiger, and H. Heisig. 2019. Where are the avalanches? Rapid mapping of a large snow avalanche period with optical satellites. *Cryosphere Discuss.* DOI: 10.5194/tc-2019-119

Bühler, Y., A. Hüni, M. Christen, R. Meister, and T. Kellenberger. 2009. Automated detection and mapping of avalanche deposits using airborne optical remote sensing data. *Cold Reg. Sci. Technol.* 57(2–3): 99–106. DOI: 10.1016/j.coldregions.2009.02.007

Deems, J.S., P.J. Gadomski, D. Vellone, R. Evanczyk, A.L. LeWinter, K.W. Birkeland, and D.C. Finnegan. 2015. Mapping starting zone snow depth with a ground-based LiFAR to assist avalanche control and forecasting. *Cold Reg. Sci. Technol.* 120: 197–204. DOI: 10.1016/j.coldregions.2015.09.002

Eckerstorfer, M., Y. Bühler, R. Frauenfelder, and E. Malnes. 2015. Remote sensing of snow avalanches: Recent advances, potential, and limitations. *Cold Reg. Sci. Technol.* 121: 126–140. DOI: 10.1016/j.coldregions.2015.11.001

Eckerstorfer, M., and E. Malnes. 2015. Manual detection of snow avalanche debris using high resolution Radarsat-2 SAR images. *Cold Reg. Sci. Technol.* 120: 205–218. DOI: 10.1016/j.coldregions.2015.08.016

Eckerstorfer, M., H. Vickers, E. Malnes, and J. Grahn. 2019. Near-real time automatic snow avalanche activity monitoring system using Sentinel-1 SAR data in Norway. *Remote Sens.* 11(23): 2863. DOI: 10.3390/rs11232863

Eckerstorfer, M., E. Malnes, R. Frauenfelder, U. Domass, and K. Brattlien. 2014. Avalanche debris detection using satellite-borne radar and optical remote sensing. In *Proc. of the Int. Snow Sci. Workshop*, Banff, AB, Canada, 29 September–3 October 2014.

Eckerstorfer, M., H. Vickers, and F. Malnes. 2016. Snow avalanche activity monitoring from space: Creating a complete avalanche activity dataset for a Norwegian forecasting region. In *Proc. of the Int. Snow Sci. Workshop*, Breckenridge, CO, USA, 3–7, pp. 199–204.

Gauthier, F., D. Germain, and B. Hétu. 2017. Logistic models as a forecasting tool for snow avalanches in a cold maritime climate: Northern Gaspésie, Québec, Canada. *Nat. Hazards* 89: 201–232. DOI: 10.1007/s11069-017-2959-3

Hafner, E.D., F. Techel, S. Leinss, and Y. Bühler. 2021. Mapping avalanches with satellites – Evaluation of performance and completeness. *Cryosphere* 15: 983–1004. DOI: 10.5194/tc-15-983-2021

Huggel, C., J. Caplan-Auerbach, C.F. Waythomas, and R.L. Wessels. 2007. Monitoring and modelling ice-rock avalanches from ice-capped Volcanoes: A case study of frequent large avalanches on Iliamna Volcano, Alaska. *J. Volcan. Geotherm. Res.* 168: 114–136.

Korzeniowska, K., Y. Bühler, M. Marty, and O. Korup. 2017. Regional snow avalanche detection using object-based image analysis of near-infrared aerial imagery. *Nat. Hazards Earth Syst. Sci.* 17: 1823–1836. DOI: 10.5194/nhess-17-1823-2017

Kumar, S., P. Srivastava, and K. Snehmani. 2017. GIS-based MCDA–AHP modelling for avalanche susceptibility mapping of Nubra valley region, Indian Himalaya. *Geocarto Int.* 32(11): 1254–1267. DOI: 10.1080/10106049.2016.1206626

Lato, M.J., R. Frauenfelder, and Y. Bühler. 2012. Automated detection of snow avalanche deposits: Segmentation and classification of optical remote sensing imagery. *Nat. Hazards Earth Syst. Sci.* 12(9): 2893–2906. DOI: 10.5194/nhess-12-2893-2012

Leinss, S., R. Wicki, S. Holenstein, S. Baffelli, and Y. Bühler. 2020. Snow avalanche detection and mapping in multitemporal and multiorbital radar images from TerraSAR-X and Sentinel-1. *Nat. Hazards Earth Syst. Sci.* 20: 1783–1803. DOI: 10.5194/nhess-20-1783-2020

Liu, Y., X. Chen, Y. Qiu, J. Hao, J. Yang, and L. Li. 2021. Mapping snow avalanche debris by object-based classification in mountainous regions from Sentinel-1 images and causative indices. *Catena* 206: 105559. DOI: 10.1016/j.catena.2021.105559

Malnes, E., M. Eckerstorfer, and H. Vickers. 2015. First Sentinel-1 detections of avalanche debris. *Cryosphere Discuss.* 9: 1943–1963. DOI: 10.5194/tcd-9-1943-2015

Martinez-Vazquez, A., and J. Fortuny-Guasch. 2006. Feasibility of snow avalanche volume retrieval by GB-SAR imagery. In *Geosci. and Remote Sens. Sympos.*, 2006. IEEE International Conference on IGARSS, pp. 743–746.

Picard, G., L. Brucker, A. Roy, F. Dupont, M. Fily, A. Royer, and C. Harlow. 2013. Simulation of the microwave emission of multi-layered snowpacks using the Dense Media Radiative transfer theory: The DMRT-ML model. *Geosci. Model Dev.* 6: 1061–1078. DOI: 10.5194/gmd-6-1061-2013

Prokop, A., P. Schön, F. Singer, P. Gaëtan, M. Naaim, and E. Thibert. 2013a. Determining avalanche modelling input parameters using terrestrial laser scanning technology. In *Proc. of the Inter. Snow Sci. Workshop*, 2013, Chamonix Mont-Blanc, France, pp. 770–774.

Prokop, A., A. Wirbel, and M. Jungmayr. 2013b. The "Avalanche Detector", a new avalanche monitoring tool using distributed acoustic fibre optic sensing. In *Proc. of the Int. Snow Sci. Workshop*, 2013, Chamonix Mont-Blanc, France, pp. 1027–1032.

Rahmati, O., O. Ghorbanzadeh, T. Teimurian, F. Mohammadi, J.P. Tiefenbacher, F. Falah, S. Pirasteh, P.T.T. Ngo, and D.T. Bui. 2019. Spatial modeling of snow avalanche using machine learning models and geo-environmental factors: Comparison of effectiveness in two mountain regions. *Remote Sens.* 11: 2995. DOI: 10.3390/rs11242995

Singh, D.K., V.D. Mishra, H.S. Gusain, N. Gupta, and A.K. Singh. 2019. Geo-spatial modeling for automated demarcation of snow avalanche hazard areas using Landsat-8 satellite images and in situ data. *J. Indian Soc. Remote Sens.* 47: 513–526. DOI: 10.1007/s12524-018-00936-w

van Herwijnen, A., and J. Schweizer. 2011. Monitoring avalanche activity using a seismic sensor. *Cold Reg. Sci. Technol.* 69(23): 165–176. DOI: 10.1016/j.coldregions.2011.06.008

Vickers, H., M. Eckerstorfer, E. Malnes, Y. Larsen, and H. Hindberg. 2016. A method for automated snow avalanche debris detection through use of synthetic aperture radar (SAR) imaging. *Earth and Space Sci.* 3: 446–462. DOI: 10.1002/2016EA000168

Wesselink, D.S., E. Malnes, M. Eckerstorfer, and R.C. Lindenbergh. 2017. Automatic detection of snow avalanche debris in central Svalbard using C-band SAR data. *Polar Res.* 36(1): 1333236. DOI: 10.1080/17518369.2017.1333236

Yang, J., C. Li, L. Li, J. Ding, R. Zhang, T. Han, and Y. Liu. 2020. Automatic detection of regional snow avalanches with scattering and interference of C-band SAR data. *Remote Sens.* 12: 2781. DOI: 10.3390/rs12172781

Index

N

O

P

Q

R

S

T

For Product Safety Concerns and Information please contact our EU
representative GPSR@taylorandfrancis.com
Taylor & Francis Verlag GmbH, Kaufingerstraße 24, 80331 München, Germany

www.ingramcontent.com/pod-product-compliance
Lightning Source LLC
Chambersburg PA
CBHW080131220326
41598CB00032B/5031